AF323996

Asymptotic Theory of Anisotropic Plates and Shells

Asymptotic Theory of Anisotropic Plates and Shells

Lenser Aghalovyan

National Academy of Sciences, Armenia

Translated by D. Prikazchikov

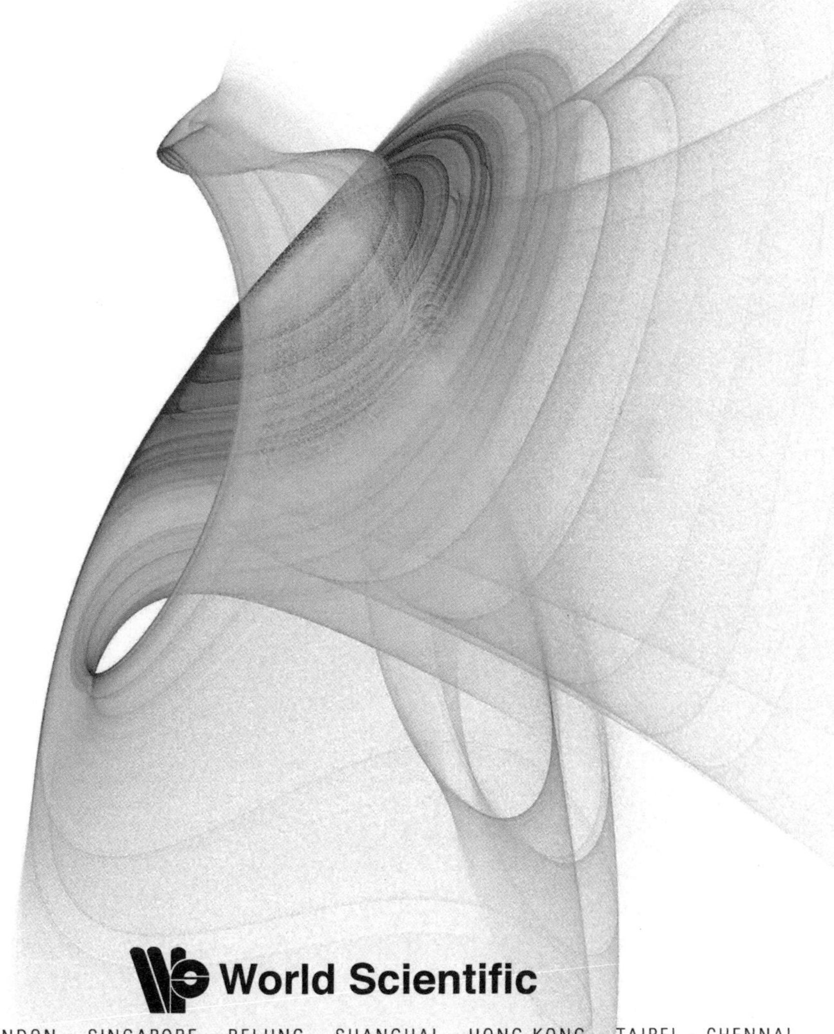

World Scientific

NEW JERSEY · LONDON · SINGAPORE · BEIJING · SHANGHAI · HONG KONG · TAIPEI · CHENNAI

Published by

World Scientific Publishing Co. Pte. Ltd.

5 Toh Tuck Link, Singapore 596224

USA office: 27 Warren Street, Suite 401-402, Hackensack, NJ 07601

UK office: 57 Shelton Street, Covent Garden, London WC2H 9HE

Library of Congress Cataloging-in-Publication Data
Agalovian, L. A., author.
   [Asimptoticheskaia teoriia anizotropnykh plastin i obolochek. English]
  Asymptotic theory of Anisotropic plates and shells / by Lenser Aghalovyan (National Academy of Sciences, Armenia) ; translated by D. Prikazchikov.
    pages cm
  Translation of: Asimptoticheskaia teoriia anizotropnykh plastin i obolochek.
  Includes bibliographical references and index.
  ISBN 978-9814579025 (hardcover : alk. paper)
  1. Elastic plates and shells. 2. Anisotropy. 3. Asymptotic expansions. 4. Plates (Engineering)--Mathematical models. 5. Shells (Engineering)--Mathematical models. I. Title.
  QA935.A29513 2015
  530.4'12--dc23
                         2014040553

British Library Cataloguing-in-Publication Data
A catalogue record for this book is available from the British Library.

Printed in Singapore

Preface

Asymptotic methods seem to be one of the natural approaches to analyze the stress-strain field arising in beams, plates and shells, justified by the geometric features. However, despite the long history of asymptotic methods, until recent decades the majority of results in the area has been obtained through *ad hoc* assumptions and expansions along the transverse coordinate. One of the reasons, in our opinion, is the fact that introduction of a small asymptotic parameter led to singularly perturbed governing equations having the small parameter in the coefficients of senior derivatives, with the mathematical theory of such equations arising relatively recently. For the same reason a number of problems for validation of technical theories of beams, plates and shells, especially, for anisotropic structures, remained open, though it has been clear that these problems may be resolved through the full analysis within the three-dimensional elasticity theory.

This monograph contains results on the theory of anisotropic beams, plates and shells relying on a unified approach, namely, on the asymptotic method of integration of the governing equations of 3D elasticity. The underlying links of the proposed methodology with existing applied theories and with the well-known Saint-Venant principle are revealed. Special attention is paid to the boundary layer solutions, along with associated velocities of attenuation. A class of boundary value problems is then pointed out, for which the Saint-Venant principle is satisfied exactly, i.e. it is shown that the principle is a generic property of the solution obtained through rigorous mathematics.

The proposed asymptotic approach is later applied to non-classical boundary value problems for plates and shells, when the face boundary conditions are different from that adopted in the classical theories for plates and shells. This leads to a progress in analysis of beams, plates and shells resting on elastic foundations and enables novel formulations for the modulus of a foundation for layered and inhomogeneous foundations.

The asymptotic method of this monograph is also applied to a number of dynamic boundary value problems for thin walled structures, including these interacting with various physical fields.

The equations and figures in this book are numbered within each chapter. In case of a reference to a formula from another chapter a triple numeration is employed, with the first number denoting the number of the chapter. The numeration of tables is continuous.

The author has had close collaborative links with Prof. A.L. Goldenveizer for a number of years, which are very gratefully recognized. Fruitful discussions of some of the results obtained by the author with Prof. S.A. Ambartsumyan, are acknowledged. The author is also indebted to Prof. P.E. Tovstik for useful remarks and comments.

L. A. Aghalovyan

Preface
for the English Edition

Every author is pleased when his research is honored and becomes available to a wider range of readers.

The Russian edition of the monograph has become known for demonstration of the capacity of asymptotic methods for solving new classes of problems for beams, plates and shells composed of both traditional elastic and composite materials. The author is therefore hopeful that the English edition to be of interest, stimulating further developments in the area of thin walled structures, especially, those interacting with various physical-mechanical fields.

A new Chapter 11 has been added to this English edition, dealing with 3D dynamic problems for anisotropic plates. Some novel classes of problems which may be successfully treated by asymptotic methods are pointed out. The literature has also been updated.

It is a pleasure for the author to express his sincere gratitude to Prof. J. Kaplunov, without whom this book would hardly become possible. His idea was supported by Prof. Yang Chen, and Dr. D.A. Prikazchikov, who has acted as a translator. Kind attitude of the editors of World Scientific Publishing Mr. Roh-Suan Tung and Ms. E.H. Chionh is also gratefully acknowledged. Finally, the author feels indebted to Dr. Lusine Ghulghazaryan for help with technical preparation of the manuscript.

L. A. Aghalovyan
September 2014

Foreword

Beams, plates, and shells are components of nearly all modern structures. Their geometry is typically characterized by having one of the dimensions significantly deviating from the others. In case of the beams one of the dimensions usually exceeding the other two, whereas in case of plates and shells the situation is opposite. This geometric feature should have general effect on the methods of analysis of these solid structures. A comprehensive analysis of the stress-strain field should rely on the formulation of 3D elasticity in the narrow region. Since this analysis is often hardly possible, and at the same time motivated by the above-mentioned geometric feature of beams, plates, and shells, a number of approximate applied theories has arisen, relying on certain assumptions. Adoption of a hypothesis leads at the end to some form of variation of the sought-for quantities over the transverse coordinate (thickness). This may be interpreted as an approximation of the associated stress and displacement components. Therefore, the current state of art (within analytical framework) is two-fold. It includes on the one hand a number of well-established technical theories of beams, plates and shells, with the range of applicability of these typically subject of refinement. On the other hand, there is always a correctly formulated problem of 3D elasticity, which often does not have a straightforward solution. Among the methods applied to this problem, and also to justification of the applied theories, we mention the power series, expansion along the Legendre polynomials, variational approach and combinations of methods. Surprisingly, despite their relatively long history, the asymptotic methods have not been applied until relatively recently, even though the presence of a small parameter is trivially observed. This could possibly be explained by the fact that the perturbation by a small parameter occurred to be singular, with the mathematical theory of such equations developed intensively only from late 1940s. Similar types of singularly perturbed equations were arising earlier in hydromechanics, celestial mechanics, electrical engineering, and other areas of natural sciences. Though all of the early treatments of such problems were intuitive, it should be mentioned that most of the adopted assumptions have been verified later through rigorous mathematics. It therefore seems logical that most of the terminology involving boundary layers, matching and others, was transferred to mathematical literature. Let us discuss briefly some of the specific distinctions between regularly and singularly perturbed equations.

The singularity of the perturbation leads to certain peculiarities of the properties of the solution as a function of a small parameter. In case of regular perturbation the resulting solution is a continuous function of a small parameter, whereas in case of singular perturbation the solution is discontinuous. For example, in case of a regularly perturbed equation, e.g.

$$u'' - \varepsilon u' = 0 \tag{0.1}$$

the solution $u = C_1 + C_2 e^{\varepsilon x}$ is continuous in ε, whereas in case of a singularly perturbed equation (containing small parameter as a coefficient within senior derivative)

$$\varepsilon u'' - u' = 0 \tag{0.2}$$

the solution $u = C_1 + C_2 e^{x/\varepsilon}$ exhibits a discontinuity at $\varepsilon = 0$, $x \neq 0$.

The solution of initial or boundary value problems for a regularly perturbed problem may be found as series in small parameter, i.e. $u = \sum_{s=0}^{\infty} \varepsilon^s u_s$. It is also known from the very beginning that in view of certain conditions, the solution for small ε is close to the solution at $\varepsilon = 0$, or the so-called degenerating or unperturbed solution. The problem is therefore reduced to further refinement of this solution using higher order approximations. In order to determine the solution for a regularly perturbed equation, the original operator has to be split only once. This is achieved by substitution of the sought-for quantities as series in small parameter into the governing equations along with the boundary and initial conditions. For example, in case of the first equation (0.1) one obtains $u_s'' = u_{s-1}'$. It should be mentioned that direct correspondence holds between the order of the equation and the number of conditions. In case of singular perturbation the operator is usually split several times, with the corresponding solutions being both continuous and discontinuous (of boundary layer type). Since the solution of a singularly perturbed problem contains regular component and the boundary layer part, the procedure usually involves several steps: 1) analysis of the degenerated problem including higher order approximations, corresponding to the first splitting of the operator (this part is usually referred to as the outer problem or outer solution within the framework of beams, plates and shells); 2) determination of the boundary layer solutions using the second splitting of the original perturbed operator; 3) matching of the obtained qualitatively different solutions through boundary and initial conditions.

At the moment the theory of singularly perturbed equations with the perturbation caused by the presence of a small parameter at senior derivatives is relatively well-established, see e.g. the monographs of Friedrichs (1955); Vishik and Lusternik (1957); Vishik and Lyusternik (1960); Goldenveizer (1959, 1960); Vazov (1968); Trenogin (1970); Nayfeh (1973); Vasilieva and Butuzov (1973); Lomov (1981). It should be noted that singularly perturbed boundary value problems for plates and shells contain the small parameter as a coefficient of only a part of senior derivative (problems for singularly degenerating areas).

In the classical case the degenerated system is of lower order than the perturbed one. As an example, we mention the classical shell theory. However, in case of plates it is no longer the case, when the degenerated system is of lower spatial dimension. This effect implies infinite number of boundary value functions, forming a countable set, which leads to novel mathematical problems, e.g. the problem of basic properties of the boundary layer solutions.

The development of asymptotic techniques for theories of plates and shells was initially carried out by several eminent scientists including K.O. Friedrichs, A. Green, A.L. Goldenveizer, E. Reiss, I.I. Vorovich, within the framework of isotropic plates and shells. As will be shown below, the approach may be extended to anisotropic plates and shells. The asymptotic method allows verification of existing applied theories, and also provides further insights on refinements along with tackling novel classes of boundary value problems.

The above-mentioned novel problems include the case of general anisotropy (containing 21 material constants), the classes of boundary value problems involving the face boundary conditions with prescribed values of displacements or mixed boundary conditions of elasticity, etc.

This monograph is concerned with development of theories for anisotropic beams, plates and shells relying on the method of asymptotic integration of equations of 3D elasticity. The underlying links of asymptotic theories with technical theories and the Saint-Venant principle are revealed. The limiting transition from the 3D formulation of elasticity to 2D approximate theories is studied for several classes of boundary conditions. The distribution of stress field in the near-edge vicinity is investigated. The corresponding dependence of the velocity of attenuation on elastic parameters is discussed. Interaction of the outer solution with the boundary layer solution is analyzed along with the cases of degeneration of edge effects.

Solutions for a new class of mixed boundary value problems for anisotropic beams, plates and shells are presented. Using the same asymptotic method, some more novel results for elastic foundations have been established.

In view of the vast literature on justification of asymptotic methods, this issue is only partly discussed in the current monograph with the main attention drawn to the results following from the method.

Contents

Chapter 1

Plane Problem for a Rectangular Elastic Strip

Technical Theory for Bernoulli-Coulomb-Euler Beams

1.1   Outer Solution for a Rectangular Isotropic Strip

Plane problems of elasticity for infinite or semi-infinite layers of finite thickness and for rectangular area have been widely studied by many researchers. This was motivated by numerous practical engineering applications and also by the fact that these relatively simple problems are serving as a basis for verification of some more advanced approaches and methods of mathematical theory of elasticity allowing their further development. A vast variety of methods including the trigonometric series, integral transforms, orthogonal polynomials, biorthogonal functions, complex variable functions, energy methods, the Schiff-Papkovich-Lourier method of homogeneous solutions along with many others were used for solution of these basic problems, see e.g. Abramyan (1957); Vorovich (1966a); Babloyan and Gulkanyan (1969); Babloyan and Mkrtchyan (1972). For simplicity we start with presenting the asymptotic method for a finite rectangular isotropic strip and later generalize that to the case of anisotropic media (Aghalovyan, 1977; Aghalovyan and Khachatryan, 1977a).

Consider a rectangular strip of length a and height $2h$. Let us solve the plane problem of elasticity in the area $\Omega = \{(x, y) : x \in [0, a], |y| \leq h, 2h \ll a\}$, with the traction components prescribed at the faces $y = \pm h$

$$\sigma_{xy} = \pm \frac{a}{h} X^{\pm}(x), \quad \sigma_y = \pm Y^{\pm}(x) \quad \text{at} \quad y = \pm h \tag{1.1}$$

along with one of the following boundary conditions at the edges $x = 0, x = a$ (Aghalovyan, 1977)

$$\sigma_x = \varphi_1(\zeta), \quad \sigma_{xy} = \varphi_2(\zeta) \qquad \text{(Problem 1)} \tag{1.2}$$

$$u = \varphi_1(\zeta), \quad \sigma_{xy} = \varphi_2(\zeta) \qquad \text{(Problem 2)} \tag{1.3}$$

$$\sigma_x = \varphi_1(\zeta), \quad \text{v} = \varphi_2(\zeta) \qquad \text{(Problem 3)} \tag{1.4}$$

$$u = \varphi_1(\zeta), \quad \text{v} = \varphi_2(\zeta) \qquad \text{(Problem 4)} \tag{1.5}$$

where $\zeta = \dfrac{y}{h}$.

Here we assume that the strip as an elastic body is in the equilibrium state, and that the following elasticity relations are satisfied

$$\frac{\partial \sigma_x}{\partial x} + \frac{\partial \sigma_{xy}}{\partial y} = 0, \qquad \frac{\partial \sigma_{xy}}{\partial x} + \frac{\partial \sigma_y}{\partial y} = 0,$$

$$\frac{\partial u}{\partial x} = \frac{1}{E}(\sigma_x - \nu\sigma_y), \qquad \frac{\partial v}{\partial y} = \frac{1}{E}(\sigma_y - \nu\sigma_x), \tag{1.6}$$

$$\frac{\partial u}{\partial y} + \frac{\partial v}{\partial x} = \frac{1}{G}\sigma_{xy}.$$

Some comments on the inequality $2h \ll a$ and the presence of the factor $\dfrac{a}{h}$ in (1.1) will be given later.

Proceeding with the solution of the above-stated plane problems of elasticity including the equilibrium equations (1.6) and the elasticity relations, we introduce the dimensionless coordinates $\xi = \dfrac{x}{a}$, $\zeta = \dfrac{y}{h}$. We will therefore result in the following system, containing a small parameter $\varepsilon = \dfrac{h}{a}$ as a factor of derivatives

$$\frac{\partial \sigma_x}{\partial \xi} + \varepsilon^{-1}\frac{\partial \sigma_{xy}}{\partial \zeta} = 0, \qquad \frac{\partial \sigma_{xy}}{\partial \xi} + \varepsilon^{-1}\frac{\partial \sigma_y}{\partial \zeta} = 0,$$

$$\frac{\partial U}{\partial \xi} = \frac{1}{E}(\sigma_x - \nu\sigma_y), \qquad \varepsilon^{-1}\frac{\partial V}{\partial \zeta} = \frac{1}{E}(\sigma_y - \nu\sigma_x), \tag{1.7}$$

$$\varepsilon^{-1}\frac{\partial U}{\partial \zeta} + \frac{\partial V}{\partial \xi} = \frac{1}{G}\sigma_{xy},$$

where E is the Young's modulus, ν is the Poisson ratio, G is the shear modulus, and $U = \dfrac{u}{a}$, $V = \dfrac{v}{a}$ are the dimensionless displacements. The obtained system (1.7) may be referred to as a singularly perturbed one, since its degenerate system is of lower dimension. Here and below we are using the term "singularly perturbed" in the same meaning. Hence, solution of (1.7) is formed by two types of solutions - the outer solution, i.e. not decaying away from the boundary and the boundary layer type solution, see Vishik and Lusternik (1957); Goldenveizer (1962); Vasilieva and Butuzov (1973). Solution of each of the problems (1-4) is also decomposed into solutions of symmetric (extensional) and anti-symmetric (bending) problems, with the corresponding boundary conditions written as:

(a) *symmetric problem*

$$\sigma_{xy} = \pm\frac{a}{h}X_1, \quad \sigma_y = Y_1 \qquad \text{at} \qquad y = \pm h \tag{1.8}$$

(b) *anti-symmetric problem*

$$\sigma_{xy} = \frac{a}{h}X_2, \quad \sigma_y = \pm Y_2 \qquad \text{at} \qquad y = \pm h, \tag{1.9}$$

where

$$X_i = \frac{1}{2}(X^+ \pm X^-), \quad Y_i = \frac{1}{2}(Y^+ \mp Y^-) \quad (i = 1, 2). \tag{1.10}$$

In case of a symmetric problem the quantities u, σ_x, σ_y are even and v, σ_{xy} are odd with respect to y, and vice versa for the anti-symmetric problem.

Let us search for the outer solution for the strip in the general form (Aghalovyan, 1977)

$$Q = \varepsilon^{-q} \sum_{s=0}^{S} \varepsilon^s Q^{(s)}. \tag{1.11}$$

Here Q is any of the traction or dimensionless displacement components, $Q^{(s)} \equiv 0$ at $s < 0$, S is the number of iterations, and q is an integer parameter which may vary for different displacement or stress components, also known as the variability index, see also Goldenveizer (1962) and Kaplunov et al. (1998).

It has to be chosen in such a way that after substitution of (1.11) into the equations of elasticity (1.7) the latter give a consistent asymptotic system providing a recurrent procedure with respect to $Q^{(s)}$. For σ_x, u, σ_{xy} and σ_y the values are $q = 2, 1, 0$, respectively, whereas for the vertical displacement v we have $q = 1$ in symmetric problem and $q = 3$ in the anti-symmetric one. As a result, we arrive at the following system

$$\frac{\partial \sigma_x^{(s)}}{\partial \xi} + \frac{\partial \sigma_{xy}^{(s)}}{\partial \zeta} = 0, \quad \frac{\partial \sigma_{xy}^{(s)}}{\partial \xi} + \frac{\partial \sigma_y^{(s)}}{\partial \zeta} = 0,$$

$$\frac{\partial U^{(s)}}{\partial \xi} = \frac{1}{E}(\sigma_x^{(s)} - \nu \sigma_y^{(s-2)}), \quad \varepsilon^{-1}\frac{\partial V^{(s)}}{\partial \zeta} = \frac{1}{E}(\sigma_y^{(m)} - \nu \sigma_x^{(n)}), \tag{1.12}$$

$$\frac{\partial U^{(s)}}{\partial \zeta} + \frac{\partial V^{(k)}}{\partial \xi} = \frac{1}{G}\sigma_{xy}^{(s-2)}.$$

In case of the symmetric problem $m = s - 2$, $n = s$, $k = s - 2$, whereas for the anti-symmetric one $m = s - 4$, $n = s - 2$, $k = s$. It may be shown that any other combinations of parameters leads to an inconsistent asymptotic system, i.e. either having non-closed form or contradicting the boundary conditions.

The system (1.12) may now be integrated with respect to ζ. Taking into account (1.8) and (1.9), we obtain for *symmetric problem*

$$U^{(s)} = u^{(s)}(\xi) + u^{*(s)}, \quad \sigma_x^{(s)} = E\frac{du^{(s)}}{d\xi} + \sigma_x^{*(s)},$$

$$V^{(s)} = -\nu\frac{du^{(s)}}{d\xi}\zeta + v^{*(s)}, \quad \sigma_{xy}^{(s)} = -p^{(s)}\zeta + \sigma_{xy}^{*(s)}, \tag{1.13}$$

$$\sigma_y^{(s)} = \frac{1}{2}(\zeta^2 - 1)\frac{dp^{(s)}}{d\xi} + Y_1^{(s)} + \sigma_y^{*(s)} - \sigma_y^{*(s)}(\zeta = 1),$$

$$p^{(s)} = -X_1^{(s)} + \sigma_{xy}^{*(s)}(\zeta = 1),$$

where $u^{(s)}$ is determined from

$$E\frac{d^2 u^{(s)}}{d\xi^2} = p^{(s)} \quad \text{i.e.} \quad Eu^{(s)} = \int_0^\xi d\xi \int_0^\xi p^{(s)} d\xi + C_1^{(s)}\xi + C_2^{(s)}, \tag{1.14}$$

$$X_1^{(0)} = X_1, \quad Y_1^{(0)} = Y_1, \quad X_1^{(s)} = Y_1^{(s)} = 0, \quad (s > 0),$$

$$u^{*(s)} = \int_0^\zeta \left(\frac{1}{G}\sigma_{xy}^{(s-2)} - \frac{dV^{(s-2)}}{d\xi}\right) d\zeta, \quad \sigma_x^{*(s)} = E\frac{du^{*(s)}}{d\xi} + \nu\sigma_y^{(s-2)},$$

$$v^{*(s)} = \int_0^\zeta \frac{1}{E}\left(\sigma_y^{(s-2)} - \nu\sigma_x^{*(s)}\right) d\zeta, \tag{1.15}$$

$$\sigma_{xy}^{*(s)} = -\int_0^\zeta \frac{\partial\sigma_x^{*(s)}}{\partial\xi} d\zeta, \quad \sigma_y^{*(s)} = -\int_0^\zeta \frac{\partial\sigma_{xy}^{*(s)}}{\partial\xi} d\zeta.$$

We remark that quantities with the asterisk are obtained from the previous order asymptotic results, namely $Q^{*(0)} = Q^{*(1)} \equiv 0$.

In case of the *anti-symmetric problem*

$$V^{(s)} = w^{(s)}(\xi) + v^{*(s)}, \quad u^{(s)} = -\zeta\frac{dw^{(s)}}{d\xi} + u^{*(s)},$$

$$\sigma_x^{(s)} = \zeta\tau_{xx}^{(s)} + \sigma_x^{*(s)},$$

$$\sigma_{xy}^{(s)} = \frac{1}{2}(\zeta^2 - 1)\tau_{xy}^{(s)} + X_2^{(s)} + \sigma_{xy}^{*(s)} - \sigma_{xy}^{*(s)}(\zeta = 1), \tag{1.16}$$

$$\sigma_y^{(s)} = \frac{1}{2}(1 - \zeta^2)\zeta q^{(s)} + \zeta Y_2^{(s)} + \sigma_y^{*(s)} - \zeta\sigma_y^{*(s)}(\zeta = 1),$$

where

$$\tau_{xx}^{(s)} = -E\frac{d^2 w^{(s)}}{d\xi^2}, \quad \tau_{xy}^{(s)} = E\frac{d^3 w^{(s)}}{d\xi^3},$$

$$q^{(s)} = Y_2^{(s)} + \frac{dX_2^{(s)}}{d\xi} - \sigma_y^{*(s)}(\zeta = 1) - \frac{d\sigma_{xy}^{*(s)}}{d\xi}(\zeta = 1),$$

$$X_2^{(0)} = X_2, \quad Y_2^{(0)} = Y_2, \quad X_2^{(s)} = Y_2^{(s)} = 0, \quad (s > 0),$$

$$v^{*(s)} = \frac{1}{E}\int_0^\zeta \left(\sigma_y^{(s-4)} - \nu\sigma_x^{(s-2)}\right) d\zeta,$$

$$u^{*(s)} = \int_0^\zeta \left(-\frac{\partial v^{*(s)}}{\partial\xi} + \frac{1}{G}\sigma_{xy}^{(s-2)}\right) d\zeta, \tag{1.17}$$

$$\sigma_x^{*(s)} = E\frac{\partial u^{*(s)}}{\partial\xi} + \nu\sigma_y^{(s-2)}, \quad \sigma_{xy}^{*(s)} = -\int_0^\zeta \frac{\partial\sigma_x^{*(s)}}{\partial\xi} d\zeta,$$

$$\sigma_y^{*(s)} = -\int_0^\zeta \frac{\partial\sigma_{xy}^{*(s)}}{\partial\xi} d\zeta.$$

The function $w^{(s)}$ satisfies the following equation

$$\frac{1}{3}E\frac{d^4 w^{(s)}}{d\xi^4} = q^{(s)}, \tag{1.18}$$

i.e.

$$\frac{1}{3}Ew^{(s)} = \int_0^\xi d\xi \int_0^\xi d\xi \int_0^\xi d\xi \int_0^\xi q^{(s)}d\xi + C_1^{(s)}\frac{\xi^3}{3!} - C_2^{(s)}\frac{\xi^2}{2!} + C_3^{(s)}\xi + C_4^{(s)}. \quad (1.19)$$

Thus, we obtained a closed solution for the inner region for any s, given by (1.13) - (1.19). The quantities X_1 and X_2 appear in (1.13), (1.16) starting from the leading order at $s = 0$. If the large parameter a/h was not introduced in the boundary conditions (1.1), (1.8), (1.9), then the contribution of X_i $(i = 1, 2)$ would start from the next order $s = 1$. These asymptotic results possess a clear physical interpretation, namely, that in the considered problems the normal and tangential loading lead to stresses of different asymptotic orders. In order to make these commensurable, the intensity of tangential loading should be greater than that of the normal loading. This fact may be seen from the corresponding scaling factor in (1.1), (1.8), and (1.9).

It may be observed from (1.14) and (1.19) that the integral of the outer problem contains four arbitrary constants in the anti-symmetric problem along with two constants in the symmetric case. Obviously, this is not enough to satisfy the edge boundary conditions (1.2)-(1.5), which provides another evidence of a singularly perturbed system, since in case of a regular perturbation analogous asymptotic solution should contain appropriate number of constants in order to satisfy the boundary conditions. Therefore, in case of a singular perturbation we require a different, qualitatively new type of solution, namely, the boundary layer, which is decaying rapidly away from the edges and satisfying the stress free edge boundary conditions.

The boundary layer solution is localized near the edges $x = 0, a$. Within the adopted asymptotic method it is assumed that the influence of the boundary layer constructed at $x = 0$ is not reaching another edge $x = a$, which is also in line with $h \ll a$. In the next section we present more details on the estimates of the decay zone for the boundary layer. We also remark that in general, though the boundary layer is localized near the edges, it still affects the whole solution through the boundary conditions at $x = 0, a$. This influence may become particularly significant in case of anisotropic plates and shells.

1.2 Boundary Layer Solution. Decay of Edge Effects

In order to construct the boundary layer solution in the vicinity of the edge $\xi = 0$, we introduce the asymptotic scaling $t = \xi/\varepsilon$, which leads to extracting the leading order terms of the derivatives along the longitudinal coordinate.

Solution of the obtained equations is sought in the form (Aghalovyan, 1973a, 1977)

$$R_b = \sum_{s=0}^N \varepsilon^{\chi_b+s} R_b^{(s)}(\zeta) \exp(-\lambda t), \quad (2.1)$$

where R_b is the appropriate displacement or stresses component. Here and below all the quantities related to boundary layer are denoted with a subscript b. The real parameters χ_b are the intensity indices, which have to be chosen in order to obtain a consistent asymptotic system after substitution of (2.1) into the equation of the plane problem and stress free boundary conditions at $\zeta = \pm 1$. It may be shown that this is only possible provided $\chi_{\sigma_i} = \chi$, $\chi_{u_i} = \chi + 1$, where χ is to be determined from the interaction between the outer solution and the boundary layer and λ is characterizing the variability of the displacement and stresses components of the boundary layer. The nature of the boundary layer solution implies decay of the solutions at $t \to +\infty$, hence, $\mathrm{Re}\lambda > 0$. In general $\lambda = \lambda(\zeta)$, however, it may be shown that in case of the considered plane problem $\lambda = const$ (Aghalovyan, 1973a). As a result, we get

$$-\lambda\sigma_{xb}^{(s)} + \frac{d\sigma_{xyb}^{(s)}}{d\zeta} = 0, \quad -\lambda\sigma_{xyb}^{(s)} + \frac{d\sigma_{yb}^{(s)}}{d\zeta} = 0,$$

$$-\lambda u_b^{(s)} = \frac{1}{E}\left(\sigma_{xb}^{(s)} - \nu\sigma_{yb}^{(s)}\right), \quad \frac{dv_b^{(s)}}{d\zeta} = \frac{1}{E}\left(\sigma_{yb}^{(s)} - \nu\sigma_{xb}^{(s)}\right), \tag{2.2}$$

$$\frac{du_b^{(s)}}{d\zeta} - \lambda v_b^{(s)} = \frac{1}{G}\sigma_{xyb}^{(s)}.$$

Since the non-homogeneous boundary conditions imposed on the faces of the strip are satisfied by the solution over the interior, system (2.2) should now be solved subject to

$$\sigma_{xyb} = \sigma_{yb} = 0 \quad \text{at} \quad \zeta = \pm 1. \tag{2.3}$$

All of the quantities in (2.2) may be expressed through $\sigma_{yb}^{(s)}$, which is determined from the fourth order ordinary differential equation. Calculating σ_{xyb} and σ_{yb} and satisfying conditions (2.3), we arrive at a homogeneous algebraic system of equations with respect to the arbitrary constants of the boundary layer solution. The existence of non-trivial solution of the latter leads to a relation, from which the constant λ is determined, namely

$$\sin 2\lambda \pm 2\lambda = 0, \tag{2.4}$$

with these two transcendental equations corresponding to the symmetric and anti-symmetric problems.

The resulting solution of the boundary value problem (2.2), (2.3) is given by

$$\sigma_{yb}^{(s)} = F_n A_n^{(s)}, \quad \sigma_{xyb}^{(s)} = \frac{F_n'}{\lambda_n} A_n^{(s)}, \quad \sigma_{xb}^{(s)} = \frac{F_n''}{\lambda_n^2} A_n^{(s)},$$

$$u_b^{(s)} = \frac{1}{E}\left(-\frac{F_n''}{\lambda_n^3} + \frac{\nu}{\lambda_n}F_n\right) A_n^{(s)}, \tag{2.5}$$

$$v_b^{(s)} = -\frac{1}{E\lambda_n^2}\left(\frac{F_n'''}{\lambda_n^2} + (2+\nu)F_n'\right) A_n^{(s)}.$$

In case of the *symmetric* problem

$$F_n(\zeta) = \zeta \sin \lambda_n \zeta - tg\lambda_n \cos \lambda_n \zeta, \qquad (2.6)$$

with λ_n being the root of $\sin 2\lambda + 2\lambda = 0$,

whereas in the *anti-symmetric* case

$$F_n(\zeta) = \sin \lambda_n \zeta - \zeta \, tg\lambda_n \cos \lambda_n \zeta, \qquad (2.7)$$

with λ_n being the solution of $\sin 2\lambda - 2\lambda = 0$.

It should be noted that we adopt summation over the repeated index n corresponding to λ_n satisfying $\mathrm{Re}\lambda_n > 0$ in (2.5) and below in solutions for the boundary layer. We also remark that the only real root of (2.4) is $\lambda = 0$. As follows from (2.2), (2.3), it corresponds to zero solution associated physically with rigid motion. Equations (2.4) are often arising in various problems of mathematical physics and elasticity, and are therefore well-studied (Hillman and Salzer, 1943; Mittelman and Hillman, 1946; Vorovich, 1966b; Uflyand, 1967). The roots of equations (2.4) except $\lambda = 0$ are complex and form a countable set, being symmetric in respect of the origin and the real axe. The eigenfunctions corresponding to $\mathrm{Re}\lambda_n > 0$ are forming the doubly complete system according to the Keldysh theorem, see Vorovich and Kopasenko (1966); Keldysh (1951). Table 1 contains data for the first ten roots of equations (2.4) presented in ascending order of their real parts.

Table 1

$2\lambda_n = X_n \pm iY_n$			
$\sin 2\lambda + 2\lambda = 0$		$\sin 2\lambda - 2\lambda = 0$	
n	X_n	Y_n	X_n	Y_n
1	4.212392	2.250728	7.497676	2.768678
2	10.712537	3.103148	13.899960	3.352210
3	17.073365	3.551087	20.238518	3.716768
4	23.398355	3.858808	26.554547	3.983142
5	29.708120	4.093704	32.859741	4.193251
6	36.009866	4.283780	39.158817	4.366795
7	42.306827	4.443445	45.454071	4.514640
8	48.600684	4.581103	51.746768	4.643428
9	54.892406	4.702095	58.037662	4.757515
10	61.182590	4.810024	64.327234	4.859917

The real part of the first root of (2.4) is in fact characterizing the speed of attenuation. Therefore, the decomposition of the full solution for rectangular strip into the outer solution and the boundary layer is only possible when the length a is such that the quantities of order $O\left[\exp\left(-\dfrac{a}{h}\mathrm{Re}\lambda_1\right)\right]$ are negligible compared to unity.

$$O\left[\exp\left(-\frac{a}{h}\mathrm{Re}\lambda_1\right)\right] \ll 1.$$

Here and below we refer to the rectangle satisfying this inequality as a strip. In case of a symmetric problem $\mathrm{Re}\lambda_1 \approx 2.1062$, whereas in the anti-symmetric case $\mathrm{Re}\lambda_1 \approx 3.75$. Hence, the above mentioned conditions hold when the distance from the edges of the strip is about $1.5-2$ times the thickness of the strip. We also remark that the values of real parts of the first root indicate that the flexural boundary layer is decaying faster than the extensional one.

The displacement and stress components defined by (2.5) are real since they are expressed through both complex conjugates λ_n and $\overline{\lambda}_n$. Denoting $\lambda_n = x_n + iy_n$, $(x_n > 0, y_n > 0)$, and $2A_n^{(s)} = A_{1n}^{(s)} - iA_{2n}^{(s)}$, the solution may be presented in general form as $Q^{(s)} = A_{1n}^{(s)}\mathrm{Re}Q^{(s)} + A_{2n}^{(s)}\mathrm{Im}Q^{(s)}$, where $A_{1n}^{(s)}$ and $A_{2n}^{(s)}$ are real arbitrary constants. In particular, the stress components are

(a) *in the symmetric case*

$$
\sigma_{xb} = \varepsilon^\chi \sum_{s=0}^{S} \varepsilon^s \{[(\delta_n - 2\omega_2)(\varphi_1 + \varphi_2) + (\gamma_n + 2\omega_1)(\varphi_3 - \varphi_4)
$$

$$
-\zeta(\varphi_7 + \varphi_8)]A_{1n}^{(s)} + [(\delta_n - 2\omega_2)(\varphi_3 - \varphi_4) - (\gamma_n + 2\omega_1)(\varphi_1 + \varphi_2)
$$

$$
+\zeta(\varphi_5 - \varphi_6)]A_{2n}^{(s)}\} \exp(-x_n t),
$$

$$
\sigma_{xyb} = \varepsilon^\chi \sum_{s=0}^{S} \varepsilon^s \{[(\gamma_n + \omega_1)(\varphi_7 + \varphi_8) + (\delta_n - \omega_2)(\varphi_5 - \varphi_6)
$$

$$
+\zeta(\varphi_3 - \varphi_4)]A_{1n}^{(s)} + [(\gamma_n + \omega_1)(\varphi_6 - \varphi_5) + (\delta_n - \omega_2)(\varphi_7 + \varphi_8)
$$

$$
-\zeta(\varphi_1 + \varphi_2)]A_{2n}^{(s)}\} \exp(-x_n t), \qquad (2.8)
$$

$$
\sigma_{yb} = \varepsilon^\chi \sum_{s=0}^{S} \varepsilon^s \{[\zeta(\varphi_7 + \varphi_8) - \gamma_n(\varphi_3 - \varphi_4) - \delta_n(\varphi_1 + \varphi_2)]A_{1n}^{(s)}
$$

$$
+[\zeta(\varphi_6 - \varphi_5) + (\varphi_1 + \varphi_2)\gamma_n - \delta_n(\varphi_3 - \varphi_4)]A_{2n}^{(s)}\} \exp(-x_n t),
$$

(b) *in the anti-symmetric case*

$$
\sigma_{xb} = \varepsilon^\chi \sum_{s=0}^{S} \varepsilon^s \{[\zeta(\varphi_1 + \varphi_2)\delta_n + \zeta(\varphi_3 - \varphi_4)\gamma_n
$$

$$
+2(\omega_1\delta_n - \omega_2\gamma_n)(\varphi_5 - \varphi_6) + (2(\omega_1\gamma_n + \omega_2\delta_n) - 1)(\varphi_7 + \varphi_8)]A_{1n}^{(s)}
$$

$$
+[-\zeta(\varphi_1 + \varphi_2)\gamma_n + \zeta(\varphi_3 - \varphi_4)\delta_n - (2(\omega_1\gamma_n + \omega_2\delta_n) - 1)(\varphi_5 - \varphi_6)
$$

$$
+2(\omega_1\delta_n - \omega_2\gamma_n)(\varphi_7 + \varphi_8)]A_{2n}^{(s)}\} \exp(-x_n t),
$$

$$
\sigma_{xyb} = \varepsilon^\chi \sum_{s=0}^{S} \varepsilon^s \{[(\omega_1\gamma_n + \omega_2\delta_n - 1)(\varphi_4 - \varphi_3) - (\omega_1\delta_n - \omega_2\gamma_n)(\varphi_1 + \varphi_2) \quad (2.9)
$$

$$
+\zeta(\varphi_5 - \varphi_6)\delta_n + \zeta(\varphi_7 + \varphi_8)\gamma_n]A_{1n}^{(s)} + [(\omega_1\gamma_n + \omega_2\delta_n - 1)(\varphi_1 + \varphi_2)
$$

$$-(\omega_1\delta_n - \omega_2\gamma_n)(\varphi_3 - \varphi_4) - \zeta(\varphi_5 - \varphi_6)\gamma_n + \zeta(\varphi_7 + \varphi_8)\delta_n]A_{2n}^{(s)}\}\exp(-x_n t),$$

$$\sigma_{yb} = \varepsilon^X \sum_{s=0}^{S} \varepsilon^s \{[\varphi_7 + \varphi_8 - \zeta(\gamma_n(\varphi_3 - \varphi_4) + \delta_n(\varphi_1 + \varphi_2))]A_{1n}^{(s)}$$

$$+[\varphi_6 - \varphi_5 + \zeta((\varphi_1 + \varphi_2)\gamma_n - \delta_n(\varphi_3 - \varphi_4))]A_{2n}^{(s)}\}\exp(-x_n t),$$

where

$$\omega_1 = \frac{x_n}{x_n^2 + y_n^2}, \qquad \omega_2 = \frac{y_n}{x_n^2 + y_n^2},$$

$$\gamma_n = \frac{\sin 2x_n}{\cos 2x_n + \mathrm{ch}2y_n}, \qquad \delta_n = \frac{\mathrm{sh}2y_n}{\cos 2x_n + \mathrm{ch}2y_n},$$

$$\varphi_1 = \sin x_n \zeta \,\mathrm{sh} y_n \zeta \cos y_n t, \qquad \varphi_2 = \cos x_n \zeta \,\mathrm{ch} y_n \zeta \sin y_n t,$$

$$\varphi_3 = \cos x_n \zeta \,\mathrm{ch} y_n \zeta \cos y_n t, \qquad \varphi_4 = \sin x_n \zeta \,\mathrm{sh} y_n \zeta \sin y_n t, \qquad (2.10)$$

$$\varphi_5 = \sin x_n \zeta \,\mathrm{ch} y_n \zeta \sin y_n t, \qquad \varphi_6 = \cos x_n \zeta \,\mathrm{sh} y_n \zeta \cos y_n t,$$

$$\varphi_7 = \sin x_n \zeta \,\mathrm{ch} y_n \zeta \cos y_n t, \qquad \varphi_8 = \cos x_n \zeta \,\mathrm{sh} y_n \zeta \sin y_n t.$$

The displacements may be treated similarly by making use of (2.5)-(2.7). Since the analysis has been carried out for a homogeneous system with zero boundary conditions, the solution is obtained to within a constant factor ε^X.

We remark that solution (2.1), (2.5) is an exact solution for any s and may be thought of as a homogeneous Papkovich solution. Functions F_n acting in the boundary layer solution satisfy the generalized orthogonality conditions,

$$\int_0^1 [F_n'' F_k'' - \lambda_n^2 \lambda_k^2 F_n F_k]d\zeta = 0, \quad (n \neq k) \qquad (2.11)$$

see Papkovich (1940).

It is worth noting that the stress tensor components of the boundary layer σ_{xb}, σ_{xyb} are self-equilibrated along the height, i.e.

$$\int_{-1}^{+1} \sigma_{xb}d\zeta = 0, \qquad \int_{-1}^{+1} \zeta\sigma_{xb}d\zeta = 0, \qquad \int_{-1}^{+1} \sigma_{xyb}d\zeta = 0. \qquad (2.12)$$

Indeed, noting that

$$F_n(\pm 1) = F_n'(\pm 1) = 0, \qquad (2.13)$$

the conditions (2.12) may be easily justified by direct integration.

The displacement components in general are violating this property, since the displacement and rotation angle at mid-height are non-zero. Thus, the self-equilibrated stresses σ_{xb}, σ_{xyb} correspond to non-self-equilibrated displacements (i.e. not satisfying (2.12)). Therefore, at the ends of the strip the stresses σ_{xb}, σ_{xyb} may be caused by self-equilibrated loads only, while the displacements appearing at the edge boundary conditions may correspond to self-equilibrated stress components only. These properties of displacements and stresses will be used later for matching the solutions of the inner problem and that of the boundary layer.

The boundary layer in the vicinity $x = a$ is constructed in a similar manner. In fact, it could be obtained from that derived for another edge $x = 0$ by scaling $t_1 = \varepsilon^{-1} - t = (a - x)/h$. We also remark that the solution of this subsection is given for a plane stress problem. The corresponding results for plane strain case may be obtained from the latter by a standard substitution of $E/(1 - \nu^2)$ and $\nu/(1 - \nu)$ instead of E and ν, respectively.

1.3   Interaction between the Boundary Layer and Outer Solution

The two types of solutions have been constructed above, namely the outer solution and the boundary layer. Clearly, their sum

$$I = Q + R_b^{(1)} + R_b^{(2)} \tag{3.1}$$

is the asymptotic solution of original singularly perturbed boundary value problem. Here Q is the outer solution, and $R_b^{(1)}$ and $R_b^{(2)}$ are boundary layers constructed in the vicinity of the edges $x = 0$ and $x = a$, respectively.

After a number of iterations it is possible to satisfy the governing equations of elasticity along with the boundary conditions at $y = \pm h$ to give asymptotic accuracy. We remark that in the above treated problems for a strip the boundary layer is satisfying all the equations and boundary conditions exactly, however it is not the case for plates and shells. At the moment the conditions (1.2)-(1.5) are not satisfied. On the other hand, the solution (3.1) contains sufficient amount of arbitrary constants in order to match these boundary conditions. Let us present some further details of the procedure.

Consider the first boundary value problem (Problem 1), see (1.2). Substituting (3.1) in the latter and noting (1.11) and (2.1), along with the fact that the only acting boundary layer solution at $\xi = 0$ is $R_b^{(1)}$, and the effect of $R_b^{(2)}$ may be neglected at the edge $\xi = 0$, we result in

$$\sigma_{xb}^{(s)} + \sigma_x^{(s+2+\chi)} = \varphi_1^{(s+\chi)},$$
$$\sigma_{xyb}^{(s)} + \sigma_{xy}^{(s+1+\chi)} = \varphi_2^{(s+\chi)}, \tag{3.2}$$
$$\text{at} \quad \xi = 0 \ (t = 0).$$

The value of χ should now be chosen in order to resolve the sequence of boundary conditions following from (3.2) in respect of the arbitrary constants of the outer solution and the boundary layer. In other words, conditions implied from (3.2) should not contradict the iterative differential equations for quantities of the outer problem and the boundary layer. Let us refer to such value of χ as a *consistent* value, which may be found in our case as $\chi = -2$. Indeed, if $|\chi| > 2$ then at $s = 0$ we obtain homogeneous boundary conditions for both the outer solution and the boundary layer, which means asymptotic degeneration. The other case of $|\chi| < 2$ leads to an inconsistent system. Thus, the only consistent value is $\chi = -2$. Then, the system (3.2) may be rewritten as

$$\sigma_{xb}^{(s)} = \varphi_1^{(s-2)} - \sigma_x^{(s)}(\xi = 0),$$
$$\sigma_{xyb}^{(s)} = \varphi_2^{(s-2)} - \sigma_{xy}^{(s-1)}(\xi = 0), \qquad (3.3)$$
$$\text{at} \quad t = 0.$$

We note that the functions on the right-hand side of (3.3) cannot be arbitrary, since they are bounded by conditions (2.12), more specifically, the last two conditions in case of a symmetric problem and the first condition in case of the anti-symmetric problem. Now, taking into account (3.3), the constants of the outer solution may be determined as

(a) *for symmetric problem*

$$C_1^{(s)} = \int_0^1 [\varphi_1^{(s-2)} - \sigma_x^{*(s)}(\xi = 0)]d\zeta, \qquad (3.4)$$

(b) *for anti-symmetric problem*

$$C_1^{(s)} = [X_2^{(s)} - \sigma_{xy}^{*(s)}(\zeta = 1)]_{\xi=0} + \int_0^1 \sigma_{xy}^{*(s)}(0,\zeta)d\zeta - \int_0^1 \varphi_2^{(s-1)}d\zeta,$$

$$C_2^{(s)} = \int_0^1 \zeta[\varphi_1^{(s-2)} - \sigma_x^{*(s)}(\xi = 0)]d\zeta, \qquad (3.5)$$

$$(\varphi_1^{(0)} = \varphi_1, \ \varphi_2^{(0)} = \varphi_2, \ \varphi_1^{(s)} = \varphi_2^{(s)} \equiv 0, \ s \neq 0).$$

The remaining constants of the outer solution are found from the boundary conditions at the opposite edge of the strip. In particular, if the conditions at the opposite edge are analogous to (2.2), and the strip is in the equilibrium state, the expressions for C_i are similar to (3.4) and (3.5), the difference would be only in the right-hand side of (3.3), i.e. between the boundary layers. Hence, in case of the first boundary value problem, as could be expected, the constants of the outer solution are determined from the conditions on just one edge. It should be mentioned that integral conditions (2.12) are the only relations required for evaluation of the constants. Substituting (3.4) and (3.5) into (1.11), (1.13)-(1.19), we obtain the stress components, while the displacement components are defined to within the rigid body motion.

Now since the outer solution is known, the right-hand sides of (3.3) are fully determined. It has been stated earlier that even though the set of functions $F_n(\zeta)$ is not orthogonal, it is bi-complete. This ensures that there are enough arbitrary constants in the boundary layer solution in order to satisfy conditions (3.3). There are several ways in which these constants may be determined. Unfortunately, only approximate treatment is available. For example, this problem may be reduced to an infinite algebraic system after expanding the coefficients of $A_n^{(s)}$ as a Fourier series in respect of a certain complete orthonormal set of functions. The matrix of the system does not depend on the order of approximation, which reveals the influence the right-hand sides only. As alternative methods, we mention the boundary collocation or the Treftz method, see e.g. Lourier (1970).

As follows from (3.3), the values of the sought-for constants depend only on the self-equilibrated part of the prescribed edge loading. From (1.13)-(1.19), (3.4), and (3.5), it may be verified that in general the right-hand side of (3.3) is non-zero at $s = 1$ leading to a non-zero boundary layer solution starting from $s = 1$. Therefore, in view of (2.1), the orders of stress and displacement components of the boundary layer are $O(\varepsilon^{-1})$ and $O(\varepsilon^{0})$, respectively. Comparing these with the orders of the appropriate quantities for the outer solution (1.11), we deduce that the displacements of the boundary layer may be neglected in comparison with the outer displacements, while the contribution of stresses of the boundary layer cannot be ignored. In particular, the orders of the outer stresses and the tangential boundary layer stress coincide.

Here and below we exclude from our consideration the corner angular points and the points corresponding to transition between the types of boundary conditions. It is well known that this matter can be investigated separately, see e.g. Williams (1952, 1959).

Investigation of the stress field in the vicinity of the corner points reduces to searching for the root of the associated transcendental equation with the least positive real part depending on various physical and geometrical parameters of the problem. The secular transcendental equation (equations) may be derived by various methods, including the Fourier method, integral transforms etc. The problems for the homogeneous isotropic body, when the contour in the vicinity of the corner point is either stress free, fixed or subject to mixed boundary conditions, were addressed by Williams (1952, 1959); Uflyand (1967); Arutyunyan and Abramyan (1968); Kalandiya (1969) and others.

Similar problems for non-homogeneous solids are also well-studied (Aksentyan, 1967; Bogy, 1968, 1970, 1971).

The distribution of stress field in the vicinity of the corner points in case of anisotropic bodies or points at the joints between parts of compound solids have been studied by Gevorkyan (1968); Chobanyan and Aleksanyan (1971); Avetisyan and Chobanyan (1972); Chobanyan (1987). Description of the wide range of practical applications of these results may be found in the works by Chobanyan and his co-authors, see e.g. Aksentyan and Luschik (1978).

1.4   Relation to the Saint-Venant's Principle

It has been shown in the previous section that the solution of the first boundary value problem for a rectangular strip is formed by the internal solution and the boundary layer. The constants of the internal problem are defined from integral conditions, i.e. the propagating stress field is not influenced by the self-equilibrated edge loading. The rapidly decaying part of the stress field is fully governed by the self-equilibrated part of the load.

Therefore, it has been shown that two statically equivalent loads applied at the edge lead to identical internal stress-strain state; the difference occurs only at the boundary layer. Within the framework of classical elasticity this reflects the validity of the Saint-Venant's principle for the considered class of problems. Thus, the Saint-Venant's principle follows immediately from the properties of solutions obtained by asymptotic integration of the governing equations of elasticity. Let us now consider several examples in order to illustrate the above stated.

Example 1.1. Let the strip $\Omega = \{(x, y) : 0 \le x \le a, |y| \le h\}$ be subjected to extension load, see Fig. 1.1. The load is distributed uniformly along one of the edges with the density P, whereas the law of distribution at the other edge is given by $\varphi(\zeta)$.

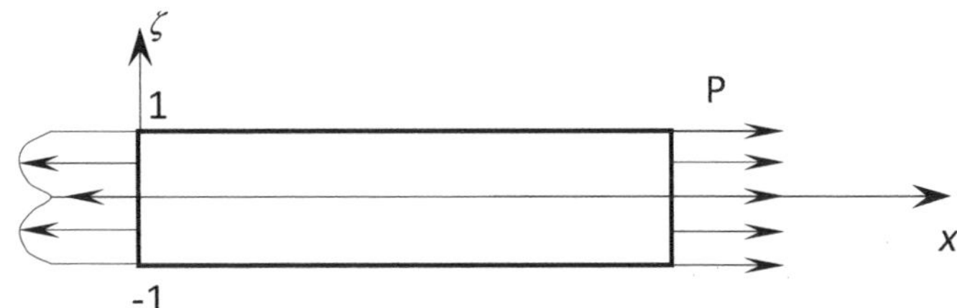

Fig. 1.1

The fact that $\varphi(\zeta)$ is even clearly implies

$$\int_{-1}^{1} \varphi(\zeta)d\zeta = 2P. \tag{4.1}$$

Therefore we get from (1.11), (1.13) and (3.4) that

$$C_1^{(0)} = C_1^{(1)} = 0, \ \Rightarrow Q^{(0)} = Q^{(1)} \equiv 0,$$

$$C_1^{(2)} = \int_0^1 \varphi d\zeta, \ \ C_1^{(s)} = 0, \ \Rightarrow Q^{(s)} \equiv 0, \quad (s \ge 3). \tag{4.2}$$

As a result, the exact solution of the internal problem is governed by

$$\sigma_x^{out} = \int_0^1 \varphi d\zeta = P, \ \ \sigma_{xy}^{out} = \sigma_y^{out} = 0. \tag{4.3}$$

The boundary layer solution may now be determined subject to conditions (3.3). As follows from (3.3), (4.2) and (4.3), the only non-zero boundary layer solution corresponds to $s = 2$ and is given by

$$\sigma_{xb}^{(2)}(t = 0) = \varphi - \int_0^1 \varphi d\zeta, \ \ \sigma_{xyb}^{(2)}(t = 0) = 0. \tag{4.4}$$

If $\varphi = const$, then (2.8) and (4.4) dictate that $R_b^{(1)} \equiv 0$, i.e. solution (4.3) is valid up to the edge and coincides with the well-known solution often presented in lecture courses on strength of materials and elasticity. Now if $\varphi \ne const$, i.e. the distribution of the load is not uniform with height, then the first function in (4.4) is not zero and the boundary layer appears at $x = 0$. Consider now several particular cases:

(a) $\varphi = 2P|\zeta|$

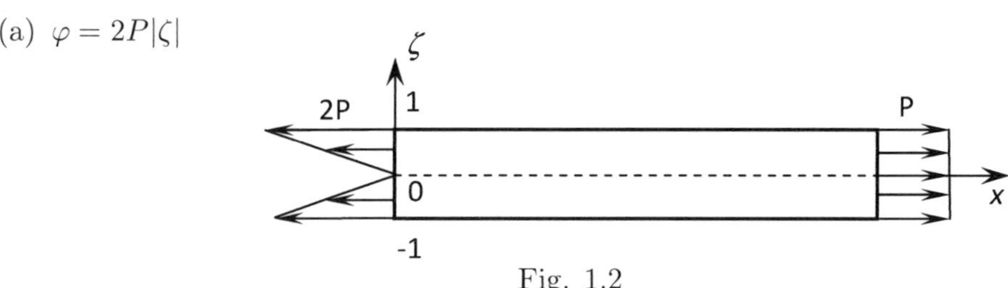

Fig. 1.2

(b) $\varphi = 2P(1 - |\zeta|)$

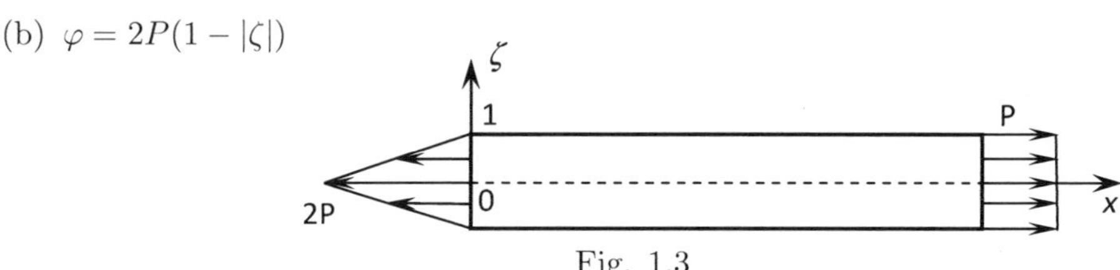

Fig. 1.3

(c) $\varphi = \frac{3}{2}P(1 - \zeta^2)$

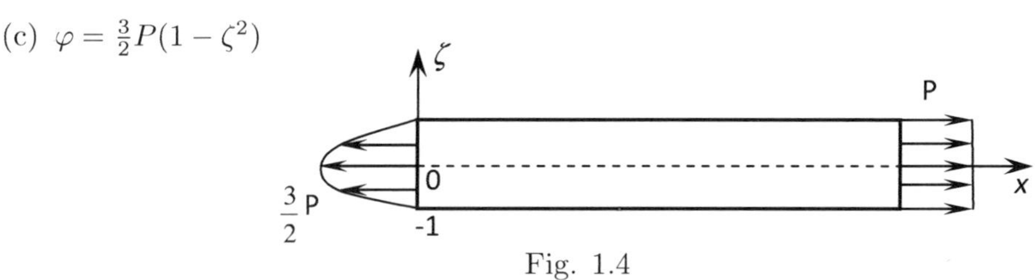

Fig. 1.4

We remark that the outer solution given by expression (4.3) is the same for all three considered cases, however, the boundary layers following from the boundary conditions are quite different.

(a)

$$\sigma_{xb}^{(2)}(t = 0) = 2P(|\zeta| - 1/2), \quad \sigma_{xyb}^{(2)}(t = 0) = 0 \qquad (4.5)$$

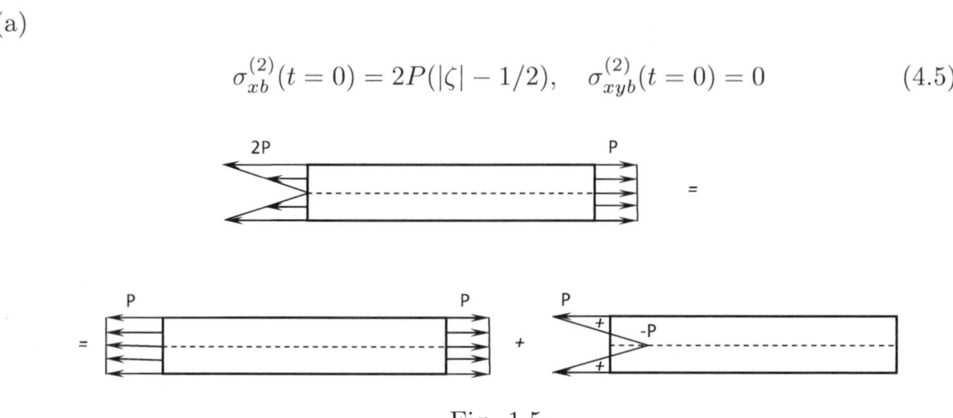

Fig. 1.5

(b)

$$\sigma_{xb}^{(2)}(t=0) = -2P(|\zeta| - 1/2), \quad \sigma_{xyb}^{(2)}(t=0) = 0 \qquad (4.6)$$

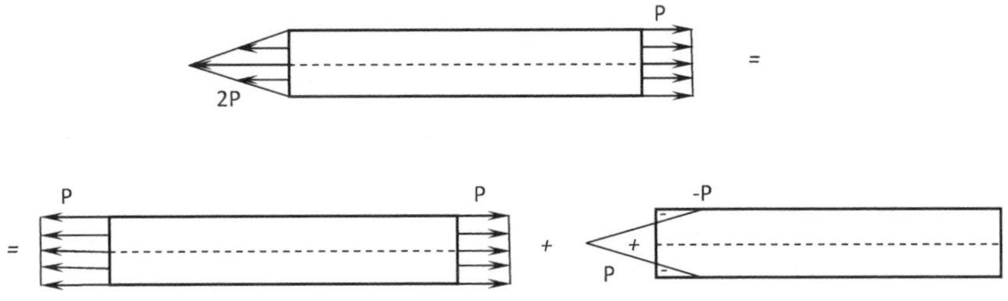

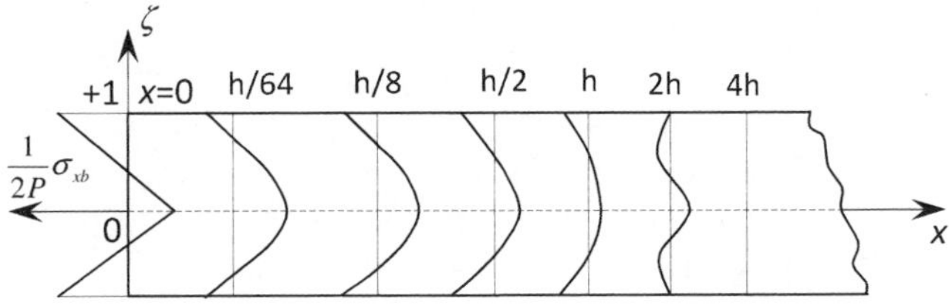

Fig. 1.6

(c)

$$\sigma_{xb}^{(2)}(t=0) = \frac{P}{2}(1 - 3\zeta^2), \quad \sigma_{xyb}^{(2)}(t=0) = 0 \qquad (4.7)$$

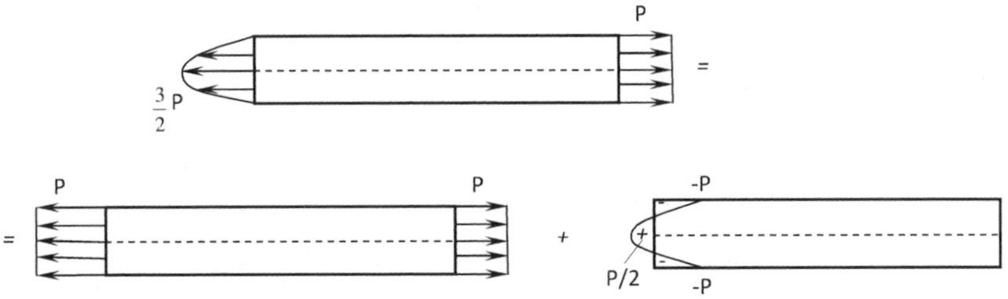

Fig. 1.7

This decomposition of the stress-strain state is illustrated by Figs. 1.5-1.7.

The boundary layer solution is given by (2.8), expressed through coefficients A_{1k} and A_{2k} which may be determined from conditions (4.1)-(4.7) by the boundary collocation method.

Fig. 1.8

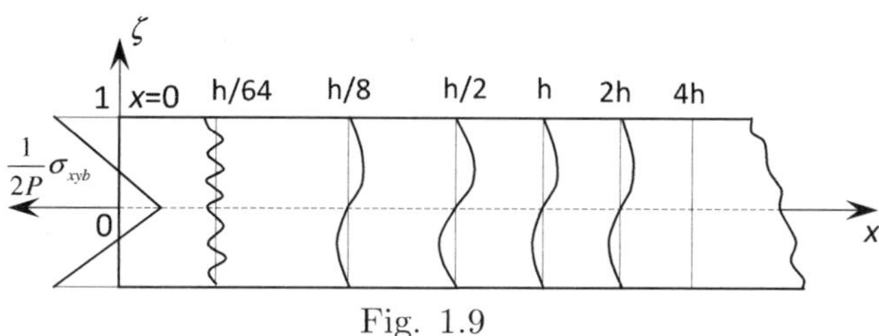

Fig. 1.9

The idea of the method is the following: the boundary conditions are satisfied for a discrete set of N points of collocation (coincidence) along the boundary, thus leading to an algebraic system of $2N$ equations, from which the coefficients A_{1k} and A_{2k} are obtained. So we may use the solution (2.8) and impose the boundary conditions (4.5), (4.6) or (4.7) at N points and then solve the system of $2N$ algebraic equations in respect of the coefficients A_{1k} and A_{2k}. The necessary accuracy of the method, as a rule, may be achieved by increasing the number N and appropriate location of the points. In general the distribution of the points may be non-uniform. Usually, in addition to the chosen points of collocation there is also a set of checkpoints which are used for another functions given along the boundary, which may be compared to the corresponding approximate values obtained through the determined coefficients A_{1k} and A_{2k}. More details on the boundary collocation method along with its applications to plates and shells may be found in Kornishin (1964). Often the set of collocation points is chosen through Chebyshev distribution, leading to uniform approximations (Lantsosh, 1961; Krasnoselsky et al., 1969). In our case the corresponding algebraic system for boundary conditions (4.5) has been solved numerically for A_{1k} and A_{2k}, then using (2.8) the stress components were calculated, see Figs. 1.8-1.9. In addition to collocation, 10 checkpoints have been chosen at $x = 0$, since at $x = 0, t = 0$ we know the values of $\sigma_{xb} = P(2\zeta - 1)$ and $\sigma_{xyb} = 0$. The maximal error for σ_{xb} is within the limits of 3-5 percent, while the stress σ_{xyb} is approximate zero up to 2-3 significant figures. The accuracy may be increased by increasing the number of collocation points. As may be seen from Fig. 1.8, the stress σ_{xb} almost keeps its initial shape and decays rapidly away from the edge of the plate. It may be neglected at the distance close to thickness $2h$ from the edge. As for the tangential stress σ_{xyb} (Fig. 1.9), some fast oscillations may be observed in the near edge vicinity $x = 0$ (since the amplitude is small, this is not very important or dangerous), and then as we move away from the edge, it takes a rather stable form. Approximately at $x = h/2$ its amplitude has a maximum of about 20% - 25% of P, and then decays rapidly. Thus, due to non-uniform nature of the edge loading, the crucial cross section could not be the edge but that of $x \approx h/2$ where the maximal tangential stress occur.

The results for the boundary conditions (4.6) are similar to the above presented apart from the sign, with the associated numerical illustrations easily obtained from Figs. 1.8-1.9 by reflection against the $O\xi$ axis.

The resulting stress for the boundary conditions (4.7) are shown on Fig. 1.10, 1.11. Their behaviour for cases a) and c) are very similar.

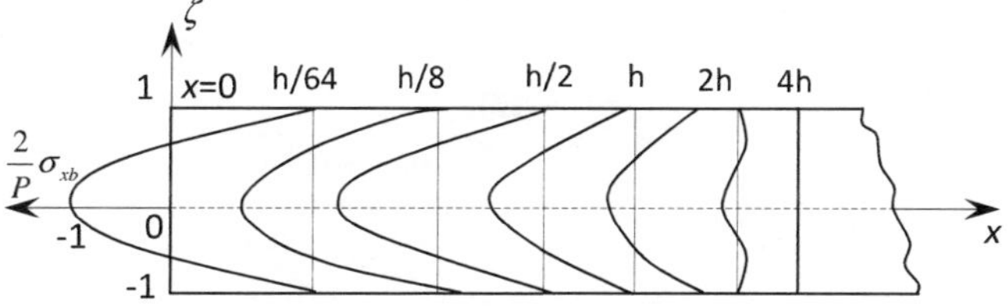

Fig. 1.10

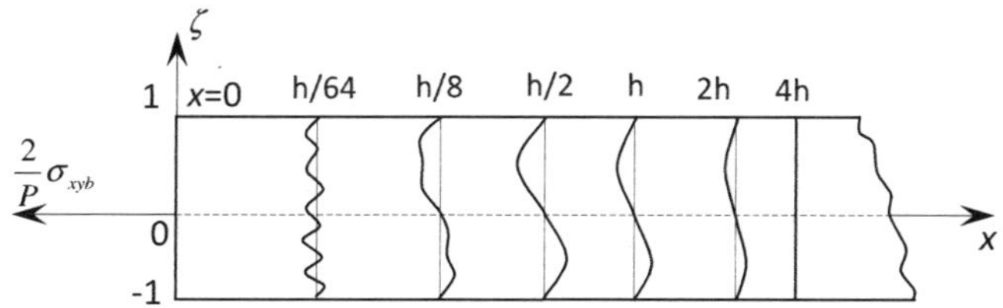

Fig. 1.11

Example 1.2. (The Navier example) Let the strip be loaded along the edge $x = 0$ as follows

$$\sigma_x = 0, \quad \sigma_{xy} = \varphi(\zeta), \quad \varphi(-\zeta) = -\varphi(\zeta) \qquad \text{at} \quad x = 0. \qquad (4.8)$$

In this case (Figs. 1.12, 1.13) conditions (1.11) and (3.4) imply

$$C_1^{(s)} \equiv 0, \quad Q^{(s)} = 0,$$

i.e.

$$\sigma_x^{out} = \sigma_{xy}^{out} = \sigma_y^{out} = 0,$$

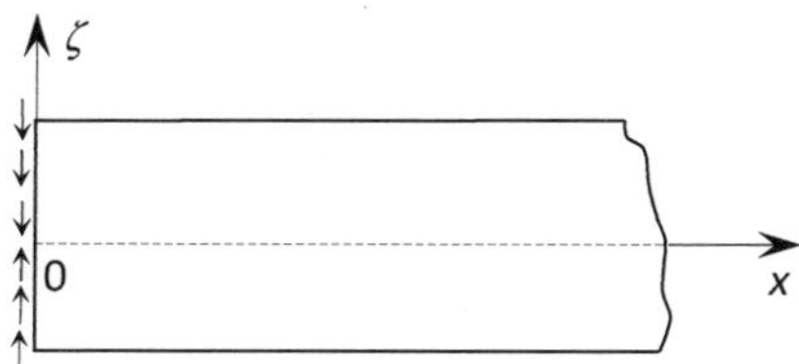

Fig. 1.12

The non-zero boundary layer solution corresponds to $s = 2$ and is governed by conditions (3.3), which take the form $\sigma_{xb}^{(2)} = 0$, $\sigma_{xyb}^{(2)} = \varphi(\zeta)$ at $t = 0$.

The boundary layer solution may then be obtained through the procedure described in Section 1.1. In the previous example of extension of the strip due to a uniformly distributed load, there was no boundary layer.

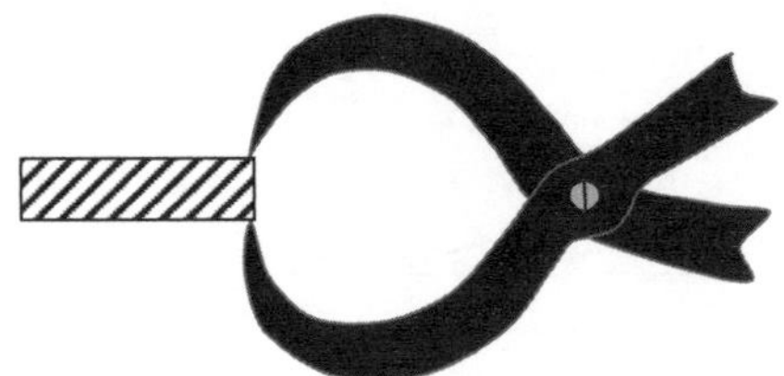

Fig. 1.13

The current example 1.2 illustrates the type of external forcing that excites the boundary layer solution only, i.e. a rapidly decaying solution.

Example 1.3. Consider now pure bending of a strip by a normal load distributed along the edges $x = 0; x = a$

$$\sigma_x = \varphi(\zeta), \quad \sigma_{xy} = 0 \quad \text{at} \quad x = 0$$
$$\sigma_x = -P\zeta, \quad \sigma_{xy} = 0 \quad \text{at} \quad x = a, \tag{4.9}$$

where $\varphi(\zeta)$ is an odd function satisfying the static equilibrium conditions

$$\int_{-1}^{+1} \zeta\varphi(\zeta)d\zeta = \frac{2}{3}P, \quad P = \frac{M}{I}h, \tag{4.10}$$

M is the bending moment, I is the moment of inertia of the cross section. Solution of the outer problem (1.11), (1.16), (3.5) leads to

$$C_1^{(2)} = 0, \quad C_2^{(2)} = \int_0^1 \zeta\varphi d\zeta, \quad C_1^{(s)} = C_2^{(s)} = 0, \quad Q^{(s)} = 0 \ (s \neq 2) \tag{4.11}$$

and

$$\sigma_x^{out} = \sigma_x^{(2)} = -\frac{M}{I}y, \quad \sigma_{xy}^{out} = 0, \quad \sigma_y^{out} = 0, \tag{4.12}$$

i.e. in case of edge loading only the outer solution within the asymptotic procedure is closed, coinciding with the classical solution for pure bending. The non-zero boundary layer solution corresponds to $s = 2$ and is governed by

$$\sigma_{xb}^{(2)}(t = 0) = \varphi(\zeta) - 3\zeta \int_0^1 \zeta\varphi(\zeta)d\zeta = \varphi(\zeta) - P\zeta, \quad \sigma_{xyb}^{(2)}(t = 0) = 0. \tag{4.13}$$

In particular, if $\varphi(\zeta) = P\zeta$, then there is no boundary layer, and we reveal a well-known result - solution (4.12) is valid till the edge. However, if $\varphi(\zeta) \neq P\zeta$, then the boundary layer emerges at $x = 0$. Consider now two special cases

$$a) \ \varphi(\zeta) = \frac{5}{3}P\zeta^3, \quad b) \ \varphi(\zeta) = \frac{\pi^2}{12}P \sin\frac{\pi}{2}\zeta \tag{4.14}$$

satisfying conditions (4.10). The outer solution in both cases is described by (4.12). However, the boundary layers given by (2.9) and (4.13) are quite different. The edge values for σ_{xb} are $\dfrac{P}{3}(5\zeta^3 - 3\zeta)$ and $\dfrac{P}{3}\left(\dfrac{\pi^2}{4}\sin\dfrac{\pi}{2}\zeta - 3\zeta\right)$, respectively.

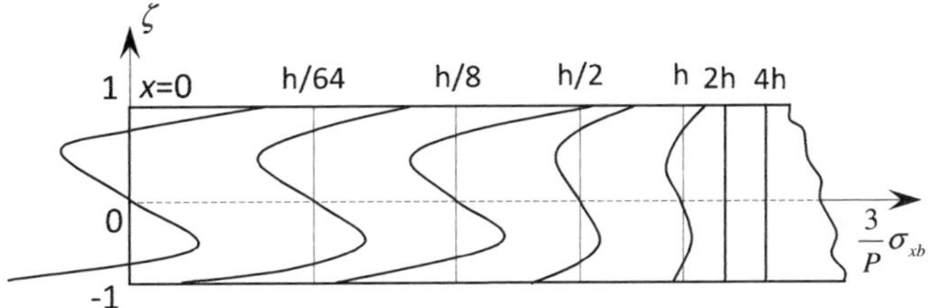

Fig. 1.14

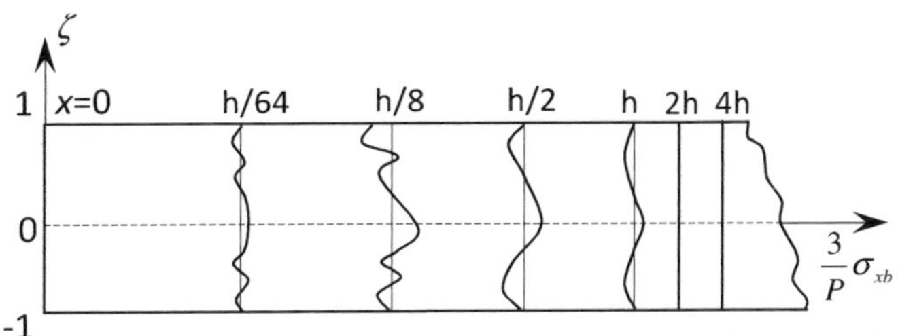

Fig. 1.15

The constants and the stress components of the boundary layer solution (2.9) may be obtained through the boundary collocation method. Numerical illustrations for stress components σ_{xb} and σ_{xyb} are given in Figs. 1.14, 1.15 (case a), Figs. 1.16, 1.17 (case b). Comparison of the values of σ_{xb} calculated from (2.9), at $t = 0$, revealed that accuracy is within 5-6%, which could clearly be enhanced by increasing the number of collocation points. Even though the internal solutions in both cases are identical, there is a principal difference between the boundary layer solutions, as seen from Figs. 1.14-1.17. Similarly to the symmetric case, the normal stress σ_{xb} almost keeps the initial form, decays rapidly, and may be neglected starting from the half-thickness distance away from the edge. The tangential stress component σ_{xyb} exhibits small-amplitude oscillations in the near-edge vicinity, and forms a stable shape as moving away from the edge. At $x \approx h/2$ the amplitude reaches its maximum (of about 10-15% of external forcing value), and then decays rapidly, keeping its shape.

In the anti-symmetric case (bending) both tangential and normal stress components may be neglected at the distance of about half-thickness, which is another
confirmation of the earlier established results in Section 1.2.

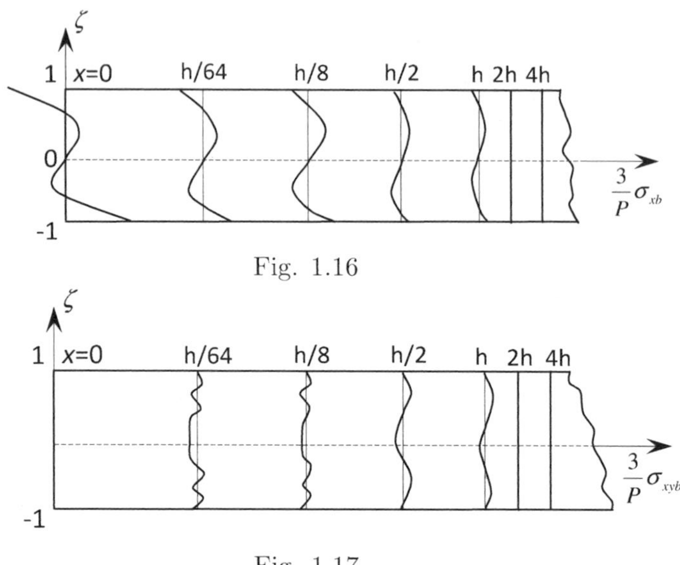

Fig. 1.16

Fig. 1.17

1.5   Relation to the Mesnager-Timoshenko Solution

It is well-known (Timoshenko and Goodier, 1970), that within the plane problem
of elasticity it is possible to determine surface load by a given bi-harmonic function
for stress components (inverse problem of elasticity). In case of a rectangular strip
Mesnager suggested to present stress components through polynomials, which provided solution to several practical problems, see Timoshenko and Goodier (1970);
Filin (1975).

All of these solutions may also be obtained as special cases from the more general
asymptotic solutions (1.11), (1.13), (1.16), (3.4) and (3.5). Let us consider some
more examples in addition to extension and pure bending analyzed in the previous
Section 1.4. In what follows we restrict ourselves to analysis of stress components
only, with the displacements easily computed through these.

a) shear - tangential force of density P is uniformly distributed along the edge
(Fig. 1.18). In our case

$$X^+ = P\varepsilon, \quad X^- = -P\varepsilon, \quad X_2^{(0)} = P\varepsilon, \quad \varphi_1 = 0, \quad \varphi_2 = P$$

$$C_1^{(0)} = P\varepsilon, \quad C_1^{(1)} = -P, \ C_2^{(0)} = C_2^{(1)} = C_1^{(s)} = C_2^{(s)} = 0 \quad (s \geq 2), \qquad (5.1)$$

therefore it may be inferred from (1.11)-(1.18), and (3.5), that

$$\sigma_x = \sigma_y = 0, \quad \sigma_{xy} = P \qquad (5.2)$$

coinciding with the well-known classical solution, when the Airy function is chosen as a second order polynomial. As follows from (3.3), (5.1), there is no boundary layer in this problem.

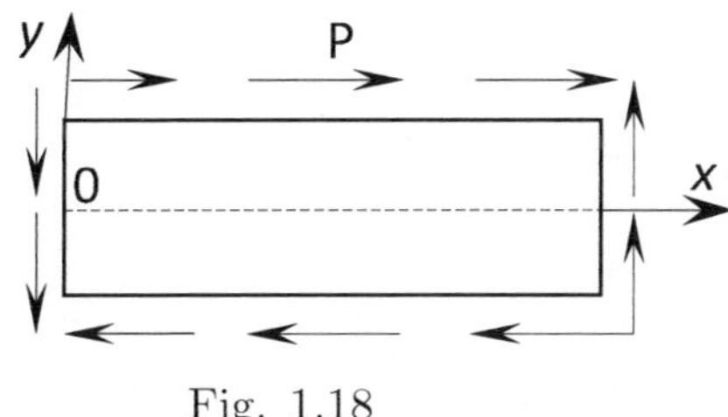

Fig. 1.18

b) bending of a cantilever beam by a normal uniformly distributed load with density q (Fig. 1.19)

$$X^+ = X^- = 0, \quad Y^+ = -q, \quad Y^- = 0, \quad \varphi_1 = \varphi_2 = 0. \tag{5.3}$$

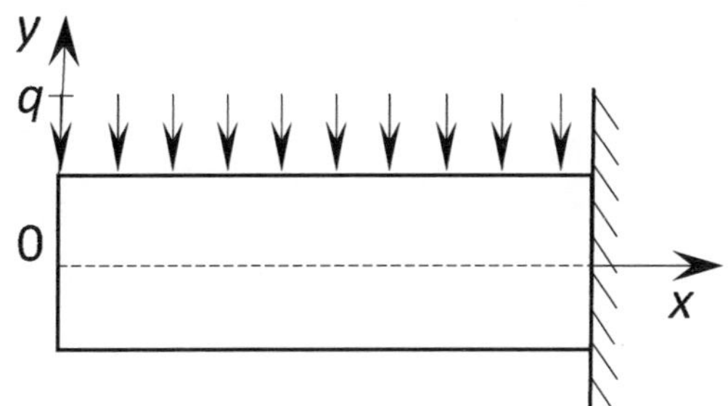

Fig. 1.19

From the outer symmetric solution (1.11), (1.13), (3.4) we deduce

$$\sigma_x = 0, \quad \sigma_{xy} = 0, \quad \sigma_y = -\frac{q}{2}. \tag{5.4}$$

From the anti-symmetric solution (1.11), (1.16), (3.5), taking note of

$$X_2^{(0)} = 0, \quad q^{(0)} = Y_2^{(0)} = \frac{1}{2}q, \quad C_1^{(s)} = 0, \quad C_2^{(2)} \neq 0, \quad C_2^{(s)} = 0, \tag{5.5}$$

ensuring the non-zero approximations for $s = 0$ and $s = 2$, we obtain

$$\sigma_x = \frac{3}{4}q\left[\varepsilon^{-2}\xi^2\zeta - \left(\frac{2}{3}\zeta^3 - \frac{2}{5}\zeta\right)\right] = \frac{q}{2I}x^2y - \frac{q}{2I}\left(\frac{2}{3}y^3 - \frac{2}{5}yh^2\right),$$

$$\sigma_{xy} = -\frac{3}{4}q\varepsilon^{-1}(\zeta^2 - 1)\xi = -\frac{q}{2I}(y^2 - h^2)x,$$

$$\sigma_y^b = -\frac{q\zeta}{4}(3 - \zeta^2) = -\frac{q}{2I}\left(yh^2 - \frac{1}{3}y^3\right), \tag{5.6}$$

$$\sigma_y = \sigma_y^{sym} + \sigma_y^b = -\frac{q}{2I}\left(yh^2 - \frac{1}{3}y^3 + \frac{2}{3}h^3\right),$$

22   *Asymptotic Theory of Anisotropic Plates and Shells*

where $I = \dfrac{2}{3}h^3 \cdot 1$ is the moment of inertia of the cross section. The formulae (5.6) coincide with the known solution provided that the Airy function of stress is chosen as a fifth order polynomial. The peculiarity of solution (5.6) is that the stress σ_x is non-zero at the edge $x = 0$ and has a self-equilibrated non-zero component. Indeed, as may be seen from (5.6), the first term of σ_x is leading order by absolute value, and is identically zero at $x = 0$. The same result follows from the technical Bernoulli-Euler beam theory for bending. The second non-zero term is self-equilibrated, smaller than the leading first term by two orders. That explains why the boundary layer solution emerges at $x = 0$, effectively taking over the self-equilibrated part of the normal stress. The boundary layer solution may be found from (2.9), (3.3) by one of the methods described in Section 1.3. Thus, the considered example already demonstrates the advantage of asymptotic method, since the solution in terms of algebraic polynomials cannot take into account the boundary layer solution.

c) A vertical cantilever loaded by hydrostatic pressure (Fig. 1.20).

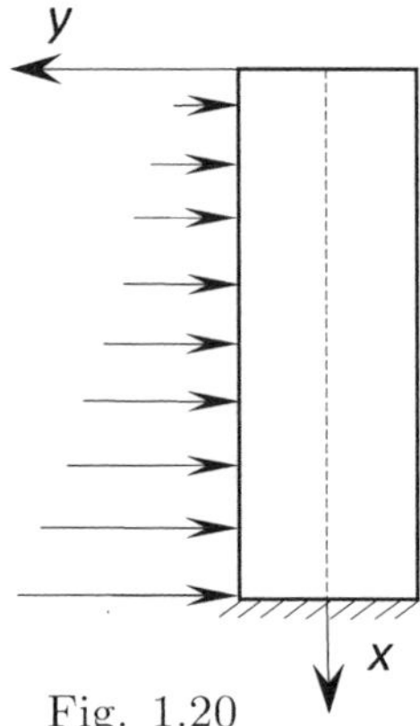

Fig. 1.20

$$X^+ = X^- = 0, \quad Y^+ = -qx, \quad Y^- = 0, \quad \varphi_1 = \varphi_2 = 0. \tag{5.7}$$

In view of (5.6), solutions of the symmetric problem (1.11), (1.13) and (3.4) lead to

$$\sigma_x^{sym} = \sigma_{xy}^{sym} = 0, \quad \sigma_y^{sym} = -\frac{1}{2}qx. \tag{5.8}$$

In case of the bending problem, taking into account

$$q^{(0)} = -\frac{1}{2}qx, \quad Y_2^{(0)} = -\frac{1}{2}qx, \quad X_2^{(s)} = Y_2^{(s)} = q^{(s)} = 0 \ (s > 0)$$

$$C_1^{(2)} = \frac{1}{4}qa\left(\frac{8}{5} + \nu\right), \quad C_1^{(s)} = C_2^{(s)} = 0 \tag{5.9}$$

we deduce the non-zero approximations for $s = 0$ and $s = 2$. Therefore, using

(1.11), (1.16) and (3.5), the solution is given by

$$\sigma_x^b = \frac{1}{4}qa\left[\varepsilon^{-2}\xi^3\zeta + 2\xi\left(\frac{3}{5}\zeta - \zeta^3\right)\right]$$

$$= \frac{1}{4}\frac{qx^3y}{h^3} + \frac{1}{4}\frac{q}{h^3}\left(-2xy^3 + \frac{6}{5}h^2xy\right)$$

$$\sigma_{xy}^b = -\frac{3}{8}qa\varepsilon^{-1}\xi^2(\zeta^2 - 1) + \frac{1}{40}qa\varepsilon(1 - 6\zeta^2 + 5\zeta^4)$$

$$= \frac{3}{8}\frac{q}{h^3}x^2(h^2 - y^2) + \frac{1}{40}\frac{q}{h^3}\left(h^4 - 6h^2y^2 + 5y^4\right) \qquad (5.10)$$

$$\sigma_y^b = -\frac{1}{4}qx(3\zeta - \zeta^3)$$

$$\sigma_y = \sigma_y^{sym} + \sigma_y^b = -\frac{1}{2}qx + \frac{1}{4}qx(\zeta^3 - 3\zeta)$$

$$= -\frac{1}{2}qx + \frac{1}{4}q\frac{x}{h^3}(y^3 - 3yh^2),$$

which coincides with a known solution, when the bi-harmonic function is chosen in the form of a sixth order polynomial, see Timoshenko and Goodier (1970). The tangential stress σ_{xy} in (5.10) is not identically zero at the edge $x = 0$, even in the absence of shear force. It is also clear from (5.10) that the self-equilibrated part is two orders smaller than the non-self-equilibrated one. It may be shown that the boundary layer solution (which may be computed through (2.9) and (3.3)) occurs both at $x = 0$ and $x = a$, and that these solutions should be added to solution (5.10).

Solutions for more complicated form of load could be obtained from (1.11), (3.4) and (3.5). For example, if the external loading is specified as a polynomial of degree n, then the iteration procedure should be stopped at $s = n + 1$ or $s = n + 2$ for odd and even n, respectively. Thus, for polynomial loading of the long edges of the strip the asymptotic approach provides an outer solution of closed form along with the associated boundary layer solutions obtained from (2.8), (2.9) and (3.3) in accordance with the procedure described in Sections 1.3 and 1.4.

It then follows that the asymptotic method is valid for a large class of external loads, since any continuous function can be uniformly approximated by an algebraic polynomial (Weierstrass theorem), giving a closed solution for a corresponding problem of elasticity.

It should also be mentioned that within the traditional approach the coefficients of a bi-harmonic polynomial should satisfy certain conditions, whereas the solutions of the asymptotic approach satisfy these automatically.

1.6   Comparison with the Bernoulli-Coulomb-Euler Theory

We remark that formulae (1.11), (1.13)-(1.19) are related directly to the Bernoulli-Coulomb-Euler beam theory based on the plane-sections hypothesis. Indeed, representing (1.18) in terms of dimensional parameters and assuming $w_0^{(s)} = a\varepsilon^{-3}w^{(s)}$, we get

$$EI\frac{d^4 w_0^{(s)}}{dx^4} = q_0^{(s)}, \tag{6.1}$$

where $q_0^{(s)} = 2q^{(s)}$, and $I = \frac{2}{3}h^3 \cdot 1$ is the moment of inertial of the cross section.

At $s = 0$ equation (6.1) coincides with the classical bending equation for a beam along with the boundary conditions, see Rabotnov (1962); Feodos'ev (1970). Thus, the outer solution based on the asymptotic method is identical to that provided by the technical theories for beams. Moreover, the asymptotic approach is more informative, since the formulae (1.11), (1.13) and (1.16) determine the stress components σ_{xy}, σ_y whereas the last one is usually neglected within the technical theories for beams.

In the symmetric problem, assuming $u_0^{(s)} = a\varepsilon^{-2}u^{(s)}$, equation (1.14) takes the form

$$EF\frac{d^2 u_0^{(s)}}{dx^2} = -q_x^{(s)}, \tag{6.2}$$

where $q_x^{(s)} = -2\varepsilon^{-1}p^{(s)}$, EF is stiffness for extension-compression, $F = 2h \cdot 1$ is the cross section area. We also note that $q_x^{(0)} = 2\varepsilon^{-1}X_1^{(0)} = \varepsilon^{-1}(X^+ + X^-)$ is the density of the external distributed tangential load. At $s = 0$, equation (6.2) coincides with the classical equation of the beam extension, see Filin (1975). Similarly to the problem for bending of the beam, the conditions (1.11), (1.13) and (1.14) at $s = 0$ provide more information than the classical theory of beam extension (compression) based on the plane section assumption, since the asymptotic approach provides results for the stress components σ_{xy}, σ_x which are usually neglected within the classical theory.

At $s > 0$, the right-hand sides in (6.1) and (6.2) are changed corresponding to different external loading, therefore providing straightforward approach to refinement of the classical results through asymptotic method. As mentioned earlier in Section 1.1, it is not possible to satisfy the edge boundary conditions by the outer solution (1.11), (1.13)-(1.19), i.e. the boundary layer solution should be introduced into consideration. Thus, the technical beam theory can only provide a leading order approximation of the outer solution. Adopting the plane section hypothesis leads to ignoring both the higher order terms in the outer solution and the boundary layer solution. Taking into account higher order terms (of order $O(\varepsilon^2)$) in the analysis of the outer problem provides refined beam theories which lead to reasonable results in most of the practical problems, however these will not be valid in the near-edge vicinity, as shown in Sections 1.4 and 1.5.

1.7 Matching of the Outer Solution with the Boundary Layer for Mixed Edge Boundary Conditions

Consider now the procedure of matching of the outer solution with the boundary layer solution in case of the mixed boundary value problems (1.3), (1.4).

In case of the problem (1.3) (problem 2), substituting (3.1) into (1.3), we deduce from the solvability that $\chi = -3$, then equations (1.3) become

$$au^{(s)} + au_b^{(s)} = \varphi_1^{(s-2)}$$

$$\sigma_{xy}^{(s-2)} + \sigma_{xyb}^{(s)} = \varphi_2^{(s-3)} \qquad \text{at} \qquad \xi = 0 \ (t = 0), \qquad (7.1)$$

As shown in Section 1.2, the displacements of the boundary layer are not self-equilibrated (2.12), i.e.

$$\int_{-1}^{+1} u_b d\zeta \neq 0, \qquad \int_{-1}^{+1} \zeta u_b d\zeta \neq 0, \qquad \int_{-1}^{+1} v_b d\zeta \neq 0 \qquad (7.2)$$

therefore matching of the outer solution and the boundary layer by means of the method described in Section 1.3 is not possible. For the same reason we deduce the invalidity of the Saint-Venant's principle for displacements. The sought-for boundary values $au_b^{(s)}(t = 0)$ and $\sigma_{xyb}^{(s)}(t = 0)$ should correspond to decaying solutions. It follows from (7.1) that

$$
\begin{aligned}
au_b^{(s)} &= f_1^{(s)}, \quad f_1^{(s)} = \varphi_1^{(s-2)} - au^{(s)}(\xi = 0) \\
\sigma_{xyb}^{(s)} &= f_2^{(s)}, \quad f_2^{(s)} = \varphi_2^{(s-3)} - \sigma_{xy}^{(s-2)}(\xi = 0).
\end{aligned}
\qquad t = 0 \qquad (7.3)
$$

The functions $f_1^{(s)}$, $f_2^{(s)}$ should be representable as series (2.1), and (2.5), i.e. along the non-orthogonal system $F_n(\zeta)$. We remark that though the system of functions $F_n(\zeta)$ is not orthogonal in traditional sense, they satisfy the generalized orthogonality condition (2.11).

The property of generalized orthogonality of homogeneous solutions has been first pointed out by P. Schiff in analysis of axisymmetric deformation of cylinder (Schiff, 1883; Prokopov, 1966, 1967; Nuller, 1969; Lourier, 1970), and later noted by Papkovich (1940, 1941) for bending of a rectangular plate and also for a plane strain problem of elasticity for a rectangular area. G.A. Grinberg was the first to present a mathematical study of representation of two functions as series along the homogeneous solutions of P.F. Papkovich, see Grinberg (1953) for conditions in respect of f_1, f_2 necessary for convergence of the series of the type (2.1) and (2.5).

Further progress was delivered by M.I. Gusein-Zade, who obtained the conditions for several new types of boundary value problems for elastic strip and also corrected some of the results by G.A. Grinberg, see Gusein-Zade (1965). The conditions that should be satisfied by functions in order to be expanded as series along the system of homogeneous solutions, or, equivalently, the conditions of existence of a decaying solution in a semi-strip, will be further referred to as compatibility conditions between the boundary layer and the internal solution. Some of these, corresponding to the boundary value problems (1.2)-(1.4) for orthotropic strip, have been obtained by the author, see Aghalovyan and Khachatryan (1975), establishing generalized orthogonality of homogeneous solutions, thus developing further the results of P.F. Papkovich.

The case of a symmetric plane-stress problem isotropic media contains one condition

$$E \int_0^1 f_1 d\zeta + \nu\, h \int_0^1 \zeta\, f_2 d\zeta = 0, \qquad (7.4)$$

while the anti-symmetric case resulted in two conditions

$$\int_0^1 f_2 d\zeta = 0, \quad 2E \int_0^1 \zeta\, f_1 d\zeta + \nu\, h \int_0^1 \zeta^2 f_2 d\zeta = 0. \qquad (7.5)$$

Therefore, within the framework of the outer problem, the equivalent displacements should satisfy the conditions (7.4) and (7.5), however they are not necessarily satisfying the integral conditions of the same structure as the forces and moments, see Section 1.3. Thus, the attempt of extension of the Saint-Venant's principle for displacements is not valid. This leads to a natural question, what will be the accuracy of approach still trying to satisfy the edge boundary condition in integral form? The latter may also be reformulated as a problem of effect of the second terms in (7.4) and (7.5) on the outer solution.

The conditions (7.4) and (7.5) may be satisfied by making use of (7.3), (1.13), (1.16). In case of the *symmetric* problem we obtain

$$C_2^{(s)} = E \int_0^1 \frac{\varphi_1^{(s-2)}}{a} d\zeta + \nu \int_0^1 \zeta(\varphi_2^{(s-4)} - \sigma_{xy}^{(s-3)}(\xi = 0))d\zeta$$

$$-E \int_0^1 u^{*(s)}(\xi = 0)d\zeta, \qquad (7.6)$$

while in case of the *anti-symmetric* one

$$C_1^{(s)} = \left(X_2^{(s)} - \sigma_{xy}^{*(s)}(\zeta = 1) \right)_{\xi=0} + \int_0^1 (\sigma_{xy}^{*(s)}(0,\zeta) - \varphi_2^{(s-1)})d\zeta$$

$$C_3^{(s)} = E \int_0^1 \zeta \left(u^{*(s)}(\xi = 0) - \frac{\varphi_1^{(s-2)}}{a} \right) d\zeta \qquad (7.7)$$

$$-\frac{\nu}{2} \int_0^1 \zeta^2(\varphi_2^{(s-4)} - \sigma_{xy}^{(s-3)}(\xi = 0))d\zeta.$$

The remaining undetermined constants including $C_1^{(s)}$ of the symmetric problem and $C_2^{(s)}$, $C_4^{(s)}$ in case of the bending problem are found from the conditions at the opposite edge. For example, if the conditions at $\xi = 1$ are given by

$$u(1,\zeta) = \psi_1(\zeta), \quad \sigma_{xy}(1,\zeta) = \psi_2(\zeta), \qquad (7.8)$$

then (for symmetric case)

$$C_1^{(s)} = E \int_0^1 \zeta \left(\frac{\psi_1^{(s-2)}}{a} - u^{*(s)}(1,\zeta) \right) d\zeta$$

$$+ \nu \int_0^1 \zeta(\psi_2^{(s-4)} - \sigma_{xy}^{(s-3)}(\xi = 1)) d\zeta$$

$$- \int_0^1 d\xi \int_0^\xi p^{(s)} d\xi - E \int_0^1 \left(\frac{\varphi_1^{(s-2)}}{a} - u^{*(s)}(\xi = 0) \right) d\zeta \qquad (7.9)$$

$$- \nu \int_0^1 \zeta(\varphi_2^{(s-4)} - \sigma_{xy}^{(s-3)}(\xi = 0)) d\zeta.$$

In the anti-symmetric case the coefficients $C_2^{(s)}$ may be determined in a similar manner from (7.8), while $C_4^{(s)}$ correspond to rigid body motion.

Returning to (7.3), which now contain known quantities in their right-hand sides, using (2.5) along with the Papkovich generalized orthogonality conditions (2.11), we are in a position to determine the coefficients of the boundary layer

$$A_n^{(s)} = \frac{\lambda_n^3}{\Delta_n} \int_0^1 (\psi_1^{(s)} F_n'' - \lambda_n^2 F_n \psi_2^{(s)}) d\zeta, \qquad (7.10)$$

where

$$\Delta_n = \int_0^1 ((F_n'')^2 - \lambda_n^4 F_n^2) d\zeta, \qquad (7.11)$$

and

$$\Psi_1^{(s)}(\zeta) = -\frac{E}{a} f_1^{(s)} + \nu \Psi_2^{(s)}, \quad \Psi_2^{(s)} = \int_0^\zeta f_2^{(s)} d\zeta - \int_0^1 f_2^{(s)} d\zeta. \qquad (7.12)$$

Hence, in case of problem 2 for any s there are closed solutions for both the outer problem and the boundary layer. Following the procedure from Grinberg (1953), the convergence of the series (2.1) and (2.5) may be verified.

As follows from (7.10), $A_n^{(0)} = A_n^{(1)} \equiv 0$, $A_n^{(2)} \neq 0$, hence the stresses and displacements of the boundary layer would be of order $O(\varepsilon^{-1})$ and $O(\varepsilon^0)$, respectively. Comparing these with the corresponding orders of the outer solution, one may deduce that in real-life problems the displacements of the boundary layer solution are negligible.

Ignoring the second terms in (7.4), (7.5) is in a sense equivalent to formal extension of the Saint-Venant's principle for displacements. Let us investigate the effect of these second terms on the coefficients of the outer solution. In case of $C_2^{(s)}$ and $C_3^{(s)}$ they lead to second integrals in (7.6), (7.7), which give non-zero contribution starting from $s = 3$. If $s < 3$ then the value of the coefficients is found from equation involving integrals of displacements; on the other hand the case $s < 3$ corresponds to traditional methods of material strength, while $s \geq 3$ is associated with more advanced methods taking into account in-plane shear and compression. Thus, in case of leading order calculations within the plane section assumption, the boundary conditions in respect of displacements may be satisfied in integral form, however, if more accurate calculations are required, then the boundary conditions should be refined. Thus, the Saint-Venant's principle is not valid for displacements in the same form as it was formulated for stresses, however, in some cases its application to displacements may still provide results of reasonable accuracy.

In problem 3, after substitution of (3.1) into (1.4) we deduce from the solvability that $\chi = -2$ and $\chi = -4$ for the symmetric and anti-symmetric cases, respectively. As a result, we get for the

symmetric problem

$$\sigma_{xb}^{(s)} = f_1^{(s)}, \quad f_1^{(s)} = \varphi_1^{(s-2)} - \sigma_x^{(s)}(\xi = 0)$$
$$av_b^{(s)} = f_2^{(s)}, \quad f_2^{(s)} = -av^{(s)}(\xi = 0) + \varphi_2^{(s-1)} \qquad \text{at} \qquad t = 0 \qquad (7.13)$$

whereas for the *anti-symmetric problem*

$$\sigma_{xb}^{(s)} = \psi_1^{(s)}, \quad \psi_1^{(s)} = -\sigma_x^{(s-2)}(\xi = 0) + \varphi_1^{(s-4)}$$
$$av_b^{(s)} = \psi_2^{(s)}, \quad \psi_2^{(s)} = -av^{(s)}(\xi = 0) + \varphi_2^{(s-3)} \qquad \text{at} \qquad t = 0. \qquad (7.14)$$

Using the right-hand side of (7.13) and the compatibility condition

$$\int_{-1}^{+1} \sigma_{xb} d\zeta = 0 \qquad (7.15)$$

we obtain

$$C_1^{(s)} = \int_0^1 (\varphi_1^{(s-2)} - \sigma_x^{*(s)}(\xi = 0))d\zeta \qquad (7.16)$$

while similar conditions for the anti-symmetric problem are

$$\int_0^1 \zeta \sigma_{xb} d\zeta = 0, \qquad (2+\nu)h \int_0^1 \zeta^3 f_1 d\zeta + 3E \int_0^1 (1 - \zeta^2) f_2 d\zeta = 0, \qquad (7.17)$$

and formulae (7.14) for $\psi_i^{(s)}$ lead to

$$C_2^{(s)} = \int_0^1 \zeta \varphi_1^{(s-2)} d\zeta - \int_0^1 \zeta \sigma_x^{*(s)}(0, \zeta)d\zeta$$

$$C_4^{(s)} = \frac{E}{2} \int_0^1 (1 - \zeta^2)\left(\frac{\varphi_2^{(s-3)}}{a} - v^{*(s)}(0, \zeta)\right)d\zeta \qquad (7.18)$$

$$-\frac{2+\nu}{10}C_2^{(s-3)} + \frac{2+\nu}{6}\int_0^1 \zeta^3(\varphi_1^{(s-5)} - \sigma_x^{*(s-3)}(0, \zeta))d\zeta.$$

In particular,

$$C_2^{(0)} = C_2^{(1)} = C_4^{(0)} = C_4^{(1)} = C_4^{(2)} = 0,$$

$$C_2^{(2)} = \int_0^1 \zeta \varphi_1 d\zeta, \quad C_4^{(3)} = \frac{E}{2} \int_0^1 (1 - \zeta^2)\left(\frac{\varphi_2}{a}\right)d\zeta. \qquad (7.19)$$

We remark that the influence of the first term of the second condition in (7.14) on the outer solution occurs only at higher orders ($s \geq 3$). However, value of the coefficient of the outer solution at $s < 3$ is no longer governed by a simple integral of the displacement v, but that with the integrand multiplied by $(1 - \zeta^2)$, therefore in this case the formal extension of the Saint-Venant's principle for displacements would not be valid even at the leading order. At the same time, the outer solution may be obtained in integral formulation within the plane section assumption, however, as mentioned above, the displacement v in the integrand should be multiplied by $(1 - \zeta^2)$. Once the coefficients of the outer solution are determined, reconsideration of (7.13) and (7.14) along with use of (2.5) and (2.11) leads to

$$A_n^{(s)} = \frac{\lambda_n^4}{\Delta_n} \int_0^1 (F_n'' \psi_2^{(s)} - \lambda_n^2 F_n \psi_1^{(s)})d\zeta \tag{7.20}$$

where

$$\Psi_1^{(s)}(\zeta) = \int_0^\zeta \bar{f}_1^{(s)} d\zeta - \int_0^1 \bar{f}_1^{(s)} d\zeta, \quad \bar{f}_1^{(s)}(\zeta) = \int_0^\zeta f_1^{(s)} d\zeta - \int_0^1 f_1^{(s)} d\zeta; \quad (f_1, \psi_1)$$

$$\Psi_2^{(s)}(\zeta) = \int_0^\zeta \bar{f}_2^{(s)} d\zeta, \quad \bar{f}_2^{(s)} = -\frac{E}{a} f_2^{(s)} - (2 + \nu)\bar{f}_1^{(s)}; \quad (f_i, \psi_i). \tag{7.21}$$

It follows then from (7.20) and (7.21) that $R_b^{(0)} \equiv 0$, $R_b^{(1)} \neq 0$ for symmetric problem, and $R_b^{(0)} = R_b^{(1)} = R_b^{(2)} \equiv 0$, $R_b^{(3)} \neq 0$ for the anti-symmetric one. Hence, for both cases the stresses and displacements of the boundary layer are of orders $O(\varepsilon^{-1})$ and $O(\varepsilon^0)$, respectively. These qualitative estimations are certainly valid for general type of loading, however, some special cases associated with various degenerations could have different results.

Finally, we remark that the described method of matching the outer solution with the boundary layer is not unique, there is also an option of treatment through variational principles.

1.8 Displacements Prescribed at the Edges of a Rectangle

Consider the case when the conditions (1.5) are imposed at $\xi = 0$, with similar conditions

$$u = \psi_1(\zeta), \quad v = \psi_2(\zeta) \quad \text{at} \quad \xi = 1 \tag{8.1}$$

prescribed at the opposite edge $\xi = 1$.

Since the existence conditions for decaying solutions cannot be expressed in simple form (Gusein-Zade, 1965), matching of the outer solution with the boundary layer may be performed through variational approach. In order to determine the arbitrary constants of the outer solution and the boundary layer, we will utilize the Castigliano variational equation. Provided that the plane equations of elasticity and static boundary conditions are satisfied, the Castigliano equation is written as

$$\int_{-1}^{+1} \left((u - \varphi_1)\delta\sigma_x + (v - \varphi_2)\delta\sigma_{xy}\right)_{\xi=0} d\zeta$$

$$+ \int_{-1}^{+1} \left((u - \psi_1)\delta\sigma_x + (v - \psi_2)\delta\sigma_{xy}\right)_{\xi=1} d\zeta = 0. \tag{8.2}$$

Let us discuss the symmetric problem. Substituting the values of $\delta\sigma_x$, $\delta\sigma_{xy}$ into (8.2) along with equating to zero the coefficients within $\delta C_1^{(k)}$, $\delta A_n^{(k)}$, $\delta B_n^{(k)}$ and taking into account the conditions $u = v = \dfrac{\partial u}{\partial \zeta} = 0$ at $\xi = 1$, $\zeta = 0$ fixing the center of gravity and the vertical element $\xi = 1$, we result in a closed algebraic system with respect to the unknown coefficients, which is solvable at $\chi = -3$

$$C_1^{(s)} + a_n(A_n^{(s)} + B_n^{(s)}) + a_0^{(s)} = 0$$

$$C_1^{(s)} + C_2^{(s)} + EU_n(\zeta = 0)B_n^{(s)} + Eu^{*(s)}(1,0) + \int_0^1 d\xi \int_0^\xi p^{(s)}d\xi = 0 \tag{8.3}$$

$$b_{nm}A_n^{(s)} + b_{0m}^{(s)} = 0 \quad (m = 1, 2, ...) \tag{8.4}$$

$$b_{nm}B_n^{(s)} + c_{0m}^{(s)} = 0 \quad (m = 1, 2, ...) \tag{8.5}$$

where

$$a_n = E\int_0^1 u_n d\zeta, \quad u_0^{(s)} = \frac{1}{E}\int_0^\xi d\xi \int_0^\xi p^{(s)}d\xi$$

$$a_0^{(s)} = E\int_0^1 \left(u^{*(s)}(\xi = 0) + u^{*(s)}(\xi = 1) + u_0^{(s)}(\xi = 1)\right) d\zeta$$

$$- \frac{E}{a}\int_0^1 \left(\varphi_1^{(s-2)} + \psi_1^{(s-2)}\right) d\zeta,$$

$$b_{nm} = \int_0^1 (u_n\sigma_m + v_n\tau_m)d\zeta, \quad v_0^{(s)} = \int_0^\xi p^{(s)}d\xi, \tag{8.6}$$

$$b_{0m}^{(s)} = E\int_0^1 \left(u^{*(s)}(\xi = 0)\sigma_m + \tau_m v^{*(s-1)}(\xi = 0)\right) d\zeta$$

$$- \frac{\nu}{E}\int_0^1 \zeta\tau_m d\zeta \cdot C_1^{(s-1)} - \int_0^1 \left(\sigma_m\frac{\varphi_1^{(s-2)}}{a} + \tau_m\frac{\varphi_2^{(s-2)}}{a}\right)d\zeta,$$

$$c_{0m}^{(s)} = \int_0^1 \left(\left(u^{*(s)}(\xi = 1) + u_0^{(s)}(\xi = 1)\right)\sigma_m\right.$$

$$\left. + \left(v^{*(s-1)}(\xi = 1) + \zeta v_0^{(s-1)}(\xi = 1)\right)\tau_m\right) d\zeta$$

$$- \frac{\nu}{E}\int_0^1 \zeta\tau_m d\zeta \cdot C_1^{(s-1)} - \int_0^1 \left(\sigma_m\frac{\psi_1^{(s-2)}}{a} + \tau_m\frac{\psi_2^{(s-2)}}{a}\right)d\zeta,$$

and σ_m, τ_n, u_n, v_n denote the coefficients at $A_n^{(s)}$ in the corresponding formulae (2.5) for the quantities σ_{xb}, σ_{xyb}, u_b, v_b, and $B_n^{(s)}$ is the arbitrary constant of the boundary layer solution, corresponding to the edge $\xi = 1$. If $u_0^{(0)}(1) = 0$, then $A_n^{(0)} = B_n^{(0)} = 0$, and the boundary layer is not affecting the coefficients of the outer solution at leading order, however, this does not imply that the boundary layer solution is generally negligible. Indeed, the boundary layer solution in the near-edge vicinity is of the same asymptotic order as the general solution (stresses of $O(\varepsilon^{-2})$, displacements of $O(\varepsilon^{-1})$), so it cannot be ignored. Moreover, if $u_0^{(0)}(1) \neq 0$, then as follows from (8.3), the boundary layer effects the values of arbitrary constants of the outer solution. It may be deduced from (8.3), that

$$C_1^{(s)} = -a^{(s)}, \quad C_2^{(s)} = a^{(s)} - b^{(s)}, \tag{8.7}$$

where

$$a^{(s)} = a_0^{(s)} + a_n(A_n^{(s)} + B_n^{(s)}),$$

$$b^{(s)} = E\left(u_n(0)B_n^{(s)} + u^{*(s)}(1,0) + \frac{1}{E}\int_0^1 d\xi \int_0^\xi p^{(s)}d\xi\right), \tag{8.8}$$

and $A_n^{(s)}$ and $B_n^{(s)}$ are determined from (8.4) and (8.5), respectively.

The anti-symmetric problem may now be treated similarly, resulting in

$$\frac{2}{3}C_1^{(s)} - C_2^{(s)} + a^{(s)} = 0,$$

$$C_1^{(s)} - 2C_2^{(s)} + b^{(s)} = 0,$$

$$C_1^{(s)} - 2C_2^{(s)} + 2C_3^{(s)} + d^{(s)} = 0, \tag{8.9}$$

$$C_1^{(s)} - 3C_2^{(s)} + 6C_3^{(s)} + 6C_4^{(s)} + c^{(s)} = 0,$$

$$c_{nm}A_n^{(s)} + c_{0m}^{(s)} = 0 \quad (m = 1, 2, ...) \tag{8.10}$$

$$c_{nm}B_n^{(s)} + d_{0m}^{(s)} = 0 \quad (m = 1, 2, ...) \tag{8.11}$$

where

$$a^{(s)} = a_0^{(s)} + a_n(A_n^{(s-1)} + B_n^{(s-1)}) - b_n B_n^{(s)},$$

$$b^{(s)} = b_0^{(s)} + b_n(A_n^{(s)} + B_n^{(s)}),$$

$$c^{(s)} = 2E\left(v_n(0)B_n^{(s-1)} + v^{*(s)}(1,0) + \frac{3}{E}\int_0^1 d\xi \int_0^\xi d\xi \int_0^\xi d\xi \int_0^\xi q^{(s)}d\xi\right),$$

$$d^{(s)} = -\frac{2}{3}E\left(u'_n(0)B_n^{(s)} + \frac{\partial u^{*(s)}(1,0)}{\partial \zeta} - \frac{3}{E}\int_0^1 d\xi \int_0^\xi d\xi \int_0^\xi q^{(s)}d\xi\right),$$

$$a_n = E\int_0^1 (\zeta^2 - 1)v_n d\zeta, \qquad b_n = -2E\int_0^1 \zeta u_n d\zeta,$$

$$c_{nm} = \int_0^1 (u_n\sigma_m + v_n\tau_m)d\zeta, \tag{8.12}$$

$$a_0^{(s)} = 2E \int_0^1 \left(\frac{1}{2}(\zeta^2 - 1)\left(v^{*(s)}(\xi = 0) + v^{*(s)}(\xi = 1) + v_0^{(s)} \right) \right.$$

$$\left. -\zeta u^{*(s)}(\xi = 1) + \zeta^2 u_0^{(s)} \right) d\zeta$$

$$-E \int_0^1 (\zeta^2 - 1)\left(\varphi_2^{(s-3)} + \psi_2^{(s-3)} \right) d\zeta + 2E \int_0^1 \zeta \psi_1^{(s-2)} d\zeta,$$

$$b_0^{(s)} = 2E \int_0^1 \zeta \left(\varphi_1^{(s-2)} + \psi_1^{(s-2)} \right) d\zeta$$

$$-\int_0^1 \zeta \left(u^{*(s)}(\xi = 0) + u^{*(s)}(\xi = 1) - \zeta u_0^{(s)} \right) d\zeta,$$

$$u_0^{(s)} = \frac{3}{E} \int_0^1 d\xi \int_0^\xi d\xi \int_0^\xi q^{(s)} d\xi, \quad v_0^{(s)} = \frac{3}{E} \int_0^1 d\xi \int_0^\xi d\xi \int_0^\xi d\xi \int_0^\xi q^{(s)} d\xi,$$

$$c_{0m}^{(s)} = \int_0^1 \left(\sigma_m u^{*(s)}(\xi = 0) + \tau_m v^{*(s+1)}(\xi = 0) \right) d\zeta$$

$$-\int_0^1 \left(\sigma_m \varphi_1^{(s-2)} + \tau_m \varphi_2^{(s-2)} \right) d\zeta,$$

$$d_{0m}^{(s)} = \int_0^1 \left(\sigma_m u^{*(s)}(\xi = 1) + \tau_m v^{*(s+1)}(\xi = 0) \right) d\zeta$$

$$-\int_0^1 \left(\sigma_m \psi_1^{(s-2)} + \tau_m \psi_2^{(s-2)} \right) d\zeta.$$

It follows from (8.12) that

$$C_1^{(s)} = -6a^{(s)} + 3b^{(s)}, \quad C_2^{(s)} = -3a^{(s)} + 2b^{(s)},$$

$$C_3^{(s)} = -\frac{1}{2}(d^{(s)} - b^{(s)}), \quad C_4^{(s)} = -\frac{1}{2}\left(a^{(s)} + \frac{1}{3}c^{(s)} - d^{(s)} \right). \tag{8.13}$$

The coefficients of the boundary layer solution are determined from (8.10)-(8.11), in particular we obtain $A_n^{(0)} = B_n^{(0)} = 0$, hence, the stress and displacement components of the boundary layer solution are of orders $O(\varepsilon^{-2})$ and $O(\varepsilon^{-1})$, respectively, and the boundary layer does not affect the outer solution at leading order.

Concluding this section, we remark that according to Mikhlin (1964), the Castigliano principle may be obtained through the method of orthogonal projections. The convergence issues for these methods have been analyzed in Mikhlin (1964) and Krasnoselsky et al. (1969), and therefore are not discussed here.

1.9   Asymptotic Analysis of an Orthotropic Rectangle

As mentioned earlier, one of the advantages of the described asymptotic approach is its applicability to more complicated objects, for example, anisotropic thin structures. As an illustration we present briefly the results of asymptotic analysis of 2D problems for orthotropic rectangle Ω satisfying conditions (1.1)-(1.5). For full details of analysis the reader is referred to Aghalovyan and Khachatryan (1977a,b). Solution of the plane problem of elasticity for the orthotropic rectangle Ω may be written in the form (3.1). It is also assumed that the principal directions of anisotropy coincide with the coordinate axes. The quantities Q may then be determined through a procedure described in Section 1.1.

In case of the *symmetric* problem we obtain

$$U^{(s)} = u^{*(s)} + u^{(s)}(\xi), \quad V^{(s)} = \zeta \mathrm{v}^{(s)} + \mathrm{v}^{*(s)},$$

$$\sigma_x^{(s)} = \tau_x^{(s)} + \sigma_x^{*(s)}, \quad \sigma_{xy}^{(s)} = -\zeta p^{(s)} + \sigma_{xy}^{*(s)}, \tag{9.1}$$

$$\sigma_y^{(s)} = \frac{1}{2}(\zeta^2 - 1)\frac{dp^{(s)}}{d\xi} + Y_1^{(s)} + \sigma_y^{*(s)} - \sigma_y^{*(s)}(\zeta = 1),$$

$$p^{(s)} = -X_1^{(s)} + \sigma_{xy}^{*(s)}(\zeta = 1),$$

where

$$\tau_x^{(s)} = \frac{1}{a_{11}}\frac{du^{(s)}}{d\xi}, \quad \mathrm{v}^{(s)} = \frac{a_{12}}{a_{11}}\frac{du^{(s)}}{d\xi},$$

$$u^{*(s)} = \int_0^\zeta \left(a_{66}\sigma_{xy}^{(s-2)} - \frac{\partial V^{(s-2)}}{\partial \xi}\right) d\zeta,$$

$$\sigma_x^{*(s)} = \frac{1}{a_{11}}\left(\frac{\partial u^{*(s)}}{\partial \xi} - a_{12}\sigma_y^{(s-2)}\right), \tag{9.2}$$

$$\mathrm{v}^{*(s)} = \int_0^\zeta \left(a_{12}\sigma_x^{*(s)} + a_{22}\sigma_y^{(s-2)}\right) d\zeta,$$

$$\sigma_{xy}^{*(s)} = -\int_0^\zeta \frac{\partial \sigma_x^{*(s)}}{\partial \xi}d\zeta, \quad \sigma_y^{*(s)} = -\int_0^\zeta \frac{\partial \sigma_{xy}^{*(s)}}{\partial \xi}d\zeta.$$

Here a_{ik} are the elasticity coefficients, and $u^{(s)}$ is the solution of the following equation

$$\frac{1}{a_{11}}\frac{d^2 u^{(s)}}{d\xi^2} = p^{(s)}, \tag{9.3}$$

i.e.

$$\frac{1}{a_{11}}u^{(s)} = \int_0^\xi d\xi \int_0^\xi p^{(s)} d\xi + C_1^{(s)} \cdot \xi + C_2^{(s)}.$$

In the *anti-symmetric* (flexural) case the outer solution is governed by (1.11),

(1.16), where

$$\tau_{xx}^{(s)} = -\frac{1}{a_{11}}\frac{d^2 w^{(s)}}{d\xi^2}, \quad \tau_{xy}^{(s)} = \frac{1}{a_{11}}\frac{d^3 w^{(s)}}{d\xi^3},$$

$$v^{*(s)} = \int_0^\zeta \left(a_{12}\sigma_x^{(s-2)} + a_{22}\sigma_y^{(s-4)} \right) d\zeta,$$

$$u^{*(s)} = \int_0^\zeta \left(a_{66}\sigma_{xy}^{(s-2)} - \frac{\partial v^{*(s)}}{\partial \xi} \right) d\zeta,$$

$$\sigma_x^{*(s)} = \frac{1}{a_{11}} \left(\frac{\partial u^{*(s)}}{\partial \xi} - a_{12}\sigma_y^{(s-2)} \right), \qquad (9.4)$$

$$\sigma_{xy}^{*(s)} = -\int_0^\zeta \frac{\partial \sigma_x^{*(s)}}{\partial \xi}d\zeta, \qquad \sigma_y^{*(s)} = -\int_0^\zeta \frac{\partial \sigma_{xy}^{*(s)}}{\partial \xi}d\zeta,$$

$$q^{(s)} = Y_2^{(s)} + \frac{dX_2^{(s)}}{d\xi} - \sigma_y^{*(s)}(\zeta = 1) - \frac{d\sigma_{xy}^{*(s)}(\zeta = 1)}{d\xi},$$

$$X_2^{(0)} = X_2, \ Y_2^{(0)} = Y_2, \ X_2^{(s)} = Y_2^{(s)} = 0, \ (s > 0),$$

and $w^{(s)}$ satisfies

$$\frac{1}{3a_{11}}\frac{d^4 w^{(s)}}{d\xi^4} = q^{(s)}, \qquad (9.5)$$

hence,

$$\frac{1}{3a_{11}}w^{(s)} = \int_0^\xi d\xi \int_0^\xi d\xi \int_0^\xi d\xi \int_0^\xi q^{(s)}d\xi + C_1^{(s)}\frac{\xi^3}{6} + \frac{1}{2}C_2^{(s)}\xi^2 + C_3^{(s)}\xi + C_4^{(s)}. \quad (9.6)$$

Returning to the actual displacement $w_0^{(s)} = a\varepsilon^{-3}w^{(s)}$, we obtain

$$\frac{1}{a_{11}}I\frac{d^4 w_0^{(s)}}{dx^4} = q_0^{(s)}, \qquad (9.7)$$

where $q_0^{(s)} = 2q^{(s)}$, and I is the moment of inertia of the cross section. It may be observed that equation (9.7) coincides with the well-known bending equation for orthotropic beam provided $s = 0$, with the rest of the formulae matching within the plane sections assumption. At $s > 0$ the loading on the right-hand side of (9.7) changes to

$$q^{(1)} = 0, \quad q^{(2)} = \left(\frac{1}{30}\frac{a_{12}}{a_{11}} - \frac{1}{15}\frac{a_{66}}{a_{11}} \right)\frac{d^3 X_2}{d\xi^3} - \left(\frac{3}{10}\frac{a_{12}}{a_{11}} + \frac{2}{5}\frac{a_{66}}{a_{11}} \right)\frac{d^2 Y_2}{d\xi^2}. \qquad (9.8)$$

It follows from (9.8) that the correction depends both on the the variation of the external loading and the material parameters. In case of real materials the ratio $\dfrac{a_{12}}{a_{11}}$ cannot be large, which is not the case for the ratio $\dfrac{a_{66}}{2a_{11}} = \dfrac{E_1}{2G_{12}}$. If the in-plane bending stiffness is sufficiently large, then $\dfrac{E_1}{2G_{12}} \ll 1$, and the classical beam theory is applicable, moreover, it would provide more accuracy for such type of anisotropy, than it would in isotropic case. However, problems may arise when the bending stiffness in the associated plane is small, then $\dfrac{E_1}{2G_{12}} \gg 1$, which can often happen in practical applications. Even for slow varying load $\dfrac{E_1}{2G_{12}} = O(\varepsilon^{-1})$, therefore the correction would be of order $O(\varepsilon^{-1})$, and then (depending on the value of ε) it may become necessary to take the next order into consideration, therefore the classical theories will no longer be able to provide the desired accuracy. In this case an iterative process described above may be constructed in order to refine the results of the classical theories. At $\dfrac{E_1}{2G_{12}} = O(\varepsilon^{-2})$ the refinement will be of the same order as the leading order terms, and hence the problem would become essentially 2D one. In such limiting cases of strong anisotropy, some special asymptotic technique may be required or even neither asymptotic nor classical methods would be applicable. The 2D formulation of such problems is discussed in Chapter 5. Thus, the formal extension of the plane section hypothesis to anisotropic beams may lead to significant computational errors. At the same time, the technical theory of Bernoulli-Euler-Coulomb for isotropic beams is asymptotically consistent in case of slow varying external load.

As for the previously considered case of isotropic rectangular strip, the outer solution does not contain enough constants to satisfy the boundary conditions (1.2)-(1.5), with these provided by the boundary layer. The latter is given by

$$\sigma_{xb}^{(s)} = \frac{F_n''}{\lambda_n^2} A_n^{(s)}, \quad \sigma_{xyb}^{(s)} = \frac{F_n'}{\lambda_n} A_n^{(s)}, \quad \sigma_{yb}^{(s)} = A_n^{(s)} F_n(\zeta),$$

$$u_b^{(s)} = -\left(a_{11}\frac{F_n''}{\lambda_n^3} + a_{12}\frac{F_n}{\lambda_n} \right) A_n^{(s)}, \tag{9.9}$$

$$v_b^{(s)} = -\left(a_{11}\frac{F_n'''}{\lambda_n^4} + (a_{12} + a_{66})\frac{F_n'}{\lambda_n^2} \right) A_n^{(s)}$$

with $F_n(\zeta)$ denoting the solution of the following boundary value problem

$$a_{11}F_n^{IV} + \lambda_n^2(a_{66} + 2a_{12})F_n'' + \lambda_n^4 a_{22}F_n = 0, \tag{9.10}$$

$$F_n(\pm 1) = F_n'(\pm 1) = 0. \tag{9.11}$$

The characteristic equation of (9.10) has the following roots depending on the elastic material parameters $(\alpha, \beta, \beta_1, \beta_2 > 0)$

(a) $i\beta, \ -i\beta,$
(b) $\pm i\beta_1, \ \pm i\beta_2,$
(c) $\alpha \pm i\beta, \ -\alpha \pm i\beta.$

The first case a) takes place when

$$D = \left(\frac{1}{2}a_{66} + a_{12}\right)^2 - a_{11}a_{22} = 0 \quad \Rightarrow \quad \beta = \sqrt[4]{\frac{a_{22}}{a_{11}}} = \sqrt[4]{\frac{E_1}{E_2}}. \tag{9.12}$$

The second case b) occurs when $D > 0$, hence

$$\beta_1 = \sqrt{\frac{1}{a_{11}}\left(0.5a_{66} + a_{12} - \sqrt{(0.5a_{66} + a_{12})^2 - a_{11}a_{22}}\right)},$$

$$\beta_2 = \sqrt{\frac{1}{a_{11}}\left(0.5a_{66} + a_{12} + \sqrt{(0.5a_{66} + a_{12})^2 - a_{11}a_{22}}\right)}. \tag{9.13}$$

Finally, in the third case c) $D < 0$ and

$$\alpha = \sqrt{\frac{1}{2a_{11}}\left(\sqrt{a_{11}a_{22}} - \left(\frac{1}{2}a_{66} + a_{12}\right)\right)},$$

$$\beta = \sqrt{\frac{1}{2a_{11}}\left(\sqrt{a_{11}a_{22}} + \left(\frac{1}{2}a_{66} + a_{12}\right)\right)}. \tag{9.14}$$

In reality, case a) corresponds to isotropic materials, case b) emerges when the bending stiffness $2G_{12}$ is less than the average extension stiffness $\sqrt{E_1 E_2}$, while the third case c) corresponds to $2G_{12} > \sqrt{E_1 E_2}$. Most anisotropic materials will be classified to case b). As will be shown below, the boundary layers associated with these three cases are quite different. For example, in case of isotropic material the decay is described by a harmonic exponential function, while for anisotropic beams, plates and shells the decay is often exponential, but not harmonic. The solutions of the boundary value problem (9.10)-(9.11), corresponding to the cases (9.12)-(9.14) are given below.

For *symmetric* problem

(a)

$$F_n(\zeta) = \sin\beta\lambda_n \cos\beta\lambda_n\zeta - \zeta\cos\beta\lambda_n \sin\beta\lambda_n\zeta, \tag{9.15}$$

with λ_n denoting the root of $\sin 2\beta\lambda_n + 2\beta\lambda_n = 0$.

(b)

$$F_n(\zeta) = \cos\beta_2\lambda_n \cos\beta_1\lambda_n\zeta - \cos\beta_1\lambda_n \cos\beta_2\lambda_n\zeta, \tag{9.16}$$

where λ_n is a root of transcendental equation

$$\beta_2 \sin\beta_2\lambda_n \cos\beta_1\lambda_n - \beta_1 \sin\beta_1\lambda_n \cos\beta_2\lambda_n = 0, \tag{9.17}$$

which may be transformed to a more convenient form

$$\omega\sin z_n + \sin\omega z_n = 0,$$

$$z_n = (\beta_1 + \beta_2)\lambda_n, \quad \omega = \frac{\beta_2 - \beta_1}{\beta_1 + \beta_2}, \quad 0 < \omega < 1. \tag{9.18}$$

(c)

$$F_n(\zeta) = \sin\beta\lambda_n \text{sh}\alpha\lambda_n \cos\beta\lambda_n\zeta \text{ch}\alpha\lambda_n\zeta$$
$$- \cos\beta\lambda_n \text{ch}\alpha\lambda_n \sin\beta\lambda_n\zeta \text{sh}\alpha\lambda_n\zeta, \tag{9.19}$$

with

$$\alpha\sin 2\beta\lambda_n + \beta\text{sh}2\alpha\lambda_n = 0$$

or

$$\omega\sin z_n + \text{sh}\omega z_n = 0 \quad z_n = 2\beta\lambda_n, \quad \omega = \frac{\alpha}{\beta} \quad (0 < \omega < 1). \tag{9.20}$$

In case of the *anti-symmetric* problem similar solutions reduce to

(a)

$$F_n(\zeta) = \cos\beta\lambda_n \sin\beta\lambda_n\zeta - \zeta\sin\beta\lambda_n \cos\beta\lambda_n\zeta, \tag{9.21}$$

with $\sin 2\beta\lambda_n - 2\beta\lambda_n = 0$

(b)

$$F_n(\zeta) = \sin\beta_2\lambda_n \sin\beta_1\lambda_n\zeta - \sin\beta_1\lambda_n \sin\beta_2\lambda_n\zeta, \tag{9.22}$$

where now

$$\beta_2 \sin\beta_1\lambda_n \cos\beta_2\lambda_n - \beta_1 \sin\beta_2\lambda_n \cos\beta_1, \lambda_n = 0,$$

or

$$\omega\sin z_n - \sin\omega z_n = 0, \tag{9.23}$$

$$z_n = (\beta_1 + \beta_2)\lambda_n, \quad \omega = \frac{\beta_2 - \beta_1}{\beta_1 + \beta_2}, \quad 0 < \omega < 1,$$

and, finally,

(c)

$$F_n(\zeta) = \cos\beta\lambda_n \mathrm{sh}\alpha\lambda_n \sin\beta\lambda_n\zeta \mathrm{ch}\alpha\lambda_n\zeta$$
$$- \sin\beta\lambda_n \mathrm{ch}\alpha\lambda_n \cos\beta\lambda_n\zeta \mathrm{sh}\alpha\lambda_n\zeta, \tag{9.24}$$

with λ_n satisfying

$$\omega\sin z_n - \mathrm{sh}\omega z_n = 0, \qquad z_n = 2\beta\lambda_n, \quad \omega = \frac{\alpha}{\beta}, \qquad 0 < \omega < 1. \tag{9.25}$$

We remark that summation among repeated indices n associated with the roots λ_n satisfying $Re\lambda_n > 0$ is assumed in (9.9), ensuring decay of the solution.

Here we note several properties of the transcendental equations presented above, since a more general discussion will be held later in Chapter 4. According to the Pikard theorem the sets of roots of equations (9.15)-(9.25) are countable, the roots cannot be purely imaginary, and they are symmetric in respect of the origin and the real axe. We shall be sorting these roots in increasing order for their real parts. The first root satisfying $Re\lambda_n > 0$ is characterizing the attenuation coefficient of the boundary layer. The values of the first ten roots of equations (9.18), (9.23) are presented in Tables 3 and 4 respectively, while the material parameters are displayed in Table 2.

As seen from (9.18), (9.23) and the corresponding Tables 3 and 4, the attenuation parameter λ_n depends on the elastic material parameters, which is a novel feature caused by anisotropy (no such dependence was observed earlier in isotropic case, see Table 1). A tendency of slower attenuation may be pointed out for most anisotropic materials compared to isotropic, however, some materials demonstrate faster decay. Some more details on these may be found in Chapter 3 (Tables 12 and 13). Large values of ω would typically correspond to strong anisotropy of a material. One more effect of strong anisotropy, especially for materials with small shear stiffness, is a tendency of slower decay of plane boundary layer.

Table 2

Elastic charac-teristics	Materials				
	Fiberglass STET	Unidirectional fiberglass winding	Fiberglass ASTT(b) - C_2-O and PN-3	SVAM 10:1	Plywood BC-1, first grade
$E_1 \cdot 10^{-9} Pa$	35.218	55.917	17.56	38.259	10.791
$E_2 \cdot 10^{-9} Pa$	28.743	13.734	12.851	17.64	8.339
$G_{12} \cdot 10^{-9} Pa$	7.456	5.592	2.747	5.199	0.883
ν_{12}	0.177	0.277	0.15	0.22	0.076
$a_{11} \cdot 10^{11} Pa^{-1}$	2.839	1.788	5.695	2.614	9.267
$a_{22} \cdot 10^{11} Pa^{-1}$	3.479	7.281	7.781	5.669	11.993
$a_{12} \cdot 10^{11} Pa^{-1}$	-0.503	-0.495	-0.854	-0.575	-0.704
$a_{66} \cdot 10^{11} Pa^{-1}$	13.413	17.884	36.406	19.234	113.263
β_1	0.549	0.673	0.483	0.573	0.231
β_2	2.017	2.999	2.421	2.567	4.923
$\beta_1 + \beta_2$	2.566	3.671	2.903	3.140	5.154
ω	0.57	0.634	0.667	0.635	0.91

Solution (2.1), (9.9) satisfies all of the equations of elasticity for an orthotropic solid along with stress free boundary conditions at $y = \pm h$ of the rectangle, so it is an exact solution for arbitrary N. It may be shown from (9.10) and (9.11) according to Aghalovyan and Khachatryan (1975) that the functions $F_n(\zeta)$ satisfy the orthogonality condition

$$\int_0^1 \left(F_n'' F_k'' - \frac{a_{22}}{a_{11}} \lambda_n^2 \lambda_k^2 F_n F_k \right) d\zeta = 0 \quad (n \neq k) \tag{9.26}$$

generalizing the famous condition of P.F. Papkovich. Similarly to isotropic case, the stress components σ_{xp}, σ_{xyp} are self-equilibrated (2.12). We also remark that a well-known substitution $\beta_{ij} = a_{ij} - \dfrac{a_{i3}a_{j3}}{a_{33}}$, $(i, j = 1, 2, 6)$ delivers the results of the associated plane strain problem provided that all coefficients a_{ij} would be replaced by β_{ij}.

Solution of the outer problem (1.11), (9.1)-(9.6) does not involve enough arbitrary constants in order to satisfy the boundary conditions at the edges $x = 0, a$ and, therefore, there is a necessity in the boundary layer solution. Matching the boundary layer with the outer solution through a procedure explained in Section 1.3, we may now obtain the values of arbitrary constants.

Table 3

	Roots of the equation $\omega \sin z_n + \sin \omega\, z_n = 0,\quad z_n = x_n + iy_n$				
	Eigenvalues $\lambda_n = \frac{z_n}{\beta_1 + \beta_2}$				
	Fiberglass STET	Unidirectional fiberglass winding	Fiberglass ASTT(b) -C_2-O and PN-3	SVAM 10:1	Plywood BC-1, first grade
ω	0.57	0.634	0.667	0.635	0.91
$\beta_1 + \beta_2$	2.56592	3.67164	2.903582	3.140545	5.1540
z_1	4.48	4.09695	3.95303	4.0922	3.297
λ_1	1.746	1.1158	1.3614	1.303	0.6397
z_2	7.520 +1.496 i	8.319 +1.368 i	9.05701	8.335 +1.358 i	6.5944
λ_2	2.930 +0.583 i	2.266 +0.373 i	3.1193	2.654 +0.432 i	1.2795
z_3	11.764	10.9935	9.612 +0.320 i	10.97802	9.8924
λ_3	4.5847	2.9942	3.310 +0.110 i	3.4956	1.9194
z_4	16.1123	15.2828	14.89127	15.2708	13.1918
λ_4	6.2793	4.1624	5.1286	4.8625	2.5595
z_5	21.580 +0.680 i	19.3413	18.84483	19.32506	16.4934
λ_5	8.410 +0.265 i	5.2678	6.4902	6.1534	3.2001
z_6	22.7655	23.7062	22.7973	23.65963	19.7987
λ_6	8.8723	6.4566	7.8514	7.5336	3.8414
z_7	27.9134	26.081 +1.267 i	27.7476	26.132 +1.302 i	23.1113
λ_7	10.879	7.103 +0.345 i	9.5563	8.321 +0.415 i	4.4841
z_8	32.2452	30.8362	28.549 +0.463 i	30.50909	26.4396
λ_8	12.567	8.3985	9.832 +0.159 i	9.7146	5.1299
z_9	36.372 +1.530 i	34.6219	33.7355	34.59462	29.8154
λ_9	14.1750 +0.596 i	9.4295	11.619	11.015	5.7849
z_{10}	39.4832	38.7329	37.6897	38.6944	33.6052
λ_{10}	15.0388	10.549	12.98	12.321	6.5202

          Asymptotic Theory of Anisotropic Plates and Shells

Table 4

	Roots of the equation $\omega \sin z_n - \sin \omega\, z_n = 0,\quad z_n = x_n + iy_n$				
	Eigenvalues $\lambda_n = \frac{z_n}{\beta_1 + \beta_2}$				
	Fiberglass STET	Unidirectional fiberglass winding	Fiberglass ASTT(b) -C_2-O and PN-3	SVAM 10:1	Plywood BC-1, first grade
ω	0.57	0.634	0.667	0.635	0.91
$\beta_1 + \beta_2$	2.56592	3.67164	2.903582	3.140545	5.154
z_1	5.89401	5.6025	5.469	5.5983	4.7086
λ_1	2.297	1.5259	1.8835	1.7826	0.91358
z_2	10.2586	9.6683	9.4224	9.6604	8.0953
λ_2	3.998	2.6332	3.2451	3.076	1.5707
z_3	14.426 +1.515 i	13.8388	13.3752	13.8216	11.4266
λ_3	5.622 +0.590 i	3.7692	4.6064	4.401	2.217
z_4	17.5535	17.209 +1.491 i	18.3876	17.239 +1.492 i	14.7433
λ_4	6.841	4.687 +0.406 i	6.3327	5.489 +0.475 i	2.8605
z_5	22.0187	20.8095	19.087 +0.403 i	20.7883	18.0462
λ_5	8.5812	5.6676	6.574 +0.139 i	6.6193	3.5014
z_6	26.5153	24.9541	24.3134	24.9345	21.3481
λ_6	10.334	6.7964	8.3736	7.9395	4.142
z_7	29.464 +1.475 i	29.0254	28.2673	28.9996	24.6477
λ_7	11.483 +0.575 i	7.9053	9.7353	9.2339	4.7822
z_8	33.771	33.9512 +0.933 i	32.2194	34.069 +0.781 i	27.9458
λ_8	13.161	9.247 +0.254 i	11.096	10.848 +0.249 i	5.4222
z_9	38.0966	35.6008	37.1206	35.436	31.2432
λ_9	14.847	9.6962	12.784	11.283	6.0619
z_{10}	43.294	40.2292	38.004 +0.509 i	40.1959	34.5403
λ_{10}	16.873	10.957	13.089 +0.175 i	12.798	6.7016

In case of the first boundary value problem (1.2) the *symmetric* problem reduces to

$$C_1^{(s)} = \int_0^1 \left(\varphi_1^{(s-2)} - \sigma_x^{*(s)}(\xi = 0) \right) d\zeta, \tag{9.27}$$

while the corresponding results for the *anti-symmetric* problem are given by

$$C_1^{(s)} = \left(X_2^{(s)} - \sigma_{xy}^{*(s)}(\zeta = 1) \right)_{\xi=0} + \int_0^1 \sigma_{xy}^{*(s)}(0,\zeta)d\zeta - \int_0^1 \varphi_2^{(s-1)}d\zeta$$

$$C_2^{(s)} = \int_0^1 \zeta \left(\varphi_1^{(s-2)} - \sigma_x^{*(s)}(\xi = 0) \right) d\zeta \tag{9.28}$$

$$(\varphi_i^{(0)} = \varphi_i, \ \varphi_i^{(s)} \equiv 0, \ s \neq 0, \ i = 1,2).$$

Then the solution is written as (9.1)-(9.6). We remark that the solutions of Lekhnitskii (1968) through polynomials may be obtained from these as particular cases. The boundary layer solution is found from (2.1), (9.9) and conditions (3.3) through one of the approximation methods described in Section 1.3.

It may be observed that similarly to isotropic case the statically equivalent edge loads lead to the same outer solutions including the constants $C_i^{(s)}$, with the only difference on the right-hand side of (3.3), which corresponds to the boundary layer solution. The self-equilibrated solution (along height) is decaying exponentially at a speed associated with $Re\lambda_1 > 0$ for equations (9.15)-(9.25). In other words, the Saint-Venant's principle is true for the considered class of boundary value problems for an orthotropic strip.

We remark that there has been an ongoing interest to various generalizations of the Saint-Venant's principle, see e.g. Von Mises (1945); Sternberg (1954); Toupin (1965); Cherepanov (1970); Berdichevsky (1974, 1978); Oleynik and Iosif'yan (1976, 1977a,b); Choi and Horgan (1977). Most of the cited works rely on energy approach. In particular it has been proved by energy method in Toupin (1965), that the stresses in a cylindrical rod caused by a self-equilibrated load applied at the edge of the road decay exponentially, with an estimate of attenuation speed obtained through the first eigenfrequency of the cylinder. Various generalizations of the Saint-Venant's principle to other physical problems along with other types of problems of mathematical physics including elliptic were presented in Berdichevsky (1974); Oleynik and Iosif'yan (1976, 1977b). An extension of the Saint-Venant's principle to dynamic problems have been performed recently by Babenkova and Kaplunov (2005).

The asymptotic approach described above allows some qualitative results for exponential decay of the boundary layer and also determination of the speed of attenuation associated with the appropriate root (with minimal $Re\lambda_1 > 0$) of the corresponding transcendental equation (9.15)-(9.25), see Tables 3 and 4. In addition, the sought-for boundary layer solution can clearly be obtained.

In case of the boundary value problem (1.3) for an orthotropic strip the conditions of existence of a decaying solution may be written as (Aghalovyan and Khachatryan, 1975):

for *symmetric* problem

$$\int_0^1 f_1(\zeta)d\zeta - a_{12}h \int_0^1 \zeta f_2(\zeta)d\zeta = 0; \tag{9.29}$$

for *anti-symmetric* problem

$$\int_0^1 f_2(\zeta)d\zeta = 0; \quad 2\int_0^1 \zeta f_1(\zeta)d\zeta - a_{12}h \int_0^1 \zeta^2 f_2(\zeta)d\zeta = 0, \tag{9.30}$$

where f_1, f_2 are the values of au_b and σ_{xyb}, respectively, evaluated at the edge. Rewriting conditions (1.3) in the form (7.3), we may obtain the constants of the outer solution from (9.29) and (9.30), namely

for *symmetric* problem

$$C_2^{(s)} = \frac{1}{a_{11}} \left(\int_0^1 \left(\frac{1}{a}\varphi_1^{(s-2)} - u^{*(s)}(\xi = 0) \right) d\zeta \right.$$

$$\left. - a_{12} \int_0^1 \zeta \left(\varphi_2^{(s-4)} - \sigma_{xy}^{(s-3)}(\xi = 0) \right) d\zeta \right), \tag{9.31}$$

for *anti-symmetric* problem

$$C_1^{(s)} = \int_0^1 \left(\sigma_{xy}^{*(s)}(0,\zeta) - \varphi_2^{(s-1)} \right) d\zeta + X_2^{(s)}(0) - \sigma_{xy}^{*(s)}(0,1), \tag{9.32}$$

$$C_3^{(s)} = \frac{1}{a_{11}} \left(\int_0^1 \zeta \left(u^{*(s)}(\xi = 0) - \frac{1}{a}\varphi_1^{(s-2)} \right) d\zeta \right.$$

$$\left. - \frac{a_{12}}{2} \int_0^1 \zeta^2 \left(\varphi_2^{(s-4)} - \sigma_{xy}^{(s-3)}(\xi = 0) \right) d\zeta \right).$$

The remaining of the constants of the outer solution, namely $C_1^{(s)}$ for the symmetric case and $C_2^{(s)}$, $C_4^{(s)}$ for the anti-symmetric one are determined from the boundary conditions at the opposite edge. After all of the constants of the outer solution are obtained, the values of $f_1^{(s)}$, $f_2^{(s)}$ in (7.3) are known, and therefore, using the generalized orthogonality condition (9.26), the constant $A_n^{(s)}$ of the boundary layer solution (2.2), (9.9) may be evaluated

$$A_n^{(s)} = \frac{\lambda_n^3}{\Delta_n} \int_0^1 \left(F_n'' \Psi_1^{(s)} - \frac{a_{22}}{a_{11}}\lambda_n^2 F_n \Psi_2^{(s)} \right) d\zeta, \tag{9.33}$$

with

$$\Delta_n = \int_0^1 \left((F_n'')^2 - \frac{a_{22}}{a_{11}}\lambda_n^4 F_n^2 \right) d\zeta, \tag{9.34}$$

$$\Psi_1^{(s)} = -\frac{1}{a_{11}} \left(\frac{1}{a}f_1^{(s)} + a_{12}\Psi_2^{(s)} \right), \quad \Psi_2^{(s)} = \int_0^\zeta f_2^{(s)}d\zeta - \int_0^1 f_2^{(s)}d\zeta.$$

Thus, closed solutions for both the outer problem and the boundary layer have been established for all values of s. It may be shown that $A_n^{(0)} = A_n^{(1)} \equiv 0$, $A_n^{(2)} \neq 0$, therefore the stress and displacement components would be of orders $O(\varepsilon^{-1})$ and $O(\varepsilon^0)$, respectively.

In case of the boundary value problem, corresponding to (1.4), the compatibility conditions are written as:

for *symmetric* problem

$$\int_0^1 \sigma_{xb}d\zeta = 0, \tag{9.35}$$

for *anti-symmetric* problem

$$\int_0^1 \zeta\sigma_{xb}d\zeta = 0, \tag{9.36}$$

$$(a_{12} + a_{66})h\int_0^1 \zeta^3\psi_1 d\zeta + 3\int_0^1 (1 - \zeta^2)\psi_2 d\zeta = 0,$$

where ψ_1, ψ_2 stand for the quantities σ_{xb} and av_b respectively, evaluated at the edge. Following the same procedure as above, we obtain the constants of the outer solution

$$C_1^{(s)} = \int_0^1 \left(\varphi_1^{(s-2)} - \sigma_x^{*(s)}(\xi = 0)\right) d\zeta, \tag{9.37}$$

for the *symmetric* case and

$$C_2^{(s)} = -\int_0^1 \zeta\left(\varphi_1^{(s-2)} - \sigma_x^{*(s)}(\xi = 0)\right) d\zeta, \tag{9.38}$$

$$C_4^{(s)} = \frac{1}{2a_{11}}\left(\int_0^1 (1 - \zeta^2)\left(\frac{1}{a}\varphi_2^{(s-3)} - v^{*(s)}(\xi = 0)\right) d\zeta\right.$$

$$\left. +\frac{1}{3}(a_{12} + a_{66})\int_0^1 \zeta^3\left(\varphi_1^{(s-5)} - \sigma_x^{(s-3)}(\xi = 0)\right) d\zeta\right),$$

for the *anti-symmetric* case. Now returning to (7.13)-(7.14), with $f_i^{(s)}$, and $\psi_i^{(s)}$ evaluated, and making use of (9.26), we obtain

$$A_n^{(s)} = \frac{\lambda_n^4}{\Delta_n}\int_0^1 \left(F_n'' \Psi_1^{(s)} - \frac{a_{22}}{a_{11}}\lambda_n^2 F_n\Psi_2^{(s)}\right) d\zeta, \tag{9.39}$$

where

$$\Psi_1^{(s)} = -\frac{1}{a_{11}}\int_0^\zeta \left(f_2^{(s)} + (a_{12} + a_{66})\bar{f}_1^{(s)}\right) d\zeta \quad (f_i, \psi_i) \tag{9.40}$$

$$\Psi_2^{(s)} = \int_0^\zeta \bar{f}_1^{(s)} d\zeta - \int_0^1 \bar{f}_1^{(s)} d\zeta, \quad \bar{f}_1^{(s)} = \int_0^\zeta f_1^{(s)} d\zeta - \int_0^1 f_1^{(s)} d\zeta \quad (f_i, \psi_i).$$

As follows from (9.39)-(9.40), $R_b^{(0)} \equiv 0$, $R_b^{(1)} \neq 0$ and $R_b^{(0)} = R_b^{(1)} = R_b^{(2)} \equiv 0$, $R_b^{(3)} \neq 0$ for symmetric and anti-symmetric problems, respectively, hence for both cases the stresses are of order $O(\varepsilon^{-1})$, while the displacements are of order $O(\varepsilon^0)$.

Now let us consider the boundary value problem for conditions (1.5). Using the variational Castigliano approach, the constants of the outer solution and the boundary layer are found from algebraic systems. In case of symmetric problem these are written as

$$C_1^{(s)} + a_n(A_n^{(s)} + B_n^{(s)}) + a_0^{(s)} = 0 \tag{9.41}$$

and

$$b_{nm}A_n^{(s)} + b_{0m}^{(s)} = 0 \quad (m = 1, 2, ...),$$
$$b_{nm}B_n^{(s)} + c_{0m}^{(s)} = 0 \quad (m = 1, 2, ...), \tag{9.42}$$

where $B_n^{(s)}$ is a constant of the boundary layer corresponding to the edge $\xi = 1$, σ_n, τ_n, u_n, v_n are the coefficients at $A_n^{(s)}$ in the appropriate formulae (9.9) for σ_{xb}, σ_{xyb}, u_b, v_b, and

$$a_n = \frac{1}{a_{11}} \int_0^1 u_n d\zeta, \quad u_0^{(s)} = a_{11} \int_0^\xi d\xi \int_0^\xi p^{(s)} d\xi,$$

$$a_0^{(s)} = \frac{1}{a_{11}} \int_0^1 \left(u^{*(s)}(\xi = 0) + u^{*(s)}(\xi = 1) + u_0^{(s)}(\xi = 1) \right) d\zeta$$

$$- \frac{1}{a\, a_{11}} \int_0^1 \left(\varphi_1^{(s-2)} + \psi_1^{(s-2)} \right) d\zeta,$$

$$b_{nm} = \int_0^1 (u_n \sigma_m + v_n \tau_m) d\zeta, \quad v_0^{(s)} = a_{11} \int_0^\xi p^{(s)} d\xi, \tag{9.43}$$

$$b_{0m}^{(s)} = \int_0^1 \left(u^{*(s)}(\xi = 0)\sigma_m + v^{*(s-1)}(\xi = 0)\tau_m \right) d\zeta$$

$$+ C_1^{(s-1)} a_{12} \int_0^1 \zeta \tau_m d\zeta - \int_0^1 \left(\sigma_m \frac{\varphi_1^{(s-2)}}{a} + \tau_m \frac{\varphi_2^{(s-2)}}{a} \right) d\zeta,$$

$$c_{0m}^{(s)} = \int_0^1 \left(\left(u^{*(s)}(\xi = 1) + u_0^{(s)}(\xi = 1) \right) \sigma_m \right.$$

$$\left. + \left(v^{*(s-1)}(\xi = 1) + \zeta v_0^{(s-1)}(\xi = 1) \right) \tau_m \right) d\zeta$$

$$+ C_1^{(s-1)} a_{12} \int_0^1 \zeta \tau_m d\zeta - \int_0^1 \left(\sigma_m \frac{\psi_1^{(s-2)}}{a} + \tau_m \frac{\psi_2^{(s-2)}}{a} \right) d\zeta.$$

The constants $A_n^{(s)}$ and $B_n^{(s)}$ may now be evaluated from (9.42), and then (from (9.41))

$$C_1^{(s)} = -a_0^{(s)} - a_n(A_n^{(s)} + B_n^{(s)}).$$

It is clearly observed that the boundary value solution affects significantly the outer solution, which was not the case for the first boundary value problem.

For the anti-symmetric case we obtain

$$\frac{2}{3}C_1^{(s)} + C_2^{(s)} + a^{(s)} = 0, \qquad C_1^{(s)} + 2C_2^{(s)} + b^{(s)} = 0 \tag{9.44}$$

and

$$c_{nm}A_n^{(s)} + c_{0m}^{(s)} = 0, \qquad c_{nm}B_n^{(s)} + d_{0m}^{(s)} = 0 \quad (m = 1, 2, ...), \tag{9.45}$$

where

$$a^{(s)} = a_0^{(s)} + a_n(A_n^{(s-1)} + B_n^{(s-1)}) + b_n B_n^{(s)},$$
$$b^{(s)} = b_0^{(s)} + b_n(A_n^{(s)} + B_n^{(s)}),$$
$$a_n = \frac{1}{a_{11}} \int_0^1 (\zeta^2 - 1)\mathrm{v}_n d\zeta, \quad b_n = -\frac{2}{a_{11}} \int_0^1 \zeta u_n d\zeta,$$
$$c_{nm} = 2 \int_0^1 (u_n\sigma_m + \mathrm{v}_n\tau_m)d\zeta,$$
$$a_0^{(s)} = \frac{2}{a_{11}} \int_0^1 \left(\frac{1}{2}(\zeta^2 - 1) \left(\mathrm{v}^{*(s)}(\xi = 0) + \mathrm{v}^{*(s)}(\xi = 1) + \mathrm{v}_0^{(s)} \right) \right.$$
$$\left. - \zeta u^{*(s)}(\xi = 1) + \zeta^2 u_0^{(s)} \right)d\zeta$$
$$-\frac{1}{a_{11}} \int_0^1 (\zeta^2 - 1) \left(\frac{\varphi_2^{(s-3)}}{a} + \frac{\psi_2^{(s-3)}}{a} \right) d\zeta + \frac{2}{aa_{11}} \int_0^1 \zeta\psi_1^{(s-2)} d\zeta, \tag{9.46}$$
$$b_0^{(s)} = -\frac{2}{a_{11}} \int_0^1 \zeta \left(u^{*(s)}(\xi = 0) + u^{*(s)}(\xi = 1) + u_0^{(s)} \right) d\zeta$$
$$+\frac{2}{a_{11}} \int_0^1 \zeta \left(\frac{\varphi_1^{(s-2)}}{a} + \frac{\psi_1^{(s-2)}}{a} \right) d\zeta,$$
$$c_{0m}^{(s)} = 2 \int_0^1 \left(\sigma_m u^{*(s-1)}(\xi = 0) + \tau_m \mathrm{v}^{*(s)}(\xi = 0) \right) d\zeta$$
$$-2 \int_0^1 \left(\sigma_m \frac{\varphi_1^{(s-3)}}{a} + \tau_m \frac{\varphi_2^{(s-3)}}{a} \right) d\zeta,$$
$$d_{0m}^{(s)} = 2 \int_0^1 \left(\sigma_m u^{*(s-1)}(\xi = 1) + \tau_m \mathrm{v}^{*(s)}(\xi = 1) \right) d\zeta$$
$$-2 \int_0^1 \left(\sigma_m \frac{\psi_1^{(s-3)}}{a} + \tau_m \frac{\psi_2^{(s-3)}}{a} \right) d\zeta,$$
$$u_0^{(s)} = 3a_{11} \int_0^1 d\xi \int_0^\xi d\xi \int_0^\xi q^{(s)} d\xi,$$
$$\mathrm{v}_0^{(s)} = 3a_{11} \int_0^1 d\xi \int_0^\xi d\xi \int_0^\xi d\xi \int_0^\xi q^{(s)} d\xi,$$

implying

$$C_1^{(s)} = -6a^{(s)} + 3b^{(s)}, \quad C_2^{(s)} = 3a^{(s)} - 2b^{(s)}. \tag{9.47}$$

The stress and displacement components of the boundary layer are of orders $O(\varepsilon^{-1})$ and $O(\varepsilon^0)$, respectively. The rest of the unknown constants, namely, $C_2^{(s)}$ for symmetric case, and $C_3^{(s)}$, $C_4^{(s)}$ for the anti-symmetric one, may be determined, for example from the following conditions of absence of rigid body motion

$$u = \mathrm{v} = \frac{\partial u}{\partial \zeta} = 0 \qquad \text{at} \quad \xi = 1,\ \zeta = 0.$$

Finally in this chapter, we remark that Khachatryan (1976) has solved plane problems for strips of general anisotropy, and Ponyatovsky (1968) has presented a solution for the outer problem for an elastic cylinder subject to arbitrary loading at the lateral surface (the Almansi's problem). The developed iteration approach allowed a solution satisfying the equations of elasticity and the boundary conditions on the lateral surfaces. The conditions at the edges are satisfied within the framework of the Saint-Venant's approach. It is clear that in order to refine the solution and satisfy the edge boundary conditions in a more accurate manner, a boundary layer solution is required.

Chapter 2

The Winkler-Fuss Model

Mixed Boundary Value Problems for Single and Two-Layer Rectangular Beam

2.1 The Second and the Mixed Boundary Value Problems

The asymptotic method can be used to solve a new class of problems, which include mixed two- and three-dimensional problems of beams and plates. In particular, such problems may arise in calculations of layered structures, hydraulic structures, airfield pavements, bases and foundations, described by the compressible layer model, and in other areas. In order to solve plane problems the equations of plane stress state are employed. In this case the rectangular area has unit thickness and the forces are applied in the plane of the rectangle. On the other hand, if the body has a rectangular cross section, which at the same time is the plane of elastic symmetry, infinite in a perpendicular direction to the rectangle plane, and the forces are acting in cross-sections and do not change towards the infinite stretch of the body, then the equations of plane strain deformation should be taken. Hereinafter we will solve the equations of plane stress state. Solution for the plane strain problem can be obtained from the plane stress solution by the well-known replacement of elastic coefficients a_{ij} by $\beta_{ij} = a_{ij} - (a_{i3}a_{j3})/a_{33}$ $(i, j = 1, 2, 4, 5, 6)$.

Let us consider the following class of mixed problems in a rectangular area $\Omega = \{(x, y) : x \in [0, a], |y| \le h, h << a\}$: displacement values are specified on the lower face of the rectangle, in particular, the lower face is fixed, while on the upper face displacements, stresses or mixed conditions are imposed. Static, kinematic or mixed conditions are set on the edges $x = 0, a$. A plane problem is assumed for the generally anisotropic rectangle. The boundary conditions are given in the form

$$u(-h) = u^-(\xi), \quad v(-h) = v^-(\xi) \qquad \text{in particular} \quad u^- = v^- = 0 \qquad (1.1)$$

along with one of the following conditions imposed at $y = +h$

$$u(h) = u^+(\xi), \quad v(h) = v^+(\xi), \tag{1.2}$$

$$v(h) = v^+(\xi), \quad \sigma_{xy}(h) = \varepsilon^{-1}\sigma_{xy}^+(\xi), \quad (\varepsilon = h/a) \tag{1.3}$$

$$u(h) = u^+(\xi), \quad \sigma_y(h) = \varepsilon^{-1}\sigma_y^+(\xi), \tag{1.4}$$

$$\sigma_{xy}(h) = \varepsilon^{-1}\sigma_{xy}^+(\xi), \quad \sigma_y(h) = \varepsilon^{-1}\sigma_y^+(\xi). \tag{1.5}$$

It can be shown that for this class of problems the boundary layers correspond to the conditions at $x = 0, a$. We consider a long rectangular strip ($2h \ll a$), so that the effect of a certain edge may be neglected in the vicinity of the opposite one. A more delicate case when these can not be neglected, will be discussed further in Section 2.4.

As before, we introduce the dimensionless variables $\xi = x/a$, $\zeta = y/h$ in the equations of plane elasticity and seek for the dimensionless stress and displacement components specified as $U = u/a$, $V = v/a$. As a result, we obtain a system which is singularly perturbed by a small parameter $\varepsilon = h/a$. As in the case of the first boundary value problem of elasticity, solutions of mixed boundary value problems stated above are composed of the outer solution and the boundary layers.

The outer solution may be written as (Aghalovyan, 1982)

$$Q = \varepsilon^{q+s} Q^{(s)}, \quad s = \overline{0, N}, \tag{1.6}$$

where Q - is any of the stresses and non-dimensional displacements, and notation $s = \overline{0, N}$ hereinafter means summation over the mute index s from zero to N, where N - is number of approximations. Following the procedure presented in Chapter 1 for the first boundary value problem, we reveal the solvability of the system in respect of $Q^{(s)}$ as

$$q = -1 \quad \text{for} \quad (\sigma_x, \sigma_{xy}, \sigma_y), \quad q = 0 \quad \text{for} \quad (U, V). \tag{1.7}$$

Comparing (1.6) and (1.7) with (1.1.10) shows that the asymptotic behaviour of the sought for quantities, corresponding to (1.7), is fundamentally different from that of the same quantities when conditions of the first boundary-value problem of elasticity theory are set on the faces of the strip or plate. In case of general loading all of the stress components here contribute equally, so none of these could be neglected. This means that the hypothesis of a beam theory is not applicable. However, there is no need for any hypothesis, since the problem has a rather straightforward analytical solution.

Substituting (1.6) into the equations of the plane problem and equating the corresponding coefficients of the same powers of ε, and making use of (1.7) leads to a system for $Q^{(s)}$, namely

$$\frac{\partial \sigma_x^{(s-1)}}{\partial \xi} + \frac{\partial \sigma_{xy}^{(s)}}{\partial \zeta} = 0, \quad \frac{\partial \sigma_{xy}^{(s-1)}}{\partial \xi} + \frac{\partial \sigma_y^{(s)}}{\partial \zeta} = 0,$$

$$\frac{\partial U^{(s-1)}}{\partial \xi} = a_{11}\sigma_x^{(s)} + a_{12}\sigma_y^{(s)} + a_{16}\sigma_{xy}^{(s)},$$

$$\frac{\partial V^{(s)}}{\partial \zeta} = a_{12}\sigma_x^{(s)} + a_{22}\sigma_y^{(s)} + a_{26}\sigma_{xy}^{(s)}, \tag{1.8}$$

$$\frac{\partial U^{(s)}}{\partial \zeta} + \frac{\partial V^{(s-1)}}{\partial \xi} = a_{16}\sigma_x^{(s)} + a_{26}\sigma_y^{(s)} + a_{66}\sigma_{xy}^{(s)}.$$

It follows from (1.8), that

$$\sigma_{xy}^{(s)} = \sigma_{xy0}^{(s)}(\xi) + \sigma_{xy}^{*(s)}(\xi,\zeta), \qquad \sigma_y^{(s)} = \sigma_{y0}^{(s)}(\xi) + \sigma_y^{*(s)}(\xi,\zeta),$$

$$\sigma_x^{(s)} = -\frac{1}{a_{11}}\left(a_{12}\sigma_{y0}^{(s)} + a_{16}\sigma_{xy0}^{(s)}\right) + \sigma_x^{*(s)}(\xi,\zeta),$$

$$U^{(s)} = \left(A_{16}\sigma_{y0}^{(s)} + A_{66}\sigma_{xy0}^{(s)}\right)\zeta + u^{(s)}(\xi) + u^{*(s)}(\xi,\zeta), \tag{1.9}$$

$$V^{(s)} = \left(A_{11}\sigma_{y0}^{(s)} + A_{16}\sigma_{xy0}^{(s)}\right)\zeta + v^{(s)}(\xi) + v^{*(s)}(\xi,\zeta),$$

where

$$\sigma_{xy}^{*(s)} = -\int_0^\zeta \frac{\partial \sigma_x^{(s-1)}}{\partial \xi}\,d\zeta, \qquad \sigma_y^{*(s)} = -\int_0^\zeta \frac{\partial \sigma_{xy}^{(s-1)}}{\partial \xi}\,d\zeta,$$

$$\sigma_x^{*(s)} = -\left(a_{12}\sigma_y^{*(s)} + a_{16}\sigma_{xy}^{*(s)}\right)a_{11}^{-1} + \frac{1}{a_{11}}\frac{\partial U^{(s-1)}}{\partial \xi},$$

$$u^{*(s)} = \int_0^\zeta \left(a_{16}\sigma_x^{*(s)} + a_{26}\sigma_y^{*(s)} + a_{66}\sigma_{xy}^{*(s)} - \frac{\partial V^{(s-1)}}{\partial \xi}\right)d\zeta,$$

$$v^{*(s)} = \int_0^\zeta \left(a_{12}\sigma_x^{*(s)} + a_{22}\sigma_y^{*(s)} + a_{26}\sigma_{xy}^{*(s)}\right)d\zeta, \tag{1.10}$$

$$A_{11} = (a_{11}a_{22} - a_{12}^2)a_{11}^{-1}, \qquad A_{16} = (a_{11}a_{26} - a_{12}a_{16})a_{11}^{-1},$$

$$A_{66} = (a_{11}a_{66} - a_{16}^2)a_{11}^{-1}.$$

The solution (1.9) contains four arbitrary functions $\sigma_{xy0}^{(s)}(\xi)$, $\sigma_{y0}^{(s)}(\xi)$, $u^{(s)}(\xi)$, $v^{(s)}(\xi)$, which are found from the boundary conditions (1.1)-(1.5). For example, in case of the problem (1.1)-(1.2), satisfying the conditions at $y = \pm h$, one arrives at

$$\sigma_{xy0}^{(s)} = \left(A_{11}\varphi^{(s)} - A_{16}f^{(s)}\right)/\Omega, \qquad \Omega = A_{11}A_{66} - A_{16}^2,$$

$$\sigma_{y0}^{(s)} = \left(A_{66}f^{(s)} - A_{16}\varphi^{(s)}\right)/\Omega,$$

$$u^{(s)}(\xi) = \frac{1}{2}\left(u^{+(s)} + u^{-(s)}\right) - \frac{1}{2}\left(u^{*(s)}(\xi,1) + u^{*(s)}(\xi,-1)\right), \tag{1.11}$$

$$v^{(s)}(\xi) = \frac{1}{2}\left(v^{+(s)} + v^{-(s)}\right) - \frac{1}{2}\left(v^{*(s)}(\xi,1) + v^{*(s)}(\xi,-1)\right),$$

where

$$u^{\pm(0)} = u^\pm/a, \quad u^{\pm(s)} = 0, \quad s \neq 0, \quad (u,v),$$

$$\varphi^{(s)} = \frac{1}{2}\left(u^{+(s)} - u^{-(s)}\right) - \frac{1}{2}\left(u^{*(s)}(\xi,1) - u^{*(s)}(\xi,-1)\right), \tag{1.12}$$

$$f^{(s)} = \frac{1}{2}\left(v^{+(s)} - v^{-(s)}\right) - \frac{1}{2}\left(v^{*(s)}(\xi,1) - v^{*(s)}(\xi,-1)\right).$$

Then, using (1.6), (1.7), (1.9)-(1.12) the components of the outer solution for (1.1)-(1.2) are determined.

Satisfying the conditions (1.1), (1.3), we have

$$\sigma_{xy0}^{(s)} = \sigma_{xy}^{+(s)} - \sigma_{xy}^{*(s)}(\xi, 1),$$

$$\sigma_{y0}^{(s)} = \left(f^{(s)} - A_{16}\sigma_{xy}^{+(s)} + A_{16}\sigma_{xy}^{*(s)}(\xi, 1) \right) A_{11}^{-1},$$

$$v^{(s)}(\xi) = \frac{1}{2}\left(v^{+(s)} + v^{-(s)} - v^{*(s)}(\xi, 1) - v^{*(s)}(\xi, -1) \right), \tag{1.13}$$

$$u^{(s)}(\xi) = u^{-(s)} + A_{16}\sigma_{y0}^{(s)} + A_{66}\sigma_{xy0}^{(s)} - u^{*(s)}(\xi, -1),$$

$$\sigma_{xy}^{+(0)} = \sigma_{xy}^{+}, \qquad \sigma_{xy}^{+(s)} = 0, \quad s \neq 0,$$

which gives the full solution of the outer problem in conjunction with (1.6) and (1.9).

The obtained solutions of the boundary value problems (1.1)-(1.2), (1.1)-(1.3) clarify the reasons of including the large parameter ε^{-1} into the boundary conditions (1.3)-(1.5). These followed from the fact that the stresses and displacements corresponding to the conditions (1.1)-(1.2) are of orders $0(\varepsilon^{-1})$ and $0(\varepsilon^{0})$, respectively (as may be observed from (1.6), (1.7), (1.11)), allowing to satisfy the conditions for displacements (1.1), (1.2). Then, in order for the representation (1.6) to satisfy the conditions for stresses, these should be of the form (1.3)-(1.5). A clear physical interpretation of the reasoning is that formulae (1.1), (1.2) imply the stresses are of order $0(\varepsilon^{-1})$, therefore the loading causing stress should be of the same order. Certainly, the conditions of the form $\sigma_y(+h) = \sigma_y^{+}$ may also be satisfied, be presented as $\sigma_y(+h) = \varepsilon^{-1} \cdot 0 + \varepsilon^{0} \cdot \sigma_y^{+}$.

The constructed iteration process allows treatment of more general boundary conditions for displacements $u(\pm h) = \varepsilon^k u^{\pm(k)}$, (u, v) and stresses $\sigma_y(\pm h) = \varepsilon^{-1+k}\sigma_y^{+(k)}$, (σ_y, σ_{xy}), $k = \overline{0, N}$. As above the notation $k = \overline{0, N}$ represents summation along the index k from zero to N. The formulae above remain valid, if we assume in the appropriate places that $Q_k^{\pm(s)} \neq 0$, $s = 0, 1, ..., N$.

In the case of the boundary conditions (1.1), (1.5), the values $Q^{(s)}$ are given by

$$\sigma_{y0}^{(s)}(\xi) = \sigma_y^{+(s)} - \sigma_y^{*(s)}(\xi, 1), \qquad \sigma_y^{+(0)} = \sigma_y^{+}, \qquad \sigma_y^{+(s)} = 0 \quad (s \neq 0),$$

$$\sigma_y^{(s)} = \sigma_y^{+(s)} + \sigma_y^{*(s)}(\xi, \zeta) - \sigma_y^{*(s)}(\xi, 1),$$

$$\sigma_{xy0}^{(s)} = \frac{1}{A_{66}}\varphi^{(s)} - \frac{A_{16}}{A_{66}}\sigma_{y0}^{(s)}, \qquad \sigma_{xy}^{(s)} = \sigma_{xy0}^{(s)} + \sigma_{xy}^{*(s)}, \tag{1.14}$$

$$v^{(s)}(\xi) = v^{-(s)} - v^{*(s)} + \frac{\Omega}{A_{66}}\sigma_{y0}^{(s)} + \frac{A_{16}}{A_{66}}\varphi^{(s)},$$

$$u^{(s)}(\xi) = \frac{1}{2}(u^{+(s)} + u^{-(s)}) - \frac{1}{2}(u^{*(s)}(\xi, 1) + u^{*(s)}(\xi, -1)),$$

$$\sigma_x^{(s)} = -\frac{1}{a_{11}A_{66}}\left((a_{12}A_{66} - a_{16}A_{16})\sigma_{y0}^{(s)} + a_{16}\varphi^{(s)} \right) + \sigma_x^{*(s)}(\xi, \zeta).$$

In case of the boundary value problem corresponding to (1.1), (1.5)

$$\sigma_{xy0}^{(s)} = \sigma_{xy}^{+(s)} - \sigma_{xy}^{*(s)}(\xi, 1), \quad \sigma_{y0}^{(s)} = \sigma_{y}^{+(s)} - \sigma_{y}^{*(s)}(\xi, 1),$$

$$(\sigma_{xy}^{+(0)} = \sigma_{xy}^{+}, \quad \sigma_{y}^{+(0)} = \sigma_{y}^{+}, \quad \sigma_{xy}^{+(s)} = \sigma_{y}^{+(s)} \equiv 0, \quad (s \neq 0)),$$

$$u^{(s)}(\xi) = A_{16}\sigma_{y0}^{(s)} + A_{66}\sigma_{xy0}^{(s)} + u^{-(s)} - u^{*(s)}(\xi, -1), \qquad (1.15)$$

$$v^{(s)}(\xi) = A_{11}\sigma_{y0}^{(s)} + A_{16}\sigma_{xy0}^{(s)} + v^{-(s)} - v^{*(s)}(\xi, -1).$$

The outer solution is then computed using (1.6), (1.7), (1.9), (1.10).

The properties of solutions of the class of problems considered above is fundamentally different from that solved in the first chapter. This difference lies in the fact that the solution of the outer problem is completely determined by the conditions on the faces $y = \pm h$. Hence the boundary layer have to match the conditions at the ends $x = 0, a$. This also means that the boundary layer does not affect the outer stresses through boundary conditions, while the outer solution directly affects the boundary layer. Physically, this is explainable, because in the case when one of the faces is fixed rigidly or a known displacement is prescribed, the system is equilibrated and the end effect cannot disturb its equilibrium. In case of the first boundary problem the system is not self-equilibrated from the very beginning, hence the edge loading should be applied in order to achieve equilibrium state. Clearly, such edge loading will affect the interior as well.

2.2 Special Cases

In all cases where the functions on the right of (1.1) - (1.5) are polynomials in ξ, the corresponding iterative process stops at a certain approximation and we obtain the outer solution in closed form. This promises certain benefits because, according to the Weierstrass theorem an arbitrary continuous function can be approximated with any desired accuracy by the appropriate polynomial. Hence, for a wide class of external loading one can obtain a closed solution of the outer problem. Let us consider some special cases. Let $u^- = v^- = 0$, $u^+ = const$, $v^+ = const$ then in the problem (1.1), (1.2) $Q^{(0)} \neq 0$, $Q^{(s)} \equiv 0$ $(s \geq 1)$, and the solution is written in the form

$$\sigma_x = -\left((a_{16}A_{11} - a_{12}A_{16})u^+ + (a_{12}A_{66} - a_{16}A_{16})v^+\right)\frac{1}{2ha_{11}\Omega_1}, \qquad (2.1)$$

$$\sigma_{xy} = \left(A_{11}u^+ - A_{16}v^+\right)/2h\Omega_1, \quad \sigma_y = \left(A_{66}v^+ - A_{16}u^+\right)/2h\Omega_1,$$

$$u = \frac{u^+}{2h}(y + h), \quad v = \frac{v^+}{2h}(y + h), \quad \Omega_1 = A_{11}A_{66} - A_{16}^2.$$

For orthotropic solid $(a_{16} = A_{16} = 0)$

$$\sigma_{xy} = G_{12}\frac{u^+}{2h}, \quad \sigma_y = \frac{E_2}{1 - \nu_{12}\nu_{21}}\frac{v^+}{2h}, \quad \sigma_x = \frac{E_1\nu_{12}}{1 - \nu_{12}\nu_{21}}\frac{v^+}{2h}, \qquad (2.2)$$

where G_{12} is the shear modulus, E_1, E_2 are the Young's moduli and ν_{ik} the Poisson's ratio.

In case of the problem (1.1), (1.3), if

$$u^- = v^- = 0, \quad v^+ = const, \qquad \varepsilon^{-1}\sigma_{xy}^+ = \tau^+ = const, \tag{2.3}$$

the exact outer solution is given by

$$\sigma_x = -\frac{1}{a_{11}A_{11}}\left(a_{12}\frac{v^+}{2h} + (a_{16}A_{11} - a_{12}A_{16})\tau^+\right),$$

$$\sigma_{xy} = \tau^+, \quad \sigma_y = \frac{1}{A_{11}}\frac{v^+}{2h} - \frac{A_{16}}{A_{11}}\tau^+, \tag{2.4}$$

$$u = \left(\frac{A_{16}}{A_{11}}\frac{v^+}{2h} + \frac{\Omega_1}{A_{11}}\tau^+\right)(y + h), \quad v = \frac{v^+}{2h}(y + h).$$

In case of the following boundary conditions at $y = \pm h$

$$u(-h) = v(-h) = 0, \quad u(h) = u^+ = const, \qquad \sigma_y(h) = \sigma^+ = const \tag{2.5}$$

the corresponding solution is written as

$$\sigma_y = \sigma^+, \qquad \sigma_{xy} = -\frac{A_{16}}{A_{66}}\sigma^+ + \frac{1}{A_{66}}\frac{u^+}{2h},$$

$$\sigma_x = -\frac{a_{12}A_{66} - a_{16}A_{16}}{a_{11}A_{66}}\sigma^+ - \frac{a_{16}}{a_{11}A_{66}}\frac{u^+}{2h}, \tag{2.6}$$

$$u = \frac{u^+}{2h}(y + h), \qquad v = \left(\frac{\Omega_1}{A_{66}}\sigma^+ + \frac{A_{16}}{A_{66}}\frac{u^+}{2h}\right)(y + h).$$

When the edge $y = -h$ is fixed by $u(-h) = v(-h) = 0$, and loading of constant density $\sigma_{xy}(h) = \tau^+ = const$, $\sigma_y(h) = \sigma_2^+ = const$ is prescribed at the opposite edge $y = h$, the solution takes the form

$$\sigma_x = -(a_{12}\sigma_2^+ + a_{16}\tau^+)a_{11}^{-1}, \qquad \sigma_{xy} = \tau^+, \quad \sigma_y = \sigma_2^+, \tag{2.7}$$

$$u = (A_{66}\tau^+ + A_{16}\sigma_2^+)(y + h), \qquad v = (A_{11}\sigma_2^+ + A_{16}\tau^+)(y + h).$$

The formulae (2.1), (2.2), (2.4), (2.6), (2.7) demonstrate that if the external edge loading is constant along the length of a rectangle, then the emerging stresses are also constants while the displacements are linear functions along the height of a rectangle.

The uneven distribution of stresses in thickness is connected with non-uniformity of external load along the length of the rectangle. The following illustrative example gives an idea of the nature of any non-uniform distribution. Let the bottom face of the rectangle be rigidly fixed ($u^- = v^- = 0$) and the normal load (hydrostatic pressure) applies to the upper face $\tau^+ = 0$, $\sigma_y(h) = -q^+\xi$, ($\xi = x/a$). Then

$Q^{(0)} \neq 0,\; Q^{(1)} \neq 0,\; Q^{(s)} \equiv 0,\; (s \geq 2)$, and the outer solution is given by

$$\sigma_{xy} = \varepsilon^{-1}\sigma_{xy}^{(0)} + \varepsilon^{0}\sigma_{xy}^{(1)} = \frac{a_{12}}{a_{11}}\frac{h-y}{a}q^{+}, \quad \sigma_{y} = \varepsilon^{-1}\sigma_{y}^{(0)} + \varepsilon^{0}\sigma_{y}^{(1)} = -\frac{x}{a}q^{+},$$

$$\sigma_{x} = \varepsilon^{-1}\sigma_{x}^{(0)} + \varepsilon^{0}\sigma_{x}^{(1)} = \frac{a_{12}}{a_{11}}\frac{x}{a}q^{+} - \frac{a_{12}a_{16}}{a_{11}^{2}}\frac{h-y}{a}q^{+} - \frac{A_{16}}{a_{11}}\frac{h+y}{a}q^{+}, \quad (2.8)$$

$$u = aU^{(0)} + a\varepsilon U^{(1)} = -A_{16}(y+h)\frac{x}{a}q^{+} - (a_{16}A_{16} - a_{11}A_{11} - a_{12}A_{66})$$

$$\times a_{11}^{-1}\frac{h}{a}(y+h)q^{+} + \frac{1}{2a_{11}}(a_{16}A_{16} - a_{11}A_{11} + a_{12}A_{66})(h^{2} - y^{2})\frac{q^{+}}{a},$$

$$v = aV^{(0)} + a\varepsilon V^{(1)} = -A_{11}(y+h)\frac{x}{a}q^{+} + \frac{a_{12}A_{16}}{a_{11}}(h^{2} - y^{2})\frac{q^{+}}{a}.$$

For loads of more complicated form, the non-uniform distribution of stresses and displacements in transverse variable is even more evident, but the arising additional terms typically do not contribute to the leading order terms. For example, in expression (2.8) for σ_x the second and third terms are small in comparison to the first one. The appearance of such terms also contributes to the overall anisotropy in the plane of the rectangle ($a_{16} \neq 0,\; A_{16} \neq 0$).

The formulae (2.7), (2.8) and similar expressions, following from (1.15), can be used to calculate the deformation of a compressible layer in the theory of elastic foundations. They are more accurate than those proposed in Vlasov and Leont'ev (1960), since the proposed solution is obtained without making any hypotheses regarding the distribution of the displacement field and exactly satisfy all of the equations and relations of elasticity.

2.3 Boundary Layer

The solutions constructed in Sections 2.1, 2.2 are valid in the entire rectangular area provided that the forces and displacements imposed at the edges $x = 0, a$ are the same as those arising from the outer problem. However, this is possible in some rather special cases, therefore, the question of satisfying the boundary conditions at $x = 0, a$ so far remains open. On the other hand, the solution of the outer problem no longer contains arbitrary constants, so we cannot use it for this purpose. It is therefore necessary to have a second solution, namely the boundary layer solution.

The boundary layer near the edge $x = 0$ is constructed as in the case of the first boundary value problem. The scaling $t = \xi/\varepsilon$ is introduced in the equations of the elasticity theory, and the solutions are found again in the form (1.6). The resulting system will be consistent if the sought-for quantities have the following asymptotic representations

$$\sigma_b = \varepsilon^{-1+s}\sigma_b^{(s)}(t,\zeta), \quad u_b = \varepsilon^{s}u_b^{(s)}(t,\zeta), \quad s = \overline{0, N}, \quad (3.1)$$

where σ_b and u_b as previously stand for the stress and displacement components.

The coefficients in (3.1) may be written as ([Aghalovyan (1984)])

$$\left(\sigma_{xb}^{(s)},\,\sigma_{xyb}^{(s)},\,\sigma_{yb}^{(s)}\right) = \left(\sigma_1^{(s)}(\zeta),\,\sigma_{12}^{(s)}(\zeta),\,\sigma_2^{(s)}(\zeta)\right)\exp(-\lambda t),$$

$$\left(u_b^{(s)},\,v_b^{(s)}\right) = \left(u_1^{(s)}(\zeta),\,u_2^{(s)}(\zeta)\right)\exp(-\lambda t). \tag{3.2}$$

Following the procedure described in details in Chapter 1, the quantities $\sigma_1^{(s)}$, $\sigma_{12}^{(s)}$, $\sigma_2^{(s)}$, $u_1^{(s)}$, $u_2^{(s)}$, λ are determined from

$$-\lambda\sigma_1^{(s)} + \frac{d\,\sigma_{12}^{(s)}}{d\zeta} = 0, \quad -\lambda\sigma_{12}^{(s)} + \frac{d\sigma_2^{(s)}}{d\zeta} = 0,$$

$$-\lambda u_1^{(s)} = a_{11}\sigma_1^{(s)} + a_{12}\sigma_2^{(s)} + a_{16}\sigma_{12}^{(s)},$$

$$\frac{du_1^{(s)}}{d\zeta} - \lambda u_2^{(s)} = a_{16}\sigma_1^{(s)} + a_{26}\sigma_2^{(s)} + a_{66}\sigma_{12}^{(s)}, \tag{3.3}$$

$$\frac{du_2^{(s)}}{d\zeta} = a_{12}\sigma_1^{(s)} + a_{22}\sigma_2^{(s)} + a_{26}\sigma_{12}^{(s)}.$$

Expressing all of the unknown quantities through $\sigma_2^{(s)}$, λ, for the sake of brevity we present the following results for an orthotropic rectangle ($a_{16} = a_{26} = 0$)

$$\sigma_1^{(s)} = \frac{1}{\lambda^2}\frac{d^2\sigma_2^{(s)}}{d\zeta^2}, \; \sigma_{12}^{(s)} = \frac{1}{\lambda}\frac{d\sigma_2^{(s)}}{d\zeta},$$

$$u_1^{(s)} = -\frac{1}{\lambda}\left(\frac{a_{11}}{\lambda^2}\frac{d^2\sigma_2^{(s)}}{d\zeta^2} + a_{12}\sigma_2^{(s)}\right), \tag{3.4}$$

$$u_2^{(s)} = -\frac{a_{11}}{\lambda^4}\frac{d^3\sigma_2^{(s)}}{d\zeta^3} - (a_{12} + a_{66})\frac{1}{\lambda^2}\frac{d\sigma_2^{(s)}}{d\zeta}.$$

Then $\sigma_2^{(s)}$ is found from

$$a_{11}\sigma_2^{(s)\,IV} + (a_{66} + 2a_{12})\lambda^2\sigma_2^{(s)\,II} + a_{22}\lambda^4\sigma_2^{(s)} = 0. \tag{3.5}$$

The characteristic equation associated with (3.5), could possess roots of the following types

(a) $i\beta$, $-i\beta$,

(b) $\pm i\beta_1$, $\pm i\beta_2$,

(c) $\alpha \pm i\beta$, $-\alpha \pm i\beta$,

with the corresponding solutions

$$a) \; \sigma_2^{(s)}(\zeta) = (A^{(s)} + B^{(s)}\zeta)\cos\lambda\beta\zeta + (C^{(s)} + D^{(s)}\zeta)\sin\lambda\beta\zeta, \tag{3.6}$$

$$b) \; \sigma_2^{(s)}(\zeta) = A^{(s)}\cos\lambda\beta_1\zeta + B^{(s)}\sin\lambda\beta_1\zeta + C^{(s)}\cos\lambda\beta_2\zeta + D^{(s)}\sin\lambda\beta_2\zeta, \tag{3.7}$$

$$c) \; \sigma_2^{(s)}(\zeta) = A^{(s)}\varphi_1 + B^{(s)}\varphi_2 + C^{(s)}\varphi_3 + D^{(s)}\varphi_4,$$

$$\varphi_1 = \operatorname{ch}\alpha\lambda\zeta\cos\lambda\beta\zeta, \quad \varphi_2 = \operatorname{sh}\alpha\lambda\zeta\sin\lambda\beta\zeta, \tag{3.8}$$

$$\varphi_3 = \operatorname{ch}\alpha\lambda\zeta\sin\lambda\beta\zeta, \quad \varphi_4 = \operatorname{sh}\alpha\lambda\zeta\cos\lambda\beta\zeta.$$

The first case a) corresponds to $\Delta = (a_{66} + 2a_{12})^2 - 4a_{11}a_{22} = 0$, then $\beta = \sqrt[4]{E_1/E_2}$.

In case of b) $\Delta > 0$

$$\beta_1 = \sqrt{\frac{a_{66} + 2a_{12} - \sqrt{\Delta}}{2a_{11}}}, \qquad \beta_2 = \sqrt{\frac{a_{66} + 2a_{12} + \sqrt{\Delta}}{2a_{11}}}, \tag{3.9}$$

whereas the third case c) $\Delta < 0$ gives

$$\alpha = \sqrt{\frac{\sqrt{a_{11}a_{22}} - \left(\frac{1}{2}a_{66} + a_{12}\right)}{2a_{11}}}, \qquad \beta = \sqrt{\frac{\sqrt{a_{11}a_{22}} + \left(\frac{1}{2}a_{66} + a_{12}\right)}{2a_{11}}}. \tag{3.10}$$

Since the outer solution satisfies the boundary conditions (1.1)- (1.5), the boundary layer solution (3.1), (3.4), (3.6)-(3.8) should satisfy the corresponding homogeneous conditions (due to linearity of the boundary value problem)

$$u_1 = u_2 = 0 \quad \text{at} \quad \zeta = -1 \tag{3.11}$$

along with one of the following conditions at $\zeta = 1$

$$\begin{aligned}
&1)\, u_1 = u_2 = 0, \quad &2)\, u_2 = \sigma_{12} = 0 \quad &\text{at} \quad \zeta = 1 \\
&3)\, u_1 = \sigma_2 = 0, \quad &4)\, \sigma_{12} = \sigma_2 = 0 \quad &\text{at} \quad \zeta = 1.
\end{aligned} \tag{3.12}$$

In addition, the boundary layer solution should decay away from the edge $x = 0$ into the interior of the domain Ω.

Matching solutions (3.4), (3.6)-(3.8) with conditions (3.11)-(3.12), we obtain

a) $\Delta = 0$, $\sigma_2^{(s)}(\zeta) = \left(\left(\dfrac{2a_{11}\beta}{\sqrt{a_{11}a_{22}} - a_{12}} - \tan\lambda\beta \right) \cos\lambda\beta\zeta + \zeta \sin\lambda\beta\zeta \right) D^{(s)}$ (3.13)

where λ is the root of a transcendental equation

$$\sin 2\lambda\beta - \frac{a_{66}}{3a_{66} + 8a_{12}} 2\lambda\beta = 0. \tag{3.14}$$

The obtained solution corresponds to symmetric case (extension/compression). In case of the bending mode

$$\sigma_2^{(s)}(\zeta) = \left(\zeta \cos\lambda\beta\zeta + \left(\tan\lambda\beta + \left(1 + \frac{4a_{12}}{a_{66}}\right) \frac{1}{\lambda\beta} \right) \sin\lambda\beta\zeta \right) B^{(s)}$$

$$\sin 2\lambda\beta + \frac{a_{66}}{3a_{66} + 8a_{12}} 2\lambda\beta = 0. \tag{3.15}$$

We note that in case of isotropic rectangle $\beta = 1$. Introduction of the notation $2\lambda = zi$ allows the following representation of the transcendental equations (3.14)-(3.15)

$$\sinh z \mp \frac{1+\nu}{1-\nu} z = 0, \tag{3.16}$$

coinciding with that given in Uflyand (1967) obtained through integral transforms in case of the second boundary value problem of elasticity for isotropic strip.

b) $\Delta > 0$

For symmetric problem (extension/compression) we get

$$\sigma_2^{(s)}(\zeta) = \left(\cos \beta_1 \lambda \zeta - \frac{a_{11}\beta_1^2 - a_{12}}{a_{11}\beta_2^2 - a_{12}} \frac{\cos \beta_1 \lambda}{\cos \beta_2 \lambda} \cos \beta_2 \lambda \zeta \right) A^{(s)}, \qquad (3.17)$$

with λ denoting the root of

$$m \sin z + m_1 \sin mz = 0, \qquad (3.18)$$

whereas for the anti-symmetric problem (bending)

$$\sigma_2^{(s)}(\zeta) = \left(\sin \beta_1 \lambda \zeta - \frac{a_{11}\beta_1^2 - a_{12}}{a_{11}\beta_2^2 - a_{12}} \frac{\sin \beta_1 \lambda}{\sin \beta_2 \lambda} \sin \beta_2 \lambda \zeta \right) C^{(s)}, \qquad (3.19)$$

with λ satisfying

$$m \sin z - m_1 \sin mz = 0, \qquad (3.20)$$

where

$$z = (\beta_1 + \beta_2)\lambda, \quad m = \frac{\beta_2 - \beta_1}{\beta_1 + \beta_2}, \quad m_1 = \frac{1 - \delta}{1 + \delta}, \quad \delta = \frac{a_{66}\sqrt{a_{11}a_{22}}}{a_{11}a_{22} - a_{12}^2}. \qquad (3.21)$$

We remark that in case of anisotropic materials the quantities m and m_1 are in the range $0 < m < 1$, $-1 < m_1 < 1$.

c) $\Delta < 0$

For symmetric problem we obtain

$$\sigma_2^{(s)}(\zeta) = \left(\varphi_1(\zeta) - \frac{((\alpha^2 - \beta^2)a_{11} + a_{12})\varphi_1(1) - 2a_{11}\alpha\beta\varphi_2(1)}{((\alpha^2 - \beta^2)a_{11} + a_{12})\varphi_2(1) + 2a_{11}\alpha\beta\varphi_1(1)} \varphi_2(\zeta) \right) A^{(s)} \quad (3.22)$$

with λ satisfying $\omega \sin z + m_1 \sinh \omega z = 0$, $z = 2\beta\lambda$, $\omega = \alpha/\beta$, whereas in case of the anti-symmetric problem

$$\sigma_2^{(s)}(\zeta) = \left(\varphi_4(\zeta) - \frac{((\alpha^2 - \beta^2)a_{11} + a_{12})\varphi_4(1) - 2a_{11}\alpha\beta\varphi_3(1)}{((\alpha^2 - \beta^2)a_{11} + a_{12})\varphi_3(1) + 2a_{11}\alpha\beta\varphi_4(1)} \varphi_3(\zeta) \right) D^{(s)} \quad (3.23)$$

$$\omega \sin z - m_1 \sinh \omega z = 0.$$

The boundary layer solution, satisfying (3.11)-(3.12), is then given by (3.4), (3.6)-(3.8), with the arbitrary constants $A^{(s)}$, $B^{(s)}$, $C^{(s)}$, $D^{(s)}$ still to be determined. Using (3.11)-(3.12), we obtain a homogeneous system in respect of the arbitrary constants, with the solvability of the latter, being expressed through an appropriate determinant, giving the transcendental equation in respect of λ. The constants $A^{(s)}$, $B^{(s)}$, $C^{(s)}$, $D^{(s)}$ are linearly dependent, therefore all of the solutions (3.6)-(3.8) may be expressed through just one arbitrary constant, say $\bar{C}^{(s)}$. Then in case a) the corresponding solution (3.6) implies

$$A^{(s)} = (b_1\lambda\beta \cos \lambda\beta + b_2 \sin \lambda\beta) \cos^2 \lambda\beta \cdot \bar{C}^{(s)},$$

$$B^{(s)} = -\frac{1}{2}b_1\lambda\beta \sin 2\lambda\beta \sin \lambda\beta \cdot \bar{C}^{(s)},$$

$$C^{(s)} = (b_2 \cos \lambda\beta - b_1\lambda\beta \sin \lambda\beta) \sin^2 \lambda\beta \cdot \bar{C}^{(s)}, \qquad (3.24)$$

$$D^{(s)} = \frac{1}{2}b_1\lambda\beta \sin 2\lambda\beta \cos \lambda\beta \cdot \bar{C}^{(s)},$$

$$b_1 = a_{11}\beta^2 - (a_{12} + a_{66}), \quad b_2 = 3a_{11}\beta^2 - (a_{12} + a_{66}),$$

with λ found from

$$\sin 4\lambda\beta - \frac{a_{66}}{3a_{66} + 8a_{12}} 4\lambda\beta = 0. \qquad (3.25)$$

In case b) the associated solution is given by (3.7), where now

$$A^{(s)} = (\gamma_2\alpha_1\beta_2\psi_2\psi_3 - \gamma_1\alpha_2\beta_1\psi_1\psi_4)\,\gamma_2\beta_2\psi_4 \cdot \bar{C}^{(s)},$$
$$B^{(s)} = (\gamma_1\alpha_2\beta_1\psi_1\psi_4 - \gamma_2\alpha_1\beta_2\psi_2\psi_3)\,\gamma_1\beta_1\psi_3 \cdot \bar{C}^{(s)},$$
$$C^{(s)} = (\gamma_2\alpha_1\beta_2\psi_1\psi_4 - \gamma_1\alpha_2\beta_1\psi_2\psi_3)\,\gamma_2\beta_2\psi_2 \cdot \bar{C}^{(s)},$$
$$D^{(s)} = (\gamma_1\alpha_2\beta_1\psi_2\psi_3 - \gamma_2\alpha_1\beta_2\psi_1\psi_4)\,\gamma_1\beta_1\psi_1 \cdot \bar{C}^{(s)}, \qquad (3.26)$$
$$\alpha_1 = a_{11}\beta_1^2 - a_{12}, \quad \alpha_2 = a_{11}\beta_2^2 - a_{12},$$
$$\gamma_1 = a_{11}\beta_1^2 - (a_{12} + a_{66}), \qquad \gamma_2 = a_{11}\beta_2^2 - (a_{12} + a_{66}),$$
$$\psi_1 = \cos\beta_1\lambda, \quad \psi_2 = \cos\beta_2\lambda, \quad \psi_3 = \sin\beta_1\lambda, \quad \psi_4 = \sin\beta_2\lambda.$$

Here λ should satisfy

$$m\sin z + m_1\sin mz = 0, \quad z = 2(\beta_1 + \beta_2)\lambda. \qquad (3.27)$$

In case c) the solution is given by (3.8), with the constants $A^{(s)}$, $B^{(s)}$, $C^{(s)}$, $D^{(s)}$ expressed through $\bar{C}^{(s)}$ as

$$A^{(s)} = (\alpha\gamma_2\psi_3 + \gamma_1\beta\psi_4)\left(\psi_1^2 + \psi_2^2\right) \cdot \bar{C}^{(s)},$$
$$B^{(s)} = (\beta\gamma_1\psi_3 - \alpha\gamma_2\psi_4)\left(\psi_1^2 + \psi_2^2\right) \cdot \bar{C}^{(s)},$$
$$C^{(s)} = (\alpha\gamma_2\psi_1 - \gamma_1\beta\psi_2)\left(\psi_3^2 + \psi_4^2\right) \cdot \bar{C}^{(s)},$$
$$D^{(s)} = -(\beta\gamma_1\psi_1 + \alpha\gamma_2\psi_2)\left(\psi_3^2 + \psi_4^2\right) \cdot \bar{C}^{(s)}, \qquad (3.28)$$
$$\gamma_1 = a_{11}(3\alpha^2 - \beta^2) + a_{12} + a_{66},$$
$$\gamma_2 = a_{11}(\alpha^2 - 3\beta^2) + a_{12} + a_{66}, \quad \psi_i = \varphi_i(\zeta = 1),$$

where λ is determined from

$$\omega\sin z + m_1\sinh\omega z = 0, \quad z = 4\beta\lambda. \qquad (3.29)$$

The considered problem is not allowing separation of the boundary layer solution into symmetric and anti-symmetric cases. It is noticed that the values of the roots for λ of (3.25), (3.27), (3.29) are twice less in comparison with the corresponding roots of (3.14), (3.18), (3.22) for the first boundary value problem. Therefore, as $Re(\lambda_1)$ characterizes the attenuation speed of the boundary layer $\exp(-Re\lambda_1 t)$, it may be deduced that the boundary layer of the first boundary value problem decays twice faster than that of the second problem. Thus, replacement of condition $u_1(1) = 0$ by $\sigma_{12}(1) = 0$ causes a 50% decrease of the attenuation speed of the boundary layer. There is a clear correspondence with physical nature of the problem, since the edges are less fixed, so the boundary effects propagate deeper.

Let us now state some of the properties of the transcendental equations (3.14)-(3.27), (3.29). Their solution sets are countable, and symmetric along the coordinate axes Ox, Oy. The roots cannot be purely imaginary. For the majority of composite materials, equations (3.18), (3.20) are taking place, having both complex conjugate pairs and real roots. Introducing $z = x + iy$, it is possible to rewrite (3.18) as

$$m \sin x \ \cosh y + m_1 \sin mx \ \cosh my = 0, \tag{3.30}$$

$$m \cos x \ \sinh y + m_1 \cos mx \ \sinh my = 0,$$

whereas for equation (3.20) we get

$$m \sin x \ \cosh y - m_1 \sin mx \ \cosh my = 0, \tag{3.31}$$

$$m \cos x \ \sinh y - m_1 \cos mx \ \sinh my = 0.$$

The system (3.30) possesses non-trivial solutions provided

$$\sin x \sin mx < 0, \quad \cos x \cos mx < 0, \tag{3.32}$$

whereas the solvability of (3.31) dictates

$$\sin x \sin mx > 0, \quad \cos x \cos mx > 0. \tag{3.33}$$

It also follows from (3.30) that

$$(m \sin x - m_1 \sin mx)(m \sin x + m_1 \sin mx)$$

$$= (m \sinh y + m_1 \sinh my)(m_1 \sinh my - m \sinh y). \tag{3.34}$$

Since the absolute value of the left-hand side of (3.34) is less than unity, so is the right-hand side. Using that $|m| < 1$, $|m_1| < 1$ we deduce

$$|y| < \text{arcsinh} \, \frac{1}{m}. \tag{3.35}$$

The same condition follows from (3.31). Therefore, the roots of equations (3.18), (3.19) are located within the strip (3.35), satisfying conditions (3.32) and (3.33), respectively. This property proves to be rather helpful for computation of the first roots. Table 5 contains the values of several first roots, while the elastic parameters of the used materials have been presented earlier in Table 2 of Chapter 1.

The solutions corresponding to other types of conditions (3.12), may be written in a similar way. Due to importance of practical applications, we also present a solution of (3.11), (3.12)$_4$ here. It is given by (3.1), (3.2), (3.4), where the case corresponding to (3.6) reduces to

$$A^{(s)} = (2a_{11}\beta \cos 2\lambda\beta \sin \lambda\beta - 2a_{11}\lambda\beta^2 \cos \lambda\beta$$

$$+ 2b\lambda^2\beta \sin \lambda\beta - b\lambda \sin 2\lambda\beta \sin \lambda\beta)D_\lambda^{(s)},$$

$$B^{(s)} = (2a_{11}\lambda\beta^2 \cos \lambda\beta - 2b\lambda^2\beta \sin \lambda\beta - b\lambda \sin 2\lambda\beta \sin \lambda\beta)\,D_\lambda^{(s)}, \tag{3.36}$$

$$C^{(s)} = (2b\lambda^2\beta \cos \lambda\beta + 2a_{11}\lambda\beta^2 \sin \lambda\beta$$

$$- 2a_{11}\beta \cos \lambda\beta \cos 2\lambda\beta + b\lambda \sin 2\lambda\beta \cos \lambda\beta)\,D_\lambda^{(s)},$$

$$D^{(s)} = (b\lambda \sin 2\lambda\beta \cos \lambda\beta - 2b\lambda^2\beta \cos \lambda\beta - 2a_{11}\lambda\beta^2 \sin \lambda\beta)D_\lambda^{(s)},$$

Table 5

	Fiberglass STET	Unidirectional fiberglass winding	Fiberglass ASTT(b) - C_2-O and PN-3	SVAM 10:1	Plywood BC-1, first grade
Roots of the equation $m\sin z + m_1\sin m\,z = 0,\ z_n = x_n + iy_n$					
m	0.57	0.634	0.667	0.635	0.91
m_1	0.628	0.67	0.695	0.673	0.83
$\beta_1 + \beta_2$	2.5659	3.67164	2.90358	3.14055	5.154
z_1	4.076	3.8743	3.7872	3.8733	3.2825
z_2	7.6344+ 0.4098 i	7.9983	7.6172	7.9968	6.5646
z_3	11.9667	11.4782	9.4355	11.467	9.846
z_4	20.1652	15.3696	11.2276	15.3592	13.126
z_5	21.8984	23.1368	15.0585	23.1237	16.4044
z_6	23.87486	25.8855	18.8457	25.9957	19.6789
z_7	27.9976	26.7323	22.6328	26.6776	22.9469
Roots of the equation $m\sin z - m_1\sin m\,z = 0,\ z_n = x_n + iy_n$					
z_1	1.8554	1.8255	1.8054	1.8188	2.1148
z_2	5.985	5.7486	5.6357	5.7439	5.1365
z_3	10.0463	9.6197	9.4228	9.6139	8.3456
z_4	14.327+ 0.45 i	13.501	13.20998	13.4943	11.5972
z_5	17.9382	17.2328+ 0.4335 i	17.0396	17.2521+ 0.4498 i	14.8642
z_6	29.596+ 0.3584 i	21.1107	18.871	21.0946	18.1382
z_7	33.9796	32.8019	20.6498	32.779	21.416

where $D_\lambda^{(s)}$ is an arbitrary constant, and the eigenvalue λ is determined from

$$\cos 4\lambda\beta + \frac{bc}{bd - 2a_{11}\beta^2 c}\frac{(4\lambda\beta)^2}{2} + \frac{(4a_{11}d + 8bc)\beta^2 - bd}{bd - 2a_{11}\beta^2 c} = 0 \qquad (3.37)$$

with

$$b = a_{11}\beta^2 - a_{12}, \quad c = a_{11}\beta^2 - a_{12} - a_{66}, \quad d = 3a_{11}\beta^2 - a_{12} - a_{66}.$$

In case of isotropic solids the equation (3.37) takes a simpler form

$$\cos 4\lambda - \frac{1+\nu}{3-\nu}\frac{(4\lambda)^2}{2} + \frac{4 + (1-\nu)^2}{4 - (1-\nu)^2} = 0. \qquad (3.38)$$

We remark that the last equation (3.38) coincides with that provided by Uflyand (1967) for an infinite strip with the same boundary conditions with $4\lambda = zi$.

The case (3.7) corresponds to

$$A^{(s)} = (-\alpha_1\beta_2 \sin\beta_1\lambda + \alpha_2(\beta_1 \sin 2\beta_2\lambda \cos\beta_1\lambda - \beta_2 \cos 2\beta_2\lambda \sin\beta_1\lambda))\, D_\lambda^{(s)},$$

$$B^{(s)} = (\alpha_2(\beta_2 \cos 2\beta_2\lambda \cos\beta_1\lambda + \beta_1 \sin 2\beta_2\lambda \sin\beta_1\lambda) - \alpha_1\beta_2 \cos\beta_1\lambda)\, D_\lambda^{(s)}, \quad (3.39)$$

$$C^{(s)} = (\alpha_1(\beta_2 \sin 2\beta_1\lambda \cos\beta_2\lambda - \beta_1 \cos 2\beta_1\lambda \sin\beta_2\lambda) - \alpha_2\beta_1 \sin\beta_2\lambda)\, D_\lambda^{(s)},$$

$$D^{(s)} = (\alpha_1(\beta_2 \sin 2\beta_1\lambda \sin\beta_2\lambda + \beta_1 \cos 2\beta_1\lambda \cos\beta_2\lambda) - \alpha_2\beta_1 \cos\beta_2\lambda)\, D_\lambda^{(s)},$$

with λ satisfying

$$(\beta_1 + \beta_2)(\alpha_1\gamma_2\beta_2 + \alpha_2\gamma_1\beta_1)\cos 2(\beta_1 - \beta_2)\lambda + (\beta_1 - \beta_2)$$
$$\times(\alpha_1\gamma_2\beta_2 - \alpha_2\gamma_1\beta_1)\cos 2(\beta_1 + \beta_2)\lambda - 2\beta_1\beta_2(\gamma_1\alpha_1 + \gamma_2\alpha_2) = 0, \quad (3.40)$$

and

$$\alpha_1 = a_{11}\beta_1^2 - a_{12}, \quad \alpha_2 = a_{11}\beta_2^2 - a_{12},$$
$$\gamma_1 = a_{11}\beta_1^2 - (a_{12} + a_{66}), \quad \gamma_2 = a_{11}\beta_2^2 - (a_{12} + a_{66}). \quad (3.41)$$

For the third case (3.8) we have

$$A^{(s)} = (\gamma_1 \mathrm{sh}\,\alpha\lambda \sin\beta\lambda(\alpha \sin 2\beta\lambda - \beta\,\mathrm{sh}\,2\alpha\lambda)$$
$$+ 2a_{11}\alpha\beta(\alpha\,\mathrm{ch}\alpha\lambda \sin\beta\lambda \cos 2\beta\lambda - \beta\,\mathrm{sh}\alpha\lambda \cos\beta\lambda\,\mathrm{ch}2\alpha\lambda))D_\lambda^{(s)},$$

$$B^{(s)} = (\gamma_1 \mathrm{ch}\,\alpha\lambda \cos\beta\lambda(\beta\,\mathrm{sh}\,2\alpha\lambda - \alpha \sin 2\beta\lambda)$$
$$- 2a_{11}\alpha\beta(\alpha\,\mathrm{sh}\alpha\lambda \cos\beta\lambda \cos 2\beta\lambda + \beta\,\mathrm{ch}\alpha\lambda \sin\beta\lambda\,\mathrm{ch}2\alpha\lambda))D_\lambda^{(s)}, \quad (3.42)$$

$$C^{(s)} = (-\gamma_1 \mathrm{sh}\,\alpha\lambda \cos\beta\lambda(\alpha \sin 2\beta\lambda + \beta\,\mathrm{sh}\,2\alpha\lambda)$$
$$+ 2a_{11}\alpha\beta(\beta\,\mathrm{sh}\alpha\lambda \sin\beta\lambda\,\mathrm{ch}2\alpha\lambda - \alpha\,\mathrm{ch}\alpha\lambda \cos\beta\lambda \cos 2\beta\lambda))D_\lambda^{(s)},$$

$$D^{(s)} = (\gamma_1 \mathrm{ch}\,\alpha\lambda \sin\beta\lambda(\alpha \sin 2\beta\lambda + \beta\,\mathrm{sh}\,2\alpha\lambda)$$
$$+ 2a_{11}\alpha\beta(\alpha\,\mathrm{sh}\alpha\lambda \sin\beta\lambda \cos 2\beta\lambda + \beta\,\mathrm{ch}\alpha\lambda \cos\beta\lambda\,\mathrm{ch}2\alpha\lambda))D_\lambda^{(s)},$$

with the equation for determining λ given by

$$(1 + \cos 2\beta\lambda)(\beta^2(2\alpha^2\gamma_3 a_{11} - \gamma_1\gamma_2)(1 + \mathrm{ch}4\alpha\lambda)$$
$$- \alpha^2(2\beta^2 a_{11}\gamma_2 + \gamma_1\gamma_3)(1 + \cos 4\beta\lambda) + 2\gamma_1(\alpha^2\gamma_3 + \beta^2\gamma_2)) = 0, \quad (3.43)$$

where
$$\gamma_1 = a_{11}(\alpha^2 - \beta^2) + a_{12},$$
$$\gamma_2 = a_{11}(3\alpha^2 - \beta^2) + a_{12} + a_{66},$$
$$\gamma_3 = a_{11}(\alpha^2 - 3\beta^2) + a_{12} + a_{66}.$$

We note that summation over λ with positive real parts is assumed within the formulae (3.1), (3.2) (taking into account (3.6)-(3.8), (3.36), (3.39), (3.42)). Hence, since every complex λ is complemented by its conjugate $\bar{\lambda}$, the solution is real and is expressed through two countable sets of arbitrary constants. It should also be noted that for every s the solution (3.1), (3.2) is exact, being essentially the homogeneous Papkovich-Lourier-Vorovich type solution for the considered class of boundary value problem. Since the boundary layer solution is determined from the linear homogeneous equations subject to homogeneous boundary conditions, it is defined to within an arbitrary constant factor, which may be represented as ε^χ, where χ is a real number.

2.4 Matching of the Outer Solution and the Boundary Layer

Solutions constructed in the previous sections 2.1-2.3 satisfy the equations of elasticity and the constitutive relations along with the boundary conditions at $y = \pm h$, but not the edge boundary conditions at $x = 0, a$. In order to satisfy the latter we write the general integral of the problem as

$$J = Q^{out} + \varepsilon^{\chi} R_b^{(1)} + \varepsilon^{\mu} R_b^{(2)}, \qquad (4.1)$$

where Q^{out} is the outer solution, $R_b^{(1)}$, $R_b^{(2)}$ are the boundary layer solutions at $x = 0$ and $x = a$, respectively. It is worth noticing that the solution $R_b^{(2)}$ may be obtained from $R_b^{(1)}$ by a formal substitution of $t_1 = 1/\varepsilon - t = (a - x)/h$ replacing t. As previously, the whole numbers χ and μ characterizing the intensity of the boundary layer solutions, are defined uniquely by the matching process of the outer solution and the boundary layers from the boundary conditions at $x = 0, a$. Let us consider the first type of boundary conditions at $x = 0$

$$\sigma_x = \bar{\sigma}_x(\zeta), \quad \sigma_{xy} = \bar{\sigma}_{xy}(\zeta) \quad \text{at} \quad x = 0. \qquad (4.2)$$

Substituting (4.1) into (4.2) and taking into account the asymptotic results (1.6) and (1.7) for the outer stresses along with approximations (3.1) for similar quantities of the boundary layer solution, and neglecting the effect of another boundary layer $R_b^{(2)}$ in the vicinity of $x = 0$, we obtain

$$\sigma_x^{(s)} + \sigma_1^{(s-\chi)} = \bar{\sigma}_x^{(s-1)}(\zeta), \quad \sigma_{xy}^{(s)} + \sigma_{12}^{(s-\chi)} = \bar{\sigma}_{xy}^{(s-1)}(\zeta) \quad \text{at} \quad x = 0 \qquad (4.3)$$

where $\sigma_x^{(s)}$, $\sigma_{xy}^{(s)}$ are the quantities of the outer problem which may be calculated using the known formulae from Sections 2.1, 2.2, and $\bar{\sigma}_x^{(0)} = \bar{\sigma}_x$, $\bar{\sigma}_{xy}^{(0)}(\zeta) = \bar{\sigma}_{xy}$, $(\bar{\sigma}_x^{(s)}, \bar{\sigma}_{xy}^{(s)}) = 0$, $s \neq 0$. It is clear that the conditions (4.3) are consistent (i.e. enable to determine sequentially the boundary layer quantities) only if $\chi = 0$. Indeed, if $\chi > 0$ then at $s = 0$ the boundary layer quantities are not taken into account, since $s - \chi < 0$, hence equation (4.3) becomes inconsistent. Similarly, if $\chi < 0$ then the components of the outer solution do not appear in the solution, implying that the boundary layer is identical zero. Thus, $\chi = 0$ is the only value leading to consistent solution. Therefore, (4.3) is rewritten in the form

$$\sigma_1^{(s)} = \bar{\sigma}_x^{(s-1)} - \sigma_x^{(s)}(0, \zeta), \quad \sigma_{12}^{(s)} = \bar{\sigma}_{xy}^{(s-1)} - \sigma_{xy}^{(s)}(0, \zeta) \quad \text{at} \quad t = 0. \qquad (4.4)$$

Since the functions on the right-hand sides of (4.4) are known, the constants of the boundary layer solution may now be determined through boundary collocation, least squares and expansion of the functions in (4.4) along the orthogonal functional system, etc. Since in general the right-hand sides of (4.4) at $s = 0$ are non-zero, the stresses of the boundary layer solution will be of the same order as those of the outer solution. It can be easily shown that the stresses on the edges (4.4) are non-self-equilibrated, however, they correspond to a decaying solution. The reason for that is that in the considered problem a decaying solution corresponds to arbitrary, even non-self-equilibrated edge loading. Therefore the Saint-Venant's principle is valid in a broader sense. This principle may be formulated as follows: the solution consists of the outer (main) part and the boundary layer component, with the outer solution defined completely through the face boundary conditions at $y = \pm h$. This structure or principle, within which the solution is a sum of the outer solution and the boundary layers, holds for all thin spatial structures. The Saint-Venant's principle may hold in a stronger sense as just observed or to be violated depending on the type of the boundary value problem, as discussed in Chapter 1, when the principle is extrapolated for other types of the boundary conditions. Thus, the formulated principle is more general, including the classical Saint-Venant's principle as a particular case, as illustrated in Chapter 1. It is also worth noting that the formulated more general principle enables qualitative analysis in virtually any physical problem for thin structures, leading to construction of the corresponding iterative processes. Clearly, provided that the classical Saint-Venant's principle is applicable, the solution procedure is simplified.

Consider now the boundary conditions of second type at $x = 0$

$$u = \bar{u}(\zeta), \qquad v = \bar{v}(\zeta) \quad \text{at} \quad x = 0. \tag{4.5}$$

It may be deduced that $\chi = 0$. Then the boundary layer is governed by

$$u_1^{(s)} = \bar{u}^{(s)} - U^{(s)}(0, \zeta), \qquad u_2^{(s)} = \bar{v}^{(s)} - V^{(s)}(0, \zeta) \qquad \text{at} \qquad t = 0 \tag{4.6}$$

where $\bar{u}^{(0)} = \bar{u}/a$, $\bar{v}^{(0)} = \bar{v}/a$, $\bar{u}^{(s)} = \bar{v}^{(s)} \equiv 0$, $(s > 0)$, quantities $U^{(s)}$, $V^{(s)}$ are determined from (1.9), and $u_1^{(s)}$, $u_2^{(s)}$ are obtained from (3.4). If zero boundary conditions are prescribed on the faces $y = \pm h$ both for stresses and displacements, then $U^{(s)} = V^{(s)} \equiv 0$, hence $Q^{out} \equiv 0$, so the general solution is given by the boundary layer only, arising from the edge boundary conditions (4.4) or (4.6). For example, in case of fixed faces $y = \pm h$, a decaying solution would correspond to any edge boundary conditions at $x = 0, a$, which is also confirmed by experimental studies, see e.g. Baikov and Strongin (1980).

For mixed boundary conditions

$$\sigma_{xy} = \bar{\sigma}_{xy}(\zeta), \quad u = \bar{u}(\zeta) \qquad \text{at} \quad x = 0 \tag{4.7}$$

we result in $\chi = 0$ and the following conditions for the boundary layer

$$\sigma_{12}^{(s)} = \bar{\sigma}_{xy}^{(s-1)} - \sigma_{xy}^{(s)}(0, \zeta), \quad u_1^{(s)} = \bar{u}^{(s)} - U^{(s)}(0, \zeta) \quad \text{at} \quad t = 0. \tag{4.8}$$

Other types of conditions at $t = 0$ may be treated in a similar way.

When matching the outer solution and the boundary layer solution at the edge $x = 0$, the influence of the boundary layer associated with the opposite edge, has been neglected, due to its rapid decay. However, this imposes a constraint on the length of the rectangle. Essentially, the length a should satisfy $1 + \exp[(-Re\lambda_1 \cdot a)/h] \approx 1$, where $Re\lambda_1$ is the real part of the first root of the appropriate transcendental equation with positive real part $(Re\lambda > 0)$. For the second boundary value problem these equations are (3.14) or (3.16); (3.18); (3.20) or (3.23), whereas in case of the first boundary value problem these are (3.37), (3.38),

(3.40) or (3.43). If the size of a is small, such that it is not possible to ignore the contribution of the boundary layer solution associated with the opposite edge, then we need to keep the terms $R_b^{(2)}$ in (4.2) and (4.5), bearing in mind that the boundary layers are exact solutions of the plane problem of elasticity. Hence, summation in the boundary layer solution will also involve that along λ with $Re\lambda < 0$, which means that the general solution cannot be split into propagating and localized parts.

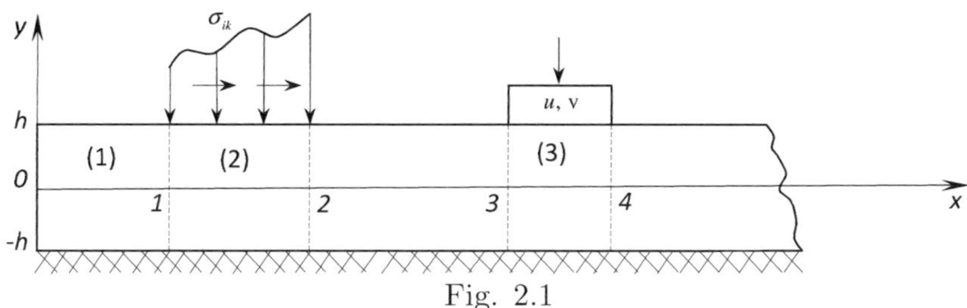

Fig. 2.1

Using the solutions of the previously considered boundary value problems, it is possible to investigate partially loaded and mixed boundary problems (Fig. 2.1). We assume for simplicity that the distance between the areas of loading is bigger than the zone of influence of the associated boundary effects. The domain is then split into sub-domains (1), (2), (3). In case of sub-domain (1) the outer solution is zero, since $\sigma_{xy}(+h) = \sigma_y(+h) = 0$; for sub-domain (2) the outer solution follows from (1.6), (1.9), and (1.15), whereas in case of sub-domain (3) the outer solution is given by (1.6), (1.9), and (1.11). The boundary layer solutions would emerge in the vicinity of the vertical cross sections 1, 2, 3, 4. The boundary layer solutions for the first two cases 1 and 2 are given by (3.1), (3.2), (3.4), and (3.36)-(3.42), with the corresponding results for cross sections 3 and 4 provided by (3.1), (3.2), (3.4), and (3.13)-(3.22). The outer solutions in all of the areas are then matched with the corresponding boundary layer solutions, then the general solution is presented as a sum of the appropriate outer solution and boundary layers on the sections 1, 2, 3, 4, due to the required continuity of the displacements (u, v) and stresses σ_x, σ_{xy} along each cross section. The boundary layers arising for cross sections 1 and 2, correspond to the problem (3.11), (3.12)$_4$, whereas in the other two cases of cross sections 3 and 4 the associated boundary layers could be of the types of the first and second boundary value problems, namely (3.11), (3.12)$_4$, and (3.11), (3.12)$_1$, respectively. In the latter case the rates of decay of the boundary layers between the stamp and outside the stamp will be different. The number of arbitrary constants in the boundary layer solutions coincides with the number of continuity conditions for each cross section. However, if the distance between sections 2 and 3 is relatively small, then the difficulty of the problem rises significantly, since all of the boundary layer solutions should be treated simultaneously.

2.5 Asymptotic Solution of the Outer Problem for a Two-Layer Anisotropic Rectangular Beam

The proposed asymptotic method is applicable to more advanced problems, in particular, for mixed boundary value problems for two-layer and multi-layer strips and beams. These theories have very important practical applications for structures resting on elastic foundations. Indeed, according to the model of a compressible layer the foundation could be thought of as a two-layer strip or plate, with the lower face of it being rigidly fixed, and the upper face of the first layer subjected to given normal and tangential forces.

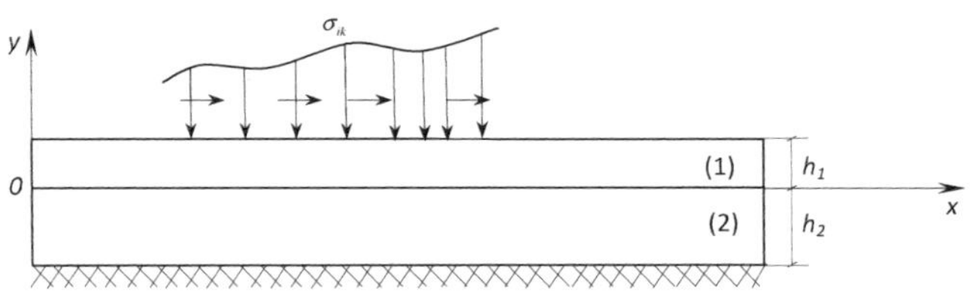

Fig. 2.2

Let us concentrate on the asymptotic analysis of the plane problem of elasticity for a two-layer rectangle $\Omega = \{(x,y) : x \in [0,a], -h_2 \le y \le h_1, h_1 + h_2 < a\}$. Let the acting volume forces be $F_x^{(i)}(x,y)$, $F_y^{(i)}(x,y)$, more specifically, we are taking into account the weights of the layers and thermal effects using the Duhamel-Neumann law, see Novatsky (1970), thus defining the temperature dependence within each layer. We shall denote the quantities associated with the upper and lower layers by the indices 1 and 2, respectively. The axis Ox is directed along the contact line between the layers (Fig. 2.2). Let us consider the case of general anisotropy for both layers. The boundary conditions include

$$\sigma_{xy}^{(1)} = \varepsilon^{-1}\sigma_{xy}^{+}(\xi), \qquad \sigma_{y}^{(1)} = \varepsilon^{-1}\sigma_{y}^{+}(\xi), \qquad \xi = x/a \qquad \text{at} \qquad y = h_1 \qquad (5.1)$$

$$u^{(2)} = u^{-}(\xi), \qquad v^{(2)} = v^{-}(\xi) \qquad \text{at} \qquad y = -h_2 \qquad (5.2)$$

(in particular $u^{-} = v^{-} = 0$), along with conditions at the edges $x = 0, a$, which could be either specified displacements, stresses or mixed boundary conditions. We will present more details on the edge boundary conditions later, since these are causing the disturbances in the near edge vicinity only, i.e. they lead to boundary layer solutions similarly to the previously analyzed case of one layer strip. Thus, the problem statement is given by the thermoelasticity equations

$$\frac{\partial \sigma_x}{\partial x} + \frac{\partial \sigma_{xy}}{\partial y} + F_x = 0, \qquad \frac{\partial \sigma_{xy}}{\partial x} + \frac{\partial \sigma_y}{\partial y} + F_y = 0,$$

$$\frac{\partial u}{\partial x} = a_{11}\sigma_x + a_{12}\sigma_y + a_{16}\sigma_{xy} + \alpha_{11}\theta, \qquad (5.3)$$

$$\frac{\partial u}{\partial y} + \frac{\partial v}{\partial x} = a_{16}\sigma_x + a_{26}\sigma_y + a_{66}\sigma_{xy} + \alpha_{12}\theta,$$

$$\frac{\partial v}{\partial y} = a_{21}\sigma_x + a_{22}\sigma_y + a_{26}\sigma_{xy} + \alpha_{22}\theta,$$

where a_{ik} are elastic parameters, α_{ik} are the coefficients of thermal expansion, $\theta = T - T_0$ is temperature change, and T_0 is absolute temperature, subject to boundary conditions (5.1), (5.2), and the edge boundary conditions at $x = 0, a$.

We begin with rescaling to dimensionless variables $\xi = x/a$, $\zeta = y/h_2$. As previously, the unknown quantities are the stress components along the layers $\sigma_{jk}^{(i)}$, and the dimensionless displacements $U^{(i)} = u^{(i)}/a$, $V^{(i)} = v^{(i)}/a$, $(i = 1, 2)$. If $h_1 > h_2$, then it is convenient to introduce $y = h_1\zeta$. The quantities $F_x^{(i)}(x, y)$, $F_y^{(i)}(x, y)$, $\theta^{(i)}(x, y)$ are also transformed

$$F_x^{(i)}(x, y) = a^{-1}\varepsilon^{-2+s}F_{xs}^{(i)}(\xi, \zeta), \quad (x, y), \quad s = \overline{0, N}$$

$$\theta^{(i)}(x, y) = \varepsilon^{-1+s}\theta^{(i,s)}(\xi, \zeta). \qquad (5.4)$$

The meaning of representation (5.4) will be explained later. The system (5.3) expressed in terms of dimensionless variables ξ, ζ, is singularly perturbed with small parameter $\varepsilon = h_2/a$. As before, the general solution is composed by the outer solution and the boundary layer. The outer solution is sought for in the form (Aghalovyan, 1983)

$$\sigma_{jk}^{(i)}(x, y) = \varepsilon^{-1+s}\sigma_{jk}^{(i,s)}, \quad U^{(i)} = \varepsilon^s u^{(i,s)} \quad (u, v), \quad s = \overline{0, N}. \qquad (5.5)$$

Following a known procedure, we obtain

$$\frac{\partial \sigma_x^{(i,s-1)}}{\partial \xi} + \frac{\partial \sigma_{xy}^{(i,s)}}{\partial \zeta} + F_{xs}^{(i)}(\xi, \zeta) = 0, \qquad \frac{\partial \sigma_{xy}^{(i,s-1)}}{\partial \xi} + \frac{\partial \sigma_y^{(i,s)}}{\partial \zeta} + F_{ys}^{(i)}(\xi, \zeta) = 0,$$

$$\frac{\partial u^{(i,s-1)}}{\partial \xi} = a_{11}^{(i)}\sigma_x^{(i,s)} + a_{12}^{(i)}\sigma_y^{(i,s)} + a_{16}^{(i)}\sigma_{xy}^{(i,s)} + \alpha_{11}^{(i)}\theta^{(i,s)}(\xi, \zeta), \quad (5.6)$$

$$\frac{\partial u^{(i,s)}}{\partial \zeta} + \frac{\partial v^{(i,s-1)}}{\partial \xi} = a_{16}^{(i)}\sigma_x^{(i,s)} + a_{26}^{(i)}\sigma_y^{(i,s)} + a_{66}^{(i)}\sigma_{xy}^{(i,s)} + \alpha_{12}^{(i)}\theta^{(i,s)}(\xi, \zeta),$$

$$\frac{\partial v^{(i,s)}}{\partial \zeta} = a_{21}^{(i)}\sigma_x^{(i,s)} + a_{22}^{(i)}\sigma_y^{(i,s)} + a_{26}^{(i)}\sigma_{xy}^{(i,s)} + \alpha_{22}^{(i)}\theta^{(i,s)}(\xi, \zeta) \quad (i = 1, 2)$$

in respect of the unknown coefficients $\sigma_{jk}^{(i,s)}$, $u^{(i,s)}, v^{(i,s)}$ $(i = 1, 2)$. The system (5.6) may now be integrated with respect to ζ. Satisfying the boundary conditions (5.1), (5.2) along with the conditions of elastic contact between layers

$$\sigma_{xy}^{(1)} = \sigma_{xy}^{(2)}, \quad \sigma_y^{(1)} = \sigma_y^{(2)}, \quad u^{(1)} = u^{(2)}, \quad v^{(1)} = v^{(2)} \quad \text{at} \quad y = 0, \qquad (5.7)$$

we arrive at the following recurrent relations for all of the sought-for quantities $\sigma_{jk}^{(i,s)}$, $u^{(i,s)}$, $v^{(i,s)}$:

for the upper layer ($i = 1$, $0 \le \zeta \le \bar{h}$, $\bar{h} = h_1/h_2$)

$$\sigma_{xy}^{(1,s)} = \sigma_{xy0}^{(1,s)}(\xi) + \sigma_{xy*}^{(1,s)}(\xi,\zeta), \quad \sigma_{xy0}^{(1,s)} = \sigma_{xy}^{+(s)} - \sigma_{xy*}^{(1,s)}(\xi,\bar{h}),$$

$$\sigma_{y}^{(1,s)} = \sigma_{y0}^{(1,s)}(\xi) + \sigma_{y*}^{(1,s)}(\xi,\zeta), \quad \sigma_{y0}^{(1,s)} = \sigma_{y}^{+(s)} - \sigma_{y*}^{(1,s)}(\xi,\bar{h}),$$

$$\sigma_{xy}^{+(0)} = \sigma_{xy}^{+}(\xi), \quad \sigma_{y}^{+(0)} = \sigma_{y}^{+}(\xi), \quad \sigma_{xy}^{+(s)} = \sigma_{y}^{+(s)} \equiv 0, \quad (s \ne 0)$$

$$\sigma_{x}^{(1,s)} = - \left(a_{12}^{(1)} \sigma_{y0}^{(1,s)} + a_{16}^{(1)} \sigma_{xy0}^{(1,s)} \right) \left(a_{11}^{(1)} \right)^{-1} + \sigma_{x*}^{(1,s)}(\xi,\zeta), \qquad (5.8)$$

$$u^{(1,s)} = \left(A_{16}^{(1)} \sigma_{y0}^{(1,s)} + A_{66}^{(1)} \sigma_{xy0}^{(1,s)} \right) \zeta + A_{16}^{(2)} \sigma_{y0}^{(2,s)} + A_{66}^{(2)} \sigma_{xy0}^{(2,s)}$$

$$+u^{-(s)} - u_{*}^{(2,s)}(\xi,-1) + u_{*}^{(2,s)}(\xi,0) + u_{*}^{(1,s)}(\xi,\zeta) - u_{*}^{(1,s)}(\xi,0),$$

$$v^{(1,s)} = \left(A_{11}^{(1)} \sigma_{y0}^{(1,s)} + A_{16}^{(1)} \sigma_{xy0}^{(1,s)} \right) \zeta + A_{11}^{(2)} \sigma_{y0}^{(2,s)} + A_{16}^{(2)} \sigma_{xy0}^{(2,s)}$$

$$+v^{-(s)} + v_{*}^{(1,s)}(\xi,\zeta) - v_{*}^{(1,s)}(\xi,0) + v_{*}^{(2,s)}(\xi,0) - v_{*}^{(2,s)}(\xi,-1),$$

$$u^{-(0)} = u^{-}/a, \quad v^{-(0)} = v^{-}/a, \quad u^{-(s)} = v^{-(s)} \equiv 0 \quad (s \ne 0)$$

$$\sigma_{jk}^{(i,s)} \equiv 0, \quad u^{(i,s)} \equiv 0, \quad (i = 1,2) \quad \text{if} \quad s < 0$$

for the lower layer ($-1 \le \zeta \le 0$)

$$\sigma_{xy}^{(2,s)} = \sigma_{xy0}^{(2,s)}(\xi) + \sigma_{xy*}^{(2,s)},$$

$$\sigma_{xy0}^{(2,s)} = \sigma_{xy}^{+(s)} - \sigma_{xy*}^{(2,s)}(\xi,0) + \sigma_{xy*}^{(1,s)}(\xi,0) - \sigma_{xy*}^{(1,s)}(\xi,\bar{h}),$$

$$\sigma_{y0}^{(2,s)} = \sigma_{y}^{+(s)} - \sigma_{y*}^{(2,s)}(\xi,0) + \sigma_{y*}^{(1,s)}(\xi,0) - \sigma_{y*}^{(1,s)}(\xi,\bar{h}),$$

$$\sigma_{y}^{(2,s)} = \sigma_{y0}^{(2,s)}(\xi) + \sigma_{y*}^{(2,s)}, \quad \sigma_{x}^{(2,s)} = \sigma_{x0}^{(2,s)}(\xi) + \sigma_{x*}^{(2,s)},$$

$$\sigma_{x0}^{(2,s)} = - \left(a_{12}^{(2)} \sigma_{y0}^{(2,s)} + a_{16}^{(2)} \sigma_{xy0}^{(2,s)} \right) \left(a_{11}^{(2)} \right)^{-1}, \qquad (5.9)$$

$$u^{(2,s)} = \left(A_{16}^{(2)} \sigma_{y0}^{(2,s)} + A_{66}^{(2)} \sigma_{xy0}^{(2,s)} \right) (1+\zeta) + u^{-(s)} + u_{*}^{(2,s)}(\xi,\zeta) - u_{*}^{(2,s)}(\xi,-1),$$

$$v^{(2,s)} = \left(A_{11}^{(2)} \sigma_{y0}^{(2,s)} + A_{16}^{(2)} \sigma_{xy0}^{(2,s)} \right) (1+\zeta) + v^{-(s)} + v_{*}^{(2,s)}(\xi,\zeta) - v_{*}^{(2,s)}(\xi,-1),$$

where

$$\sigma_{xy*}^{(i,s)}(\xi,\zeta) = - \int_{0}^{\zeta} \left(F_{xs}^{(i)} + \frac{\partial \sigma_{x}^{(i,s-1)}}{\partial \xi} \right) d\zeta,$$

$$\sigma_{y*}^{(i,s)}(\xi,\zeta) = - \int_{0}^{\zeta} \left(F_{ys}^{(i)} + \frac{\partial \sigma_{xy}^{(i,s-1)}}{\partial \xi} \right) d\zeta,$$

$$\sigma_{x*}^{(i,s)} = - \left(a_{12}^{(i)} \sigma_{y*}^{(i,s)} + a_{16}^{(i)} \sigma_{xy*}^{(i,s)} - \frac{\partial u^{(i,s-1)}}{\partial \xi} + \alpha_{11}^{(i)} \theta^{(i,s)}(\xi,\zeta) \right) \left(a_{11}^{(i)} \right)^{-1},$$

$$u_{*}^{(i,s)} = \int_{0}^{\zeta} \left(a_{16}^{(i)} \sigma_{x*}^{(i,s)} + a_{26}^{(i)} \sigma_{y*}^{(i,s)} + a_{66}^{(i)} \sigma_{xy*}^{(i,s)} + \alpha_{12}^{(i)} \theta^{(i,s)}(\xi,\zeta) - \frac{\partial v^{(i,s-1)}}{\partial \xi} \right) d\zeta,$$

$$v_*^{(i,s)} = \int_0^\zeta \left(a_{21}^{(i)} \sigma_{x*}^{(i,s)} + a_{22}^{(i)} \sigma_{y*}^{(i,s)} + a_{26}^{(i)} \sigma_{xy*}^{(i,s)} + \alpha_{22}^{(i)} \theta^{(i,s)}(\xi,\zeta) \right) d\zeta, \qquad (5.10)$$

$$A_{11}^{(i)} = \left(a_{11}^{(i)} a_{22}^{(i)} - (a_{12}^{(i)})^2 \right) \left(a_{11}^{(i)} \right)^{-1}, \quad A_{16}^{(i)} = \left(a_{11}^{(i)} a_{26}^{(i)} - a_{12}^{(i)} a_{16}^{(i)} \right) \left(a_{11}^{(i)} \right)^{-1},$$

$$A_{66}^{(i)} = \left(a_{11}^{(i)} a_{66}^{(i)} - (a_{16}^{(i)})^2 \right) \left(a_{11}^{(i)} \right)^{-1} \qquad (i = 1, 2).$$

The obtained solution physically corresponds to the situation when the contribution of the volume forces and temperature field is of the same order with that of surface forces and displacements. Therefore, the associated components should be present in the equations and solution, provided that (5.4) is satisfied. The latter implies that the volume forces should possess a relatively high intensity. In order to achieve the required balance, a large parameter ε^{-1} has been introduced in the boundary conditions. However, if the acting volume forces and temperature field do not satisfy (5.4), and are of less asymptotic order, then their influence will take effect only at the next order of approximation.

As for the case of one layer, conditions (5.1) and (5.2) are sufficient to determine fully the outer solution. The boundary layer solution would then be found, allowing the general solution to satisfy the edge boundary conditions at $x = 0, a$.

2.6 Special Cases

There is a wide variety of problems for which it is possible to find a closed-form outer solution from the asymptotic iterative process. Here we will mention several of these.

Let the lower face of the foundation be fixed rigidly ($u^- = v^- = 0$), with the specified normal and tangential loads of constant intensity $\sigma_y^{(1)} = -\sigma_2^+$, $\sigma_{xy}^{(1)} = \tau^+$ at $y = h_1$ (σ_2^+, $\tau^+ = \text{const}$), acting on the upper layer. Let us take into account the influence of weights of the layer as volume forces

$$F_y^{(1)} = -\rho_1 g, \quad F_x^{(1)} = 0, \quad 0 \leq y \leq h_1$$
$$F_y^{(2)} = -\rho_2 g, \quad F_x^{(2)} = 0, \quad -h_2 \leq y \leq 0 \qquad (6.1)$$

where ρ_1, ρ_2 are the corresponding densities of the layers. Since the equations of equilibrium contain scaled volume forces, i.e. forces per unit volume, the formulae (6.1) describe fully the effect of weights of the layers.

The exact solution of the outer problem is delivered by the leading order approximation $s = 0$ of (5.5), since $Q^{(s)} \equiv 0$ for $s \geq 1$. We obtain for the upper layer ($0 \le y \le h_1$)

$$\sigma_{xy}^{(1)} = \tau^+, \quad \sigma_y^{(1)} = -\sigma_2^+ + \rho_1 g(y - h_1),$$

$$\sigma_x^{(1)} = \frac{1}{a_{11}^{(1)}}[a_{12}^{(1)}\sigma_2^+ - a_{16}^{(1)}\tau^+ - a_{12}^{(1)}\rho_1 g(y - h_1)], \qquad (6.2)$$

$$u^{(1)} = \left(A_{66}^{(1)}\tau^+ - A_{16}^{(1)}\sigma_2^+\right)y - A_{16}^{(2)}\sigma_2^+ h_2 + A_{66}^{(2)}\tau^+ h_2$$

$$-A_{16}^{(1)}\rho_1 gh_1 y - A_{16}^{(2)}\rho_1 gh_1 h_2 + \frac{1}{2}b_{16}^{(1)}\rho_1 gy^2 - \frac{1}{2}b_{16}^{(2)}\rho_2 gh_2^2,$$

$$v^{(1)} = \left(A_{66}^{(1)}\tau^+ - A_{11}^{(1)}\sigma_2^+\right)y - A_{11}^{(2)}\sigma_2^+ h_2 + A_{16}^{(2)}\tau^+ h_2$$

$$-A_{11}^{(1)}\rho_1 gh_1 y - A_{11}^{(2)}\rho_1 gh_1 h_2 + \frac{1}{2}b_{11}^{(1)}\rho_1 gy^2 - \frac{1}{2}b_{11}^{(2)}\rho_2 gh_2^2,$$

$$b_{11}^{(i)} = \frac{a_{11}^{(i)}a_{22}^{(i)} - (a_{12}^{(i)})^2}{a_{11}^{(i)}}, \quad b_{16}^{(i)} = \frac{1}{a_{11}^{(i)}}\left(a_{11}^{(i)}a_{26}^{(i)} - a_{12}^{(i)}a_{16}^{(i)}\right) \qquad (i = 1, 2)$$

whereas for the lower layer ($-h_2 \le y \le 0$)

$$\sigma_{xy}^{(2)} = \tau^+, \quad \sigma_y^{(2)} = -\sigma_2^+ - \rho_1 gh_1 + \rho_2 gy,$$

$$\sigma_x^{(2)} = \frac{1}{a_{11}^{(2)}}[a_{12}^{(2)}\sigma_2^+ - a_{16}^{(2)}\tau^+ + a_{12}^{(2)}(\rho_1 gh_1 - \rho_2 gy)],$$

$$u^{(2)} = \left(A_{66}^{(2)}\tau^+ - A_{16}^{(2)}\sigma_2^+\right)(y + h_2) - A_{16}^{(2)}\rho_1 gh_1(y + h_2)$$

$$+\frac{1}{2}b_{16}^{(2)}\rho_2 g(y^2 - h_2^2), \qquad (6.3)$$

$$v^{(2)} = \left(A_{16}^{(2)}\tau^+ - A_{11}^{(2)}\sigma_2^+\right)(y + h_2) - A_{11}^{(2)}\rho_1 gh_1(y + h_2)$$

$$+\frac{1}{2}b_{11}^{(2)}\rho_2 g(y^2 - h_2^2).$$

In case of isotropic layers $A_{11} = (1 - \nu^2)/E$, $A_{66} = a_{66} = 2(1 + \nu)/E$, $A_{16} = a_{16} = a_{26} = 0$, (with appropriate indices of the layers added to the Young's modulus E and the Poisson's ratio ν), the solution (6.2), (6.3) is simplified drastically.

Let us consider now the same problem, but without taking into account the effect of weights of the layers, when the upper layer is subject to a normal load, changing linearly along ξ (e.g. hydrostatic pressure)

$$\sigma_y^{(1)} = -q^+\xi, \quad \sigma_{xy}^{(1)} = 0 \text{ at } y = h_1.$$

For the sake of brevity we are presenting below the solution for a two-layer orthotropic strip/beam.

For the upper layer $(0 \leq y \leq h_1)$

$$\sigma_{xy}^{(1)} = \frac{a_{12}^{(1)}}{a_{11}^{(1)}} \frac{q^+}{a}(h_1 - y), \quad \sigma_y^{(1)} = -q^+\xi, \quad \sigma_x^{(1)} = \frac{a_{12}^{(1)}}{a_{11}^{(1)}} q^+\xi,$$

$$u^{(1)} = \frac{a_{12}^{(1)}}{a_{11}^{(1)}} a_{66}^{(2)} q^+ \frac{h_1 h_2}{a} - \frac{1}{2}\left(3A_{11}^{(2)} - \frac{a_{12}^{(2)}}{a_{11}^{(2)}} a_{66}^{(2)}\right)\frac{h_2^2}{a}q^+$$

$$+\frac{1}{2}\left(A_{11}^{(1)} - \frac{a_{12}^{(1)}}{a_{11}^{(1)}} a_{66}^{(1)}\right)\frac{y^2}{a}q^+ + A_{11}^{(2)} q^+ y\frac{h_2}{a}, \tag{6.4}$$

$$v^{(1)} = -\left(A_{11}^{(2)} h_2 + A_{11}^{(1)} y\right) q^+\xi.$$

For the lower layer $(-h_2 \leq y \leq 0)$

$$\sigma_{xy}^{(2)} = \frac{a_{12}^{(1)}}{a_{11}^{(1)}} q^+ \frac{h_1}{a} - \frac{a_{12}^{(2)}}{a_{11}^{(2)}} q^+ \frac{y}{a}, \quad \sigma_y^{(2)} = -q^+\xi, \quad \sigma_x^{(2)} = \frac{a_{12}^{(2)}}{a_{11}^{(2)}} q^+\xi,$$

$$u^{(2)} = \left(\frac{a_{12}^{(1)}}{a_{11}^{(1)}} a_{66}^{(2)} h_1 - A_{11}^{(2)} h_2\right)(y + h_2)\frac{q^+}{a}$$

$$+\frac{1}{2}\left(A_{11}^{(2)} - \frac{a_{12}^{(2)}}{a_{11}^{(2)}} a_{66}^{(2)}\right)(y^2 - h_2^2)\frac{q^+}{a}, \tag{6.5}$$

$$v^{(2)} = -A_{11}^{(2)}(y + h_2) q^+\xi.$$

Let us now present one more solution for a problem for a two-layer orthotropic strip (Fig. 2.3) for which

$$\sigma_y(h_1) = -q^+(1 - 2\xi)^2, \quad \sigma_{xy}(h_1) = 0$$

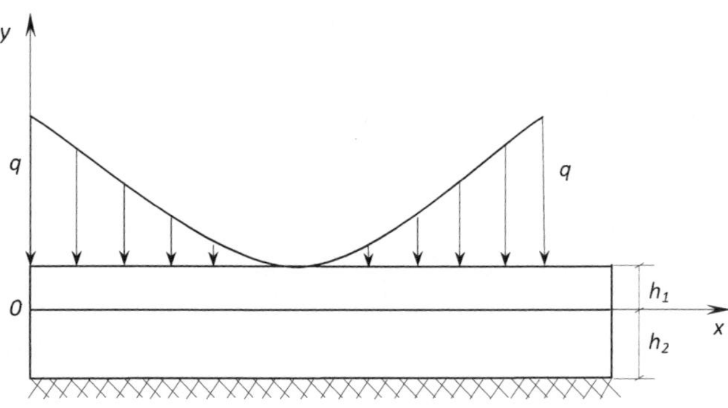

Fig. 2.3

The upper layer $(0 \leq y \leq h_1)$ quantities are given by

$$\sigma_{xy}^{(1)} = \frac{4q^+}{a}\frac{a_{12}^{(1)}}{a_{11}^{(1)}}(1-2\xi)(y-h_1), \quad \sigma_y^{(1)} = -q^+(1-2\xi)^2 + \frac{4q^+}{a^2}\frac{a_{12}^{(1)}}{a_{11}^{(1)}}(y-h_1)^2,$$

$$\sigma_x^{(1)} = q^+\frac{a_{12}^{(1)}}{a_{11}^{(1)}}(1-2\xi)^2 - 4q^+\left(\frac{a_{12}^{(1)}}{a_{11}^{(1)}}\right)^2\left(\frac{h_1-y}{a}\right)^2$$

$$+\frac{a_{66}^{(2)}a_{12}^{(2)}}{a_{11}^{(1)}a_{11}^{(2)}}4q^+\left(\frac{h_2}{a}\right)^2 + \frac{8q^+}{a^2}\frac{a_{12}^{(1)}}{(a_{11}^{(1)})^2}\left(a_{66}^{(1)}\left(h_1 y - \frac{1}{2}y^2\right) + a_{66}^{(2)}h_1 h_2\right)$$

$$+\frac{4q^+}{a^2 a_{11}^{(1)}}\left(A_{11}^{(1)}y^2 + A_{11}^{(2)}(h_2^2 + 2h_2 y)\right), \quad (6.6)$$

$$u^{(1)} = -\frac{4q^+}{a}\left(\frac{a_{12}^{(1)}}{a_{11}^{(1)}}h_1(h_2 a_{66}^{(2)} + a_{66}^{(1)}y) + \frac{1}{2}\left(\frac{a_{66}^{(2)}a_{12}^{(2)}}{a_{11}^{(2)}} + A_{11}^{(2)}\right)h_2^2\right.$$

$$\left. +A_{11}^{(2)}h_2 y - \frac{1}{2}\left(\frac{a_{66}^{(1)}a_{12}^{(1)}}{a_{11}^{(1)}} - A_{11}^{(1)}\right)y^2\right)(1-2\xi),$$

$$v^{(1)} = -q^+\left(A_{11}^{(2)}h_2 + A_{11}^{(1)}y\right)\left((1-2\xi)^2 - \frac{4a_{12}^{(1)}}{a_{11}^{(1)}}\left(\frac{h_1}{a}\right)^2\right)$$

$$+\frac{4}{3}q^+\frac{a_{12}^{(2)}}{a_{11}^{(2)}}\left(\frac{a_{12}^{(2)}}{a_{11}^{(2)}}(2a_{66}^{(2)} - a_{12}^{(2)}) + A_{11}^{(2)} + a_{22}^{(2)}\right)\frac{h_2^3}{a^2}$$

$$+\frac{8q^+}{a^2}\frac{a_{12}^{(1)}}{a_{11}^{(1)}}\left(\left(\frac{a_{12}^{(1)}a_{66}^{(2)}}{a_{11}^{(1)}}h_1 + \frac{1}{2}h_2\left(\frac{a_{12}^{(2)}}{a_{11}^{(2)}}a_{66}^{(2)} + A_{11}^{(2)}\right)\right)\right)yh_2$$

$$+\frac{1}{2}\left(\frac{a_{12}^{(1)}}{a_{11}^{(1)}}(a_{12}^{(1)} + a_{66}^{(1)})h_1 + A_{11}^{(2)}h_2 - a_{22}^{(1)}h_1\right)y^2$$

$$-\frac{1}{6}\left(\frac{a_{12}^{(1)}}{a_{11}^{(1)}}(a_{12}^{(1)} + a_{66}^{(1)}) - A_{11}^{(1)} - a_{22}^{(1)}\right)y^3 + \frac{1}{2}\left(\frac{a_{12}^{(2)}}{a_{11}^{(2)}}(a_{66}^{(2)} - a_{12}^{(2)}) + a_{22}^{(2)}\right)h_1 h_2^2\right).$$

For the lower layer $(-h_2 \leq y \leq 0)$

$$\sigma_{xy}^{(2)} = \frac{4q^+}{a}(1-2\xi)\left(\frac{a_{12}^{(2)}}{a_{11}^{(2)}}y - \frac{a_{12}^{(1)}}{a_{11}^{(1)}}h_1\right),$$

$$\sigma_y^{(2)} = -q^+(1-2\xi)^2 + 4q^+\frac{a_{12}^{(1)}}{a_{11}^{(1)}}\left(\frac{h_1}{a}\right)^2 + \frac{8q^+}{a^2}\left(\frac{a_{12}^{(2)}}{a_{11}^{(2)}}\frac{y^2}{2} - \frac{a_{12}^{(1)}}{a_{11}^{(1)}}h_1 y\right),$$

$$\sigma_x^{(2)} = q^+\frac{a_{12}^{(2)}}{a_{11}^{(2)}}(1-2\xi)^2 - 4q^+\frac{a_{12}^{(2)}}{a_{11}^{(2)}}\frac{a_{12}^{(1)}}{a_{11}^{(1)}}\left(\frac{h_1}{a}\right)^2$$

$$+\frac{8q^+}{a^2}\left(\frac{a_{12}^{(2)}}{a_{11}^{(2)}}\left(\frac{a_{12}^{(1)}}{a_{11}^{(1)}}h_1 y - \frac{a_{12}^{(2)}}{2a_{11}^{(2)}}y^2\right) + \frac{a_{66}^{(2)}}{a_{11}^{(1)}}\frac{a_{12}^{(1)}}{a_{11}^{(2)}}h_1(y + h_2)\right)$$

$$+\frac{A_{11}^{(2)}}{a_{11}^{(2)}}h_2(h_2+y)-\frac{1}{2a_{11}^{(2)}}\left(\frac{a_{66}^{(2)}a_{12}^{(2)}}{a_{11}^{(2)}}-A_{11}^{(2)}\right)(y^2-h_2^2)\Bigg),$$

$$u^{(2)}=-\frac{4q^+}{a}(1-2\xi)\left(\frac{a_{66}^{(2)}a_{12}^{(1)}}{a_{11}^{(1)}}(y+h_2)h_1\right.$$

$$\left.+A_{11}^{(2)}(h_2+y)h_2-\frac{1}{2}\left(\frac{a_{66}^{(2)}a_{12}^{(2)}}{a_{11}^{(2)}}-A_{11}^{(2)}\right)(y^2-h_2^2)\right),\quad(6.7)$$

$$v^{(2)}=-q^+A_{11}^{(2)}(1-2\xi)^2(y+h_2)+\frac{4q^+}{a^2}\frac{a_{12}^{(1)}}{a_{11}^{(1)}}h_1(y+h_2)\left(A_{11}^{(2)}h_1-a_{22}^{(2)}(y-h_2)\right)$$

$$+\frac{8q^+}{a^2}\frac{a_{12}^{(2)}}{a_{11}^{(2)}}\left(\left(\frac{a_{12}^{(1)}a_{66}^{(2)}}{a_{11}^{(1)}}h_1+\frac{1}{2}h_2\left(\frac{a_{66}^{(2)}a_{12}^{(2)}}{a_{11}^{(2)}}+A_{11}^{(2)}\right)\right)(y+h_2)h_2\right.$$

$$+\frac{1}{2}\left(\frac{a_{12}^{(1)}}{a_{11}^{(1)}}(a_{12}^{(2)}+a_{66}^{(2)})h_1+A_{11}^{(2)}h_2\right)(y^2-h_2^2)$$

$$\left.-\frac{1}{6}\left(\frac{a_{12}^{(2)}}{a_{11}^{(2)}}(a_{12}^{(2)}+a_{66}^{(2)})-A_{11}^{(2)}-a_{22}^{(2)}\right)(y^3+h_2^3)\right).$$

Let us now evaluate the thermal stresses in a two-layer beam, when the temperature $\theta=T-T_0$ is an arbitrary integrable function of the transverse coordinate $\theta^{(i)}=\theta^{(i)}(\zeta)$ $(i=1,2)$; $\zeta=y/h_2$, assuming zero external loading and volume force

$$\sigma_{xy}(h_1)=\sigma_y(h_1)=0,\quad u(-h_2)=v(-h_2)=0.\quad(6.8)$$

Using the relations (5.8)-(5.10), it is possible to show that $Q^{(0)}\neq0$, $Q^{(s)}\equiv0$ at $s\geq1$. Therefore, the resulting solution is

$$\sigma_{xy}^{(1)}=\sigma_y^{(1)}=0,\quad\sigma_x^{(1)}=-\frac{\alpha_{11}^{(1)}}{a_{11}^{(1)}}\theta^{(1)}(\zeta),\quad0\leq\zeta\leq\bar{h}=h_1/h_2,$$

$$u^{(1)}=\alpha_{12}^{(1)}h_2\int_0^\zeta\theta^{(1)}(\zeta)d\zeta+\alpha_{12}^{(2)}h_2\int_{-1}^0\theta^{(2)}(\zeta)d\zeta,\quad(6.9)$$

$$v^{(1)}=\left(\alpha_{22}^{(1)}-\alpha_{11}^{(1)}\frac{a_{12}^{(1)}}{a_{11}^{(1)}}\right)h_2\int_0^\zeta\theta^{(1)}(\zeta)d\zeta+\left(\alpha_{22}^{(2)}-\alpha_{11}^{(2)}\frac{a_{12}^{(2)}}{a_{11}^{(2)}}\right)h_2\int_{-1}^0\theta^{(2)}(\zeta)d\zeta,$$

$$\sigma_{xy}^{(2)}=\sigma_y^{(2)}=0,\quad\sigma_x^{(2)}=-\frac{\alpha_{11}^{(2)}}{a_{11}^{(2)}}\theta^{(2)}(\zeta),\quad-1\leq\zeta\leq0,$$

$$u^{(2)}=\alpha_{12}^{(2)}h_2\int_{-1}^\zeta\theta^{(2)}(\zeta)d\zeta,\quad v^{(2)}=\left(\alpha_{22}^{(2)}-\alpha_{11}^{(2)}\frac{a_{12}^{(2)}}{a_{11}^{(2)}}\right)h_2\int_{-1}^\zeta\theta^{(2)}(\zeta)d\zeta.$$

2.7 Boundary Layer for a Two-Layer Beam

The boundary layer for two-layer beam is constructed in a similar manner to that of a single layer. We now transform to new variables $t=x/h_2=\xi/\varepsilon$, $\zeta=y/h_2$

in equations (5.3) (the effect of the temperature field is ignored now since this was addressed previously in the outer problem). Then the solution is sought in the form of an expansion along the functions of the boundary layer type

$$\sigma_x^{(i)}, \sigma_y^{(i)}, \sigma_{xy}^{(i)} = \varepsilon^{-1+s} \left(\sigma_1^{(i,s)}(\zeta), \sigma_2^{(i,s)}(\zeta), \sigma_{12}^{(i,s)}(\zeta) \right) \exp(-\lambda t) \quad s = \overline{0, N}$$

$$u^{(i)}, v^{(i)} = \varepsilon^s \left(u_1^{(i,s)}(\zeta), u_2^{(i,s)}(\zeta) \right) \exp(-\lambda t) \quad (i = 1, 2). \quad (7.1)$$

In case of orthotropic layers

$$\sigma_1^{(i,s)} = \frac{1}{\lambda^2} \frac{d^2 \sigma_2^{(i,s)}}{d\zeta^2}, \quad \sigma_{12}^{(i,s)} = \frac{1}{\lambda} \frac{d\sigma_2^{(i,s)}}{d\zeta},$$

$$u_1^{(i,s)} = -\frac{1}{\lambda} \left(\frac{a_{11}^{(i)}}{\lambda^2} \frac{d^2 \sigma_2^{(i,s)}}{d\zeta^2} + a_{12}^{(i)} \sigma_2^{(i,s)} \right), \quad (7.2)$$

$$u_2^{(i,s)} = -\frac{a_{11}^{(i)}}{\lambda^4} \frac{d^3 \sigma_2^{(i,s)}}{d\zeta^3} - (a_{12}^{(i)} + a_{66}^{(i)}) \frac{1}{\lambda^2} \frac{d\sigma_2^{(i,s)}}{d\zeta},$$

where $\sigma_2^{(i,s)}$ are governed by

$$a_{11}^{(i)} \frac{d^4 \sigma_2^{(i,s)}}{d\zeta^4} + (2a_{12}^{(i)} + a_{66}^{(i)}) \lambda^2 \frac{d^2 \sigma_2^{(i,s)}}{d\zeta^2} + a_{22}^{(i)} \lambda^4 \sigma_2^{(i,s)} = 0. \quad (7.3)$$

Since the obtained equation (7.3) is formally identical to (3.5) apart from the index (i), the associated solutions (3.6)-(3.8) for $\sigma_2^{(i,s)}$ could now be used to give

$$\sigma_2^{(i,s)} = \lambda^4 \sum_{j=1}^{4} A_{j(i)}^{(s)} \psi_{j(i)}(\zeta),$$

$$\sigma_{12}^{(i,s)} = \lambda^3 \sum_{j=1}^{4} A_{j(i)}^{(s)} \psi'_{j(i)}(\zeta), \quad \sigma_1^{(i,s)} = \lambda^2 \sum_{j=1}^{4} A_{j(i)}^{(s)} \psi''_{j(i)}(\zeta),$$

$$u_1^{(i,s)} = -\lambda \sum_{j=1}^{4} \left(a_{11}^{(i)} \psi''_{j(i)} + \lambda^2 a_{12}^{(i)} \psi_{j(i)} \right) A_{j(i)}^{(s)}, \quad (7.4)$$

$$u_2^{(i,s)} = -\sum_{j=1}^{4} \left(a_{11}^{(i)} \psi'''_{j(i)} + \lambda^2 (a_{12}^{(i)} + a_{66}^{(i)}) \psi'_{j(i)} \right) A_{j(i)}^{(s)}, \quad (i = 1, 2)$$

where $A_{j(i)}^{(s)}$ are constants to be determined, and $\psi_{1(i)}$, $\psi_{2(i)}$, $\psi_{3(i)}$, $\psi_{4(i)}$ are the coefficients with respect to $A^{(s)}$, $B^{(s)}$, $C^{(s)}$, $D^{(s)}$ respectively in (3.6)-(3.8). For example in the case corresponding to (3.6) we have

$$\psi_{1(i)} = \cos \lambda \beta_{(i)} \zeta, \quad \psi_{2(i)} = \zeta \cos \lambda \beta_{(i)} \zeta, \quad \psi_{3(i)} = \sin \lambda \beta_{(i)} \zeta, \quad \psi_{4(i)} = \zeta \sin \lambda \beta_{(i)} \zeta.$$

The solution (7.1), (7.4) should satisfy the following conditions at the boundaries $y = h_1$, $-h_2$ and contact area at $y = 0$

$$\sigma_{xy} = \sigma_y = 0 \quad \text{at} \quad y = h_1 \text{ or } \zeta = \bar{h} = h_1/h_2,$$

$$u = v = 0 \quad \text{at} \quad y = -h_2 \ (\zeta = -1), \quad (7.5)$$

$$\sigma_{xy}^{(1)} = \sigma_{xy}^{(2)}, \quad \sigma_y^{(1)} = \sigma_y^{(2)}, \quad u^{(1)} = u^{(2)}, \quad v^{(1)} = v^{(2)} \quad \text{at} \quad y = 0 \ (\zeta = 0).$$

The conditions (7.5) imply the following homogeneous algebraic system regarding the eight unknown coefficients $A_{j(i)}^{(s)}$

$$A_{j(1)}^{(s)}\psi_{j(1)}(\zeta_1) = 0, \quad , \quad A_{j(1)}^{(s)}\psi_{j(1)}'(\zeta_1) = 0, \quad j = \overline{1,4}, \quad \zeta_1 = h_1/h_2,$$

$$A_{j(1)}^{(s)}\psi_{j(1)}(0) - A_{j(2)}^{(s)}\psi_{j(2)}(0) = 0, \quad A_{j(1)}^{(s)}\psi_{j(1)}'(0) - A_{j(2)}^{(s)}\psi_{j(2)}'(0) = 0,$$

$$\left(a_{11}^{(1)}\psi_{j(1)}''(0) + \lambda^2 a_{12}^{(i)}\psi_{j(1)}(0)\right)A_{j(1)}^{(s)} = \left(a_{11}^{(2)}\psi_{j(2)}''(0) + \lambda^2 a_{12}^{(2)}\psi_{j(2)}(0)\right)A_{j(2)}^{(s)},$$

$$\left(a_{11}^{(1)}\psi_{j(1)}'''(0) + \lambda^2(a_{12}^{(1)} + a_{66}^{(1)})\psi_{j(1)}'(0)\right)A_{j(1)}^{(s)}$$

$$= \left(a_{11}^{(2)}\psi_{j(2)}'''(0) + \lambda^2(a_{12}^{(2)} + a_{66}^{(2)})\psi_{j(2)}'(0)\right)A_{j(2)}^{(s)}, \qquad (7.6)$$

$$\left(a_{11}^{(2)}\psi_{j(2)}''(-1) + \lambda^2 a_{12}^{(2)}\psi_{j(2)}(-1)\right)A_{j(2)}^{(s)} = 0,$$

$$\left(a_{11}^{(2)}\psi_{j(2)}'''(-1) + \lambda^2(a_{12}^{(2)} + a_{66}^{(2)})\psi_{j(2)}'(-1)\right)A_{j(2)}^{(s)} = 0.$$

Here the notation $j = \overline{1,4}$ means summation over dummy index j from 1 to 4. Depending on the type of anisotropy of the layers, the functions $\psi_{j(1)}$ and $\psi_{j(2)}$ could be determined from (3.6)-(3.8), for example, $\psi_{j(1)}$ and $\psi_{j(2)}$ could be defined from (3.6) and (3.7), respectively.

In order for (7.6) to possess non-trivial solutions, the associated determinant $\Delta(\lambda)$ should be equal to zero, providing an equation for λ. Since $\Delta(\lambda)$ is an entire function, the roots of the characteristic equation $\Delta(\lambda) = 0$ form a countable set. Using (7.6), all of the constants $A_{j(i)}^{(s)}$ could be expressed through one complex constant, which would have a conjugate associated solution. Then any λ with positive real part $Re\lambda > 0$ would possess a unique eigenfunction $\sigma_2^{(s)}$. Hence, in view of (7.1) the solution would require summation of all particular solutions along λ with $Re\lambda > 0$.

We note that for any s the solution (7.1) and (7.4) is an exact solution of the elasticity equations.

The procedure of matching the outer solution and the boundary layer is essentially the same described previously for a single layer.

2.8　Models of Elastic Foundation

The obtained results are closely related to applied models of elastic foundation, however, before revealing this link let us first present a brief review of typical existing models of elastic foundation.

One of the oldest, and perhaps the most popular is the Winkler-Fuss-Zimmerman model, which is also referred to as the modulus of foundation model. Within its framework if a plate or beam rests on an elastic foundation, then the reactive pressure $p(x,y)$ is proportional to the deflection $W(x,y)$, i.e. $p(x,y) = KW(x,y)$, where K is the so-called modulus of a foundation. According to the Winkler-Fuss hypothesis, provided that the pressure is created by a force acting on a small localized area, the resulting displacements (proportional to pressure) occur only in the area, however the neighboring area is not affected. This type of elastic foundation could have a physical interpretation as a system of independent springs. Due to simplicity of the model and its physical lucidity, it has been used in many applied problems. Later the Winkler-Fuss model has been modified by introducing the modulus of a foundation depending on the horizontal variable $K = K(x)$, which enables taking into account possible inhomogeneity along this direction. However, the experiments and observations from real-life civil engineering clearly showed that the displacements occur not only in the area under pressure, but in the neighboring vicinity as well, therefore, justifying further improvement of the Winkler model. The required model needed to take into account the distribution of displacements, being able to link the pressure at one point and displacement in a different point. One of the options was a model of elastic half-space (half-plane). Within this model the foundation was treated as an elastic half-space, some parts of which were loaded by acting force or a rigid stamp. The unknown reactive pressure $p(x, y)$ under stamp was calculated from the equality of the normal displacement and deflection in the contact area. However, this new model has also turned out to be not ideal, in particular, depending on the form of external load the resulting formulae appeared to be rather complicated, moreover, it has been proved that the calculations within the model of a homogeneous elastic half-space provide overestimated results for deflection, and also demonstrate dramatic increase of reactive pressure near the boundaries of applied load.

In order to develop an efficient model, which is still taking into account the distribution of displacements by relatively easier means, a two-parametric model of elastic foundation has been suggested (M.M. Filonenko-Borodich, P.L. Pasternak, V.Z. Vlasov). Within the model, the properties of elastic foundation are characterized by two parameters. The first parameter is associated with compression ability, and has a meaning similar to that of the Winkler modulus of the foundation. The second parameter reflects the work of elastic foundation on shear, and allows incorporation of distribution of displacements in the foundation. The results of calculations within this two-parametric model reveal the deflections occurring not only directly under the external loading, but also in the vicinity of the latter. One more advantage of the two-parametric model is that it gives more accurate results in comparison to the model of a homogeneous elastic half-space. However, the drawback of the two-parametric model is that it still results in concentrated forces near the boundary of loading area. One more problematic issue is related to evaluation of the values of parameters of the model.

It may be remarked in general that the Winkler model of elastic foundation along with the two-parametric model and other approaches to the problem, which are oriented to analysis of the structures resting on elastic foundation (except for the Vlasov-Leontiev model), cannot provide results for stresses and strains within the foundation itself. However, it is very important for many applications to determine not only the reactive pressure, but deflection and pressure within the lower layers of the foundation, providing comprehensive results. The model of a homogeneous elastic half-space could be an answer to this challenge, but it has some other drawbacks mentioned above.

In order to bridge the gap between the results of calculations with the experimental data for subgrades, it was suggested to use the model of inhomogeneous elastic half-space, with the Young's modulus increasing steadily with depth (Wieghardt, G.K. Klein). The form of dependence could vary including linear, parabolic, exponential etc. However, this approach contains significant complications, related to evaluation of the form of dependence and also integrating differential equations with variable coefficients.

On the other hand, the experimental data and building practice revealed that in real life only the upper layers of subgrades are compressed, so the foundation may be thought as incompressible starting from certain depth (having zero displacements). Therefore, the model of elastic foundation as a compressible layer has become increasingly popular starting from 1960s (K.E. Egorov, V.Z. Vlasov - N.N. Leontiev, I.K. Samarin, V.A. Florin, and others). The thickness of compressible layer is usually called the active zone of the foundation. The elastic modulus of the foundation could be considered either constant or variable along thickness. There are some difficulties within this approach related to evaluating the thickness of compressible layer, however, there arc also some empiric laws addressing this matter.

In this section we have briefly described some of the most typical models of elastic foundations. Some more details on the subject could be found in Vlasov and Leont'ev (1960); Gorbunov-Posadov et al. (1984); Wang et al. (2005).

2.9 The Winkler-Fuss Model

Within the model of the compressible layer, the elastic foundation is treated as a two-layer or multi-layer elastic plate (beam/strip), with fixed lower face (imposing zero displacements) with some given normal and tangential forces acting on the upper face. Some volume forces could also be taken into account (e.g. the weights of all of the layers or some particular layers) along with incorporating thermal effects. Sometimes the statement of the problem could include free lower face of the compressible layer, i.e. equating normal displacement and tangential stresses to zero. In general, the problem in question corresponds to a 3D (or 2D) problem of thermoelasticity. Earlier in this chapter we have considered a plane problem for a two-layer strip/beam. This solution could have a straightforward generalization to arbitrary number of layers. The structure of solution will still be in line with (5.5), but there will be more intermediate steps of continuity conditions on the contact areas between the layers. The obtained solution (5.5), (5.8)-(5.10) is asymptotically exact, i.e. obtained without relying on any ad-hoc assumption, which is a clear advantage in comparison to solution presented in Vlasov and Leont'ev (1960).

Since the iteration process (5.8)-(5.10) often stops at certain stage providing a close-form solution, it becomes possible to analyze the solution near the contact area of the layers, and also to investigate the validity of the Winkler-Fuss-Zimmerman and other applied models of elastic foundation.

We begin with the exact outer solution (6.2), (6.3). In the simplest case, without noting the effects of the weights of the layer, the stresses and displacements on the contact area $y = 0$ are written as (in what follows the c index is associated with the contact area)

$$\sigma_{xy}^{(c)} = \tau^+, \qquad \sigma_y^{(c)} = -\sigma_2^+, \qquad (9.1)$$

$$u^{(c)} = -A_{16}^{(2)}\sigma_2^+ h_2 + A_{66}^{(2)} h_2 \tau^+, \quad v^{(c)} = -A_{11}^{(2)}\sigma_2^+ h_2 + A_{16}^{(2)} h_2 \tau^+.$$

It follows from (9.1) that in case of anisotropic layers $u^{(c)} \neq 0$ even for the uniform normal load ($\tau^+ = 0$). This implies immediately that the Winkler-Fuss model is not valid for anisotropic foundation, since normal load would cause tangential displacements. The hypothesis would be valid for isotropic foundation or for orthotropic foundation when the principal directions of anisotropy coincide with the coordinate axes. From (9.1) for the orthotropic case ($A_{16}^{(2)} = 0$) we have $\sigma_y^{(c)} = v^{(c)}/(A_{11}^{(2)} h_2)$. Therefore, the modulus of the foundation is

$$K = \frac{1}{A_{11}^{(2)} h_2} = \frac{E_2^{(2)}}{(1 - \nu_{12}^{(2)}\nu_{21}^{(2)})h_2}. \qquad (9.2)$$

In case of isotropic layer $K = E_{(2)}/(1 - \nu_{(2)}^2)h_2$, which coincides with the well-known formula, see Gorbunov-Posadov et al. (1984), p. 51. The formulae above are obtained within the plane stress assumption. We note once again that these could be easily transformed to the plane strain case. In particular, (9.2) would be transformed to

$$K = \frac{(1 - \nu_{(2)})E_{(2)}}{(1 + \nu_{(2)})(1 - 2\nu_{(2)})h_2} \qquad (9.3)$$

which coincides with the modulus of foundation given by N.M. Gersevanov relying on the assumption that the elastic layer does not move along the surface of a rigid foundation.

If the weight of the layers are taken into account, then the stresses and displacements on the contact area for an orthotropic layer are given by

$$\sigma_y^{(c)} = -\sigma_2^+ - \rho_1 g h_1, \quad u^{(c)} = 0,$$

$$v^{(c)} = -A_{11}^{(2)} h_2(\sigma_2^+ + \rho_1 g h_1) - \frac{1}{2}b_{11}^{(2)}\rho_2 g h_2^2 \qquad (9.4)$$

implying that $\sigma_y^{(c)} \neq K v^{(c)}$. As follows from (9.4) the contact pressure and displacement would be proportional provided that we ignore the weight of the second layer (foundation) causing additional deflection.

If the acting load has both tangential and normal components ($\tau^+ \neq 0, \sigma_2^+ \neq 0$), the contact points gain both normal and tangential displacements. In particular, for an orthotropic layer

$$\sigma_{xy}^{(c)} = \tau^+, \qquad \sigma_y^{(c)} = -\sigma_2^+, \tag{9.5}$$

$$u^{(c)} = A_{66}^{(2)} h_2 \tau^+, \quad v^{(c)} = -A_{11}^{(2)} h_2 \sigma_2^+.$$

It may be deduced from the last formulae that in case of both tangential and normal loading the contact interaction between the layer and foundation is characterized by two modules of the foundation. The first modulus coincides with the Winkler one (9.2), whereas the second coefficient is

$$\beta = \frac{1}{A_{66}^{(2)} h_2} = G^{(2)}/h_2 \tag{9.6}$$

where $G^{(2)}$ is the shear modulus of the foundation (lower layer). Therefore the two-parametric model of P.L. Pasternak along with other two-parametric models is more efficient than the Winker-Fuss model. In case of general anisotropy the contact between the two layers would be elastic, therefore it will be described by three parameters.

It should be noted that all of the considerations above in this section were related to uniform loading. In order to address the case of non-uniform loading let us rewrite equations (6.4)-(6.7) on the contact area.

In case of linear normal loading it follows from (6.4)-(6.5) that

$$\sigma_{xy}^{(c)} = \frac{a_{12}^{(1)}}{a_{11}^{(1)}} q^+ \frac{h_1}{a}, \qquad \sigma_y^{(c)} = -q^+ \xi,$$

$$u^{(c)} = \left(\left(\frac{a_{12}^{(1)}}{a_{11}^{(1)}} h_1 + \frac{1}{2} \frac{a_{12}^{(2)}}{a_{11}^{(2)}} h_2 \right) a_{66}^{(2)} - \frac{3}{2} A_{11}^{(2)} h_2 \right) \frac{h_2}{a} q^+, \tag{9.7}$$

$$v^{(c)} = -A_{11}^{(2)} h_2 q^+ \xi, \quad \sigma_y^{(c)} = \frac{1}{A_{11}^{(2)} h_2} v^{(c)},$$

whereas (6.6)-(6.7) imply for the normal load with parabolic distribution that

$$\sigma_{xy}^{(c)} = -\frac{4a_{12}^{(1)}}{a_{11}^{(1)}} (1 - 2\xi) q^+ \frac{h_1}{a}, \qquad \sigma_y^{(c)} = -q^+ (1 - 2\xi)^2 + 4q^+ \frac{a_{12}^{(1)}}{a_{11}^{(1)}} \left(\frac{h_1}{a} \right)^2,$$

$$u^{(c)} = -4q^+ \left(\frac{a_{12}^{(1)} a_{66}^{(1)}}{a_{11}^{(1)}} h_1 + \frac{1}{2} \left(\frac{a_{12}^{(2)} a_{66}^{(2)}}{a_{11}^{(2)}} + A_{11}^{(2)} \right) h_2 \right) \frac{h_2}{a} (1 - 2\xi),$$

$$v^{(c)} = -q^+ A_{11}^{(2)} h_2 \left((1 - 2\xi)^2 - 4 \frac{a_{12}^{(1)}}{a_{11}^{(1)}} \left(\frac{h_1}{a} \right)^2 \right) \tag{9.8}$$

$$+ \frac{4}{3} q^+ \frac{h_2 a_{12}^{(2)}}{a_{11}^{(2)}} \left(\frac{a_{12}^{(2)}}{a_{11}^{(2)}} (2a_{66}^{(2)} - a_{12}^{(2)}) + A_{11}^{(2)} + a_{22}^{(2)} \right) \left(\frac{h_2}{a} \right)^2$$

$$+ 4q^+ h_1 \frac{a_{12}^{(1)}}{a_{11}^{(1)}} \left(\frac{a_{12}^{(2)}}{a_{11}^{(2)}} (a_{66}^{(2)} - a_{12}^{(2)}) + a_{22}^{(2)} \right) \left(\frac{h_2}{a} \right)^2.$$

The resulting expressions (6.4)-(6.7), (9.7)-(9.8) allow straightforward further observations. The first conclusion is that the tangential stresses and displacements are caused by non-uniformity of the normal loading. In case of (9.7), though the contact pressure is proportional to vertical displacement with the Winkler modulus of foundation, the original Winkler-Fuss assumption is still not valid since $u^{(c)} \neq 0$. A similar situation occurs for (9.8). Thus, it is observed that the non-uniform distribution of the normal load causes violation of the Winkler-Fuss hypothesis. This violation becomes clearer with the increase in the rate of change of the loading profile.

However, it should be noted that the additional terms arising in the solution are of the next asymptotic order in comparison to that corresponding to the Winkler-Fuss model for elastic foundation. For example, in (9.8), the second and third terms in expression for $v^{(c)}$ are of order $O(\varepsilon^2)$ and are actually refinements to the Winkler term.

It also follows from (9.7) and (9.8) that the contact tangential displacement $u^{(c)}$ at medium anisotropy is of order $O(\varepsilon)$, so it could be neglected compared to the normal displacement at small ε. However, if the shear stiffness $G^{(2)}$ is weak, then the possibility of ignoring the tangential displacement is questionable. Thus, it follows from the above that the Winkler-Fuss model is a leading order approximation of the outer problem within the proposed asymptotic approach.

In case of the model of compressible layer there is no need to adopt any assumptions of kinematic or other nature as in Vlasov and Leont'ev (1960), since the asymptotic method provides a rather straightforward solution of the appropriate 3D problem of elasticity.

The outer solution satisfies all of the equations of theory of elasticity along with the face boundary conditions at $y = h_1, -h_2$ and conditions at the contact area. However, as discussed earlier, it is not the case for the edge boundary conditions at $x = 0, a$. Therefore there is a need in the boundary layer solution, i.e. a homogeneous solution decaying rapidly away from the edge into the interior of the domain Ω. The full solution required construction of the boundary layer as shown in Section 2.7, and the procedure of matching the boundary layer with the outer solution of Section 2.4. The width of the zone containing effects of the boundary layer is typically of thickness 1-1.5 of the layer.

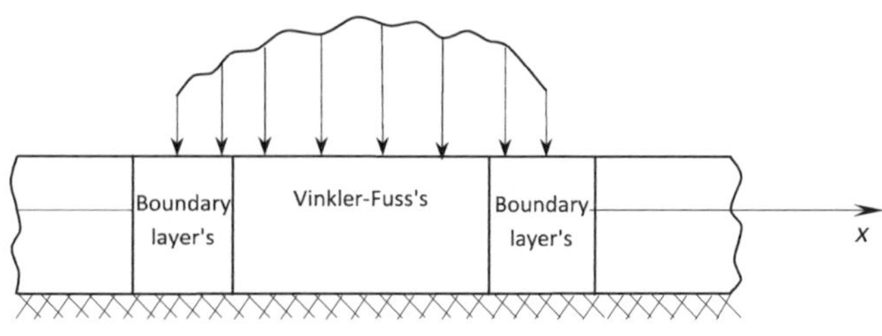

Fig. 2.4

The boundary layer also occurs around discontinuities of the load and also around points at which there is a change of type of boundary conditions. The whole area of the foundation could generally be split into two parts (Fig. 2.4). The first part generally allows application of the Winkler-Fuss-Zimmerman model, whereas the second part involves the effects of the boundary layer, where any model involving approaches through modulus of foundation, is no longer valid.

2.10 The Modulus of a Foundation with Variable Elastic Parameters

Consider now the problem for a two-layer anisotropic rectangular strip within the framework of plane problem of elasticity with variable elastic parameters. Let us perform the asymptotic analysis depending on the variation of elastic modules along the thickness of the layer. This would have implications for evaluating the modulus of the foundations in case of elastic foundations with variable elastic parameters.

Consider now a plane problem of elasticity for a two-layer rectangular strip $\Omega = \{(x,y) : x \in [0,a], -h \leq y \leq h, 2h = h_1 + h_2 \ll a\}$. The thicknesses of the layers are h_1 and h_2 so that $h_1 + h_2 = 2h$. Both layers are assumed to be anisotropic, with the elastic parameters $(a_{ik} = a_{ik}(\zeta))$ being integrable functions of the transverse coordinate. Let the boundary conditions at the lower face of the strip $y = -h$ be given by

$$u(-h) = u^-, \quad v(-h) = v^- \quad \text{in particular} \quad u^- = v^- \equiv 0 \qquad (10.1)$$

whereas on the upper face $y = h$ we impose one of the following conditions considered earlier

$$u(h) = u^+(\xi), \quad v(h) = v^+(\xi), \quad \xi = x/a \qquad (10.2)$$

$$\sigma_{xy}(h) = \varepsilon^{-1}\sigma_{xy}^+(\xi), \quad v(h) = v^+(\xi), \quad \varepsilon = h/a \qquad (10.3)$$

$$u(h) = u^+(\xi), \quad \sigma_y(h) = \varepsilon^{-1}\sigma_y^+(\xi) \qquad (10.4)$$

$$\sigma_{xy}(h) = \varepsilon^{-1}\sigma_{xy}^+(\xi), \quad \sigma_y(h) = \varepsilon^{-1}\sigma_y^+(\xi). \qquad (10.5)$$

The solution should also satisfy the continuity conditions over the contact area along with edge boundary conditions at $x = 0, a$.

We begin with scaling to dimensionless variables $\xi = x/a$, $\zeta = y/h$. Then the unknown stresses $\sigma_{jk}^{(i)}$ and dimensionless displacements $U^{(i)} = u^{(i)}/a$, $V^{(i)} = v^{(i)}/a$, $i = 1, 2$ are sought in the form

$$\sigma_{jk}^{(i)} = \varepsilon^{-1+s}\sigma_{jk}^{(i,s)}, \quad U^{(i)} = \varepsilon^s u^{(i,s)} \quad (u,v), \quad s = \overline{0, N}. \qquad (10.6)$$

The solution is then given by Aghalovyan and Adamyan (1986) as

$$\sigma_{xy}^{(s)} = \sigma_{xy0}^{(s)}(\xi) + \sigma_{xy}^{*(s)}(\xi,\zeta), \quad \sigma_y^{(s)} = \sigma_{y0}^{(s)}(\xi) + \sigma_y^{*(s)}(\xi,\zeta),$$

$$\sigma_x^{(s)} = -(a_{12}\sigma_{y0}^{(s)} + a_{16}\sigma_{xy0}^{(s)})a_{11}^{-1} + \sigma_x^{*(s)}, \quad (10.7)$$

$$U^{(s)} = \sigma_{y0}^{(s)}\int_0^\zeta A_{16}(\zeta)d\zeta + \sigma_{xy0}^{(s)}\int_0^\zeta A_{66}(\zeta)d\zeta + u^{(s)}(\xi) + u^{*(s)}(\xi,\zeta),$$

$$V^{(s)} = \sigma_{y0}^{(s)}\int_0^\zeta A_{11}(\zeta)d\zeta + \sigma_{xy0}^{(s)}\int_0^\zeta A_{16}(\zeta)d\zeta + v^{(s)}(\xi) + v^{*(s)}(\xi,\zeta),$$

where

$$u^{*(s)}(\xi,\zeta) = \int_0^\zeta \left(a_{16}\sigma_x^{*(s)} + a_{26}\sigma_y^{*(s)} + a_{66}\sigma_{xy}^{*(s)} - \frac{\partial V^{(s-1)}}{\partial \xi}\right)d\zeta,$$

$$v^{*(s)}(\xi,\zeta) = \int_0^\zeta \left(a_{12}\sigma_x^{*(s)} + a_{22}\sigma_y^{*(s)} + a_{26}\sigma_{xy}^{*(s)}\right)d\zeta,$$

$$\sigma_{xy}^{*(s)} = -\int_0^\zeta \frac{\partial\sigma_x^{(s-1)}}{\partial\xi}d\zeta, \quad \sigma_y^{*(s)} = -\int_0^\zeta \frac{\partial\sigma_{xy}^{(s-1)}}{\partial\xi}d\zeta, \quad (10.8)$$

$$\sigma_x^{*(s)} = -\left(a_{12}\sigma_y^{*(s)} + a_{16}\sigma_{xy}^{*(s)}\right)a_{11}^{-1} + a_{11}^{-1}\frac{\partial U^{(s-1)}}{\partial\xi},$$

$$A_{11} = (a_{11}a_{22} - a_{12}^2)a_{11}^{-1}, \quad A_{16} = (a_{11}a_{16} - a_{12}a_{16})a_{11}^{-1},$$

$$A_{66} = (a_{11}a_{66} - a_{16}^2)a_{11}^{-1}.$$

All of the quantities above depending on ζ should be assigned with an index (1) or (2) in case of the upper and lower layers, respectively

$$A_{ik}(\zeta) = \begin{cases} A_{ik}^{(1)}(\zeta), & \zeta_0 \le \zeta \le 1 \\ A_{ik}^{(2)}(\zeta), & -1 \le \zeta \le \zeta_0 \end{cases}, \quad \sigma_{ik}(\xi,\zeta) = \begin{cases} \sigma_{ik}^{(1)}(\xi,\zeta), & \zeta_0 \le \zeta \le 1 \\ \sigma_{ik}^{(2)}(\xi,\zeta), & -1 \le \zeta \le \zeta_0 \end{cases},$$

with $\zeta_0 = (h_2 - h_1)/(h_1 + h_2)$ standing for the dimensionless value of the contact line. The functions $\sigma_{xy0}^{(s)}(\xi)$, $\sigma_{y0}^{(s)}(\xi)$, $u^{(s)}(\xi)$, $v^{(s)}(\xi)$ are the same for both layers and have to be determined. The continuity conditions on the line $\zeta = \zeta_0$ are satisfied. Let us prove it for example for the stress σ_{xy}. The other cases may be treated similarly. From (10.8) we have

$$\sigma_{xy}^{*(1,s)}(\xi,\zeta) = -\int_0^{\zeta_0} \frac{\partial\sigma_x^{(2,s-1)}}{\partial\xi}d\zeta - \int_{\zeta_0}^\zeta \frac{\partial\sigma_x^{(1,s-1)}}{\partial\xi}d\zeta,$$

hence,

$$\sigma_{xy}^{*(1,s)}(\xi,\zeta_0) = -\int_0^{\zeta_0} \frac{\partial\sigma_x^{(2,s-1)}}{\partial\xi}d\zeta = \sigma_{xy}^{*(2,s)}(\xi,\zeta_0).$$

Therefore, in view of (10.7), we deduce

$$\sigma_{xy}^{(1,s)}(\xi,\zeta_0) = \sigma_{xy}^{(2,s)}(\xi,\zeta_0).$$

The presented proof is valid for $\zeta_0 > 0$. In the opposite case $\zeta_0 < 0$ corresponding to $h_1 > h_2$, we obtain

$$\sigma_{xy}^{*(2,s)}(\xi,\zeta) = -\int_0^{\zeta_0} \frac{\partial \sigma_x^{(1,s-1)}}{\partial \xi} d\zeta - \int_{\zeta_0}^{\zeta} \frac{\partial \sigma_x^{(2,s-1)}}{\partial \xi} d\zeta,$$

which again implies

$$\sigma_{xy}^{*(2,s)}(\xi,\zeta_0) = \sigma_{xy}^{*(1,s)}(\xi,\zeta_0),$$

and then, using (10.7), $\sigma_{xy}^{(1,s)}(\xi,\zeta_0) = \sigma_{xy}^{(2,s)}(\xi,\zeta_0)$.

Satisfying the boundary conditions (10.1), (10.2) gives

$$\sigma_{xy0}^{(s)} = (b_{11}\varphi^{(s)} - b_{16}f^{(s)})/\Delta,$$

$$\sigma_{y0}^{(s)} = (b_{66}f^{(s)} - b_{16}\varphi^{(s)})/\Delta, \qquad (10.9)$$

$$u^{(s)}(\xi) = u^{-(s)} + \left((b_{66}c_{16} - b_{16}c_{66})f^{(s)} + (b_{11}c_{66} - b_{16}c_{16})\varphi^{(s)}\right)/\Delta - u^{*(s)}(\xi,-1),$$

$$v^{(s)}(\xi) = v^{-(s)} + \left((b_{66}c_{11} - b_{16}c_{16})f^{(s)} + (b_{11}c_{16} - b_{16}c_{11})\varphi^{(s)}\right)/\Delta - v^{*(s)}(\xi,-1),$$

where

$$\varphi^{(s)} = u^{+(s)} - u^{-(s)} - u^{*(s)}(\xi,1) + u^{*(s)}(\xi,-1),$$

$$f^{(s)} = v^{+(s)} - v^{-(s)} - v^{*(s)}(\xi,1) + v^{*(s)}(\xi,-1), \qquad (10.10)$$

$$b_{ik} = \int_{-1}^{1} A_{ik}d\zeta, \quad c_{ik} = \int_{-1}^{0} A_{ik}d\zeta, \quad \Delta = b_{11}b_{66} - b_{16}^2,$$

$$u^{\pm(0)} = u^{\pm}/a, \quad v^{\pm(0)} = v^{\pm}/a, \quad u^{\pm(s)} = v^{\pm(s)} = 0 \quad (s \neq 0).$$

Then all of the unknowns may be determined from (10.6)-(10.8). When $u^{-} = v^{-} = 0$, $u^{+} = $ const, $v^{+} = $ const, we obtain the following exact solution

$$\sigma_{xy}^{(i)} = (b_{11}u^{+} - b_{16}v^{+})/(h\Delta), \quad \sigma_{y}^{(i)} = (b_{66}v^{+} - b_{16}u^{+})/(h\Delta),$$

$$\sigma_x^{(i)} = \left((b_{11}a_{16}^{(i)} - b_{16}a_{12}^{(i)})u^{+} + (b_{66}a_{12}^{(i)} - b_{16}a_{16}^{(i)})v^{+}\right)(h\Delta a_{11}^{(i)})^{-1},$$

$$u = \left((b_{11}d_{66} - b_{16}d_{16})u^{+} - (b_{66}d_{16} - b_{16}d_{66})v^{+}\right)/\Delta, \qquad (10.11)$$

$$v = \left((b_{66}d_{11} - b_{16}d_{16})v^{+} + (b_{16}d_{11} - b_{11}d_{16})u^{+}\right)/\Delta,$$

where

$$b_{jk} = \int_{-1}^{1} A_{jk}d\zeta = \int_{-1}^{\zeta_0} A_{jk}^{(2)}d\zeta + \int_{\zeta_0}^{1} A_{jk}^{(1)}d\zeta, \qquad (10.12)$$

$$d_{jk} = \int_{-1}^{\zeta} A_{jk}d\zeta = \begin{cases} \int_{-1}^{\zeta_0} A_{jk}^{(2)}d\zeta, & -1 \leq \zeta \leq \zeta_0 \\ \int_{-1}^{\zeta_0} A_{jk}^{(2)}d\zeta + \int_{\zeta_0}^{\zeta} A_{jk}^{(1)}d\zeta, & \zeta_0 \leq \zeta \leq 1. \end{cases}$$

The closed form of solution may be obtained also for the case when u^{+}, and v^{+} are polynomials.

Consider now several special cases which are often met in practical applications. Let the layers be isotropic, and

$$E^{(1)} = E_1 = \text{const}, \quad \nu^{(1)} = \nu_1 \approx \text{const} \tag{10.13}$$

$$E^{(2)} = E_2 \exp\left(-m(\zeta - \zeta_0)\right), \quad \nu^{(2)} = \nu_2 \approx \text{const}$$

then the solution (10.11) takes the form

$$\sigma_{xy}^{(i)} = u^+ E_2 / (2(1 + \nu_2)hF(n_1)),$$

$$\sigma_y^{(i)} = \frac{v^+ E_2}{(1 - \nu_2^2)hF(n_2)}, \quad \sigma_x^{(i)} = \frac{\nu_i v^+ E_2}{(1 - \nu_2^2)hF(n_2)},$$

$$u^{(1)} = \frac{u^+}{F(n_1)}\left(m_1 + (\zeta - \zeta_0)n_1\right), \quad v^{(1)} = \frac{v^+}{F(n_2)}\left(m_1 + (\zeta - \zeta_0)n_2\right),$$

$$u^{(2)} = \frac{u^+}{mF(n_1)}\left(\exp m(\zeta - \zeta_0) - \exp(-m(1 + \zeta_0))\right), \tag{10.14}$$

$$v^{(2)} = \frac{v^+}{mF(n_2)}\left(\exp m(\zeta - \zeta_0) - \exp(-m(1 + \zeta_0))\right),$$

$$n_1 = \frac{1 + \nu_1}{1 + \nu_2}\frac{E_2}{E_1}, \quad n_2 = \frac{1 - \nu_1}{1 - \nu_2}n_1, \quad m_1 = \left(1 - \exp(-m(1 + \zeta_0))\right)m^{-1},$$

$$F(n_i) = m_1 + (1 - \zeta_0)n_i.$$

The analysis of the obtained solution depending on the value of parameter m reveals that if $m > 0$, then the Young's modulus $E^{(2)}$ of the second layer increases with increase of depth counting from the contact area whereas the case of $m < 0$ demonstrate opposite behavior. At $m = 0$ it is possible to show in the limit that the solution reduces to that of constant elastic modules, namely

$$\sigma_{xy}^{(i)} = u^+ \left/ \left(\frac{2(1 + \nu_1)}{E_1}h_1 + \frac{2(1 + \nu_2)}{E_2}h_2\right)\right.,$$

$$\sigma_y^{(i)} = v^+ \left/ \left(\frac{1 - \nu_1^2}{E_1}h_1 + \frac{1 - \nu_2^2}{E_2}h_2\right)\right., \quad \sigma_x^{(i)} = \nu_i \sigma_y^{(i)}, \tag{10.15}$$

$$u^{(1)} = u^+ \left(\frac{1 + \nu_1}{E_1}\left(y - \frac{h_2 - h_1}{2}\right) + \frac{1 + \nu_2}{E_2}h_2\right) \left/ \left(\frac{1 + \nu_1}{E_1}h_1 + \frac{1 + \nu_2}{E_2}h_2\right)\right.,$$

$$v^{(1)} = v^+ \left(\frac{1 - \nu_1^2}{E_1}\left(y - \frac{h_2 - h_1}{2}\right) + \frac{1 - \nu_2^2}{E_2}h_2\right) \left/ \left(\frac{1 - \nu_1^2}{E_1}h_1 + \frac{1 - \nu_2^2}{E_2}h_2\right)\right.,$$

$$u^{(2)} = u^+ (y + h)\frac{1 + \nu_2}{E_2} \left/ \left(\frac{1 + \nu_1}{E_1}h_1 + \frac{1 + \nu_2}{E_2}h_2\right)\right.,$$

$$v^{(2)} = v^+ (y + h)\frac{1 - \nu_2^2}{E_2} \left/ \left(\frac{1 - \nu_1^2}{E_1}h_1 + \frac{1 - \nu_2^2}{E_2}h_2\right)\right..$$

When $m \gg 1$ it follows from (10.14) that

$$\sigma_{xy}^{(i)} = \frac{u^+ E_1}{2h_1(1 + \nu_1)}, \quad \sigma_y^{(i)} = \frac{v^+}{h_1}\frac{E_1}{(1 - \nu_1^2)}, \quad \sigma_x^{(i)} = \nu_i \sigma_y^{(i)},$$

$$u^{(1)} = \frac{u^+}{h_1}\left(y - \frac{h_2 - h_1}{2}\right), \quad v^{(1)} = \frac{v^+}{h_1}\left(y - \frac{h_2 - h_1}{2}\right), \tag{10.16}$$

$$u^{(2)} \approx 0, \quad v^{(2)} \approx 0.$$

The last formulae has a clear physical interpretation: the displacements of the second layer decrease as its stiffness increases. At the same time the displacements of the first layer demonstrate linear variation along thickness, becoming infinitesimal at the contact area between the layers due to high stiffness of the second layer. If $m \ll -1$ then

$$\sigma_{xy}^{(i)} \approx 0, \quad \sigma_y^{(i)} \approx 0, \quad \sigma_x^{(i)} \approx 0 \tag{10.17}$$

$$u^{(i)} = u^+, \quad v^{(i)} = v^+, \quad (\zeta \neq -1).$$

If the stiffness of the second layer is small, then as may be observed from (10.17), the stresses are small, in other words, the foundation is no longer resisting.

Consider now the case of piecewise linear variation of elastic modulus. Let

$$E^{(1)} = E_1 = \text{const}, \qquad\qquad \nu^{(1)} = \nu_1 \approx \text{const}, \tag{10.18}$$

$$E^{(2)} = (E_2 + E_3\zeta_0 + (E_2 - E_3)\zeta)/(1 + \zeta_0), \quad \nu^{(2)} = \nu_2 \approx \text{const},$$

i.e. the elastic modulus of the second layer $E^{(2)}$ changes linearly, taking its values within $[E_2; E_3]$. Taking into account (10.18), we deduce from (10.11) that the solution corresponding to conditions (10.1), (10.2) is given by

$$\sigma_{xy}^{(i)} = \frac{u^+ E_2}{2h(1+\nu_2)D(n_1)}, \quad \sigma_y^{(i)} = \frac{v^+ E_2}{(1-\nu_2^2)hD(n_2)}, \quad \sigma_x^{(i)} = \nu_i \sigma_y^{(i)},$$

$$u^{(1)} = \frac{u^+}{D(n_1)} \left((1+\zeta_0)f(c) + n_1(\zeta - \zeta_0)\right), \tag{10.19}$$

$$v^{(1)} = \frac{v^+}{D(n_2)} \left((1+\zeta_0)f(c) + n_2(\zeta - \zeta_0)\right),$$

$$u^{(2)} = \left(u^+(1+\zeta_0)\ln(E^{(2)}/E_3)\right) / \left((1-c)D(n_1)\right),$$

$$v^{(2)} = \left(v^+(1+\zeta_0)\ln(E^{(2)}/E_3)\right) / \left((1-c)D(n_2)\right),$$

$$c = E_3/E_2, \quad f(c) = \frac{\ln c}{c-1}, \quad D(n_i) = (1+\zeta_0)f(c) + n_i(1-\zeta_0),$$

$$n_1 = \frac{1+\nu_1}{1+\nu_2}\frac{E_2}{E_1}, \quad n_2 = \frac{1-\nu_1}{1-\nu_2}n_1.$$

It may be shown that at $c = 1$ the obtained solution (10.19) coincides with (10.15). When $c \gg 1$, i.e. $E_3 \gg E_2$ we re-establish (10.16), however the asymptotic results (10.16) are achieved faster. Finally, if $c \ll 1$, we result in (10.17).

Let us now present the results for the boundary conditions (10.3). Satisfying these, we obtain

$$\sigma_{xy0}^{(s)} = \sigma_{xy}^{+(s)} - \sigma_{xy}^{*(s)}(\xi, 1),$$

$$\sigma_{y0}^{(s)} = \frac{1}{b_{11}} \left(v^{+(s)} - v^{-(s)} - v^{*(s)}(\xi, 1) + v^{*(s)}(\xi, -1) - b_{16}\sigma_{xy0}^{(s)}\right), \tag{10.20}$$

$$u^{(s)}(\xi) = u^{-(s)} - u^{*(s)}(\xi, -1) + c_{66}\sigma_{xy0}^{(s)} + c_{16}\sigma_{y0}^{(s)},$$

$$v^{(s)}(\xi) = v^{-(s)} - v^{*(s)}(\xi, -1) + c_{16}\sigma_{xy0}^{(s)} + c_{11}\sigma_{y0}^{(s)}.$$

Substitution of the latter into (10.7) gives the resulting solution. In particular, when $v^+ = \mathrm{const}$, $\quad \varepsilon^{-1}\sigma_{xy}^+ = \tau^+ = \mathrm{const}$, we have

$$\sigma_{xy} = \tau^+, \quad \sigma_x = -\frac{a_{12}}{a_{11}}\sigma_y, \quad \sigma_y = \frac{1}{b_{11}}\left(\frac{v^+}{h} - b_{16}\tau^+\right),$$

$$u = \frac{1}{b_{11}}\left(\mathrm{d}_{16}v^+ + (b_{11}d_{66} - b_{16}d_{16})\tau^+h\right), \qquad (10.21)$$

$$v = \frac{1}{b_{11}}\left(\mathrm{d}_{11}v^+ + (b_{11}d_{16} - b_{16}d_{11})\tau^+h\right).$$

In case of a variable elastic modulus specified in (10.13), the solution is written as

$$\sigma_{xy}^{(i)} = \tau^+, \quad \sigma_y^{(i)} = \frac{v^+E_2}{(1 - \nu_2^2)hF(n_2)}, \quad \sigma_x^{(i)} = \nu_i\sigma_y^{(i)},$$

$$u^{(1)} = \frac{2(1 + \nu_2)}{E_1}\left(m_1 + (\zeta - \zeta_0)n_1\right)\tau^+h,$$

$$v^{(1)} = \frac{v^+}{F(n_2)}\left(m_1 + (\zeta - \zeta_0)n_2\right), \qquad (10.22)$$

$$u^{(2)} = \frac{2(1 + \nu_2)\tau^+h}{mE_2}\left(\exp m(\zeta - \zeta_0) - \exp(-m(1 + \zeta_0))\right),$$

$$v^{(2)} = \frac{v^+}{mF(n_2)}\left(\exp m(\zeta - \zeta_0) - \exp(-m(1 + \zeta_0))\right).$$

In case of (10.18) we get

$$\sigma_{xy}^{(i)} = \tau^+, \quad \sigma_y^{(i)} = \frac{v^+E_2}{(1 - \nu_2^2)hD(n_2)}, \quad \sigma_x^{(i)} = \nu_i\sigma_y^{(i)},$$

$$u^{(1)} = \left(\frac{2(1 + \nu_2)}{E_2}(1 + \zeta_0)f(c) + \frac{2(1 + \nu_1)}{E_1}(\zeta - \zeta_0)\right)\tau^+h, \qquad (10.23)$$

$$v^{(1)} = \frac{v^+}{D(n_2)}\left((1 + \zeta_0)f(c) + n_2(\zeta - \zeta_0)\right),$$

$$u^{(2)} = \frac{2(1 + \nu_2)}{E_2 - E_3}(1 + \zeta_0)\tau^+h\ln\left(E^{(2)}/E_3\right),$$

$$v^{(2)} = \frac{1}{(1 - c)D(n_2)}(1 + \zeta_0)v^+\ln\left(E^{(2)}/E_3\right).$$

Let us now discuss briefly the case of the boundary conditions (10.4). First we have

$$\sigma_{y0}^{(s)} = \sigma_y^{+(s)} - \sigma_y^{*(s)}(\xi, 1),$$

$$\sigma_{xy0}^{(s)} = \frac{1}{b_{66}}\left(u^{+(s)} - u^{-(s)} - u^{*(s)}(\xi, 1) + u^{*(s)}(\xi, -1) - b_{16}\sigma_{y0}^{(s)}\right),$$

$$u^{(s)}(\xi) = u^{+(s)} - \sigma_{y0}^{(s)}\int_0^1 A_{16}d\zeta - \sigma_{xy0}^{(s)}\int_0^1 A_{66}d\zeta - u^{*(s)}(\xi, 1), \qquad (10.24)$$

$$v^{(s)}(\xi) = v^{-(s)} + c_{11}\sigma_{y0}^{(s)} + c_{16}\sigma_{xy0}^{(s)} - v^{*(s)}(\xi, -1).$$

Then, substituting (10.24) into (10.6)-(10.7), we obtain the resulting solution. If $u^- = v^- \equiv 0$, $\varepsilon^{-1}\sigma_y^+ = \sigma_2^+ = \text{const}$, and $u^+ = \text{const}$ we get

$$\sigma_y = \sigma_2^+, \quad \sigma_{xy} = \frac{1}{b_{66}}\left(\frac{u^+}{h} - b_{16}\sigma_2^+\right),$$

$$\sigma_x = -\frac{1}{a_{11}b_{66}}\left(a_{16}\frac{u^+}{h} + (a_{12}b_{66} - a_{16}b_{16})\sigma_2^+\right), \tag{10.25}$$

$$u = \frac{1}{b_{66}}\left(d_{66}u^+ + (b_{66}d_{16} - b_{16}d_{66})\sigma_2^+ h\right),$$

$$v = \frac{1}{b_{66}}\left(d_{16}u^+ + (b_{66}d_{11} - b_{16}d_{16})\sigma_2^+ h\right).$$

In case of the boundary conditions (10.5)

$$\sigma_{xy0}^{(s)} = \sigma_{xy}^{+(s)} - \sigma_{xy}^{*(s)}(\xi,1), \quad \sigma_{y0}^{(s)} = \sigma_y^{+(s)} - \sigma_y^{*(s)}(\xi,1),$$

$$u^{(s)}(\xi) = u^{-(s)} + c_{16}\sigma_{y0}^{(s)} + c_{66}\sigma_{xy0}^{(s)} - u^{*(s)}(\xi,-1), \tag{10.26}$$

$$v^{(s)}(\xi) = v^{-(s)} + c_{11}\sigma_{y0}^{(s)} + c_{16}\sigma_{xy0}^{(s)} - v^{*(s)}(\xi,-1).$$

The closed form solution corresponding to $\varepsilon^{-1}\sigma_{xy}^+ = \tau^+ = \text{const}$, $\varepsilon^{-1}\sigma_y^+ = \sigma_2^+ = \text{const}$, $u^- = v^- \equiv 0$ takes the form

$$\sigma_{xy}^{(i)} = \tau^+, \quad \sigma_y^{(i)} = \sigma_2^+, \quad \sigma_x^{(i)} = -\left(a_{12}^{(i)}\sigma_2^+ + a_{16}^{(i)}\tau^+\right)/a_{11}^{(i)}, \tag{10.27}$$

$$u = (d_{16}\sigma_2^+ + d_{66}\tau^+)h, \quad v = (d_{11}\sigma_2^+ + d_{16}\tau^+)h.$$

In case of the variable elastic modulus (10.13) the solution (10.27) may be represented as

$$\sigma_{xy}^{(i)} = \tau^+, \quad \sigma_y^{(i)} = \sigma_2^+, \quad \sigma_x^{(i)} = \nu_i\sigma_2^{(i)}, \quad i = 1,2,$$

$$u^{(1)} = \frac{2(1+\nu_2)}{E_2}(m_1 + (\zeta - \zeta_0)n_1)\tau^+ h,$$

$$v^{(1)} = \frac{1-\nu_2^2}{E_2}(m_1 + (\zeta - \zeta_0)n_2)\sigma_2^+ h, \tag{10.28}$$

$$u^{(2)} = \frac{2(1+\nu_2)}{mE_2}(\exp m(\zeta - \zeta_0) - \exp(-m(1+\zeta_0)))\tau^+ h,$$

$$v^{(2)} = \frac{1-\nu_2^2}{mE_2}(\exp m(\zeta - \zeta_0) - \exp(-m(1+\zeta_0)))\sigma_2^+ h.$$

In the particular case of constant elastic modules ($m = 0$) the solution reduces to

$$\sigma_{xy}^{(i)} = \tau^+, \quad \sigma_y^{(i)} = \sigma_2^+, \quad \sigma_x^{(i)} = \nu_i\sigma_2^+, \quad i = 1,2,$$

$$u^{(1)} = \left(\frac{2(1+\nu_1)}{E_1}(\zeta - \zeta_0) + \frac{2(1+\nu_2)}{E_2}(1+\zeta_0)\right)\tau^+ h,$$

$$v^{(1)} = \left(\frac{1-\nu_1^2}{E_1}(\zeta - \zeta_0) + \frac{1-\nu_2^2}{E_2}(1+\zeta_0)\right)\sigma_2^+ h, \tag{10.29}$$

$$u^{(2)} = \frac{2(1+\nu_2)}{E_2}(1+\zeta)\tau^+ h, \quad v^{(2)} = \frac{1-\nu_2^2}{E_2}(1+\zeta)\sigma_2^+ h.$$

Analysis of (10.28) in a particular case, when elastic modulus of the second layer is increasing with depth and $m \gg 1$, gives

$$\sigma_{xy}^{(i)} = \tau^+, \quad \sigma_y^{(i)} = \sigma_2^+, \quad \sigma_x^{(i)} = \nu_i \sigma_2^+, \quad i = 1, 2,$$

$$u^{(1)} = \frac{2(1 + \nu_1)}{E_1}(\zeta - \zeta_0)\tau^+ h, \tag{10.30}$$

$$v^{(1)} = \frac{1 - \nu_1^2}{E_1}(\zeta - \zeta_0)\sigma_2^+ h,$$

$$u^{(2)} \approx 0, \quad v^{(2)} \approx 0.$$

The formulae (10.30) also confirm that the amplitude of displacements decay with the increase of the stiffness of the foundation. Using (10.28), it is possible to find the depth (the thickness of the active zone), starting from which the foundation could be effectively treated as incompressible.

When $m \ll -1$, it follows from (10.28) that

$$\sigma_{xy}^{(i)} = \tau^+, \quad \sigma_y^{(i)} = \sigma_2^+, \quad \sigma_x^{(i)} = \nu_i \sigma_2^+, \tag{10.31}$$

$$u^{(i)} \gg 1, \quad v^{(i)} \gg 1 \quad (\zeta \neq -1)$$

indicating clearly that the resistance of the foundation decreases with the decrease in stiffness, causing increase in the amplitudes of displacements.

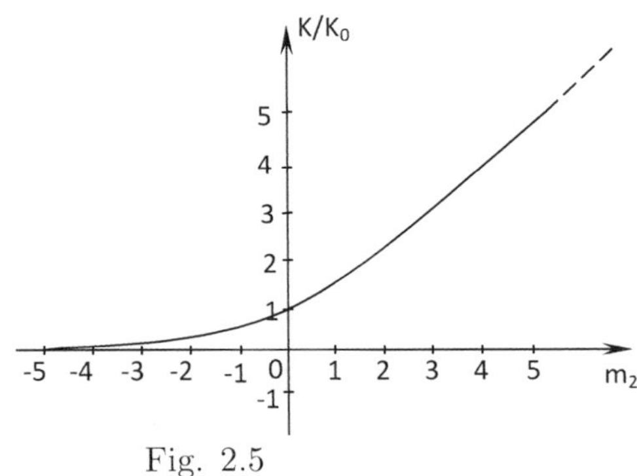

Fig. 2.5

In case of the variable elastic modulus defined by (10.18), the solution (10.27) takes the form

$$\sigma_{xy}^{(i)} = \tau^+, \quad \sigma_y^{(i)} = \sigma_2^+, \quad \sigma_x^{(i)} = \nu_i \sigma_2^+,$$

$$u^{(1)} = \left(\frac{2(1 + \nu_1)}{E_1}(\zeta - \zeta_0) + \frac{2(1 + \nu_2)}{E_2}(1 + \zeta_0)f(c) \right) \tau^+ h,$$

$$v^{(1)} = \left(\frac{1 - \nu_1^2}{E_1}(\zeta - \zeta_0) + \frac{1 - \nu_2^2}{E_2}(1 + \zeta_0)f(c) \right) \sigma_2^+ h, \tag{10.32}$$

$$u^{(2)} = \frac{2(1 + \nu_2)}{E_2 - E_3}(1 + \zeta_0)\tau^+ h \ln\left(E^{(2)}/E_3 \right),$$

$$v^{(2)} = \frac{1 - \nu_2^2}{E_2 - E_3}(1 + \zeta_0)\sigma_2^+ h \ln\left(E^{(2)}/E_3 \right).$$

Solutions (10.27)-(10.32) could be further analyzed in respect of the validity of the Winkler-Fuss hypothesis and to clarify the issue of the modulus of the foundation in case of the variable elastic modulus. Provided that the elastic modulus has exponential dependence on the transverse coordinate, it follows from (10.28) that (Aghalovyan and Adamyan, 1987)

$$\sigma_y^{(c)} = Kv^{(c)}, \quad K = K_0 \frac{m_2}{1 - \exp(-m_2)} \tag{10.33}$$

with

$$K_0 = \frac{E_2}{(1 - \nu_2^2)h_2}, \quad m_2 = \frac{mh_2}{h} \tag{10.34}$$

where K_0 coincides with the modulus of subgrade reaction for a foundation with constant modulus E_2, and m_2 characterizes the variation of the elastic modulus with thickness. Therefore, in view of (10.13) the variable modulus $E^{(2)}$ may be written as

$$E^{(2)} = E_2 \exp\left(-m_2(y - y_0)/h_2\right), \quad -h_2 \leq y - y_0 \leq 0 \tag{10.35}$$

implying the range of the modulus $E^{(2)}$ as $[E_2; E_2 \exp m_2]$.

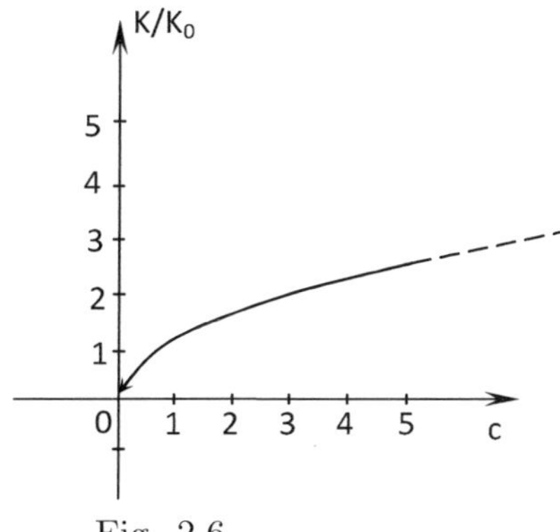

Fig. 2.6

It follows from (10.33) that in case of the variable elastic modulus the contact pressure is still proportional to the corresponding normal displacements, however, the modified modulus of subgrade reaction K depends crucially on the value of m_2. The ratio K/K_0 increases with the growth of m_2 and decreases to zero as $m_2 \rightarrow -\infty$, see Fig. 2.5.

If the elastic modulus varies as (10.18), it follows once again from (10.32) that $\sigma_y^{(c)} = Kv^{(c)}$, however, the modified modulus of subgrade reaction is now given by

$$K = K_0 \frac{c - 1}{\ln c}, \quad c = E_3/E_2. \tag{10.36}$$

The dependence of K/K_0 on parameter c is illustrated in Fig. 2.6.

Concluding this section, we remark that similarly to homogeneous foundation, in case of non-uniform normal loading the contact points will cause additional tangential displacements at the contact area, with tangential stresses arising, however, the leading order terms are still the Winkler terms.

Chapter 3

Direct Asymptotic Integration of 3D Elasticity Equations for Orthotropic Plates

3.1 Governing Equations

Before stating the governing equations and formulating and solving the boundary value problems by asymptotic method, let us first describe briefly the history of development of theories for elastic plates and shells.

The theories of plates and shells may actually be thought of as a part of a more general problem, namely, development of models in continuum mechanics (Sedov, 1965). The main goal of any plate or shell theory is reduction of the 3D boundary value problem of elasticity to a relatively simpler 2D approximate statement. The existing approaches could be generally classified into three groups including the following: a) the hypotheses methods, b) methods involving expansions along thickness, and c) asymptotic methods, which could all be in their turn separated into subgroups. The first two groups of methods are originating historically from the beginning of theories of plates and shells, whereas the asymptotic method has been developed rapidly since 1960-s.

The first significant contributions to plate theories were made by Cauchy and Poisson by means of expansions along thickness. However, a well-known contradiction between the order of four of the governing equation and the three boundary conditions derived by Poisson, has stopped further theoretical development for nearly half of a century. The method of hypotheses emerged from the famous work of Kirchhoff, see e.g. [Timoshenko (1957)]. There are numerous publications using the hypotheses methods for both isotropic and anisotropic plates and shells, including Ambartsumyan (1961, 1967, 1974); Vlasov (1949); Goldenveizer (1961); Goldenveizer et al. (1979); Darevskii (1961); Korolev (1965); Lekhnitskii (1968); Lourier (1947); Mushtari and Galimov (1957); Novozhilov (1962); Obraztsov and Onanov (1973); Ogibalov (1958); Sapondjyan (1975); Sarkisyan (1976); Filin (1987); Chernykh (1962); Green and Zerna (1954); Ciarlet (1997); Kienzler (2002); Pietraszkiewicz, Chroscielewski and Makowski (2005); Altenbach and Meenen (2008); Podio-Guidugli (2008), and many others. It should be noted that before 1940s the absolute majority of contributions in this area were relying on the Kirchhoff-Love hypothesis, whereas later some more approaches has been suggested, motivated by the observed cases of insufficient accuracy of calculations performed on the base of the Kirchhoff-Love theory, in particular in anisotropic, layered plates and shells, dynamic problems, analysis of concentration of stresses, etc. Over a rather short period of time there appeared a number of refined theories basing on improved hypotheses, including the theories of Reissner (1944, 1945), [Ambartsumyan (1958, 1961, 1967, 1974)] and various Timoshenko-type theories, see e.g. Ainola (1965, 1967); Galimov (1976); Grigorenko and Vasilenko (1981); Paimushin (1978, 1980); Pelekh (1973); Naghdi (1956).

In case of layered plates and shells there have been even more approaches, involving almost all appropriate geometrical and physical hypotheses, along with various mathematical approximations, see e.g. Ambartsumyan (1961, 1967); Bolotin (1963); Bolotin and Novichkov (1980); Grigolyuk (1957, 1958); Grigolyuk, Chulkov (1965); Chulkov and Ivanov (1969); Grigorenko (1973); Guz' (1971); Guz' and Nemish (1982); Lekhnitskii (1968); Andreev and Nemirovskii (1977); Nemirovskii (1970).

The method of expansions along the transverse coordinate γ has been developed by Kilchevskii (1939, 1962, 1963), contributing significantly to formulation of consistent boundary conditions. Another modification of the expansion method using the principle of possible displacements and variational techniques has been suggested by Mushtari (1947); Mushtari and Galimov (1957); Mushtari and Teregulov (1959); Teregulov (1962). The variational techniques have also been used for derivation of approximate theories for plates and shells (Berdichevsky, 1972, 1983), and for solving boundary value problems for shells, see Abovsky et al. (1978). Some estimates of accuracy of classical theories of plates and shells have been obtained through energy approach, see e.g. Morgenstern (1959a,b); Babuska and Prager (1960); Koiter (1970, 1971); Shoykhet (1973).

One more approach within the method of expansion is a representation of the unknown quantities through the system of special functions of the transverse coordinates, in particular the Legendre polynomials. This approach has been developed by Vekua (1955, 1982) for isotropic shells, and later generalized to anisotropic plates and shells (Ponyatovsky, 1964, 1965; Khoma, 1975). Among the technical difficulties associated with the expansion method we mention an increase of order of the governing approximate equations with the order of the approximations, which is rather important, as according to I.N. Vekua, "sometimes the overwhelming complexity of the mathematical theory may totally devalue its practical applicability".

Another way of reduction of the 3D boundary value problem of elasticity to a 2D one is based on the Lourier symbolic method (Lourier, 1942, 1955). The method enables one to determine a wide class of particular solutions satisfying the non-homogeneous face boundary conditions along with the so-called homogeneous solutions satisfying the traction-free conditions on the faces. Combining these solutions, it is usually possible to satisfy the edge boundary conditions to a certain accuracy. The symbolic Lourier method has been widely used, see e.g. Lekhnitskii (1962); Khachatryan T. (1963); Nigul (1963); Prokopov (1965); Prokopov and

Gruzdev (1968). Some further developments of the symbolic method were made by I.I. Vorovich and his group.

Reduction of the 3D problem of elasticity to the associated 2D problem for plates and shells may also be achieved through the method of initial functions, within which the sought-for displacements and stresses are represented in terms of these on the initial surface. The problem is then reformulated by means of the 6 unknown 2D functions, for more details see Volkov (1971); Vlasov (1975).

Some rather important results in the theory of plates and shells have also been obtained through the asymptotic method. There are two major research strands which could be pointed out within this general approach. The first direction is associated with direct asymptotic integration of the 3D equations of elasticity, see Aghalovyan (1966a,b,c, 1972a, 1973a); Goldenveizer (1959, 1960, 1961, 1962, 1963, 1969, 1976); Goldenveizer and Kolos (1965); Gusein-Zade (1966, 1974, 1978); Kolos (1964); Rogacheva (1974, 1975); Chernishev (1966, 1970); Novotn (1970, 1972); Friedrichs (1950, 1955); Friedrichs and Dressler (1961); Green (1962a,b); Rutten (1971); Widera (1969); Finikova and Stolyar (2011).

The asymptotic method proved to be an efficient approach for statics and free vibrations of shells within the framework of the classical 2D shell theory, see e.g. Goldenveizer (1961, 1976); Goldenveizer et al. (1979); Tovstik (1975). It has also been generalized to dynamical problems of plates and shells in Kaplunov et al. (1998).

The second major research direction associated with the asymptotic method, has been developed by I.I. Vorovich and his group, see Aksentyan and Vorovich (1963); Vorovich and Shlenev (1963); Vorovich (1966a); Vorovich and Bazarenko (1965); Vorovich and Vilenskaia (1966); Vorovich and Kadomtsev (1970); Vorovich (1975); Romenskaya and Shlenev (1973); Ustinov and Iudovich (1973); Ustinov (1974, 1976).

Here we restrict ourselves to discussing analytical approaches only, not touching any of the vast number of contributions through numerical schemes.

Some more general reviews on the plate and shell theories may be found in Alumyae (1972); Ambartsumyan (1964, 1968); Vorovich (1966b, 1975); Vorovich and Shlenev (1963); Galimov and Surkin (1967); Galin'sh (1967); Goldenveizer (1968a); Grigolyuk and Kogan (1972); Grigolyuk and Selezov (1973); Filin (1987); Koiter (1971); Koiter and Simmonds (1973); Thai and Choi (2013).

In this chapter we will illustrate the efficiency of the asymptotic method for 3D problems of elasticity for anisotropic plates. Among the advantages of the method we mention: construction of an applied 2D theory describing the stress and strain components both in the near-edge vicinity and the interior; estimation of accuracy of the Kirchhoff plate theory along with the refined theories relying on the geometrical and physical assumptions; investigate fully the effect of anisotropy; study the behavior in the near-edge zones and determine the attenuation of the boundary layers, etc.

Consider an orthotropic plate of thickness $2h$, which is relatively small compared to the other two typical scales of the plate. Let the mid-plane coincide with the coordinate plane (α, β) of the orthogonal curvilinear system, with the third axis γ directed normally to the mid-plane. Let us also assume that the principal directions of anisotropy coincide with the appropriate coordinate axes. Then the equilibrium equations are written as

$$H_\alpha \frac{\partial \sigma_{\alpha\alpha}}{\partial \alpha} + H_\beta \frac{\partial \sigma_{\alpha\beta}}{\partial \beta} + \frac{\partial \sigma_{\alpha\gamma}}{\partial \gamma} - (\sigma_{\alpha\alpha} - \sigma_{\beta\beta}) \frac{H_\alpha}{H_\beta} \frac{\partial H_\beta}{\partial \alpha} - 2 \frac{H_\beta}{H_\alpha} \frac{\partial H_\alpha}{\partial \beta} \sigma_{\alpha\beta} = 0, \quad (\alpha, \beta)$$

$$H_\alpha \frac{\partial \sigma_{\alpha\gamma}}{\partial \alpha} + H_\beta \frac{\partial \sigma_{\beta\gamma}}{\partial \beta} + \frac{\partial \sigma_{\gamma\gamma}}{\partial \gamma} - \frac{H_\alpha}{H_\beta} \frac{\partial H_\beta}{\partial \alpha} \sigma_{\alpha\gamma} - \frac{H_\beta}{H_\alpha} \frac{\partial H_\alpha}{\partial \beta} \sigma_{\beta\gamma} = 0. \tag{1.1}$$

The constitutive relations take the form

$$H_\alpha \frac{\partial u_\alpha}{\partial \alpha} - \frac{H_\beta}{H_\alpha} \frac{\partial H_\alpha}{\partial \beta} u_\beta = a_{11}\sigma_{\alpha\alpha} + a_{12}\sigma_{\beta\beta} + a_{13}\sigma_{\gamma\gamma},$$

$$H_\beta \frac{\partial u_\beta}{\partial \beta} - \frac{H_\alpha}{H_\beta} \frac{\partial H_\beta}{\partial \alpha} u_\alpha = a_{12}\sigma_{\alpha\alpha} + a_{22}\sigma_{\beta\beta} + a_{23}\sigma_{\gamma\gamma},$$

$$\frac{\partial w}{\partial \gamma} = a_{13}\sigma_{\alpha\alpha} + a_{23}\sigma_{\beta\beta} + a_{33}\sigma_{\gamma\gamma}, \tag{1.2}$$

$$H_\beta \frac{\partial u_\alpha}{\partial \beta} + H_\alpha \frac{\partial u_\beta}{\partial \alpha} + \frac{H_\beta}{H_\alpha} \frac{\partial H_\alpha}{\partial \beta} u_\alpha + \frac{H_\alpha}{H_\beta} \frac{\partial H_\beta}{\partial \alpha} u_\beta = a_{66}\sigma_{\alpha\beta},$$

$$H_\alpha \frac{\partial w}{\partial \alpha} + \frac{\partial u_\alpha}{\partial \gamma} = a_{55}\sigma_{\alpha\gamma}, \quad H_\beta \frac{\partial w}{\partial \beta} + \frac{\partial u_\beta}{\partial \gamma} = a_{44}\sigma_{\beta\gamma},$$

where σ_{ik} are the stress tensor components, u_α, u_β, w are the displacement components, H_α, H_β are the Lamè parameters and a_{ik} are elastic constants. Here and below we use (α, β) to denote that there is also a second relation which may be obtained by changing α to β and vice versa.

The conditions on the face planes $\gamma = \pm h$ are given by

$$\sigma_{\gamma\gamma} = Z^+, \quad \sigma_{\alpha\gamma} = \varepsilon^{-1}X^+, \quad \sigma_{\beta\gamma} = \varepsilon^{-1}Y^+ \quad \text{at} \quad \gamma = h$$

$$\sigma_{\gamma\gamma} = Z^-, \quad \sigma_{\alpha\gamma} = \varepsilon^{-1}X^-, \quad \sigma_{\beta\gamma} = \varepsilon^{-1}Y^- \quad \text{at} \quad \gamma = -h \tag{1.3}$$

where $\varepsilon = h/a$ is a small parameter, a is the typical scale of the plate, $X^\pm(\alpha, \beta)$, $Y^\pm(\alpha, \beta)$ and $Z^\pm(\alpha, \beta)$ are the tangential and normal components of surface loading, respectively.

The solutions of the boundary value problem (1.1) - (1.3) are sought subject to the conditions on the edges of the plate, which will be discussed later. The solution may be obtained as a sum of these for symmetric and antisymmetric problems (Lourier, 1955), which are formulated through (1.1), (1.2) and the following boundary conditions

$$\sigma_{\alpha\gamma} = \pm \frac{1}{2}\varepsilon^{-1}X_1, \quad \sigma_{\beta\gamma} = \pm \frac{1}{2}\varepsilon^{-1}Y_1, \quad \sigma_{\gamma\gamma} = \frac{1}{2}Z_1 \quad \text{at} \quad \gamma = \pm h \tag{1.4}$$

and

$$\sigma_{\alpha\gamma} = \frac{1}{2}\varepsilon^{-1}X_2, \quad \sigma_{\beta\gamma} = \frac{1}{2}\varepsilon^{-1}Y_2, \quad \sigma_{\gamma\gamma} = \pm \frac{1}{2}Z_2 \quad \text{at} \quad \gamma = \pm h \tag{1.5}$$

respectively, with

$$X_1 = X^+ + X^-, \quad (X,Y), \quad Z_1 = Z^+ - Z^-,$$
$$X_2 = X^+ - X^-, \quad (X,Y), \quad Z_2 = Z^+ + Z^-. \tag{1.6}$$

We remark that in case of the antisymmetric (bending) problem the quantities $\sigma_{\alpha\alpha}, \sigma_{\alpha\beta}, \sigma_{\beta\beta}, \sigma_{\gamma\gamma}, u_\alpha, u_\beta$ are odd functions whereas $\sigma_{\alpha\gamma}, \sigma_{\beta\gamma}, w$ are even, whereas in case of the symmetric (extension) problem the situation is opposite.

3.2 Outer Solution

Here we study the outer solution over the interior of the plate, which is characterized by rapid variation of displacements and stresses along the normal coordinate γ compared to these along the in-plane coordinates α and β. Let us scale the variables as

$$\alpha = a\xi, \quad \beta = a\eta, \quad \gamma = h\zeta. \tag{2.1}$$

Then

$$H_\alpha = aH_\xi, \quad H_\beta = aH_\eta, \quad H_\alpha \frac{\partial}{\partial \alpha} = H_\xi \frac{\partial}{\partial \xi} \quad (\alpha, \beta; \xi, \eta), \tag{2.2}$$

therefore set (1.1), (1.2) may be rewritten as

$$H_\alpha \frac{\partial \sigma_{\alpha\alpha}}{\partial \xi} + H_\beta \frac{\partial \sigma_{\alpha\beta}}{\partial \eta} + \varepsilon^{-1} \frac{\partial \sigma_{\alpha\gamma}}{\partial \zeta}$$
$$-\frac{H_\alpha}{H_\beta} \frac{\partial H_\beta}{\partial \xi} (\sigma_{\alpha\alpha} - \sigma_{\beta\beta}) - 2\frac{H_\beta}{H_\alpha} \frac{\partial H_\alpha}{\partial \eta} \sigma_{\alpha\beta} = 0, \quad (\alpha, \beta),$$

$$H_\alpha \frac{\partial \sigma_{\alpha\gamma}}{\partial \xi} + H_\beta \frac{\partial \sigma_{\beta\gamma}}{\partial \eta} + \varepsilon^{-1} \frac{\partial \sigma_{\gamma\gamma}}{\partial \zeta} - \frac{H_\alpha}{H_\beta} \frac{\partial H_\beta}{\partial \xi} \sigma_{\alpha\gamma} - \frac{H_\beta}{H_\alpha} \frac{\partial H_\alpha}{\partial \eta} \sigma_{\beta\gamma} = 0,$$

$$H_\alpha \frac{\partial u_\alpha}{\partial \xi} - \frac{H_\beta}{H_\alpha} \frac{\partial H_\alpha}{\partial \eta} u_\beta = a(a_{11}\sigma_{\alpha\alpha} + a_{12}\sigma_{\beta\beta} + a_{13}\sigma_{\gamma\gamma}),$$

$$H_\beta \frac{\partial u_\beta}{\partial \eta} - \frac{H_\alpha}{H_\beta} \frac{\partial H_\beta}{\partial \xi} u_\alpha = a(a_{12}\sigma_{\alpha\alpha} + a_{22}\sigma_{\beta\beta} + a_{23}\sigma_{\gamma\gamma}), \tag{2.3}$$

$$\varepsilon^{-1} \frac{\partial w}{\partial \zeta} = a(a_{13}\sigma_{\alpha\alpha} + a_{23}\sigma_{\beta\beta} + a_{33}\sigma_{\gamma\gamma}),$$

$$H_\alpha \frac{\partial u_\beta}{\partial \xi} + \frac{H_\alpha}{H_\beta} \frac{\partial H_\beta}{\partial \xi} u_\beta + H_\beta \frac{\partial u_\alpha}{\partial \eta} + \frac{H_\beta}{H_\alpha} \frac{\partial H_\alpha}{\partial \eta} u_\alpha = a \cdot a_{66}\sigma_{\alpha\beta},$$

$$\varepsilon^{-1} \frac{\partial u_\alpha}{\partial \zeta} + H_\alpha \frac{\partial w}{\partial \xi} = a \cdot a_{55}\sigma_{\alpha\gamma}, \quad \varepsilon^{-1} \frac{\partial u_\beta}{\partial \zeta} + H_\beta \frac{\partial w}{\partial \eta} = a \cdot a_{44}\sigma_{\beta\gamma}.$$

The obtained system (2.3) is singularly perturbed, degenerating as $\varepsilon \to 0$. However the case of degeneration differs significantly from the classical one, studied by Lusternik and Vishik, where the small parameter is a coefficient of the higher order operator. In our case the small parameter is present only within part of the higher order derivatives, therefore only partial degeneration is observed at $\varepsilon \to 0$, causing emergence of a countable set of boundary layers, implying some further mathematical problems, in particular the problem of basis properties of the boundary layers. This was first noted by Vorovich (1966a, 1975).

The solution of (2.3) is composed of the outer solution and the boundary layer. As previously, the outer solution is sought in the following form, see Aghalovyan (1965, 1972b); Goldenveizer (1962)

$$Q = \varepsilon^{-q} \sum_{s=0}^{S} \varepsilon^s Q^{(s)}, \qquad (2.4)$$

where Q is any of the displacement and stress components, q is a natural number which is found from the condition of a system being consistent, (q differs for various displacements and stresses), determined from substitution of (2.4) into (2.3) and equating the coefficients in powers of ε to zero starting from the lowest power. It may be shown that the consistency of the system for $Q^{(s)}$ requires

$$q = 2 \text{ for } \sigma_{\alpha\alpha}, \sigma_{\alpha\beta}, \sigma_{\beta\beta}, u_\alpha, u_\beta; \quad q = 1 \text{ for } \sigma_{\alpha\gamma}, \sigma_{\beta\gamma} \qquad (2.5)$$

$q = 0$ for $\sigma_{\gamma\gamma}$; $q = 1$ and $q = 3$ for w in the symmetric and antisymmetric cases, respectively. We note that so far no restrictions on the values of elastic coefficients are required since these emerge naturally through the analysis of the outer solution. The resulting system is given by

$$H_\alpha \frac{\partial \sigma_{\alpha\alpha}^{(s)}}{\partial \xi} + H_\beta \frac{\partial \sigma_{\alpha\beta}^{(s)}}{\partial \eta} + \frac{\partial \sigma_{\alpha\gamma}^{(s)}}{\partial \zeta}$$

$$-\frac{H_\alpha}{H_\beta} \frac{\partial H_\beta}{\partial \xi} \left(\sigma_{\alpha\alpha}^{(s)} - \sigma_{\beta\beta}^{(s)} \right) - 2 \frac{H_\beta}{H_\alpha} \frac{\partial H_\alpha}{\partial \eta} \sigma_{\alpha\beta}^{(s)} = 0, \quad (\alpha, \beta),$$

$$H_\alpha \frac{\partial \sigma_{\alpha\gamma}^{(s)}}{\partial \xi} + H_\beta \frac{\partial \sigma_{\beta\gamma}^{(s)}}{\partial \eta} + \frac{\partial \sigma_{\gamma\gamma}^{(s)}}{\partial \zeta} - \frac{H_\alpha}{H_\beta} \frac{\partial H_\beta}{\partial \xi} \sigma_{\alpha\gamma}^{(s)} - \frac{H_\beta}{H_\alpha} \frac{\partial H_\alpha}{\partial \eta} \sigma_{\beta\gamma}^{(s)} = 0,$$

$$\frac{1}{a} \left(H_\alpha \frac{\partial u_\alpha^{(s)}}{\partial \xi} - \frac{H_\beta}{H_\alpha} \frac{\partial H_\alpha}{\partial \eta} u_\beta^{(s)} \right) = a_{11} \sigma_{\alpha\alpha}^{(s)} + a_{12} \sigma_{\beta\beta}^{(s)} + a_{13} \sigma_{\gamma\gamma}^{(s-2)},$$

$$\frac{1}{a} \left(H_\beta \frac{\partial u_\beta^{(s)}}{\partial \eta} - \frac{H_\alpha}{H_\beta} \frac{\partial H_\beta}{\partial \xi} u_\alpha^{(s)} \right) = a_{12} \sigma_{\alpha\alpha}^{(s)} + a_{22} \sigma_{\beta\beta}^{(s)} + a_{23} \sigma_{\gamma\gamma}^{(s-2)},$$

$$\frac{1}{a} \frac{\partial w^{(s)}}{\partial \zeta} = a_{13} \sigma_{\alpha\alpha}^{(m)} + a_{23} \sigma_{\beta\beta}^{(m)} + a_{33} \sigma_{\gamma\gamma}^{(n)}, \qquad (2.6)$$

$$H_\alpha \frac{\partial u_\beta^{(s)}}{\partial \xi} + H_\beta \frac{\partial u_\alpha^{(s)}}{\partial \eta} + \frac{H_\alpha}{H_\beta} \frac{\partial H_\beta}{\partial \xi} u_\beta^{(s)} + \frac{H_\beta}{H_\alpha} \frac{\partial H_\alpha}{\partial \eta} u_\alpha^{(s)} = a \cdot a_{66} \sigma_{\alpha\beta}^{(s)},$$

$$\frac{\partial u_\beta^{(s)}}{\partial \zeta} + H_\beta \frac{\partial w^{(k)}}{\partial \eta} = a \cdot a_{44} \sigma_{\beta\gamma}^{(s-2)}, \qquad \frac{\partial u_\alpha^{(s)}}{\partial \zeta} + H_\alpha \frac{\partial w^{(k)}}{\partial \xi} = a \cdot a_{55} \sigma_{\alpha\gamma}^{(s-2)},$$

where $m = s$, $n = s - 2$, $k = s - 2$ correspond to the symmetric problem, and $m = s - 2$, $n = s - 4$, $k = s$ is for the antisymmetric one. Here and below we assume that $Q^{(s)} \equiv 0$ at $s < 0$. The system (2.6) may be integrated along ζ. In the antisymmetric case the quantities $\sigma_{\alpha\gamma}, \sigma_{\beta\gamma}, w$ are even functions of ζ, whereas $\sigma_{\alpha\alpha}, \sigma_{\alpha\beta}, \sigma_{\beta\beta}, \sigma_{\gamma\gamma}, u_{\alpha}, u_{\beta}$ are odd ones. The situation for the symmetric case is opposite. Integrating (2.6) in the antisymmetric (bending) problem, we obtain

$$W^{(s)} = w^{(s)}(\alpha, \beta) + W^{*(s)}, \quad u_{\alpha}^{(s)} = \zeta v_{\alpha}^{(s)} + u_{\alpha}^{*(s)}, \quad (\alpha, \beta),$$

$$\sigma_{\alpha\alpha}^{(s)} = \zeta \tau_{\alpha\alpha}^{(s)} + \sigma_{\alpha\alpha}^{*(s)} \quad (\alpha, \beta), \quad \sigma_{\alpha\beta}^{(s)} = \zeta \tau_{\alpha\beta}^{(s)} + \sigma_{\alpha\beta}^{*(s)}, \tag{2.7}$$

$$\sigma_{\alpha\gamma}^{(s)} = \zeta^2 \tau_{\alpha\gamma}^{(s)} + T_{\alpha\gamma}^{(s)} + \sigma_{\alpha\gamma}^{*(s)} \quad (\alpha, \beta),$$

$$\sigma_{\gamma\gamma}^{(s)} = \zeta^3 \tau_{\gamma\gamma}^{(s)} + \zeta T_{\gamma\gamma}^{(s)} + \sigma_{\gamma\gamma}^{*(s)},$$

where

$$v_{\alpha}^{(s)} = -a H_{\alpha} \frac{\partial w^{(s)}}{\partial \alpha}, \quad (\alpha, \beta),$$

$$\tau_{\alpha\alpha}^{(s)}(\alpha, \beta) = \frac{1}{\Omega} \left(H_{\alpha} \left(a_{22} \frac{\partial v_{\alpha}^{(s)}}{\partial \alpha} + \frac{a_{12}}{H_{\beta}} \frac{\partial H_{\beta}}{\partial \alpha} v_{\alpha}^{(s)} \right) \right.$$

$$\left. - H_{\beta} \left(a_{12} \frac{\partial v_{\beta}^{(s)}}{\partial \beta} + \frac{a_{22}}{H_{\alpha}} \frac{\partial H_{\alpha}}{\partial \beta} v_{\beta}^{(s)} \right) \right), \quad (\alpha, \beta; 1, 2),$$

$$\tau_{\alpha\beta}^{(s)}(\alpha, \beta) = \frac{1}{a_{66}} \left(H_{\beta} \frac{\partial v_{\alpha}^{(s)}}{\partial \beta} + H_{\alpha} \frac{\partial v_{\beta}^{(s)}}{\partial \alpha} + \frac{H_{\beta}}{H_{\alpha}} \frac{\partial H_{\alpha}}{\partial \beta} v_{\alpha}^{(s)} + \frac{H_{\alpha}}{H_{\beta}} \frac{\partial H_{\beta}}{\partial \alpha} v_{\beta}^{(s)} \right),$$

$$\tau_{\alpha\gamma}^{(s)}(\alpha, \beta) = -\frac{a}{2} \left(H_{\alpha} \frac{\partial \tau_{\alpha\alpha}^{(s)}}{\partial \alpha} + H_{\beta} \frac{\partial \tau_{\alpha\beta}^{(s)}}{\partial \beta} \right.$$

$$\left. - \frac{H_{\alpha}}{H_{\beta}} \frac{\partial H_{\beta}}{\partial \alpha} \left(\tau_{\alpha\alpha}^{(s)} - \tau_{\beta\beta}^{(s)} \right) - 2 \frac{H_{\beta}}{H_{\alpha}} \frac{\partial H_{\alpha}}{\partial \beta} \tau_{\alpha\beta}^{(s)} \right), \quad (\alpha, \beta),$$

$$\tau_{\gamma\gamma}^{(s)}(\alpha, \beta) = -\frac{a}{3} \left(H_{\alpha} \frac{\partial \tau_{\alpha\gamma}^{(s)}}{\partial \alpha} + H_{\beta} \frac{\partial \tau_{\beta\gamma}^{(s)}}{\partial \beta} - \frac{H_{\alpha}}{H_{\beta}} \frac{\partial H_{\beta}}{\partial \alpha} \tau_{\alpha\gamma}^{(s)} - \frac{H_{\beta}}{H_{\alpha}} \frac{\partial H_{\alpha}}{\partial \beta} \tau_{\beta\gamma}^{(s)} \right), \tag{2.8}$$

$$T_{\gamma\gamma}^{(s)}(\alpha, \beta) = -a \left(H_{\alpha} \frac{\partial T_{\alpha\gamma}^{(s)}}{\partial \alpha} + H_{\beta} \frac{\partial T_{\beta\gamma}^{(s)}}{\partial \beta} - \frac{H_{\alpha}}{H_{\beta}} \frac{\partial H_{\beta}}{\partial \alpha} T_{\alpha\gamma}^{(s)} - \frac{H_{\beta}}{H_{\alpha}} \frac{\partial H_{\alpha}}{\partial \beta} T_{\beta\gamma}^{(s)} \right),$$

$$\Omega = a_{11} a_{22} - a_{12}^2.$$

The quantities with asterisks are defined as

$$W^{*(s)} = a \int_0^{\zeta} \left(a_{13} \sigma_{\alpha\alpha}^{(s-2)} + a_{23} \sigma_{\beta\beta}^{(s-2)} + a_{33} \sigma_{\gamma\gamma}^{(s-4)} \right) d\zeta,$$

$$u_{\alpha}^{*(s)} = -a \int_0^{\zeta} H_{\alpha} \frac{\partial W^{*(s)}}{\partial \alpha} d\zeta + a \cdot a_{55} \int_0^{\zeta} \sigma_{\alpha\gamma}^{(s-2)} d\zeta,$$

$$u_{\beta}^{*(s)} = -a \int_0^{\zeta} H_{\beta} \frac{\partial W^{*(s)}}{\partial \beta} d\zeta + a \cdot a_{44} \int_0^{\zeta} \sigma_{\beta\gamma}^{(s-2)} d\zeta, \tag{2.9}$$

$$\sigma_{\alpha\alpha}^{*(s)} = \frac{1}{\Omega} \left(H_{\alpha} \left(a_{22} \frac{\partial u_{\alpha}^{*(s)}}{\partial \alpha} + \frac{a_{12}}{H_{\beta}} \frac{\partial H_{\beta}}{\partial \alpha} u_{\alpha}^{*(s)} \right) \right.$$

$$- H_\beta \left(a_{12} \frac{\partial u_\beta^{*(s)}}{\partial \beta} + \frac{a_{22}}{H_\alpha} \frac{\partial H_\alpha}{\partial \beta} u_\beta^{*(s)} \right) \right) - A_1 \sigma_{\gamma\gamma}^{(s-2)}, \quad (\alpha, \beta; 1, 2),$$

$$\sigma_{\alpha\beta}^{*(s)} = \frac{1}{a_{66}} \left(H_\alpha \frac{\partial u_\beta^{*(s)}}{\partial \alpha} + \frac{H_\alpha}{H_\beta} \frac{\partial H_\beta}{\partial \alpha} u_\beta^{*(s)} + H_\beta \frac{\partial u_\alpha^{*(s)}}{\partial \beta} + \frac{H_\beta}{H_\alpha} \frac{\partial H_\alpha}{\partial \beta} u_\alpha^{*(s)} \right),$$

$$\sigma_{\alpha\gamma}^{*(s)} = -a \int_0^\zeta \left(H_\alpha \frac{\partial \sigma_{\alpha a}^{*(s)}}{\partial \alpha} + H_\beta \frac{\partial \sigma_{\alpha\beta}^{*(s)}}{\partial \beta} \right.$$

$$\left. - \frac{H_\alpha}{H_\beta} \frac{\partial H_\beta}{\partial \alpha} \left(\sigma_{\alpha a}^{*(s)} - \sigma_{\beta\beta}^{*(s)} \right) - 2 \frac{H_\beta}{H_\alpha} \frac{\partial H_\alpha}{\partial \beta} \sigma_{\alpha\beta}^{*(s)} \right) d\zeta, \quad (\alpha, \beta),$$

$$\sigma_{\gamma\gamma}^{*(s)} = -a \int_0^\zeta \left(H_\alpha \frac{\partial \sigma_{\alpha\gamma}^{*(s)}}{\partial \alpha} + H_\beta \frac{\partial \sigma_{\beta\gamma}^{*(s)}}{\partial \beta} - \frac{H_\alpha}{H_\beta} \frac{\partial H_\beta}{\partial \alpha} \sigma_{\alpha\gamma}^{*(s)} - \frac{H_\beta}{H_\alpha} \frac{\partial H_\alpha}{\partial \beta} \sigma_{\beta\gamma}^{*(s)} \right) d\zeta,$$

$$A_1 = \frac{a_{13} a_{22} - a_{12} a_{23}}{\Omega}, \quad A_2 = \frac{a_{23} a_{11} - a_{12} a_{13}}{\Omega},$$

are polynomials with respect to ζ. These are zeros at $s = 0, 1$ since $Q^{(s)} \equiv 0$ at $s < 0$. The recurrent formulas (2.9) enable one to determine all of the quantities of approximation s provided that the quantities of the previous orders $0, 1, ..., s-1$ are known.

Let us require that the stresses defined through (2.4)-(2.9) satisfy the boundary conditions (1.5). We assume here that the functions X_i, Y_i, Z_i are independent of ε, otherwise these should be expanded in series along ε. Taking into account (2.7), the conditions (1.5) are written as

$$T_{\gamma\gamma}^{(s)} + \tau_{\gamma\gamma}^{(s)} = \frac{1}{2} Z_2^{(s)}, \quad T_{\alpha\gamma}^{(s)} + \tau_{\alpha\gamma}^{(s)} = \frac{1}{2} X_2^{(s)} \quad (\alpha, \beta), \ (X, Y) \tag{2.10}$$

where

$$Z_2^{(0)} = Z_2, \quad Z_2^{(1)} = 0, \quad Z_2^{(k)} = -2\sigma_{\gamma\gamma}^{*(k)}(\zeta = 1) \quad (k > 1)$$

$$X_2^{(0)} = X_2, \quad X_2^{(1)} = 0, \quad X_2^{(k)} = -2\sigma_{\alpha\gamma}^{*(k)}(\zeta = 1) \quad (\alpha, \beta), \ (X, Y). \tag{2.11}$$

It may be noticed that the equations (2.8)-(2.10) form a complete system in respect of the unknowns $v_\alpha^{(s)}, v_\beta^{(s)}, w^{(s)}, \tau_{\alpha\alpha}^{(s)}, \tau_{\beta\beta}^{(s)}, \tau_{\alpha\beta}^{(s)}, \tau_{\alpha\gamma}^{(s)}, \tau_{\beta\gamma}^{(s)}, \tau_{\gamma\gamma}^{(s)}, T_{\alpha\gamma}^{(s)}, T_{\beta\gamma}^{(s)}, T_{\gamma\gamma}^{(s)}$. As seen from (2.8), the quantities $v_\alpha^{(s)}, v_\beta^{(s)}$ are expressed through $w^{(s)}$. Also, from (2.10), (2.11) and the last two equations of (2.8) it is possible to express $T_{\alpha\gamma}^{(s)}, T_{\beta\gamma}^{(s)}, T_{\gamma\gamma}^{(s)}, \tau_{\gamma\gamma}^{(s)}$ through $\tau_{\alpha\gamma}^{(s)}, \tau_{\beta\gamma}^{(s)}, X_2, Y_2, Z_2, \sigma_{\alpha\gamma}^{*(s)}(\zeta = 1), \sigma_{\beta\gamma}^{*(s)}(\zeta = 1), \sigma_{\gamma\gamma}^{*(s)}(\zeta = 1)$. After these transformations, we arrive at the following system in respect of six unknowns $\tau_{\alpha\alpha}^{(s)}, \tau_{\beta\beta}^{(s)}, \tau_{\alpha\beta}^{(s)}, \tau_{\alpha\gamma}^{(s)}, \tau_{\beta\gamma}^{(s)}, w^{(s)}$

$$H_\alpha H_\beta \left(\frac{\partial}{\partial \alpha} \left(\frac{\tau_{\alpha\alpha}^{(s)}}{H_\beta} \right) - \tau_{\beta\beta}^{(s)} \frac{\partial}{\partial \alpha} \left(\frac{1}{H_\beta} \right) \right.$$

$$\left. + \frac{\partial}{\partial \beta} \left(\frac{\tau_{\alpha\beta}^{(s)}}{H_\alpha} \right) + \tau_{\alpha\beta}^{(s)} \frac{\partial}{\partial \beta} \left(\frac{1}{H_\alpha} \right) \right) = -\frac{2}{a} \tau_{\alpha\gamma}^{(s)}, \quad (\alpha, \beta), \tag{2.12}$$

$$H_\alpha H_\beta \left(\frac{\partial}{\partial \alpha} \left(\frac{\tau_{\alpha\gamma}^{(s)}}{H_\beta} \right) + \frac{\partial}{\partial \beta} \left(\frac{\tau_{\beta\gamma}^{(s)}}{H_\alpha} \right) \right)$$

$$= \frac{3}{4} \left(\frac{1}{a} Z_2^{(s)} + H_\alpha H_\beta \left(\frac{\partial}{\partial \alpha} \left(\frac{X_2^{(s)}}{H_\beta} \right) + \frac{\partial}{\partial \beta} \left(\frac{Y_2^{(s)}}{H_\alpha} \right) \right) \right),$$

$$\tau_{\alpha\alpha}^{(s)} = a \left(\frac{a_{22}}{\Omega} \left(-H_\alpha \frac{\partial}{\partial \alpha} \left(H_\alpha \frac{\partial w^{(s)}}{\partial \alpha} \right) - H_\alpha H_\beta \frac{\partial}{\partial \beta} \left(\frac{1}{H_\alpha} \right) \cdot H_\beta \frac{\partial w^{(s)}}{\partial \beta} \right) \right.$$

$$\left. - \frac{a_{12}}{\Omega} \left(-H_\beta \frac{\partial}{\partial \beta} \left(H_\beta \frac{\partial w^{(s)}}{\partial \beta} \right) - H_\alpha H_\beta \frac{\partial}{\partial \alpha} \left(\frac{1}{H_\beta} \right) \cdot H_\alpha \frac{\partial w^{(s)}}{\partial \alpha} \right) \right), \quad (\alpha, \beta; 1, 2),$$

$$\tau_{\alpha\beta}^{(s)} = -\frac{2a}{a_{66}} H_\alpha H_\beta \left(\frac{\partial^2 w^{(s)}}{\partial \alpha \partial \beta} + \frac{1}{H_\alpha} \frac{\partial H_\alpha}{\partial \beta} \frac{\partial w^{(s)}}{\partial \alpha} + \frac{1}{H_\beta} \frac{\partial H_\beta}{\partial \alpha} \frac{\partial w^{(s)}}{\partial \beta} \right).$$

The latter may be reduced to one equation in terms of $w^{(s)}$. After all of the unknowns of (2.12) are determined, the outer solution is given by (2.4), (2.7).

Let us now discuss the symmetric (extensional) case, for which

$$u_\alpha^{(s)} = v_\alpha^{(s)} + u_\alpha^{*(s)} \quad (\alpha, \beta), \quad W^{(s)} = \zeta w^{(s)} + W^{*(s)},$$

$$\sigma_{\alpha\alpha}^{(s)} = \tau_{\alpha\alpha}^{(s)} + \sigma_{\alpha\alpha}^{*(s)} \quad (\alpha, \beta), \quad \sigma_{\alpha\beta}^{(s)} = \tau_{\alpha\beta}^{(s)} + \sigma_{\alpha\beta}^{*(s)}, \tag{2.13}$$

$$\sigma_{\alpha\gamma}^{(s)} = \zeta \tau_{\alpha\gamma}^{(s)} + \sigma_{\alpha\gamma}^{*(s)} \quad (\alpha, \beta), \quad \sigma_{\gamma\gamma}^{(s)} = \frac{1}{2} \zeta^2 \tau_{\gamma\gamma}^{(s)} + T_{\gamma\gamma}^{(s)} + \sigma_{\gamma\gamma}^{*(s)}.$$

The unknown quantities in (2.13) may be expressed through $v_\alpha^{(s)}(\alpha, \beta)$, $v_\beta^{(s)}(\alpha, \beta)$ as

$$\tau_{\alpha\alpha}^{(s)}(\alpha, \beta) = \frac{1}{\Omega} \left(a_{22} H_\alpha \frac{\partial v_\alpha^{(s)}}{\partial \alpha} - a_{12} H_\beta \frac{\partial v_\beta^{(s)}}{\partial \beta} \right.$$

$$\left. + a_{12} \frac{H_\alpha}{H_\beta} \frac{\partial H_\beta}{\partial \alpha} v_\alpha^{(s)} - a_{22} \frac{H_\beta}{H_\alpha} \frac{\partial H_\alpha}{\partial \beta} v_\beta^{(s)} \right), \quad (\alpha, \beta; 1, 2),$$

$$\tau_{\alpha\beta}^{(s)}(\alpha, \beta) = \frac{1}{a_{66}} \left(H_\beta \frac{\partial v_\alpha^{(s)}}{\partial \beta} + H_\alpha \frac{\partial v_\beta^{(s)}}{\partial \alpha} + \frac{H_\beta}{H_\alpha} \frac{\partial H_\alpha}{\partial \beta} v_\alpha^{(s)} + \frac{H_\alpha}{H_\beta} \frac{\partial H_\beta}{\partial \alpha} v_\beta^{(s)} \right),$$

$$\tau_{\alpha\gamma}^{(s)}(\alpha, \beta) = -a \left(H_\alpha \frac{\partial \tau_{\alpha\alpha}^{(s)}}{\partial \alpha} + H_\beta \frac{\partial \tau_{\alpha\beta}^{(s)}}{\partial \beta} \right.$$

$$\left. - \frac{H_\alpha}{H_\beta} \frac{\partial H_\beta}{\partial \alpha} (\tau_{\alpha\alpha}^{(s)} - \tau_{\beta\beta}^{(s)}) - 2 \frac{H_\beta}{H_\alpha} \frac{\partial H_\alpha}{\partial \beta} \tau_{\alpha\beta}^{(s)} \right), \quad (\alpha, \beta), \tag{2.14}$$

$$\tau_{\gamma\gamma}^{(s)}(\alpha, \beta) = -a \left(H_\alpha \frac{\partial \tau_{\alpha\gamma}^{(s)}}{\partial \alpha} + H_\beta \frac{\partial \tau_{\beta\gamma}^{(s)}}{\partial \beta} - \frac{H_\alpha}{H_\beta} \frac{\partial H_\beta}{\partial \alpha} \tau_{\alpha\gamma}^{(s)} - \frac{H_\beta}{H_\alpha} \frac{\partial H_\alpha}{\partial \beta} \tau_{\beta\gamma}^{(s)} \right),$$

$$w^{(s)}(\alpha, \beta) = a \left(a_{13} \tau_{\alpha\alpha}^{(s)} + a_{23} \tau_{\beta\beta}^{(s)} \right).$$

The quantities with the asterisks are given by

$$u_\alpha^{*(s)} = a \int_0^\zeta \left(a_{55} \sigma_{\alpha\gamma}^{(s-2)} - H_\alpha \frac{\partial W^{(s-2)}}{\partial \alpha} \right) d\zeta, \quad (\alpha, \beta; 5, 4),$$

$$\sigma_{\alpha\alpha}^{*(s)} = \frac{1}{\Omega}\left(a_{22}H_\alpha\frac{\partial u_\alpha^{*(s)}}{\partial\alpha} - a_{12}H_\beta\frac{\partial u_\beta^{*(s)}}{\partial\beta}\right.$$

$$\left. +a_{12}\frac{H_\alpha}{H_\beta}\frac{\partial H_\beta}{\partial\alpha}u_\alpha^{*(s)} - a_{22}\frac{H_\beta}{H_\alpha}\frac{\partial H_\alpha}{\partial\beta}u_\beta^{*(s)}\right) - A_1\sigma_{\gamma\gamma}^{(s-2)}, \quad (\alpha,\beta;1,2),$$

$$\sigma_{\alpha\beta}^{*(s)} = \frac{1}{a_{66}}\left(H_\alpha\frac{\partial u_\beta^{*(s)}}{\partial\alpha} + H_\beta\frac{\partial u_\alpha^{*(s)}}{\partial\beta} + \frac{H_\beta}{H_\alpha}\frac{\partial H_\alpha}{\partial\beta}u_\alpha^{*(s)} + \frac{H_\alpha}{H_\beta}\frac{\partial H_\beta}{\partial\alpha}u_\beta^{*(s)}\right),$$

$$W^{*(s)} = a\int_0^\zeta\left(a_{13}\sigma_{\alpha\alpha}^{*(s)} + a_{23}\sigma_{\beta\beta}^{*(s)} + a_{33}\sigma_{\gamma\gamma}^{(s-2)}\right)d\zeta, \qquad (2.15)$$

$$\sigma_{\alpha\gamma}^{*(s)} = -a\int_0^\zeta\left(H_\alpha\frac{\partial\sigma_{\alpha\alpha}^{*(s)}}{\partial\alpha} + H_\beta\frac{\partial\sigma_{\alpha\beta}^{*(s)}}{\partial\beta}\right.$$

$$\left. -\frac{H_\alpha}{H_\beta}\frac{\partial H_\beta}{\partial\alpha}(\sigma_{\alpha\alpha}^{*(s)} - \sigma_{\beta\beta}^{*(s)}) - 2\frac{H_\beta}{H_\alpha}\frac{\partial H_\alpha}{\partial\beta}\sigma_{\alpha\beta}^{*(s)}\right)d\zeta, \quad (\alpha,\beta),$$

$$\sigma_{\gamma\gamma}^{*(s)} = -a\int_0^\zeta\left(H_\alpha\frac{\partial\sigma_{\alpha\gamma}^{*(s)}}{\partial\alpha} + H_\beta\frac{\partial\sigma_{\beta\gamma}^{*(s)}}{\partial\beta} - \frac{H_\alpha}{H_\beta}\frac{\partial H_\beta}{\partial\alpha}\sigma_{\alpha\gamma}^{*(s)} - \frac{H_\beta}{H_\alpha}\frac{\partial H_\alpha}{\partial\beta}\sigma_{\beta\gamma}^{*(s)}\right)d\zeta.$$

Satisfying the conditions (1.4), bearing in mind (2.4) and (2.13), we obtain

$$\tau_{\alpha\gamma}^{(s)} = \frac{1}{2}X_1^{(s)}, \quad \tau_{\beta\gamma}^{(s)} = \frac{1}{2}Y_1^{(s)}, \quad T_{\gamma\gamma}^{(s)} = \frac{1}{2}Z_1^{(s)} - \frac{1}{2}\tau_{\gamma\gamma}^{(s)}, \qquad (2.16)$$

where

$$X_1^{(0)} = X_1, \quad X_1^{(1)} = 0, \quad X_1^{(k)} = -2\sigma_{\alpha\gamma}^{*(k)}(\zeta=1) \quad (k>1), \quad (X,Y)$$

$$Z_1^{(0)} = Z_1, \quad Z_1^{(1)} = 0, \quad Z_1^{(k)} = -2\sigma_{\gamma\gamma}^{*(k)}(\zeta=1) \quad (k>1). \qquad (2.17)$$

In view of (2.16)

$$\sigma_{\alpha\gamma}^{(s)} = \frac{1}{2}\zeta X_1^{(s)} + \sigma_{\alpha\gamma}^{*(s)}, \quad (\alpha,\beta;X,Y),$$

$$\sigma_{\gamma\gamma}^{(s)} = \frac{1}{2}(\zeta^2-1)\tau_{\gamma\gamma}^{(s)} + \frac{1}{2}Z_1^{(s)} + \sigma_{\gamma\gamma}^{*(s)} \qquad (2.18)$$

with the system (2.14) being reduced to that for five unknowns $v_\alpha^{(s)}$, $v_\beta^{(s)}$, $\tau_{\alpha\alpha}^{(s)}$, $\tau_{\beta\beta}^{(s)}$, $\tau_{\alpha\beta}^{(s)}$, given by

$$H_\alpha\frac{\partial\tau_{\alpha\alpha}^{(s)}}{\partial\alpha} + H_\beta\frac{\partial\tau_{\alpha\beta}^{(s)}}{\partial\beta} - \frac{H_\alpha}{H_\beta}\frac{\partial H_\beta}{\partial\alpha}(\tau_{\alpha\alpha}^{(s)} - \tau_{\beta\beta}^{(s)})$$

$$-2\frac{H_\beta}{H_\alpha}\frac{\partial H_\alpha}{\partial\beta}\tau_{\alpha\beta}^{(s)} = -\frac{1}{2a}X_1^{(s)}, \quad (\alpha,\beta;X,Y),$$

$$\tau_{\alpha\alpha}^{(s)} = \frac{1}{\Omega}\left(a_{22}H_\alpha\frac{\partial v_\alpha^{(s)}}{\partial\alpha} - a_{12}H_\beta\frac{\partial v_\beta^{(s)}}{\partial\beta}\right.$$

$$\left. +a_{12}\frac{H_\alpha}{H_\beta}\frac{\partial H_\beta}{\partial\alpha}v_\alpha^{(s)} - a_{22}\frac{H_\beta}{H_\alpha}\frac{\partial H_\alpha}{\partial\beta}v_\beta^{(s)}\right), \quad (\alpha,\beta;1,2), \qquad (2.19)$$

$$\tau_{\alpha\beta}^{(s)}(\alpha,\beta) = \frac{1}{a_{66}}\left(H_\beta\frac{\partial v_\alpha^{(s)}}{\partial\beta} + H_\alpha\frac{\partial v_\beta^{(s)}}{\partial\alpha} + \frac{H_\beta}{H_\alpha}\frac{\partial H_\alpha}{\partial\beta}v_\alpha^{(s)} + \frac{H_\alpha}{H_\beta}\frac{\partial H_\beta}{\partial\alpha}v_\beta^{(s)}\right).$$

It may be observed that the obtained system (2.19) coincides with the formulation of the generalized plane-stress (for any s). Excluding the unknowns $\tau_{\alpha\alpha}^{(s)}, \tau_{\beta\beta}^{(s)}, \tau_{\alpha\beta}^{(s)}$, the system (2.19) is transformed to two equations in respect of $v_\alpha^{(s)}$, $v_\beta^{(s)}$. Solving the latter, and using (2.4), (2.13), (2.14), and (2.18), we obtain all quantities of the outer solution.

We remark that there is no strict necessity to split the problem into symmetric and anti-symmetric parts, since this division into two cases (symmetric and anti-symmetric) would still occur if we sought-for solution of (2.3) in the form of (2.4), (2.5) ($q_w = 3$) satisfying the conditions (1.3). We also note that in case of general anisotropy the separation of the problem into symmetric and antisymmetric cases is not possible, with more details presented in Chapter 5.

3.3 Relation to the Kirchhoff Plate Theory

Let us show that the construction of approximations within asymptotic theory is in some sense equivalent to analysis of plate according to classical theory. In particular, we will first demonstrate that systems (2.12), (2.19) for $s = 0$ are formally equivalent to the classical bending and extensional theories for anisotropic plates and then discuss higher orders. Let us introduce the common definitions of the moments $M_\alpha, M_\beta, H_{\alpha\beta}$, tangential forces $T_\alpha, T_\beta, T_{\alpha\beta}$ and transverse forces N_α, N_β as

$$T_\alpha = \int_{-h}^{+h} \sigma_{\alpha\alpha} d\gamma, \qquad M_\alpha = \int_{-h}^{+h} \gamma\sigma_{\alpha\alpha} d\gamma, \quad (\alpha, \beta),$$

$$S_{\alpha\beta} = \int_{-h}^{+h} \sigma_{\alpha\beta} d\gamma, \qquad H_{\alpha\beta} = \int_{-h}^{+h} \gamma\sigma_{\alpha\beta} d\gamma, \tag{3.1}$$

$$N_\alpha = \int_{-h}^{+h} \sigma_{\alpha\gamma} d\gamma, \quad (\alpha, \beta),$$

and denote by $T_\alpha^{(s)}, M_\alpha^{(s)}, S_{\alpha\beta}^{(s)}, H_{\alpha\beta}^{(s)}, N_\alpha^{(s)}$ the associated values at approximation s. Taking into account (2.4), (2.5), (2.7), (2.14), and (3.1), we have

$$M_\alpha^{(s)} = \frac{2}{3}a^2\varepsilon^s\tau_{\alpha\alpha}^{(s)} + a^2\varepsilon^s M_\alpha^{*(s)}, \quad (\alpha, \beta),$$

$$H_{\alpha\beta}^{(s)} = \frac{2}{3}a^2\varepsilon^s\tau_{\alpha\beta}^{(s)} + a^2\varepsilon^s H_{\alpha\beta}^{*(s)}, \tag{3.2}$$

$$N_\alpha^{(s)} = a\varepsilon^s\left(-\frac{4}{3}\tau_{\alpha\gamma}^{(s)} + X_2^{(s)} + N_\alpha^{*(s)}\right), \quad (\alpha, \beta; X, Y)$$

and

$$T_\alpha^{(s)} = a\varepsilon^{-1+s}\left(2\tau_{\alpha\alpha}^{(s)} + T_a^{*(s)}\right), \quad (\alpha, \beta),$$

$$S_{\alpha\beta}^{(s)} = a\varepsilon^{-1+s}\left(2\tau_{\alpha\beta}^{(s)} + S_{a\beta}^{*(s)}\right) \tag{3.3}$$

where

$$(T_\alpha^{*(s)}, S_{\alpha\beta}^{*(s)}, N_\alpha^{*(s)}) = \int_{-h}^{+h} (\sigma_{\alpha\alpha}^{*(s)}, \sigma_{\alpha\beta}^{*(s)}, \sigma_{\alpha\gamma}^{*(s)}) d\zeta$$

$$(M_\alpha^{*(s)}, H_{\alpha\beta}^{*(s)}) = \int_{-h}^{+h} \zeta(\sigma_{\alpha\alpha}^{*(s)}, \sigma_{\alpha\beta}^{*(s)}) d\zeta. \tag{3.4}$$

Using (3.2), the quantities $\tau_{\alpha\alpha}^{(s)}, \tau_{\beta\beta}^{(s)}, \tau_{\alpha\beta}^{(s)}, \tau_{\alpha\gamma}^{(s)}, \tau_{\beta\gamma}^{(s)}$ may be expressed through the transverse forces. Substitution into (2.12) gives

$$\frac{\partial}{\partial\alpha}(BM_\alpha^{(s)}) + \frac{\partial}{\partial\beta}(AH_{\alpha\beta}^{(s)}) + \frac{\partial A}{\partial\beta}H_{\alpha\beta}^{(s)} - \frac{\partial B}{\partial\alpha}M_\beta^{(s)}$$

$$= AB\left(N_\alpha^{(s)} - a\varepsilon^s X_2^{(s)}\right) + a^2\varepsilon^s\left(\frac{\partial(BM_\alpha^{*(s)})}{\partial\alpha} + \frac{\partial}{\partial\beta}(AH_{\alpha\beta}^{*(s)})\right.$$

$$\left. + \frac{\partial A}{\partial\beta}H_{\alpha\beta}^{*(s)} - \frac{\partial B}{\partial\alpha}M_\beta^{*(s)} - \frac{1}{a}ABN_\alpha^{*(s)}\right), \quad (\alpha,\beta),$$

$$\frac{\partial}{\partial\alpha}(BN_\alpha^{(s)}) + \frac{\partial}{\partial\beta}(AN_\beta^{(s)}) = -\varepsilon^s ABZ_2^{(s)} + a\varepsilon^s\left(\frac{\partial BN_\alpha^{*(s)}}{\partial\alpha} + \frac{\partial AN_\beta^{*(s)}}{\partial\beta}\right), \tag{3.5}$$

$$M_\alpha^{(s)} = \frac{2}{3}a^3\varepsilon^s B_{11}(\chi_1^{(s)} + \nu_2\chi_2^{(s)}) + a^2\varepsilon^s M_\alpha^{*(s)}, \quad (\alpha,\beta;1,2),$$

$$H_{\alpha\beta}^{(s)} = \frac{2}{3}a^3\varepsilon^s B_{66}\tau^{(s)} + a^2\varepsilon^s H_{\alpha\beta}^{*(s)},$$

where $A = \dfrac{1}{H_\alpha}, B = \dfrac{1}{H_\beta}$ are the coefficients of the first quadratic form

$$B_{11} = a_{22}/\Omega, \quad B_{22} = a_{11}/\Omega, \quad B_{66} = 1/a_{66},$$

$$\chi_1^{(s)} = -H_\alpha\frac{\partial}{\partial\alpha}\left(H_\alpha\frac{\partial w^{(s)}}{\partial\alpha}\right) + \frac{H_\beta}{H_\alpha}\frac{\partial H_\alpha}{\partial\beta}H_\beta\frac{\partial w^{(s)}}{\partial\beta}, \quad (\alpha,\beta), \quad (1,2), \tag{3.6}$$

$$\tau^{(s)} = -2H_\alpha H_\beta\left(\frac{\partial^2 w^{(s)}}{\partial\alpha\partial\beta} + \frac{1}{H_\alpha}\frac{\partial H_\alpha}{\partial\beta}\frac{\partial w^{(s)}}{\partial\alpha} + \frac{1}{H_\beta}\frac{\partial H_\beta}{\partial\alpha}\frac{\partial w^{(s)}}{\partial\beta}\right).$$

In other words, $\chi_1^{(s)}, \chi_2^{(s)}, \tau^{(s)}$ turn out to be the usual coefficients (Ambartsumyan, 1974; Goldenveizer, 1961, 1976) describing the deformation of bending and relative deformation of torsion for approximation s. If we restrict ourselves to leading order only, then it follows from (2.4), (2.5) that $w^{(0)} = \varepsilon^3 w$. Substituting this result into (3.5) and (3.6), it may be easily verified that the system (3.5) is then coinciding with that derived for plate bending relying on the classical Kirchhoff hypothesis (Ambartsumyan, 1967; Lekhnitskii, 1968). Therefore, indeed in case of construction of the leading order outer solution the governing equations are identical to the classical ones. Moreover, as shown in Chapter 4, the corresponding boundary conditions also coincide. The higher orders of approximation are essentially refining the solution of the outer problem. The L.H.S. of the system (3.5) at higher orders is also similar to the classical formulation for plate bending, with the R.H.S. being known at any order s provided that the previous approximations

$0, 1, ..., (s-1)$ have been constructed. This is rather convenient, as mathematically we operate with the well-known operators, for which there already exist a number of mathematical methods.

As mentioned above, the system (2.12) or (3.5) may be reduced to a single equation in respect of $w^{(s)}$ through representing $\tau_{\alpha\alpha}^{(s)}, \tau_{\alpha\beta}^{(s)}, \tau_{\beta\beta}^{(s)}$ as

$$\tau_{\alpha\alpha}^{(s)} = aB_{11}\left(\chi_1^{(s)} + \nu_2\chi_2^{(s)}\right), \quad (\alpha, \beta; 1, 2),$$

$$\tau_{\alpha\beta}^{(s)} = aB_{66}\tau^{(s)}. \tag{3.7}$$

Substituting the latter into the first two equations of (2.12), the values of $\tau_{\alpha\gamma}^{(s)}$ and $\tau_{\beta\gamma}^{(s)}$ may be defined, which in turn may be used in the third equation of (2.12), yielding

$$H_\alpha H_\beta \left(\frac{\partial}{\partial\alpha} H_\alpha \left(B_{11} \frac{\partial}{\partial\alpha} \frac{\chi_1^{(s)} + \nu_2\chi_2^{(s)}}{H_\beta} - B_{22}\left(\chi_2^{(s)} + \nu_1\chi_1^{(s)}\right)\frac{\partial}{\partial\alpha}\left(\frac{1}{H_\beta}\right) \right.\right.$$

$$+ B_{66}\left(\frac{\partial}{\partial\beta}\frac{\tau^{(s)}}{H_\alpha} + \tau^{(s)}\frac{\partial}{\partial\beta}\frac{1}{H_\alpha} \right) \right) + \frac{\partial}{\partial\beta} H_\beta \left(B_{22}\frac{\partial}{\partial\beta}\frac{\chi_2^{(s)} + \nu_2\chi_1^{(s)}}{H_\alpha} \right.$$

$$- B_{11}\left(\chi_1^{(s)} + \nu_2\chi_2^{(s)}\right)\frac{\partial}{\partial\beta}\frac{1}{H_\alpha} + B_{66}\left(\frac{\partial}{\partial\alpha}\frac{\tau^{(s)}}{H_\beta} + \tau^{(s)}\frac{\partial}{\partial\alpha}\frac{1}{H_\beta} \right) \right) \tag{3.8}$$

$$= -\frac{3}{2a^2}\left(\frac{1}{a}Z_2^{(s)} + H_\alpha H_\beta \left(\frac{\partial}{\partial\alpha}\left(\frac{X_2^{(s)}}{H_\beta}\right) + \frac{\partial}{\partial\beta}\left(\frac{Y_2^{(s)}}{H_\alpha}\right) \right) \right).$$

Denoting $W_0^{(s)} = \varepsilon^{s-3}w^{(s)}$, it may be observed that (3.8) takes the form of a commonly known plate bending equation for any order s in terms of deflection of the plate, with the R.H.S. varying depending on the order s. At leading order the equation coincides with the classical one. Moreover, the analysis of the leading order approximation provides more analysis compared to that available through classical approach since it enables one to define the stress $\sigma_{\gamma\gamma}$. Indeed, as follows from (2.7)-(2.10),

$$\sigma_{\alpha\gamma}^{(s)} = (\zeta^2 - 1)\tau_{\alpha\gamma}^{(s)} + \frac{1}{2}X_2^{(s)} + \sigma_{\alpha\gamma}^{*(s)}, \quad (\alpha, \beta; X, Y),$$

$$\sigma_{\gamma\gamma}^{(s)} = (\zeta^3 - \zeta)\tau_{\gamma\gamma}^{(s)} + \frac{1}{2}Z_2^{(s)}\zeta + \sigma_{\gamma\gamma}^{*(s)} \tag{3.9}$$

with $\tau_{\alpha\gamma}^{(s)}, \tau_{\gamma\gamma}^{(s)}$ given by (2.8) after solving (2.12) or (3.5).

In the extensional case the quantities $\tau_{\alpha\alpha}^{(s)}, \tau_{\alpha\beta}^{(s)}, \tau_{\beta\beta}^{(s)}$ may be expressed through tangential forces in (3.3) and then substituted into (2.19), which takes the form

$$\frac{1}{AB}\frac{\partial BT_\alpha^{(s)}}{\partial\alpha} - \frac{1}{AB}\frac{\partial B}{\partial\alpha}T_\beta^{(s)} + \frac{1}{B}\frac{\partial S_{\alpha\beta}^{(s)}}{\partial\beta} + \frac{2}{AB}\frac{\partial A}{\partial\beta}S_{\alpha\beta}^{(s)}$$

$$= \varepsilon^{-1+s}\left(-X_1^{(s)} + a\left(\frac{1}{AB}\frac{\partial BT_\alpha^{*(s)}}{\partial\alpha} \right.\right. \tag{3.10}$$

$$\left.\left. -\frac{1}{AB}\frac{\partial B}{\partial\alpha}T_\beta^{*(s)} + \frac{1}{B}\frac{\partial S_{\alpha\beta}^{*(s)}}{\partial\beta} + \frac{2}{AB}\frac{\partial A}{\partial\beta}S_{\alpha\beta}^{*(s)} \right) \right), \quad (\alpha, \beta; X, Y)$$

and

$$T_\alpha^{(s)} = 2hB_{11}\left(\varepsilon_1^{(s)} + \nu_2\varepsilon_2^{(s)}\right) + a\varepsilon^{-1+s}T_\alpha^{*(s)}, \quad (\alpha,\beta;1,2),$$

$$S_{\alpha\beta}^{(s)} = 2hB_{66}\omega^{(s)} + a\varepsilon^{-1+s}S_{\alpha\beta}^{*(s)} \tag{3.11}$$

where

$$\varepsilon_1^{(s)} = \varepsilon^{-2+s}\left(\frac{1}{A}\frac{\partial v_\alpha^{(s)}}{\partial\alpha} + \frac{1}{AB}\frac{\partial A}{\partial\beta}v_\beta^{(s)}\right), \qquad (1,2;\alpha,\beta),$$

$$\omega^{(s)} = \varepsilon^{-2+s}\left(\frac{1}{B}\frac{\partial v_\alpha^{(s)}}{\partial\beta} + \frac{1}{A}\frac{\partial v_\beta^{(s)}}{\partial\alpha} - \frac{1}{AB}\frac{\partial A}{\partial\beta}v_\alpha^{(s)} - \frac{1}{AB}\frac{\partial B}{\partial\alpha}v_\beta^{(s)}\right) \tag{3.12}$$

with $\varepsilon_1^{(s)}, \varepsilon_2^{(s)}, \omega^{(s)}$ standing for the components of tangential deformation coinciding with classical results at leading order $s = 0$.

We note that equations (3.10) at the leading order approximation $s = 0$ are identical to the plane stress equations of the plate, whereas (3.11) coincide with the constitutive relations, given in Ambartsumyan (1961); Lekhnitskii (1968). At higher orders $s > 0$ these will differ formally from the classical formulation only in the R.H.S., with these known from the previous approximations. Therefore, the asymptotic method allows solution of the outer problem through formal iterative solution of the well-known operators. The issues concerning the corresponding boundary conditions will be discussed in more details in the following Chapter 4.

Excluding the quantities $T_\alpha^{(s)}, S_{\alpha\beta}^{(s)}, T_\beta^{(s)}$ from equations (3.10) and (3.11), the problem is reformulated in terms of two unknowns, namely the displacements $v_\alpha^{(s)}, v_\beta^{(s)}$, giving

$$C_{11}\frac{1}{AB}\frac{\partial}{\partial\alpha}B\left(\varepsilon_1^{(s)} + \nu_2\varepsilon_2^{(s)}\right) - C_{22}\frac{1}{AB}\frac{\partial B}{\partial\alpha}\left(\varepsilon_2^{(s)} + \nu_1\varepsilon_1^{(s)}\right)$$

$$+C_{66}\left(\frac{1}{B}\frac{\partial\omega^{(s)}}{\partial\beta} + \frac{2}{AB}\frac{\partial A}{\partial\beta}\omega^{(s)}\right) = -\varepsilon^{-1+s}X_1^{(s)}, \quad (\alpha,\beta;1,2;X,Y), \tag{3.13}$$

$$C_{ik} = 2hB_{ik}.$$

Assuming that the forces $X_1^{(s)}$ and $Y_1^{(s)}$ have potential $\Psi^{(s)}$, i.e.

$$X_1^{(s)} = -\frac{1}{AB}\left(\frac{\partial\Psi^{(s)}}{\partial\alpha} - \frac{1}{A}\frac{\partial B}{\partial\alpha}\Psi^{(s)}\right),$$

$$Y_1^{(s)} = -\frac{1}{AB}\left(\frac{\partial\Psi^{(s)}}{\partial\beta} - \frac{1}{B}\frac{\partial A}{\partial\beta}\Psi^{(s)}\right) \tag{3.14}$$

solution of (3.10), (3.11) may now be expressed through the function $\Phi^{(s)}$. The equilibrium equations (3.10) are satisfied by

$$T_\alpha^{(s)} = \frac{1}{B}\frac{\partial}{\partial\beta}\left(\frac{1}{B}\frac{\partial\Phi^{(s)}}{\partial\beta}\right) + \frac{1}{AB}\frac{\partial B}{\partial\alpha}\cdot\frac{1}{A}\frac{\partial\Phi^{(s)}}{\partial\alpha} + \frac{1}{B}\Psi^{(s)} + a\varepsilon^{-1+s}T_\alpha^{*(s)},$$

$$T_\beta^{(s)} = \frac{1}{A}\frac{\partial}{\partial\alpha}\left(\frac{1}{A}\frac{\partial\Phi^{(s)}}{\partial\alpha}\right) + \frac{1}{AB}\frac{\partial A}{\partial\beta}\cdot\frac{1}{B}\frac{\partial\Phi^{(s)}}{\partial\beta} + \frac{1}{A}\Psi^{(s)} + a\varepsilon^{-1+s}T_\beta^{*(s)}, \tag{3.15}$$

$$S_{\alpha\beta}^{(s)} = -\frac{1}{B}\frac{\partial}{\partial\beta}\left(\frac{1}{A}\frac{\partial\Phi^{(s)}}{\partial\alpha}\right) + \frac{1}{AB}\frac{\partial B}{\partial\alpha}\cdot\frac{1}{B}\frac{\partial\Phi^{(s)}}{\partial\beta} + a\varepsilon^{-1+s}S_{\alpha\beta}^{*(s)}.$$

Expressing $\varepsilon_1^{(s)}, \varepsilon_2^{(s)}, \omega^{(s)}$ through $T_\alpha^{(s)}$, $T_\beta^{(s)}, S_{\alpha\beta}^{(s)}$, it is possible to obtain an equation for the potential function $\Phi^{(s)}$ from the continuity of deformations

$$
\frac{\partial}{\partial\alpha}\left(\frac{B}{A}\frac{\partial\varepsilon_2^{(s)}}{\partial\alpha}\right) + \frac{\partial}{\partial\beta}\left(\frac{A}{B}\frac{\partial\varepsilon_1^{(s)}}{\partial\beta}\right) + \frac{\partial}{\partial\alpha}\left(\frac{1}{A}\frac{\partial B}{\partial\alpha}(\varepsilon_2^{(s)} - \varepsilon_1^{(s)})\right)
$$
$$
+ \frac{\partial}{\partial\beta}\left(\frac{1}{B}\frac{\partial A}{\partial B}(\varepsilon_1^{(s)} - \varepsilon_2^{(s)})\right) - \frac{\partial}{\partial\alpha}\left(\frac{1}{A}\frac{\partial A}{\partial\beta}\omega^{(s)}\right) \tag{3.16}
$$
$$
- \frac{\partial}{\partial\beta}\left(\frac{1}{B}\frac{\partial B}{\partial\alpha}\omega^{(s)}\right) - \frac{\partial^2\omega^{(s)}}{\partial\alpha\partial\beta} = 0.
$$

Using (3.15), the latter transforms to

$$
\left(a_{11}\frac{\partial}{\partial\beta}\frac{A}{B}\frac{\partial}{\partial\beta} + a_{12}\frac{\partial}{\partial\alpha}\frac{B}{A}\frac{\partial}{\partial\alpha} + (a_{11} - a_{12})\left(\frac{\partial}{\partial\beta}\left(\frac{1}{B}\frac{\partial A}{\partial\beta}\times\right)\right.\right.
$$
$$
\left.\left.- \frac{\partial}{\partial\alpha}\left(\frac{1}{B}\frac{\partial B}{\partial\alpha}\times\right)\right)\right)\left(\frac{1}{B}\frac{\partial}{\partial\beta}\left(\frac{1}{B}\frac{\partial\Phi^{(s)}}{\partial\beta}\right) + \frac{1}{AB}\frac{\partial B}{\partial\alpha}\cdot\frac{1}{A}\frac{\partial\Phi^{(s)}}{\partial\alpha} + \frac{1}{B}\Psi^{(s)}\right)
$$
$$
+ \left(a_{22}\frac{\partial}{\partial\alpha}\frac{B}{A}\frac{\partial}{\partial\alpha} + a_{12}\frac{\partial}{\partial\beta}\frac{A}{B}\frac{\partial}{\partial\beta} + (a_{22} - a_{12})\left(\frac{\partial}{\partial\alpha}\left(\frac{1}{A}\frac{\partial B}{\partial a}\times\right)\right.\right. \tag{3.17}
$$
$$
\left.\left.- \frac{\partial}{\partial\beta}\left(\frac{1}{B}\frac{\partial A}{\partial\beta}\times\right)\right)\right)\left(\frac{1}{A}\frac{\partial}{\partial\alpha}\left(\frac{1}{A}\frac{\partial\Phi^{(s)}}{\partial a}\right) + \frac{1}{AB}\frac{\partial A}{\partial\beta}\cdot\frac{1}{B}\frac{\partial\Phi^{(s)}}{\partial\beta} + \frac{1}{A}\Psi^{(s)}\right)
$$
$$
+ a_{66}\left(\frac{\partial^2}{\partial\alpha\partial\beta} + \frac{\partial}{\partial\alpha}\left(\frac{1}{A}\frac{\partial A}{\partial\beta}\times\right) + \frac{\partial}{\partial\beta}\left(\frac{1}{B}\frac{\partial B}{\partial\alpha}\times\right)\right)
$$
$$
\times\left(\frac{1}{B}\frac{\partial}{\partial\beta}\left(\frac{1}{A}\frac{\partial\Phi^{(s)}}{\partial\alpha}\right) - \frac{1}{AB^2}\frac{\partial B}{\partial\alpha}\cdot\frac{1}{B}\frac{\partial\Phi^{(s)}}{\partial\beta}\right) = 0.
$$

It may be shown that at $s = 0$ the last equation (3.17) turns out to be a well-known biharmonic equation (provided that the material is isotropic and body and mass forces are absent). In case of a general loading equations (2.4), (2.7), (2.12)-(2.14), (2.19), or (3.5), (3.8), (3.10), (3.11) and (3.17) provide a complete system for asymptotic analysis of the outer solution.

3.4 Relation to the Iterative Ambartsumyan Theory

Here we discuss the refined theories of S.A. Ambartsumyan, suggested by him at the end of 1950-s, along with some of their modifications for anisotropic plates and shells taking into account the cross-sectional shear, see Ambartsumyan (1957, 1958, 1961, 1967, 1974). Let us first discuss one of these known as the iterative theory, relying on the following assumptions:

a) the displacement component u_γ normal to the mid-surface of plate or shell is independent of the transverse coordinate γ;

b) when calculating the deformations $e_{\alpha\gamma}, e_{\beta\gamma}$ the tangential stresses $\tau_{\alpha\gamma}, \tau_{\beta\gamma}$ coincide with these $\tau_{\alpha\gamma}^0, \tau_{\beta\gamma}^0$ of the classical theory;

c) the normal stress $\sigma_{\gamma\gamma}$ may be neglected on the planes parallel to the mid-surface (or taken as classical $\sigma_{\gamma\gamma}^0$).

Naturally, it is interesting to investigate the iterative theory within the framework of the asymptotic method. Using the iterative procedure of Section 3.2, we construct the next order $(s = 2)$ approximation of the outer solution. It has been shown in Sections 3.2 and 3.3, that in case of bending all quantities may be expressed through $w^{(s)}$, which in turn is determined from (3.8). For second order approximation we have

$$L\left(w^{(2)}\right) = \frac{3}{a^3}\left(\sigma_{\gamma\gamma}^{*(2)} - \frac{\partial\sigma_{\gamma\gamma}^{*(2)}}{\partial\zeta}\right)\Bigg|_{\zeta=1} \tag{4.1}$$

where L is the operator of the L.H.S. of (3.8), and the quantity $\sigma_{\gamma\gamma}^{*(2)}$ may be obtained from

$$\sigma_{\gamma\gamma}^{*(2)} = -a\int_0^\zeta H_\alpha H_\beta\left(\frac{\partial}{\partial\alpha}\left(\frac{\sigma_{\alpha\gamma}^{*(2)}}{H_\beta}\right) + \frac{\partial}{\partial\beta}\left(\frac{\sigma_{\beta\gamma}^{*(2)}}{H_\alpha}\right)\right)d\zeta,$$

$$\sigma_{\alpha\gamma}^{*(2)} = -a\int_0^\zeta H_\alpha H_\beta\left(\frac{\partial}{\partial\alpha}\left(\frac{\sigma_{\alpha\alpha}^{*(2)}}{H_\beta}\right) - \sigma_{\beta\beta}^{*(2)}\frac{\partial}{\partial\alpha}\frac{1}{H_\beta} + H_\alpha\frac{\partial}{\partial\beta}\left(\frac{\sigma_{\alpha\beta}^{*(2)}}{H_\alpha^2}\right)\right)d\zeta,$$

$$\sigma_{\alpha\alpha}^{*(2)} = \frac{1}{6}a^2\zeta^3\left(B_{11}\chi_1 + B_{12}\chi_2\right)\left(a_{13}\tau_{\alpha\alpha}^{(0)} + a_{23}\tau_{\beta\beta}^{(0)}\right) + a\left(\frac{1}{3}\zeta^3 - \zeta\right)$$

$$\times\left(a_{55}\ell_{11}\left(\tau_{\alpha\gamma}^{(0)}\right) + a_{44}\ell_{21}\left(\tau_{\beta\gamma}^{(0)}\right)\right) + \frac{1}{2}a\zeta\left(a_{55}\ell_{11}(X_2) + a_{44}\ell_{21}(Y_2)\right) - A_1\sigma_{\gamma\gamma}^{(0)}$$

$$(\alpha,\beta;1,2), \tag{4.2}$$

$$\sigma_{\alpha\beta}^{*(2)} = \frac{1}{6}a^2\zeta^3 B_{66}\left(a_{13}\tau\left(\tau_{\alpha\alpha}^{(0)}\right) + a_{23}\tau\left(\tau_{\beta\beta}^{(0)}\right)\right) + aB_{66}\left(\frac{1}{3}\zeta^3 - \zeta\right)$$

$$\times\left(a_{55}\ell_{31}\left(\tau_{\alpha\gamma}^{(0)}\right) + a_{44}\ell_{32}\left(\tau_{\beta\gamma}^{(0)}\right)\right) + \frac{1}{2}a\zeta B_{66}\left(a_{55}\ell_{31}(X_2) + a_{44}\ell_{32}(Y_2)\right),$$

$$\sigma_{\gamma\gamma}^{(0)} = (\zeta^3 - \zeta)\tau_{\gamma\gamma}^{(0)} + \frac{1}{2}\zeta Z_2,$$

$$\tau_{\alpha\gamma}^{(0)} = -\frac{1}{2}aH_\alpha H_\beta\left(\frac{\partial}{\partial\alpha}\left(\frac{\tau_{\alpha\alpha}^{(0)}}{H_\beta}\right) - \tau_{\beta\beta}^{(0)}\frac{\partial}{\partial\alpha}\frac{1}{H_\beta} + H_\alpha\frac{\partial}{\partial\beta}\left(\frac{\tau_{\alpha\beta}^{(0)}}{H_\alpha^2}\right)\right), \quad (\alpha,\beta),$$

$$H_\alpha H_\beta\left(\frac{\partial}{\partial\alpha}\frac{\tau_{\alpha\gamma}^{(0)}}{H_\beta} + \frac{\partial}{\partial\beta}\frac{\tau_{\beta\gamma}^{(0)}}{H_\alpha}\right) = \frac{3}{4}\left(\frac{1}{a}Z_2 + H_\alpha H_\beta\left(\frac{\partial}{\partial\alpha}\left(\frac{X_2}{H_\beta}\right) + \frac{\partial}{\partial\beta}\left(\frac{Y_2}{H_\alpha}\right)\right)\right),$$

$$\tau_{\gamma\gamma}^{(0)} = -\frac{1}{4}\left(Z_2 + aH_\alpha H_\beta\left(\frac{\partial}{\partial\alpha}\left(\frac{X_2}{H_\beta}\right) + \frac{\partial}{\partial\beta}\left(\frac{Y_2}{H_\alpha}\right)\right)\right),$$

$$\tau_{\alpha\alpha}^{(0)} = aB_{11}\left(\chi_1\left(w^{(0)}\right) + \nu_2\chi_2(w^{(0)})\right), \quad (\alpha,\beta;1,2),$$

$$\tau_{\alpha\beta}^{(0)} = aB_{66}\tau\left(w^{(0)}\right).$$

Here the operators

$$\chi_1 = -H_\alpha \frac{\partial}{\partial \alpha}\left(H_\alpha \frac{\partial}{\partial \alpha}\right) + \frac{H_\beta}{H_\alpha}\frac{\partial H_\alpha}{\partial \beta}H_\beta\frac{\partial}{\partial \beta},$$

$$\tau = -2H_\alpha H_\beta\left(\frac{\partial^2}{\partial\alpha\partial\beta} + \frac{1}{H_\alpha}\frac{\partial H_\alpha}{\partial\beta}\frac{\partial}{\partial\alpha} + \frac{1}{H_\beta}\frac{\partial H_\beta}{\partial\alpha}\frac{\partial}{\partial\beta}\right), \qquad (4.3)$$

$$\ell_{11} = B_{11}H_\alpha\frac{\partial}{\partial\alpha} - B_{12}\frac{H_\alpha}{H_\beta}\frac{\partial H_\beta}{\partial\alpha},$$

$$\ell_{21} = B_{12}H_\beta\frac{\partial}{\partial\beta} - B_{11}\frac{H_\beta}{H_\alpha}\frac{\partial H_\alpha}{\partial\beta},$$

$$\ell_{31} = H_\beta\frac{\partial}{\partial\beta} + \frac{H_\beta}{H_\alpha}\frac{\partial H_\alpha}{\partial\beta}, \quad A_1 = a_{13}B_{11} + a_{23}B_{12} \quad (\alpha,\beta;1,2).$$

Representing the relations (4.2), (4.3) in the rectangular coordinate system, we get these in simplified form. Indeed, substituting $H_\alpha = H_\beta = 1$ and excluding $\sigma_{\alpha\gamma}^{*(2)}$, $\sigma_{\beta\gamma}^{*(2)}$ from the first three equations of (4.2), we result in

$$\sigma_{zz}^{*(2)} = a^2\int_0^\zeta d\zeta \int_0^\zeta \left(\frac{\partial^2\sigma_{xx}^{*(2)}}{\partial x^2} + 2\frac{\partial^2\sigma_{xy}^{*(2)}}{\partial x\partial y} + \frac{\partial^2\sigma_{yy}^{*(2)}}{\partial y^2}\right)d\zeta,$$

$$\sigma_{xx}^{*(2)} = -\frac{1}{6}a^2\zeta^3\left(B_{11}\frac{\partial^2}{\partial x^2} + B_{12}\frac{\partial^2}{\partial y^2}\right)\left(a_{13}\tau_{xx}^{(0)} + a_{23}\tau_{yy}^{(0)}\right)$$

$$+a\left(\frac{1}{3}\zeta^3 - \zeta\right)\left(a_{55}B_{11}\frac{\partial\tau_{xz}^{(0)}}{\partial x} + a_{44}B_{12}\frac{\partial\tau_{yz}^{(0)}}{\partial y}\right)$$

$$+\frac{1}{2}a\zeta\left(a_{55}B_{11}\frac{\partial X_2}{\partial x} + a_{44}B_{12}\frac{\partial Y_2}{\partial y}\right)$$

$$-A_1\left(\zeta^3 - \zeta\right)\tau_{zz}^{(0)} - \frac{1}{2}\zeta A_1 Z_2, \quad (x,y;1,2),$$

$$\sigma_{xy}^{*(2)} = -\frac{1}{3}a^2\zeta^3 B_{66}\left(a_{13}\frac{\partial^2\tau_{xx}^{(0)}}{\partial x\partial y} + a_{23}\frac{\partial^2\tau_{xy}^{(0)}}{\partial x\partial y}\right) + aB_{66}\left(\frac{1}{3}\zeta^3 - \zeta\right) \quad (4.4)$$

$$\times\left(a_{55}\frac{\partial\tau_{xz}^{(0)}}{\partial y} + a_{44}\frac{\partial\tau_{yz}^{(0)}}{\partial x}\right) + \frac{1}{2}a\zeta B_{66}\left(a_{55}\frac{\partial X_2}{\partial y} + a_{44}\frac{\partial Y_2}{\partial x}\right),$$

$$\tau_{xx}^{(0)} = -aB_{11}\left(\frac{\partial^2 w^{(0)}}{\partial x^2} + \nu_2\frac{\partial^2 w^{(0)}}{\partial y^2}\right) \quad (x,y,;1,2), \quad \tau_{xy}^{(0)} = -aB_{66}\frac{\partial^2 w^{(0)}}{\partial x\partial y},$$

$$B_{11}\frac{\partial^4 w^{(0)}}{\partial x^4} + 2(B_{12} + 2B_{66})\frac{\partial^4 w^{(0)}}{\partial x^2\partial y^2} + B_{22}\frac{\partial^4 w^{(0)}}{\partial y^4}$$

$$= \frac{3}{2a^2}\left(\frac{1}{a}Z_2 + \frac{\partial X_2}{\partial x} + \frac{\partial Y_2}{\partial y}\right).$$

Expressing $\sigma_{\alpha\alpha}^{*(2)}$, $\sigma_{\beta\beta}^{*(2)}$, $\sigma_{\alpha\beta}^{*(2)}$ through $w^{(0)}$ in (4.2) and substituting these into formulas for $\sigma_{\alpha\gamma}^{*(2)}$, $\sigma_{\beta\gamma}^{*(2)}$ and further into $\sigma_{\gamma\gamma}^{*(2)}$, and using the leading order approximation (3.8), it is possible to obtain the resulting expression for $\sigma_{\gamma\gamma}^{*(2)}$. The second order is then found explicitly from (4.1). The obtained expressions are rather lengthy, therefore we present here the R.H.S. of (4.1) in case of a transversely isotropic body in rectangular coordinate system. Clearly, the case of general anisotropy in curvilinear coordinates may be treated absolutely similarly. It will be shown below that the results are independent of the choice of the coordinate system. Adopting in (4.4)

$$a_{44} = a_{55} = 1/G', \quad B_{11} = B_{22} = B_{12} + 2B_{66} = E/(1 - \nu^2),$$

$$a_{13} = a_{23} = -\nu_{13}/E' = -\nu'/E', \quad A_1 = A_2 = -\frac{E}{E'}\frac{\nu'}{1 - \nu}, \tag{4.5}$$

where E and ν are the Young's modulus and the Poisson's ratio, respectively, and E', G', ν' are the Young's modulus, the shear modulus, and the Poisson's ratio in the plane perpendicular to the plane of isotropy (mid-plane of the plate), bearing in ming the value of $\sigma_{zz}^{*(2)}$ from (4.4) and using (4.1), we obtain

$$L\left(w^{(2)}\right) = \frac{3}{2a}\left(8\frac{G}{G'} - 3\nu'\frac{E}{E'}\right)\frac{\Delta Z_2}{10(1 - \nu)}$$

$$+ \frac{3}{2}\left(4\frac{G}{G'} + \nu'\frac{E}{E'}\right)\frac{1}{30(1 - \nu)}\Delta(\operatorname{div} P) \tag{4.6}$$

where $P = \{X_2, Y_2\}$ is a vector of tangential loading. Rewriting $w^{(s)} = \varepsilon^{3-s}W_0^{(s)}$ and assuming $W = W_0^{(0)} + W_0^{(1)} + W_0^{(2)} = \varepsilon^{-3}w^{(0)} + \varepsilon^{-2}w^{(1)} + \varepsilon^{-1}w^{(2)}$, we obtain the following bending equation for a transversely isotropic plate

$$D\Delta\Delta W = Z_2 - \left(2\frac{G}{G'} - 0.75\frac{E}{E'}\nu'\right)\frac{(2h)^2\Delta Z_2}{10(1 - \nu)}$$

$$+ \left(\frac{G}{G'} + 0.25\frac{E}{E'}\nu'\right)\frac{(2h)^2 a}{30(1 - \nu)}\Delta(\operatorname{div} P) \tag{4.7}$$

with $D = \frac{2}{3}\frac{Eh^3}{1 - \nu^2}$ denoting bending stiffness, and Δ the Laplace operator. The last equation (4.7) derived in rectangular coordinate system is in fact invariant of the choice of transformations of coordinate system. For example, it may be easily verified that in case of isotropic material this equation coincides with that obtained by Kolos (1964) for circular plates.

We remark that in case of $P = 0$ the obtained equation (4.7) differs from that of the iterative S.A. Ambartsumyan theory

$$D\Delta\Delta W = Z_2 - \left(2\frac{G}{G'} - \frac{E}{E'}\nu'\right)\frac{(2h)^2}{10(1 - \nu)}\Delta Z_2 \tag{4.8}$$

only by the factor E/E', which is caused by the assumption a). Indeed, a more rigorous mathematical treatment within the asymptotic method implies

$$W = W_0(\alpha, \beta) + \zeta^2 W^*(\alpha, \beta).$$

However, in case of moderate anisotropy the second term W^* is smaller by two orders than the first term $W^* = O(\varepsilon^2 W_0)$. Taking into account this term characterizing transverse compression causes decrease of the coefficient E/E' from 1 to 0.75. As seen from (4.7), (4.8), the refinement to classical theory depends on the ratios $G/G', E/E'$. If it is dominated by G/G', then the effect of transverse compression may be neglected (the fact of the Poisson's ratio ν' being relatively small should also be noted) whereas in cases of G/G' and E/E' being of the same asymptotic orders or $E/E' \gg G/G'$ the effect of transverse compression should be taken into consideration. This factor has been introduced into a refined version of the Ambartsumyan iterative theory, see Ambartsumyan (1974), which already provide equations identical to that derived above by the asymptotic method.

We note that in case of isotropic material the refinement is of order $O(\varepsilon^2)$, see also Goldenveizer (1962); Kolos (1965). As mentioned above, the effect of anisotropy causes dependence of the order of refined terms on the ratios G/G' and E/E'. If any of the values G/G' and E/E' is of order $O(\varepsilon^{-1})$, then it follows from Sections 3.2 and 3.3 that the refined terms to the classical theory are of order $\sim O(\varepsilon)$, whereas if G/G', $E/E' \sim O(\varepsilon^{-2})$, the refinement is $O(1)$ which means that the Kirchhoff theory is no longer applicable and that further analysis should be carried out relying on the elasticity theory. It should be noted that G/G', $E/E' \sim O(\varepsilon^{-1})$ is often the case, see Table 6, whereas the second case is very seldom due to the fact that ε is small. At the same time using these materials for real engineering structures would be rather difficult, since it leads to significant complications in the analysis of stability of such structures, etc.

Thus, the iterative theory of S.A. Ambartsumyan is a step forward within plates and shells theories since it takes into the factors of different resistance for extension and shear in tangential and transverse directions.

We also remark that even though the outer solutions are identical for both approaches, in the near-edge vicinity the difference is quite dramatic since adoption of the Kirchhoff hypotheses or any other refined assumptions affect significantly the boundary layer solutions, causing distortion of these (up to total neglecting of the edge effects). It should also be noted that the boundary layer solution affects the outer solution through the boundary conditions. The matter will be discussed in more details in Chapter 4.

3.5   Two Types of Boundary Layer and Related Governing Equations

As mentioned earlier, the system (2.3) is a singularly perturbed one, therefore its solution consists of the outer solution over the interior and the rapidly decaying solution localized near the edge, i.e. the boundary layer. Consider the edge $\alpha = \alpha_0$. Let us introduce the scaled variables

$$\alpha - \alpha_0 = h\xi_1, \quad \beta = a\eta, \quad \gamma = h\zeta \tag{5.1}$$

and assume that differentiation with respect to ξ_1, η, ζ does not change the order of the quantities. Then

$$\frac{\partial}{\partial\alpha} = \frac{1}{a}\varepsilon^{-1}\frac{\partial}{\partial\xi_1}, \quad \frac{\partial}{\partial\beta} = \frac{1}{a}\frac{\partial}{\partial\eta}, \quad \frac{\partial}{\partial\gamma} = \frac{1}{a}\varepsilon^{-1}\frac{\partial}{\partial\zeta}. \tag{5.2}$$

Expanding the Lamè coefficients H_α, H_β, along with the curvatures

$$k_\alpha = -\frac{H_\beta}{H_a}\frac{\partial H_a}{\partial\beta}, \quad k_\beta = -\frac{H_\alpha}{H_\beta}\frac{\partial H_\beta}{\partial\alpha}$$

of the α and β coordinate lines as the Taylor series near $\alpha = \alpha_0$ and denoting

$$k_\alpha = \sum_{n=0}^{\infty}\varepsilon^n a^n \xi_1^n k_{\alpha n}, \quad (\alpha,\beta), \quad H_\alpha = \sum_{n=0}^{\infty}\varepsilon^n a^n \xi_1^n H_{\alpha n}, \quad (\alpha,\beta),$$

$$\partial_\alpha = H_\alpha\frac{\partial}{\partial\alpha} = \frac{\varepsilon^{-1}}{a}\sum_{n=0}^{\infty}\varepsilon^n \xi_1^n d_{1n}, \quad d_{1n} = a^n H_{\alpha n}\frac{\partial}{\partial\xi_1}, \tag{5.3}$$

$$\partial_\beta = H_\beta\frac{\partial}{\partial\beta} = a^{-1}\sum_{n=0}^{\infty}\varepsilon^n \xi_1^n d_{2n}, \quad d_{2n} = a^n H_{\beta n}\frac{\partial}{\partial\eta}$$

after substitution of (5.2) and (5.3) into (1.1) and (1.2), the solution may be sought in the general form (Aghalovyan, 1972a, 1978a)

$$R = \sum_{s=0}^{N}\varepsilon^{\chi_R+s} R^{(s)} \tag{5.4}$$

where R is any of the stresses or dimensionless displacements u_α/a, u_β/a, w/a. Since the non-homogeneous boundary conditions at $\gamma = \pm h$ are satisfied by the outer solution, for solution (5.4) we have

$$\sigma_{a\gamma} = \sigma_{\beta\gamma} = \sigma_{\gamma\gamma} = 0 \quad \text{at} \ , \ \gamma = \pm h(\zeta = \pm 1). \tag{5.5}$$

Substituting (5.4) into the latter and constructing the iterative procedure, it is possible to obtain the consistent system for determining $R^{(s)}$ provided

$$\chi_\sigma = \chi, \quad \chi_u = \chi + 1$$

where the whole number χ characterizes the intensity of the boundary layer, which may be found from matching the boundary layer with the outer solution. The system is given by

$$\xi_1^m d_{1m}\sigma_{\alpha\alpha}^{(s-m)} + \xi_1^m d_{2m}\sigma_{\alpha\beta}^{(s-m-1)} + \frac{\partial\sigma_{\alpha\gamma}^{(s)}}{\partial\zeta}$$

$$+a^{m+1}\xi_1^m k_{\beta m}\left(\sigma_{\alpha\alpha}^{(s-m-1)} - \sigma_{\beta\beta}^{(s-m-1)}\right) + 2a^{m+1}\xi_1^m k_{\alpha m}\sigma_{\alpha\beta}^{(s-m-1)} = 0,$$

$$\xi_1^m d_{1m}\sigma_{\alpha\gamma}^{(s-m)} + \xi_1^m d_{2m}\sigma_{\beta\gamma}^{(s-m-1)} + \frac{\partial\sigma_{\gamma\gamma}^{(s)}}{\partial\zeta}$$

$$+a^{m+1}\xi_1^m k_{\beta m}\sigma_{\alpha\gamma}^{(s-m-1)} + a^{m+1}\xi_1^m k_{\alpha m}\sigma_{\beta\gamma}^{(s-m-1)} = 0, \tag{5.6}$$

$$\frac{1}{a}\xi_1^m d_{1m}u_\alpha^{(s-m)} + a^m\xi_1^m k_{\alpha m}u_\beta^{(s-m-1)} = a_{11}\sigma_{\alpha\alpha}^{(s)} + a_{12}\sigma_{\beta\beta}^{(s)} + a_{13}\sigma_{\gamma\gamma}^{(s)},$$

$$\frac{1}{a}\xi_1^m d_{2m}u_\beta^{(s-m-1)} + a^m\xi_1^m k_{\beta m}u_\alpha^{(s-m-1)} = a_{12}\sigma_{\alpha\alpha}^{(s)} + a_{22}\sigma_{\beta\beta}^{(s)} + a_{23}\sigma_{\gamma\gamma}^{(s)},$$

$$\frac{1}{a}\frac{\partial w^{(s)}}{\partial \zeta} = a_{13}\sigma^{(s)}_{\alpha\alpha} + a_{23}\sigma^{(s)}_{\beta\beta} + a_{33}\sigma^{(s)}_{\gamma\gamma}, \qquad \xi_1^m d_{1m}w^{(s-m)} + \frac{\partial u^{(s)}_\alpha}{\partial \zeta} = aa_{55}\sigma^{(s)}_{\alpha\gamma},$$

$$\xi_1^m d_{1m}\sigma^{(s-m)}_{\alpha\beta} + \frac{\partial \sigma^{(s)}_{\beta\gamma}}{\partial \zeta} + \xi_1^m d_{2m}\sigma^{(s-m-1)}_{\beta\beta}$$

$$+a^{m+1}\xi_1^m k_{\alpha m}\left(\sigma^{(s-m-1)}_{\beta\beta} - \sigma^{(s-m-1)}_{\alpha\alpha}\right) + 2a^{m+1}\xi_1^m k_{\beta m}\sigma^{(s-m-1)}_{\alpha\beta} = 0,$$

$$\frac{1}{a}\xi_1^m d_{1m}u^{(s-m)}_\beta + \frac{1}{a}\xi_1^m d_{2m}u^{(s-m-1)}_\alpha$$

$$-a^m\xi_1^m\left(k_{\alpha m}u^{(s-m-1)}_\alpha + k_{\beta m}u^{(s-m-1)}_\beta\right) = a_{66}\sigma^{(s)}_{\alpha\beta},$$

$$\frac{1}{a}\frac{\partial u^{(s)}_\beta}{\partial \zeta} + \frac{1}{a}\xi_1^m d_{2m}w^{(s-m-1)} = a_{44}\sigma^{(s)}_{\beta\gamma}.$$

It is assumed in (5.6) that summation is performed along the dummy index m in the limits from 0 to s. Then the system (5.6) may be split into two subsystems.
a)

$$d_{10}\sigma^{(s)}_{\alpha\alpha} + \partial_\zeta\sigma^{(s)}_{\alpha\gamma} = R^{(s-1)}_\alpha, \qquad d_{10}\sigma^{(s)}_{\alpha\gamma} + \partial_\zeta\sigma^{(s)}_{\gamma\gamma} = R^{(s-1)}_\gamma,$$

$$d_{10}u^{(s)}_\alpha = b_{11}\sigma^{(s)}_{\alpha\alpha} + b_{13}\sigma^{(s)}_{\gamma\gamma} + R^{(s-1)}_{u_\alpha} + \frac{a_{12}}{a_{22}}R^{(s-1)}_{u_\beta}, \tag{5.7}$$

$$\partial_\zeta w^{(s)} = b_{13}\sigma^{(s)}_{\alpha\alpha} + b_{33}\sigma^{(s)}_{\gamma\gamma} + \frac{a_{23}}{a_{22}}R^{(s-1)}_{u_\beta},$$

$$d_{10}w^{(s)} + \partial_\zeta u^{(s)}_\alpha = a_{55}\sigma^{(s)}_{\alpha\gamma} + R^{(s-1)}_w,$$

$$\sigma^{(s)}_{\beta\beta} = -\frac{a_{12}}{a_{22}}\sigma^{(s)}_{\alpha\alpha} - \frac{a_{23}}{a_{22}}\sigma^{(s)}_{\gamma\gamma} + \frac{1}{a_{22}}R^{(s-1)}_{u_\beta},$$

where

$$b_{ij} = a_{ij} - \frac{a_{i2}a_{j2}}{a_{22}}, \qquad (i,j = 1,3),$$

$$R^{(s-1)}_\alpha = -\xi_1^n d_{1n}\sigma^{(s-n)}_{\alpha\alpha} - \xi_1^m d_{2m}\sigma^{(s-m-1)}_{\alpha\beta}$$

$$-a^{m+1}\xi_1^m k_{\beta m}\left(\sigma^{(s-m-1)}_{\alpha\alpha} - \sigma^{(s-m-1)}_{\beta\beta}\right) - 2a^{m+1}\xi_1^m k_{\alpha m}\sigma^{(s-m-1)}_{\alpha\beta},$$

$$R^{(s-1)}_\gamma = -\xi_1^n d_{1n}\sigma^{(s-n)}_{\alpha\gamma} - \xi_1^m d_{2m}\sigma^{(s-m-1)}_{\beta\gamma}$$

$$-a^{m+1}\xi_1^m(k_{\beta m}\sigma^{(s-m-1)}_{\alpha\gamma} + k_{\alpha m}\sigma^{(s-m-1)}_{\beta\gamma}), \tag{5.8}$$

$$R^{(s-1)}_{u_\alpha} = -\xi_1^n d_{1n}u^{(s-n)}_\alpha - a^{m+1}\xi_1^m k_{\alpha m}u^{(s-m-1)}_\beta,$$

$$R^{(s-1)}_{u_\beta} = \xi_1^m d_{2m}u^{(s-m-1)}_\beta + a^{m+1}\xi_1^m k_{\beta m}u^{(s-m-1)}_\alpha,$$

$$R^{(s-1)}_w = -\xi_1^n d_{1n}w^{(s-n)}.$$

We note that equations (5.7) are that of the generalized plane stress equations in the plane $(\xi_1, \ \zeta)$, being homogeneous at $s = 0$ and non-homogeneous at $s > 0$.

The second subsystem is b)

$$d_{10}\sigma^{(s)}_{\alpha\beta} + \partial_\zeta\sigma^{(s)}_{\beta\gamma} = R^{(s-1)}_\tau, \quad \partial_\zeta = \frac{\partial}{\partial\zeta}, \tag{5.9}$$

$$d_{10}u^{(s)}_\beta = a_{66}\sigma^{(s)}_{\alpha\beta} + R^{(s-1)}_\beta, \qquad \partial_\zeta u^{(s)}_\beta = a_{44}\sigma^{(s)}_{\beta\gamma} - \xi_1^m d_{2m}w^{(s-m-1)}$$

with

$$R_\tau^{(s-1)} = -\xi_1^n d_{1n}\sigma_{\alpha\beta}^{(s-n)} - \xi_1^m d_{2m}\sigma_{\beta\beta}^{(s-m-1)}$$

$$-a^{m+1}\xi_1^m k_{\alpha m}\left(\sigma_{\beta\beta}^{(s-m-1)} - \sigma_{\alpha\alpha}^{(s-m-1)}\right) - 2a^{m+1}\xi_1^m k_{\beta m}\sigma_{\alpha\beta}^{(s-m-1)}, \qquad (5.10)$$

$$R_\beta^{(s-1)} = -\xi_1^n d_{1n}u_\beta^{(s-n)} - \xi_1^m d_{2m}u_\alpha^{(s-m-1)}$$

$$+a^{m+1}\xi_1^m\left(k_{\alpha m}u_\alpha^{(s-m-1)} + k_{\beta m}u_\beta^{(s-m-1)}\right).$$

It is assumed that in (5.8) and (5.10) summation is performed along n from 1 to s, whereas that along m is from 0 to s. The equations (5.9) are the antiplane stress (torsion) equations for anisotropic solid.

The boundary layer solution is sought in the following rectangular domain $\Omega = \{(\xi_1,\zeta) : 0 \le \xi_1 < +\infty, -1 \le \zeta \le +1\}$. The discussed solution should satisfy the conditions (5.5) on the faces and should also decay rapidly away from the edge $\alpha = \alpha_0(\xi_1 = 0)$. We note that though total decay should occur at $\xi_1 = \dfrac{\ell}{a}\varepsilon^{-1} = \dfrac{\ell}{h}$, where ℓ is the length of the α line $\alpha = \alpha_0$, since the decay is exponential and ε is small, it could be assumed that the opposite edge $\alpha = \ell$ corresponds to $\xi_1 = +\infty$, see Green (1962b); Goldenveizer (1969). It should be noted that the boundary layer solution is not defined for arbitrary edge loading, therefore additional restrictions on the edge boundary values of stresses should be discussed. These restricted conditions could be derived from (5.7) and (5.9). Let us introduce the notation $t = \xi_1/H_{\alpha 0}$ and apply operators

$$\int_{-1}^{+1} d\zeta \int_0^\infty dt, \qquad \int_{-1}^{+1} \zeta d\zeta \int_0^\infty dt \qquad (5.11)$$

and

$$\int_{-1}^{+1} d\zeta \int_0^\infty dt, \qquad \int_{-1}^{+1} d\zeta \int_0^\infty t\,dt \qquad (5.12)$$

to equations (5.7) and (5.9), respectively (the applicability of the operators is guaranteed since these correspond to the usual moments and forces). After some transformations

$$\int_{-1}^{+1} \sigma_{\alpha\alpha}^{(s)}(t=0)d\zeta = -\int_{-1}^{+1} d\zeta \int_0^\infty R_\alpha^{(s-1)}dt,$$

$$\int_{-1}^{+1} \zeta\sigma_{\alpha\alpha}^{(s)}(t=0)d\zeta = -\int_{-1}^{+1} \zeta d\zeta \int_0^\infty R_\alpha^{(s-1)}dt + \int_{-1}^{+1} d\zeta \int_0^\infty R_\gamma^{(s-1)}t\,dt, \quad (5.13)$$

$$\int_{-1}^{+1} \sigma_{\alpha\gamma}^{(s)}(t=0)d\zeta = -\int_{-1}^{+1} d\zeta \int_0^\infty R_\gamma^{(s-1)}dt.$$

Applying the first operator of (5.11) to the equilibrium equations (5.9) results in

$$\int_{-1}^{+1} \sigma_{\alpha\beta}^{(s)}(t=0)d\zeta = -\int_{-1}^{+1} d\zeta \int_0^\infty R_\tau^{(s-1)}dt. \qquad (5.14)$$

Since the faces and the opposite edge $\xi_1 = +\infty$ are free, the self-equilibrium of a semi-infinite strip Ω occurs if and only if the edge stresses at $\xi_1 = 0$ are in equilibrium with the mass forces, which is stated in (5.13), (5.14). The latter coincide with conditions of Goldenveizer (1961, 1969, 1976); Green (1962b), the difference is only in the meaning of $R_\alpha, R_\gamma, R_\tau$.

Applying the first operator of (5.12) to the second equation of (5.9) along with the second operator of (5.12) to the first equation of (5.9), we obtain the existence condition for decaying solution for the anti-plane problem in case of the displacement specified at the edge

$$\int_{-1}^{+1} u_\beta^{(s)}(t=0)d\zeta = a_{66}\int_0^\infty tdt \int_{-1}^{+1} R_\tau^{(s-1)}d\zeta - \int_{-1}^{+1} d\zeta \int_0^\infty R_\beta^{(s-1)}dt. \quad (5.15)$$

At $s = 0$ the systems (5.7), (5.9) are homogeneous and separated, therefore, the decaying solution may be found by the Fourier method, see Aghalovyan (1972a, 1973a, 1978a). Let us denote the solutions for the plane and anti-plane problems by superscript "b" and "a", respectively. Then

$$\overset{b}{\underset{(0)}{Q}} = \sum_{(\lambda_n)} \overset{b}{Q}_n(\zeta)\exp(-\lambda_n t) \quad (5.16)$$

or

$$\left(\overset{b}{\underset{\alpha\alpha}{\sigma}}{}^{(0)}, \overset{b}{\underset{\alpha\gamma}{\sigma}}{}^{(0)}, \overset{b}{\underset{\gamma\gamma}{\sigma}}{}^{(0)}\right) = \sum_{(\lambda_n)} \left(\overset{b}{\underset{\alpha}{\tau}}{}^{(0)}, \overset{b}{\underset{\alpha\gamma}{\tau}}{}^{(0)}, \overset{b}{\underset{\gamma}{\tau}}{}^{(0)}\right)_n \exp(-\lambda_n t),$$

$$\left(\overset{b}{\underset{\alpha}{u}}{}^{(0)}\Big/a, \overset{b}{w}{}^{(0)}\Big/a\right) = \sum_{(\lambda_n)} \left(\overset{b}{u}{}_{\alpha(0)}, \overset{b}{w}{}_{(0)}\right)_n \exp(-\lambda_n t). \quad (5.17)$$

Substituting (5.17) into (5.7) gives

$$-\lambda_n \overset{b}{\underset{\alpha}{\tau}}{}^{(0)} + \frac{d\overset{b}{\underset{\alpha\gamma}{\tau}}{}^{(0)}}{d\zeta} = 0, \qquad -\lambda_n \overset{b}{\underset{\alpha\gamma}{\tau}}{}^{(0)} + \frac{d\overset{b}{\underset{\gamma}{\tau}}{}^{(0)}}{d\zeta} = 0,$$

$$-\lambda_n \overset{b}{u}{}_{\alpha(0)} = b_{11}\overset{b}{\underset{\alpha}{\tau}}{}^{(0)} + b_{13}\overset{b}{\underset{\gamma}{\tau}}{}^{(0)}, \quad (5.18)$$

$$\frac{d\overset{b}{w}{}^{(0)}}{d\zeta} = b_{13}\overset{b}{\underset{\alpha}{\tau}}{}^{(0)} + b_{33}\overset{b}{\underset{\gamma}{\tau}}{}^{(0)}, \qquad -\lambda_n \overset{b}{w}{}_{(0)} + \frac{d\overset{b}{u}{}_{\alpha(0)}}{d\zeta} = a_{55}\overset{b}{\underset{\alpha\gamma}{\tau}}{}^{(0)}.$$

Satisfying the boundary conditions (5.5), we obtain

$$\overset{b}{\underset{\alpha}{\tau}}{}^{(0)} = \frac{F_n''}{\lambda_n^2}A_n^{(0)}, \qquad \overset{b}{\underset{\alpha\gamma}{\tau}}{}^{(0)} = \frac{F_n'}{\lambda_n}A_n^{(0)}, \qquad \overset{b}{\underset{\gamma}{\tau}}{}^{(0)} = F_n A_n^{(0)},$$

$$\overset{b}{u}{}_{\alpha(0)} = -\frac{1}{\lambda_n^3}\left(b_{11}F_n'' + b_{13}\lambda_n^2 F_n\right)A_n^{(0)}, \quad (5.19)$$

$$\overset{b}{w}{}_{(0)} = -\frac{1}{\lambda_n^4}\left(b_{11}F_n''' + (b_{13} + a_{55})\lambda_n^2 F_n'\right)A_n^{(0)},$$

where $F_n(\zeta)$ is the solution of

$$b_{11}F_n^{IV} + \lambda_n^2(a_{55} + 2b_{13})F_n'' + \lambda_n^4 b_{33}F_n = 0,$$

$$F_n(\pm 1) = F_n'(\pm 1) = 0. \quad (5.20)$$

Similarly to the previously analyzed case of the orthotropic strip Ω, the characteristic (secular) equation, corresponding to (5.20) could possess either

1. purely imaginary repeated roots $i\beta$, $-i\beta$;
2. purely imaginary roots $\pm i\beta_1$, $\pm i\beta_2$ $(\beta_i > 0)$
3. complex conjugate roots $\alpha \pm i\beta$, $-\alpha \pm i\beta$.

The first case takes place provided

$$D = \left(\frac{1}{2}a_{55} + b_{13}\right)^2 - b_{11}b_{33} = 0, \quad \beta = \sqrt[4]{\frac{b_{33}}{b_{11}}}. \tag{5.21}$$

For the second case $D > 0$ or $\left|\frac{1}{2}a_{55} + b_{13}\right| > \sqrt{b_{11}b_{33}}$ hence

$$\beta_1 = \sqrt{\frac{\left|\frac{1}{2}a_{55} + b_{13}\right| - \sqrt{D}}{b_{11}}}, \quad \beta_2 = \sqrt{\frac{\left|\frac{1}{2}a_{55} + b_{13}\right| + \sqrt{D}}{b_{11}}}. \tag{5.22}$$

Finally, the third case corresponds to $D < 0$, therefore

$$\alpha = \sqrt{\frac{\sqrt{b_{11}b_{13}} - \left|\frac{1}{2}a_{55} + b_{13}\right|}{2b_{11}}}, \quad \beta = \sqrt{\frac{\sqrt{b_{11}b_{13}} + \left|\frac{1}{2}a_{55} + b_{13}\right|}{2b_{11}}}. \tag{5.23}$$

In practice, the second case means that the shear stiffness in the transverse plane is less than the mean stiffness for extension/compression in the same plane, i.e. $2G_{13} < \sqrt{E_1 E_3}$, while in the third case $2G_{13} > \sqrt{E_1 E_3}$, which is a rather seldom case for real materials.

The solutions for the cases presented above are given below.

1)$i\beta$, $-i\beta$

Symmetric problem:

$$F_n(\zeta) = \sin\lambda_n\beta\cos\lambda_n\beta\zeta - \zeta\cos\lambda_n\beta\sin\lambda_n\beta\zeta \tag{5.24}$$

where λ_n is the root of $\sin 2\lambda_n\beta + 2\lambda_n\beta = 0$.

Antisymmetric problem:

$$F_n(\zeta) = \cos\lambda_n\beta\sin\lambda_n\beta\zeta - \zeta\sin\lambda_n\beta\cos\lambda_n\beta\zeta,$$
$$\sin 2\lambda_n\beta - 2\lambda_n\beta = 0. \tag{5.25}$$

2)$i\beta_1$, $i\beta_2$ $(\beta_i > 0)$

Symmetric problem:

$$F_n(\zeta) = \cos\lambda_n\beta_2\cos\lambda_n\beta_1\zeta - \cos\lambda_n\beta_1\cos\lambda_n\beta_2\zeta \tag{5.26}$$

where λ_n is determined from

$$\omega\sin z_n + \sin\omega z_n = 0,$$
$$z_n = (\beta_1 + \beta_2)\lambda_n, \quad \omega = \frac{\beta_2 - \beta_1}{\beta_1 + \beta_2} \quad 0 < \omega < 1. \tag{5.27}$$

Antisymmetric problem:

$$F_n(\zeta) = \sin\lambda_n\beta_2\sin\lambda_n\beta_1\zeta - \sin\lambda_n\beta_1\sin\lambda_n\beta_2\zeta,$$
$$\omega\sin z_n - \sin\omega z_n = 0, \quad z_n = (\beta_1 + \beta_2)\lambda_n, \quad \omega = \frac{\beta_2 - \beta_1}{\beta_1 + \beta_2}. \tag{5.28}$$

3)$\alpha + i\beta$, $-\alpha + i\beta$

Symmetric problem:

$$F_n(\zeta) = \sin\lambda_n\beta\,\mathrm{sh}\alpha\lambda_n\cos\beta\lambda_n\zeta\,\mathrm{ch}\alpha\lambda_n\zeta - \cos\lambda_n\beta\,\mathrm{ch}\alpha\lambda_n\sin\beta\lambda_n\zeta\,\mathrm{sh}\alpha\lambda_n\zeta \quad (5.29)$$

with λ_n satisfying

$$\omega\sin z_n + \mathrm{sh}\omega z_n = 0, \qquad z_n = 2\beta\lambda_n, \quad \omega = \frac{\alpha}{\beta} \quad 0 < \omega < 1. \tag{5.30}$$

Antisymmetric problem:

$$F_n(\zeta) = \cos\lambda_n\beta\,\mathrm{sh}\alpha\lambda_n\sin\beta\lambda_n\zeta\,\mathrm{ch}\alpha\lambda_n\zeta - \sin\lambda_n\beta\,\mathrm{ch}\alpha\lambda_n\cos\beta\lambda_n\zeta\,\mathrm{sh}\alpha\lambda_n\zeta,$$

$$\omega\sin z_n - \mathrm{sh}\omega z_n = 0, \qquad z_n = 2\beta\lambda_n, \quad \omega = \frac{\alpha}{\beta}. \tag{5.31}$$

The transcendental equations (5.24)-(5.31) follow from (5.5). The problem (5.20) for F_n reduces to a homogeneous algebraic system, therefore the conditions for the existence of non-trivial solutions, i.e. vanishing of the appropriate determinant leads to the corresponding transcendental equations. These equations possess a number of important features. In particular, if λ_n is the root of equation, then $-\lambda_n$, $\pm\overline{\lambda}_n$ are also the roots, in other words the roots are located symmetrically in all the four quadrants. Therefore, according to the Picard theorem, the solutions of these equations form a countable set. One more important property of the equations (5.24)-(5.31) is that all the solutions for λ_n cannot be imaginary. Let us prove it, for example, for equation (5.27). Indeed, assuming $z_n = iy_n$, $y_n \in \mathbb{R}$, we rewrite the equation as

$$\omega\,\sinh y_n + \sinh\omega\,y_n = 0.$$

The function $f(y_n) = \omega\,\sinh y_n + \sinh\omega\,y_n c$ is monotonously increasing over the domain since $f'(y_n) = \omega(\cosh y_n + \cosh\omega\,y_n) > 0$. Therefore, taking into account that $f(0) = 0$, the only solution of $f(y_n) = 0$ is trivial. Hence, equation (5.27) does not possess any imaginary roots. The other equations may be treated similarly.

We remark that even though $\lambda = 0$ is a root of equations (5.24)-(5.31), the corresponding solution for stresses is zero, i.e. the rigid body motion associated with $\lambda = 0$, has been incorporated earlier in the outer solution.

The boundary value problem (5.20) is a generalized non-self-similar problem for eigenvalues and eigenfunctions. Similar problems, including more general ones, have been considered by Tamarkin (1917); Keldysh (1951); Palant (1961); Markus (1962); Vizitei and Markus (1965); Krein and Langer (1965); Ustinov and Iudovich (1973); Ustinov (1974); Kostyuchenko and Orazov (1975, 1977); Orazov (1976). As follows from the cited works, the system of eigenfunctions of the problem (5.20) is fourfold complete, whereas the system of eigenfunctions corresponding to $Re\lambda_n > 0$ is doubly complete. Since the boundary layer solution is decaying, it corresponds to $Re\lambda_n > 0$, therefore, the system of functions of the boundary layer is doubly complete.

We note that equations (5.24) and (5.25) coincide with the transcendental equations for the plane boundary layer for isotropic rectangle, having complex roots, as was discussed in Chapter 1.

Equations (5.27) and (5.28) possess real roots (for $0 < \omega < 1$). It may be shown that the roots cannot be complex $z_n = x_n + iy_n$ such that $x_n \to \infty$, $y_n \to \infty$. Indeed, this may be demonstrated from the asymptotic expansion for the roots having large modulus. Another way of demonstrating that could be obtained through Maclaurin series expansions of the functions $\sin z_n$, $\sin \omega\, z_n$ in (5.27) and (5.28), and then apply the famous de Gua theorem for the roots of the polynomials, stating that for a polynomial

$$a_0 x^n + a_1 x^{n-1} + a_2 x^{n-2} + \cdots + a_n = 0$$

having all real coefficients and real roots, the following inequalities are true

$$a_k^2 > a_{k-1} a_{k+1} \quad (k = 1, 2, ..., n-1).$$

In order to investigate the existence of complex roots, one could apply the corollary of the de Gua theorem, namely, if there exists a coefficient a_k, $(k = 1, 2, ..., n-1)$, such that

$$a_k^2 \leq a_{k-1} a_{k+1}$$

then the polynomial has at least one couple of complex roots. The first seven roots of equations (5.27), (5.28) has been obtained numerically for several materials, see Tables 6 and 7.

The first seven eigenvalues are presented in Tables 8 and 9, with information on elastic parameters of the materials given in Chjen (1970); Lekhnitskii (1977). It may be observed that $Rez_1 > \pi$, meaning that the case of propagating plane boundary layer (or the so-called weak boundary layer) is not possible for anisotropic plates. However, as will be shown below, the effect of propagating boundary layer may occur in the anti-plane case. This phenomenon is not observed in isotropic plates.

In case of isotropic plate the decay in extensional/compressional case is slower compared to that of the bending case. As seen from Tables 8 and 9, similar effect may be noted for anisotropic materials. The attenuation velocity depends on the material properties, and for some anisotropic materials this velocity is less than that observed in the isotropic case. The opposite behavior is demonstrated by the graphite-epoxy material (for extension/compresion), due to its higher stiffness in transverse direction than that of the longitudinal one (Table 8).

Anisotropy often adds a complimentary "resistance", which causes exponential decay, however, it is not harmonic (since the first root of the characteristic equation is real).

Equations (5.30), (5.31) cannot possess real roots. Let us present a proof for equation (5.30). Similar considerations are valid in case of (5.31) as well. Let $f(z) = \omega \sin z + \sinh \omega z$, then $f'(z) = \omega (\cos z + \cosh \omega z)$, so $f'(z) > 0$ at $z > 0$, therefore, since $f(0) = 0$, $f(z)$ has no real roots apart from $z = 0$. It has been shown earlier that equations (5.30) and (5.31) cannot possess imaginary solutions, therefore their roots form a countable set of the form $z_n = \pm x_n \pm iy_n$ with the limiting point at infinity.

Table 6

Elastic characteristics	Materials				
	Unidirectional fiberglass winding	Fiberglass STET	Fiberglass ASTT(b)-C_2-O and PN-3	SVAM 10:1	Graphite-epoxy
1	2	3	4	5	6
$E_1 10^{-9}$Pa	55.917	35.2179	17.5599	38.259	7.2594
$E_2 10^{-9}$Pa	13.734	28.7433	12.8511	17.658	7.2594
$E_3 10^{-9}$Pa	13.734	17.9523	4.2183	9.6138	84.6603
$G_{12} 10^{-9}$Pa	5.592	7.4556	2.7468	5.1993	2.7468
$G_{23} 10^{-9}$Pa	4.905	6.1803	2.3544	3.1392	4.20849
$G_{13} 10^{-9}$Pa	5.592	6.4746	2.3544	3.8357	4.20849
ν_{12}	0.277	0.177	0.15	0.22	0.323
ν_{23}	0.4	0.371	0.31	0.31	0.0257
ν_{31}	0.068	0.157	0.08	0.07	0.3002
β_1	1.22365	0.646914	0.816764	0.652242	0.23067
β_2	3.002848	2.096657	2.478728	3.0113	1.34067
ω	0.421	0.528	0.504	0.644	0.706
z_n	Roots of the equation $\omega \sin z_n + \sin \omega \, z_n = 0$				
z_1	5.1154 +1.6488i	4.9223	5.55374	4.05068	3.80917
z_2	8.3587	7.0685 +1.1902i	6.661 +0.6377i	8.492 +1.2476i	7.79214
z_3	13.9185	12.2309	12.51652	10.82398	10.7968 +1.4003i
z_4	16.6884 +1.4871i	16.8852	17.8056	15.16315	14.40889
z_5	22.1894	20.1515 +1.5729i	19.4103 +0.9153i	19.18311	18.31786
z_6	26.9385	24.44628	25.03292	23.33729	22.12027
z_7	30.5937	29.04024	30.18361	26.5603 +1.4714i	25.94057

Table 7

Elastic characteristics	Materials				
	Unidirectional fiberglass winding	Fiberglass STET	Fiberglass ASTT(b)-C_2-O and PN-3	SVAM 10:1	Graphite-epoxy
1	2	3	4	5	6
β_1	1.22365	0.646914	0.816764	0.652242	0.23067
β_2	3.002848	2.096657	2.478728	3.0113	1.34067
ω	0.421	0.528	0.504	0.644	0.706
z_n	Roots of the equation $\omega \sin z_n - \sin \omega\, z_n = 0$				
z_1	6.88801	6.11631	6.258251	5.56104	5.323286
z_2	10.9171 +1.7865i	10.85764	11.65118	9.59086	9.161465
z_3	15.31192	13.6316 +1.4448i	13.0498 +0.8017i	13.67874	12.96059
z_4	21.2411 +1.2138i	18.34174	15.7573 +1.0042i	17.5111 +1.4524i	16.82544
z_5	23.40535	22.95215	18.77473	20.58245	21.0575 +1.2055i
z_6	29.10092	26.6567 +1.6216i	23.98657	24.75941	23.3184
z_7	32.7487 +1.7419i	30.5405	31.29114	28.77838	27.46158

In order to verify the existence of complex roots, it is possible to replace the functions $\sin z_n$ and $\sinh \omega z_n$ by their Maclaurin series, and then apply the de Gua theorem. Then equation (5.30) may be solved in respect of $z_n = x_n + iy_n$, giving two equations from the real and imaginary parts, yielding

$$\omega\, \sin x_n \cosh y_n + \sinh \omega x_n \cos \omega\, y_n = 0,$$
$$\omega\, \cos x_n \sinh y_n + \cosh \omega x_n \sin \omega\, y_n = 0. \tag{5.32}$$

We note that asymptotic approximations may be obtained for roots having large modulus. Indeed, assuming for large x_n and y_n

$$\sinh y_n \sim \frac{1}{2}\exp y_n, \quad \cosh y_n \sim \frac{1}{2}\exp y_n,$$
$$\sinh \omega x_n \approx \cosh \omega x_n \sim \frac{1}{2}\exp(\omega x_n). \tag{5.33}$$

Table 8

$\lambda_n = \dfrac{z_n}{\beta_1+\beta_2}$	Eigenvalues λ_n corresponding to equation $\omega \sin z_n + \sin \omega z_n = 0$					Isotropic body $\sin 2\lambda_n + 2\lambda_n = 0$
	Unidirectional fiberglass winding	Fiberglass STET	Fiberglass ASTT(b)-C_2-O and PN-3	SVAM 10:1	Graphite-epoxy	
$\beta_1+\beta_2$	4.226498	2.743571	3.295492	3.663542	1.57134	-
ω	0.421	0.528	0.504	0.644	0.706	-
λ_1	1.2103 +0.392i	1.7941	1.68525	1.10567	2.42415	2.1062 +1.125i
λ_2	1.9777	2.5764 +0.434i	2.021 +0.194i	2.3179 +0.341i	4.95891	5.3563 +1.552i
λ_3	3.2932	4.45802	3.79807	2.95451	6.8711 +0.891i	8.5367 +1.776i
λ_4	3.9485 +0.352i	6.15446	5.40302	4.13893	9.16981	11.699 +1.929i
λ_5	5.2501	7.3449 +0.573i	5.8899 +0.278i	5.23622	11.65748	14.854 +2.047i
λ_6	6.3737	8.91039	7.59611	6.37014	14.07733	18.005 +2.142i
λ_7	7.2385	10.58483	9.15906	7.2499 +0.402i	16.50857	21.153 +2.222i

Therefore, equations (5.32) imply the following asymptotic relations

$$\exp(\omega x_n - y_n) = -\frac{\omega \sin x_n}{\cos \omega y_n} = -\frac{\omega \cos x_n}{\sin \omega y_n} \tag{5.34}$$

leading to

$$\cos(x_n + \omega y_n) = 0,$$
$$x_n + \omega y_n = (2k+1)\frac{\pi}{2}, \qquad (k = 0, 1, ...), \tag{5.35}$$
$$\exp(\omega x_n - y_n) = \omega(-1)^{k+1}.$$

In the latter the L.H.S. is positive, hence $k = 2n - 1$, $(n = 1, 2, ...)$, therefore

$$x_n \approx \frac{1}{1+\omega^2}\frac{\pi}{2}(4n-1) + \frac{\omega}{1+\omega^2}\ln\omega,$$
$$y_n \approx \frac{\omega}{1+\omega^2}\frac{\pi}{2}(4n-1) - \frac{1}{1+\omega^2}\ln\omega. \tag{5.36}$$

     Asymptotic Theory of Anisotropic Plates and Shells

Table 9

$\lambda_n = \dfrac{z_n}{\beta_1+\beta_2}$	Eigenvalues λ_n corresponding to equation $\omega \sin z_n - \sin \omega z_n = 0$					Isotropic body $\sin 2\lambda_n - 2\lambda_n = 0$
	Unidirectional fiberglass winding	Fiberglass STET	Fiberglass ASTT(b) - C_2-O and PN-3	SVAM 10:1	Graphite-epoxy	
$\beta_{1+}\beta_2$	4.226498	2.743571	3.295492	3.663542	1.57134	-
ω	0.421	0.528	0.504	0.644	0.706	-
λ_1	1.6297204	2.229324	1.899034	1.517941	3.387737	3.7488 +1.3843i
λ_2	2.583 +0.4227i	3.957485	3.53549	2.61792	5.830352	6.9499 +1.6761i
λ_3	3.622839	4.9686 +0.527i	3.959 +0.2433i	3.733747	8.248113	10.119 +1.8584i
λ_4	5.0257 +0.287i	6.685353	4.782 +0.303i	4.779 +0.396i	10.7077	13.277 +1.9916i
λ_5	5.537764	8.365784	5.697095	5.6181832	13.401 +0.767i	16.4298 +2.097i
λ_6	6.885351	9.716 +0.591i	7.278601	6.758326	14.839818	19.579 +2.1834i
λ_7	7.748 +0.412i	11.13166	9.495135	7.855343	17.476536	22.727 +2.2573i

Thus, the asymptotic approximation for roots of large modulus of equation (5.30) is given by

$$z_n \sim \frac{1}{1+\omega^2}\left(\frac{\pi}{2}(4n-1) + \omega \ln \omega + i\left(\frac{\pi}{2}(4n-1)\omega - \ln \omega\right)\right). \tag{5.37}$$

Similarly, for equation (5.31) it may be deduced that

$$x_n \sim (4n+1)\frac{\pi}{2}\frac{1}{1+\omega^2} + \frac{\omega}{1+\omega^2}\ln \omega,$$

$$y_n \sim (4n+1)\frac{\pi}{2}\frac{\omega}{1+\omega^2} - \frac{1}{1+\omega^2}\ln \omega, \tag{5.38}$$

$$z_n \sim \frac{1}{1+\omega^2}\left((4n+1)\frac{\pi}{2} + \omega \ln \omega + i\left(\frac{\pi}{2}(4n+1)\omega - \ln \omega\right)\right),$$

$$(n = 0, 1, 2, ...).$$

We remark that the study of boundary layer by the Lourier-Vorovich method leads once again to equations (5.24)-(5.31). In particular, equations (5.24), (5.25) have been used within isotropic context by Aksentyan and Vorovich (1963); Vorovich and Malkina (1966). Using the same approach, equations (5.30) and (5.31) have been obtained for transversely isotropic plates by Romenskaya and Shlenev (1973, 1976).

It should be noted that the function $F_n(\zeta)$ of (5.19) satisfy the generalized orthogonality condition (Aghalovyan and Khachatryan, 1975)

$$\int_0^1 \left(F_n'' F_k'' - \frac{b_{33}}{b_{11}} \lambda_n^2 \lambda_k^2 F_n F_k \right) d\zeta = 0 \qquad (n \neq k) \tag{5.39}$$

which is used further when determining the arbitrary constants of the boundary layer solution.

One more important property of solutions (5.17), (5.19) is related to self-equilibrated nature of the stresses $\overset{b}{\sigma}\,\overset{(0)}{_\alpha}$, $\overset{b}{\sigma}\,\overset{(0)}{_{\alpha\gamma}}$ in arbitrary cross section $t = t_0$, namely

$$\int_{-1}^{+1} \overset{b}{\sigma}\,\overset{(0)}{_\alpha} d\zeta = 0, \qquad \int_{-1}^{+1} \zeta \overset{b}{\sigma}\,\overset{(0)}{_\alpha} d\zeta = 0, \qquad \int_{-1}^{+1} \overset{b}{\sigma}\,\overset{(0)}{_{\alpha\gamma}} d\zeta = 0. \tag{5.40}$$

In other words, at the edge the plane boundary layer can take over only the self-equilibrated part of the load.

At $s = 0$ the decaying solution of (5.9) satisfying the second boundary condition of (5.5), may be determined uniquely by the Fourier method, giving

$$\overset{a}{u}\,\overset{(0)}{_\beta} = \Phi(t, \zeta)$$

$$\overset{a}{\sigma}\,\overset{(0)}{_{\alpha\beta}} = \frac{1}{a_{66}} \frac{\partial \Phi}{\partial t}, \qquad \overset{a}{\sigma}\,\overset{(0)}{_{\beta\gamma}} = \frac{1}{a_{44}} \frac{\partial \Phi}{\partial \zeta}. \tag{5.41}$$

In the symmetric problem

$$\Phi = \sum_{n=1}^{\infty} A_n(\eta) \cos \pi n \zeta \exp\left(-\sqrt{\frac{a_{66}}{a_{44}}} \pi n t \right) \tag{5.42}$$

whereas in the antisymmetric case

$$\Phi = \sum_{n=1}^{\infty} A_n(\eta) \sin(2n - 1) \frac{\pi}{2} \zeta \exp\left(-\sqrt{\frac{a_{66}}{a_{44}}} (2n - 1) \frac{\pi}{2} t \right). \tag{5.43}$$

It may be noticed for solution (5.43) that for arbitrary $t = t_0$

$$\int_{-1}^{+1} \overset{a}{\sigma}\,\overset{(0)}{_{\alpha\beta}} d\zeta = 0, \tag{5.44}$$

i.e. the antiplane boundary layer can take over the part of tangential loading which does not contribute to emergence of the transverse force.

Thus, in case of spatial problem there exist two types of boundary layer solutions, namely the plane and the anti-plane solutions, which are both decaying but at different speed. The plane boundary layer decays as $\exp(-\min Re\lambda_n \cdot t)$, where

$\min Re\lambda_n$ is the real part of the first root of equations (5.24)-(5.31) such that $Re\lambda > 0$.

The antiplane boundary layer solution decays as

$$\exp\left(-\sqrt{\frac{a_{66}}{a_{44}}}\pi t\right) = \exp\left(-\sqrt{\frac{G_{23}}{G_{12}}}\pi t\right) \tag{5.45}$$

in case of symmetric problem, and as

$$\exp\left(-\sqrt{\frac{G_{23}}{G_{12}}}\frac{\pi}{2}t\right) \tag{5.46}$$

in the antisymmetric one.

The results (5.45) and (5.46) reveal that the anti-plane boundary layer decays slower for materials with less shear stiffness in the transverse plane. The effect of the boundary layer arising near the edge $\alpha = \alpha_0$ should be taken into account on the opposite edge ($t = O(\varepsilon^{-1})$) already at $\sqrt{G_{23}/G_{12}}\pi \sim O(\varepsilon^1)$, and the solution will be propagating, i.e. it will be a "weak" boundary layer at $\sqrt{G_{23}/G_{12}}\pi \sim O(\varepsilon^2)$.

The phenomenon of "weak" boundary layers is also observed in layered plates, when the elastic Lamè parameters are piecewise-continuous function of the transverse coordinate, as has been shown by Gusein-Zade (1970); Vorovich et al. (1975); Ustinov (1976). It should be noted that the "weak" boundary layer solutions are not possible within isotropic framework, therefore, the Kirchhoff plate theory is more accurate for isotropic plates, rather than anisotropic. Therefore, the anisotropy of the material is a factor to be considered before choosing the applied plate theory.

At $s > 0$ the systems (5.7) and (5.9) are non-homogeneous, so the form of solutions depends basically on the choice of system at $s = 0$. For example, if we start from (5.7) at $s = 0$, then the obtained λ_n do not satisfy the transcendental equations for the anti-plane problem ($\sin\lambda = 0$ and $\cos\lambda = 0$ in the symmetric and antisymmetric cases respectively). Hence, the determinant of the boundary value system (5.9), (5.5) at $s = 0$ is non-zero, therefore at $s = 0$ we obtain the trivial solution of (5.9). Similarly, the trivial solution (5.7) may be obtained in the opposite case.

The systems (5.7) and (5.9) should be solved for $s \geq 1$ with superscripts "b" and "a" respectively, giving the two linearly independent solutions, $\overset{b}{Q}$, $\overset{a}{Q}$. In the first case the plane stress state is dominant, being accompanied by a certain anti-plane boundary layer, with the associated quantities denoted by superscript "b". In the second case the anti-plane boundary layer is dominant, whereas a certain accompanying plane boundary layer is also present (denoted by superscript "a").

As shown above

$$\overset{b}{\sigma}{}^{(0)}_{\alpha\alpha}, \overset{b}{\sigma}{}^{(0)}_{\alpha\gamma}, \overset{b}{\sigma}{}^{(0)}_{\gamma\gamma}, \overset{b}{u}{}^{(0)}_{\alpha}, \overset{b}{w}{}^{(0)} \neq 0, \qquad \overset{b}{\sigma}{}^{(0)}_{\alpha\beta} = \overset{b}{\sigma}{}^{(0)}_{\beta\gamma} = \overset{b}{u}{}^{(0)}_{\beta} \equiv 0,$$

$$\overset{a}{\sigma}{}^{(0)}_{\alpha\beta}, \overset{a}{\sigma}{}^{(0)}_{\beta\gamma}, \overset{a}{u}{}^{(0)}_{\beta} \neq 0, \qquad \overset{a}{\sigma}{}^{(0)}_{\alpha\alpha} = \overset{a}{\sigma}{}^{(0)}_{\alpha\gamma} = \overset{a}{\sigma}{}^{(0)}_{\gamma\gamma} = \overset{a}{u}{}^{(0)}_{\alpha} = \overset{a}{w}{}^{(0)} \equiv 0. \tag{5.47}$$

We note that the major difference between $\overset{b}{Q}$ and $\overset{a}{Q}$ is the attenuation speed. For arbitrary s

$$\overset{b}{Q}{}^{(s)} = \overset{b}{\bar{Q}}{}^{(s)} + \overset{b}{Q}{}^{*(s)} \qquad (b,a) \tag{5.48}$$

where $\overset{b}{\bar{Q}}{}^{(s)}$ is the solution of the homogeneous system and $\overset{b}{Q}{}^{*(s)}$ is a particular solution of the inhomogeneous system, satisfying (5.5). Since the boundary layer is a solution of a homogeneous system of equations of 3D elasticity subject to homogeneous boundary conditions (5.5), the solution is defined up to a constant factor. Therefore if $\overset{b}{Q}$ is a solution of this boundary value problem, $\varepsilon^{\mu}\overset{b}{Q}$ is also a solution. Similarly, $\varepsilon^{\chi}\overset{a}{Q}$ is the anti-plane boundary layer solution. The resulting boundary layer solution is then represented as

$$Q = \varepsilon^{\mu}\overset{b}{Q} + \varepsilon^{\chi}\overset{a}{Q} \tag{5.49}$$

where the whole numbers χ and μ characterize the intensities of the boundary layers, which should be chosen in order to satisfy the edge boundary conditions. It may be shown that the numbers χ and μ are defined uniquely. Then

$$\overset{a}{Q} = \varepsilon^{s}\overset{a}{Q}{}^{(s)}, \quad \overset{b}{Q} = \varepsilon^{s}\overset{b}{Q}{}^{(s)}, \quad s = \overline{0,S}.$$

Here and below we assume that summation is performed with respect to repeated index s in the limits between 0 and S, where S is the number of approximations. Hence, (5.49) may be rewritten in the form

$$Q = \varepsilon^{\mu+s}\overset{b}{Q}{}^{(s)} + \varepsilon^{\chi+s}\overset{a}{Q}{}^{(s)}. \tag{5.50}$$

The outer solution constructed in Section 3.2 aligned with the boundary layer solution

$$J = Q^{out} + \varepsilon^{\mu+s}\overset{b}{Q}{}^{(s)} + \varepsilon^{\chi+s}\overset{a}{Q}{}^{(s)} \tag{5.51}$$

contains sufficient number of arbitrary constants in order to satisfy the edge boundary conditions of 3D elasticity. This question is discussed in more details in the next chapter.

Chapter 4

Matching of the Outer Solution and the Boundary Layer for an Orthotropic Plate

4.1 Boundary Conditions at the Edges

One of the crucial issues for any plate or shell approximate theory is related to satisfying boundary conditions at the edge. Indeed, the 2D approximate equations do not allow the exact 3D boundary conditions to be matched, therefore the questions of approximate 2D boundary conditions along with the accuracy estimates of these arise. In case of the variational approach the boundary conditions are embedded into the approach, however, the initial assumptions adopted should allow all possible solutions. It is well known that in case of a free edge Poisson suggested three conditions, namely vanishing of bending and torsional moments along with the transverse force, see Timoshenko (1957). However, these conditions could not match the fourth order of the bending equation for a plate. It has then been shown by Kirchhoff through energy approach that two of the conditions suggested by Poisson for torsional moment and transverse force, can be replaced by one boundary condition. The physical explanation of this has been clarified by Thomson and Tait. For some period of time issue of boundary conditions was considered as a resolved one. However, after the appearance of the Reissner's theory and other refined theories, the problem of boundary conditions has become important once again. The main idea of any refined plate theory lies in its ability to describe a wider class of problems (both in the inner and outer zones) by relatively simple formulations in terms of both governing equations and boundary conditions.

The work of Friedrichs (1950) was the first where the Kirchhoff conditions for free edge were obtained through method of direct asymptotic integration of the 3D boundary value problem. It seems that this contribution along with the works of Goldenveizer (1959, 1960, 1961, 1962, 1963, 1968a,b, 1969, 1976); Goldenveizer et al. (1979, 1990) have provided motivation for further development of asymptotic methods not only within theories for plates and shells, but also in more general scope of mathematical physics, see Friedrichs (1955).

The asymptotic analysis of the boundary conditions of the classical theory and development of novel conditions for isotropic plates were the subject of the following series of works by Friedrichs and Dressler (1961); Goldenveizer (1962);

123

Green (1962b); Aksentyan and Vorovich (1963); Kolos (1965); Vorovich and Malkina (1966).

In this chapter we consider the question of reducing boundary conditions from the original 3D elasticity to asymptotic 2D formulation for orthotropic plates including both cases of extension/compression and bending. The developed approach of representation of solution through a combination of the outer solution and the boundary layer allows a general treatment of a wide class of boundary value problems modeling various ways of real life fixing, see Aghalovyan (1966c, 1978a,b).

Consider the following types of boundary conditions which are most typically imposed on the edge of a plate $\alpha = \alpha_0$

$$\sigma_{\alpha\alpha} = \sigma_{\alpha\beta} = \sigma_{\alpha\gamma} = 0, \tag{1.1}$$

$$\sigma_{\alpha\alpha} = \sigma_{\alpha\beta} = w = 0, \tag{1.2}$$

$$\sigma_{\alpha\alpha} = v = w = 0, \tag{1.3}$$

$$u = v = w = 0, \tag{1.4}$$

$$u = \sigma_{\alpha\beta} = w = 0. \tag{1.5}$$

Similar conditions on the edge $\beta = \beta_0$ could be considered along with some others.

4.2   First Boundary Value Problem. Iterative Matching

The following asymptotic expansions hold for the outer problem and the boundary layer

$$\sigma_{\alpha\alpha} = \varepsilon^{-2+s}\sigma_{\alpha\alpha}^{(s)} + \varepsilon^{\chi+s}\overset{a}{\sigma}{}_{\alpha\alpha}^{(s)} + \varepsilon^{\mu+s}\overset{b}{\sigma}{}_{\alpha\alpha}^{(s)},$$

$$\sigma_{\alpha\beta} = \varepsilon^{-2+s}\sigma_{\alpha\beta}^{(s)} + \varepsilon^{\chi+s}\overset{a}{\sigma}{}_{\alpha\beta}^{(s)} + \varepsilon^{\mu+s}\overset{b}{\sigma}{}_{\alpha\beta}^{(s)}, \tag{2.1}$$

$$\sigma_{\alpha\gamma} = \varepsilon^{-1+s}\sigma_{\alpha\gamma}^{(s)} + \varepsilon^{\chi+s}\overset{a}{\sigma}{}_{\alpha\gamma}^{(s)} + \varepsilon^{\mu+s}\overset{b}{\sigma}{}_{\alpha\gamma}^{(s)}.$$

In order to satisfy condition (1.1), expansions (2.1) should be substituted into these, yielding consistent conditions for both outer problem and the boundary layers, giving $\chi = \mu = -2$. Clearly, if $|\chi| < 2$ or $|\mu| < 2$, then the conditions will be inconsistent, whereas $|\chi| > 2$ and $|\mu| > 2$ imply the homogeneous (zero) conditions for the boundary layers, leading to trivial boundary layer solutions. Thus, the consistent values of μ and χ are uniquely defined. As a result, we obtain

$$\overset{b}{\sigma}{}_{\alpha\alpha}^{(s)} + \overset{a}{\sigma}{}_{\alpha\alpha}^{(s)} + \sigma_{\alpha\alpha}^{(s)} = 0, \quad \overset{b}{\sigma}{}_{\alpha\gamma}^{(s)} + \overset{a}{\sigma}{}_{\alpha\gamma}^{(s)} + \sigma_{\alpha\gamma}^{(s-1)} = 0, \tag{2.2}$$

$$\sigma_{\alpha\beta}^{(s)} + \overset{a}{\sigma}{}_{\alpha\beta}^{(s)} + \overset{b}{\sigma}{}_{\alpha\beta}^{(s)} = 0 \quad \text{at} \quad \alpha = \alpha_0(t = 0).$$

Now expressing $\overset{b}{\sigma}{}^{(s)}_{\alpha\alpha}(t=0)$, $\overset{b}{\sigma}{}^{(s)}_{\alpha\gamma}(t=0)$, $\overset{a}{\sigma}{}^{(s)}_{\alpha\beta}(t=0)$ in (2.2) through other quantities and bearing in mind conditions (3.5.13) and (3.5.14), with the corresponding quantities with the added indices "b" or "a", these conditions will be written as

$$\int_{-1}^{+1} \zeta\sigma^{(s)}_{\alpha\alpha}(\alpha=\alpha_0)d\zeta = \int_{-1}^{+1}\zeta d\zeta \int_0^\infty \left(\overset{b}{R}{}^{(s-1)}_\alpha + \overset{a}{R}{}^{(s-1)}_\alpha\right)dt$$

$$- \int_{-1}^{+1} d\zeta \int_0^\infty \left(\overset{b}{R}{}^{(s-1)}_\gamma + \overset{a}{R}{}^{(s-1)}_\gamma\right)tdt,$$

$$\int_{-1}^{+1} \sigma^{(s)}_{\alpha\gamma}(\alpha=\alpha_0)d\zeta = \int_{-1}^{+1} d\zeta \int_0^\infty \left(\overset{b}{R}{}^{(s)}_\gamma + \overset{a}{R}{}^{(s)}_\gamma\right)dt, \qquad (2.3)$$

$$\int_{-1}^{+1} \sigma^{(s)}_{\alpha\alpha}(\alpha_0)d\zeta = \int_{-1}^{+1} d\zeta \int_0^\infty \left(\overset{b}{R}{}^{(s-1)}_\alpha + \overset{a}{R}{}^{(s-1)}_\alpha\right)dt,$$

$$\int_{-1}^{+1} \sigma^{(s)}_{\alpha\beta}(\alpha_0)d\zeta = \int_{-1}^{+1} d\zeta \int_0^\infty \left(\overset{b}{R}{}^{(s-1)}_\tau + \overset{a}{R}{}^{(s-1)}_\tau\right)dt. \qquad (2.4)$$

The quantities $\sigma^{(s)}_{\alpha\alpha}$, $\sigma^{(s)}_{\alpha\beta}$, $\sigma^{(s)}_{\alpha\gamma}$ of the outer problem are determined by Eq. (3.2.7), then the conditions (2.4) for antisymmetric case are satisfied, whereas conditions (2.3) at $s=0$ imply the following relations for the outer problem

$$\tau^{(0)}_{\alpha\alpha}(\alpha_0) = 0,$$

$$\int_{-1}^{+1} \sigma^{(s)}_{\alpha\gamma}(\alpha=\alpha_0)d\zeta$$

$$= -\int_{-1}^{+1} d\zeta \int_0^\infty \left(\xi_1 d_{11} \overset{b}{\sigma}{}^{(0)}_{\alpha\gamma} + d_{20}\overset{a}{\sigma}{}^{(0)}_{\beta\gamma} + ak_{\beta 0}\overset{b}{\sigma}{}^{(0)}_{\alpha\gamma} + ak_{\alpha 0}\overset{a}{\sigma}{}^{(0)}_{\beta\gamma}\right)dt. \qquad (2.5)$$

Since $\overset{a}{\sigma}{}^{(0)}_{\alpha\alpha} = \overset{a}{\sigma}{}^{(0)}_{\alpha\gamma} \equiv 0$, it follows from the first condition of (2.5), and the first two conditions (2.2) for the plane strain boundary layer that

$$\overset{b}{\sigma}{}^{(0)}_{\alpha\alpha} = 0, \quad \overset{b}{\sigma}{}^{(0)}_{\alpha\gamma} = 0 \quad \text{at} \quad t=0, \qquad (2.6)$$

hence

$$\overset{b}{Q}{}^{(0)} \equiv 0. \qquad (2.7)$$

The last condition (2.2) implies for the anti-plane boundary layer

$$\overset{a}{\sigma}{}^{(0)}_{\alpha\beta}(t=0) = -\zeta\tau^{(0)}_{\alpha\beta}(\alpha_0), \qquad (2.8)$$

where $\tau^{(0)}_{\alpha\beta}(\alpha_0)$ is to be determined from the outer solution.

In view of (2.7) the integral on the right-hand side of (2.5) takes the form

$$I_1 = -\int_{-1}^{+1} d\zeta \int_0^\infty \left(d_{20}\overset{a}{\sigma}{}^{(0)}_{\beta\gamma} + ak_{\alpha 0}\overset{a}{\sigma}{}^{(0)}_{\beta\gamma}\right)dt$$

$$= -aH_{\beta 0}\frac{\partial}{\partial\beta} - \int_{-1}^{+1} d\zeta - \int_0^\infty \overset{a}{\sigma}{}^{(0)}_{\beta\gamma}dt. \qquad (2.9)$$

Integrating by parts and using (3.5.9) along with (2.8), we get

$$\int_{-1}^{+1} d\zeta \int_{0}^{\infty} \overset{a}{\underset{\beta\gamma}{\sigma}}{}^{(0)} dt = -\int_{-1}^{+1} \zeta \overset{a}{\underset{\alpha\beta}{\sigma}}{}^{(0)} (t=0) d\zeta = \frac{2}{3} \tau_{\alpha\beta}^{(0)}(\alpha_0), \tag{2.10}$$

$$I_1 = -\frac{2}{3} a \frac{\partial}{\partial s_\beta} \tau_{\alpha\beta}^{(0)}(\alpha_0), \qquad \frac{\partial}{\partial s_\beta} = H_{\beta 0} \frac{\partial}{\partial \beta}.$$

Now using (3.2.7), and (2.10) enables representing (2.5) as

$$\tau_{\alpha\alpha}^{(0)}(\alpha_0) = 0,$$

$$a\left(-\frac{4}{3}\tau_{\alpha\gamma}^{(0)} + p_\alpha^{(0)}\right) = -\frac{2}{3} a^2 \frac{\partial \tau_{\alpha\beta}^{(0)}(\alpha_0,\beta)}{\partial s_\beta} \qquad \text{at} \quad \alpha = \alpha_0 \tag{2.11}$$

which would act as boundary conditions for the outer bending problem. We remark that in terms of the classical theory (3.3.1), (3.3.2) the conditions (2.11) are written as

$$M_\alpha^{(0)} = 0,$$

$$N_\alpha^{(0)} + \frac{\partial H_{\alpha\beta}^{(0)}}{\partial s_\beta} = 0 \qquad \text{at} \quad \alpha = \alpha_0 \tag{2.12}$$

coinciding with the case of the free face boundary conditions of the classical Kirchhoff theory. As shown earlier, the governing equations of the Kirchhoff theory may also be obtained by asymptotic method, therefore, it may be concluded that the Kirchhoff theory is effectively the leading order approximation of the asymptotic integration procedure for the equations of three dimensional elasticity in case of the outer problem, clearly not been able to take into account any boundary layer effects.

After the outer solution has been determined, the expression for $\tau_{\alpha\beta}^{(0)}(\alpha_0)$ is obtained, and the anti-plane boundary layer is found from (3.5.41), (3.5.43) and (2.8), namely

$$\overset{a}{\underset{\alpha\beta}{\sigma}}{}^{(0)} = \frac{1}{a_{66}} \frac{\partial \Phi}{\partial t}, \qquad \overset{a}{\underset{\beta\gamma}{\sigma}}{}^{(0)} = \frac{1}{a_{44}} \frac{\partial \Phi}{\partial \zeta},$$

$$\overset{a}{\underset{\beta}{u}}{}^{(0)} = \Phi(t,\zeta), \qquad \Phi = -\tau_{\alpha\beta}^{(0)}(\alpha_0,\beta)\Phi_*, \tag{2.13}$$

$$\Phi_* = \frac{16}{\pi^3} \sqrt{a_{44} a_{66}} \sum_{n=1}^{\infty} \frac{(-1)^n}{(2n-1)^3} \sin(2n-1)\frac{\pi}{2}\zeta \exp\left(-\sqrt{\frac{a_{66}}{a_{44}}}(2n-1)\frac{\pi}{2}t\right).$$

Thus, at leading order the plane boundary layer is zero. At the same time the anti-plane boundary layer at leading order is non-zero, therefore in case of the free faces the major contribution comes from the anti-plane boundary layer. Since the Kirchhoff theory does not take into account any boundary layer effects, its application would give misleading results in the vicinity of the free boundary, most importantly for the quantities $\tau_{\alpha\beta}$, $\tau_{\beta\gamma}$. We note that the results for the other components of stresses and displacements could be reasonable.

At $s = 1$ it follows from (2.2) and (2.3) that

$$\frac{2}{3}\tau_{\alpha\alpha}^{(1)}(\alpha_0) = -J_2 + J_3, \qquad (2.14)$$

$$-\frac{4}{3}\tau_{\alpha\gamma}^{(1)} + p_\alpha^{(1)} = -\left(J_4 + J_5 + J_6 + J_7 + J_8\right),$$

where

$$J_2 = \int_{-1}^{+1} \zeta d\zeta \int_0^\infty \left(d_{20}\,\overset{a}{\sigma}\,\overset{(0)}{\alpha\beta} + 2ak_{\alpha0}\,\overset{a}{\sigma}\,\overset{(0)}{\alpha\beta}\right) dt = aH_{\alpha0}H_{\beta0}\frac{\partial}{\partial\beta}\left(\frac{J^*}{H_{\alpha0}}\right),$$

$$J_3 = \int_{-1}^{+1} d\zeta \int_0^\infty \left(d_{20}\,\overset{a}{\sigma}\,\overset{(0)}{\beta\gamma} + ak_{\alpha0}\,\overset{a}{\sigma}\,\overset{(0)}{\beta\gamma}\right) t\,dt = -a\frac{H_{\beta0}}{H_{\alpha0}}\frac{\partial}{\partial\beta}\left(H_{\alpha0}J^*\right),$$

$$J_4 = \int_{-1}^{+1} d\zeta \int_0^\infty \left(\xi_1 d_{11}\,\overset{b}{\sigma}\,\overset{(1)}{\alpha\gamma} + ak_{\beta0}\,\overset{b}{\sigma}\,\overset{(1)}{\alpha\gamma}\right) dt$$

$$= a\left(k_{\beta0} - H_{\alpha1}\right)\int_{-1}^{+1} d\zeta \int_0^\infty \overset{b}{\sigma}\,\overset{(1)}{\alpha\gamma}\,dt$$

$$= a\left(H_{\alpha1} - k_{\beta0}\right)\int_{-1}^{+1} d\zeta \int_0^\infty \overset{b}{R}\,\overset{(0)}{\gamma}\,t\,dt = 0,$$

$$J_5 = \int_{-1}^{+1} d\zeta \int_0^\infty \left(d_{20}\,\overset{b}{\sigma}\,\overset{(1)}{\beta\gamma} + ak_{\alpha0}\,\overset{b}{\sigma}\,\overset{(1)}{\beta\gamma}\right) dt$$

$$= aH_{\beta0}\frac{\partial}{\partial\beta}\left(\int_{-1}^{+1} d\zeta \int_0^\infty \overset{b}{\sigma}\,\overset{(1)}{\beta\gamma}\,dt\right) \quad (2.15)$$

$$= -aH_{\beta0}\frac{\partial}{\partial\beta}\left(\int_{-1}^{+1} \zeta\,\overset{b}{\sigma}\,\overset{(1)}{\alpha\beta}(0)d\zeta\right),$$

$$J_6 = \int_{-1}^{+1} d\zeta \int_0^\infty \left(d_{20}\,\overset{a}{\sigma}\,\overset{(1)}{\beta\gamma} + ak_{\alpha0}\,\overset{a}{\sigma}\,\overset{(1)}{\beta\gamma}\right) dt$$

$$= aH_{\beta0}\frac{\partial}{\partial\beta}\left(\int_{-1}^{+1} d\zeta \int_0^\infty \overset{a}{\sigma}\,\overset{(1)}{\beta\gamma}\,dt\right)$$

$$= aH_{\beta0}\frac{\partial}{\partial\beta}\left(-\int_{-1}^{+1} \zeta\,\overset{a}{\sigma}\,\overset{(1)}{\alpha\beta}(0)d\zeta + a\left(2k_{\beta0} - H_{\alpha1}\right)J^*\right),$$

$$J_5 + J_6 = aH_{\beta0}\frac{\partial}{\partial\beta}\left(\frac{2}{3}\tau_{\alpha\beta}^{(1)}(\alpha_0) + a\left(2k_{\beta0} - H_{\alpha1}\right)J^*\right),$$

$$J_7 = \int_{-1}^{+1} d\zeta \int_0^\infty \left(\xi_1 d_{11}\,\overset{a}{\sigma}\,\overset{(1)}{\alpha\gamma} + ak_{\beta0}\,\overset{a}{\sigma}\,\overset{(1)}{\alpha\gamma}\right) dt$$

$$= a\left(k_{\beta0} - H_{\alpha1}\right)\left(\int_{-1}^{+1} d\zeta \int_0^\infty \overset{a}{\sigma}\,\overset{(1)}{\alpha\gamma}\,dt\right)$$

$$= a\left(H_{\alpha1} - k_{\beta0}\right)\int_{-1}^{+1} d\zeta \int_0^\infty \overset{a}{R}\,\overset{(0)}{\gamma}\,t\,dt$$

$$= a^2\left(H_{\alpha1} - k_{\beta0}\right)\frac{H_{\beta0}}{H_{\alpha0}}\frac{\partial}{\partial\beta}\left(H_{\alpha0}J^*\right),$$

$$J_8 = \int_{-1}^{+1} d\zeta \int_0^\infty \left(\xi_1 d_{21} \overset{a\ (0)}{\sigma}_{\beta\gamma} + a^2 \xi_1 k_{\alpha 1} \overset{a\ (0)}{\sigma}_{\beta\gamma} \right) dt$$

$$= a^2 H_{\alpha 0} \int_{-1}^{+1} d\zeta \int_0^\infty \left(H_{\beta 1} \frac{\partial \overset{a\ (0)}{\sigma}_{\beta\gamma}}{\partial \beta} + k_{\alpha 1} \overset{a\ (0)}{\sigma}_{\beta\gamma} \right) t\, dt$$

$$= -a^2 \left(\frac{H_{\beta 1}}{H_{\alpha 0}} \frac{\partial \left(H_{\alpha 0}^2 J^* \right)}{\partial \beta} + H_{\alpha 0} k_{\alpha 1} J^* \right),$$

$$J^* = \int_{-1}^{+1} \zeta d\zeta \int_0^\infty \overset{a\ (0)}{\sigma}_{\alpha\beta} dt = -\sqrt{\frac{a_{44}}{a_{66}}} \frac{A}{2} \cdot \frac{2}{3} \tau_{\alpha\beta}^{(0)}(\alpha_0)$$

and

$$A = \frac{384}{\pi^5} \sum_{n=1}^{\infty} \frac{1}{(2n-1)^5} \approx 1.26$$

is a constant value for isotropic plate investigated in Kolos (1965).

Taking into consideration (2.15) and (3.3.2), the conditions (2.14) may be rewritten as

$$M_\alpha^{(1)} - \sqrt{\frac{a_{44}}{a_{66}}} Ah H_{\beta 0} \frac{\partial H_{\alpha\beta}^{(0)}}{\partial \beta} = 0,$$

$$N_\alpha^{(1)} + H_{\beta 0} \frac{\partial H_{\alpha\beta}^{(1)}}{\partial \beta} - \sqrt{\frac{a_{44}}{a_{66}}} Ah H_{\beta 0} \frac{\partial (k_{\beta 0} H_{\alpha\beta}^{(0)})}{\partial \beta} = 0 \quad \text{at} \quad \alpha = \alpha_0 \qquad (2.16)$$

providing the asymptotic solution of the outer problem up to $O(\varepsilon^2)$. The plane boundary layer at $s = 1$ is non-zero, specified from

$$\overset{b\ (1)}{\sigma}_{\alpha\alpha} = -\zeta \tau_{\alpha\alpha}^{(1)} - \overset{a\ (1)}{\sigma}_{\alpha\alpha}(t = 0),$$

$$\overset{b\ (1)}{\sigma}_{\alpha\gamma} = -\sigma_{\alpha\gamma}^{(0)} - \overset{a\ (1)}{\sigma}_{\alpha\gamma}(t = 0) \quad \text{at} \quad t = 0 \ (\alpha = \alpha_0). \qquad (2.17)$$

The corresponding expressions for the anti-plane boundary layer are given by

$$\overset{a\ (1)}{\sigma}_{\alpha\beta} = -\zeta \tau_{\alpha\beta}^{(1)}(\alpha_0) - \overset{b\ (1)}{\sigma}_{\alpha\beta}(t = 0) \quad \text{at} \quad t = 0 \ (\alpha = \alpha_0). \qquad (2.18)$$

The right-hand sides of (2.17) and (2.18) are known functions after solution of the outer problem has been determined $\left(\overset{b}{Q}{}^{(1)} \neq 0, \ \overset{a}{Q}{}^{(1)} \neq 0 \right)$. If the attention is restricted to the first two approximations $s = 0, 1$, the expressions (2.12) and (2.16) may be united to yield

$$M_\alpha - \sqrt{\frac{a_{44}}{a_{66}}} Ah \frac{\partial H_{\alpha\beta}^0}{\partial s_\beta} = 0, \qquad (2.19)$$

$$N_\alpha + \frac{\partial H_{\alpha\beta}}{\partial s_\beta} - \sqrt{\frac{a_{44}}{a_{66}}} Ah \frac{\partial (k_{\beta 0} H_{\alpha\beta}^0)}{\partial s_\beta} = 0 \quad \text{at} \quad \alpha = \alpha_0.$$

Here the underlined terms are refinement to the classical boundary conditions, and $H^0_{\alpha\beta}$ is the rotation moment according to the Kirchhoff theory. The obtained conditions (2.19) are generalizing the Kirchhoff boundary conditions for the free boundary, taking into account the variation of material properties in the transverse direction. In case of isotropic material conditions (2.19) coincide with that obtained in Kolos (1965) following the method proposed in Goldenveizer (1962). In case of isotropic media the refinement is of order $\sim O(\varepsilon)$, whereas for anisotropic materials the order of refinement depends on the degree of anisotropy, being more significant for plates with smaller shear stiffness in the transverse direction. It follows that the Kirchhoff conditions are satisfied exactly for plates with infinitely large shear stiffness in the transverse direction. If $G_{12}/G_{23} \sim O(\varepsilon^{-2})$, the refined terms are of the first order, hence, the Kirchhoff theory is no longer valid. As discussed above, in this case a "weakly localized" anti-plane boundary layer could arise. It should be noted that this case is relatively rare for real materials, however, the cases of $G_{12}/G_{23} \sim O(\varepsilon^{-1/2})$ or $\sim O(\varepsilon^{-1})$ are rather typical. Then the contribution of the refinement could reach 25% - 50%, hence, the importance of conditions (2.19) becomes crucial. We remark that (2.19) take into account both material properties and geometry of the plate (possible curvature).

The approximations above were constructed for $s = 0, 1$. Similarly, the procedure could be continued for arbitrary s, however, significant contribution of the next order refinements is doubtful.

Consider now symmetric case (extension/compression), when $\sigma_{\alpha\alpha}, \sigma_{\alpha\beta}$, $\sigma_{\beta\beta}$, $\sigma_{\gamma\gamma}$, u_α, u_β are even functions of ζ and $\sigma_{\alpha\gamma}, \sigma_{\beta\gamma}, w$ are odd functions. Hence, conditions (2.3) are satisfied automatically, whereas (2.4) imply

$$\tau^{(0)}_{\alpha\alpha} = 0, \quad \tau^{(0)}_{\alpha\beta} = 0 \quad \text{at} \quad \alpha = \alpha_0. \tag{2.20}$$

Using the latter, consider the outer problem. It follows from (2.2) that

$$\overset{b}{\sigma}{}^{(0)}_{\alpha\alpha} = \overset{b}{\sigma}{}^{(0)}_{\alpha\gamma} = 0; \quad \overset{a}{\sigma}{}^{(0)}_{\alpha\beta} = 0 \quad \text{at} \quad t = 0 \tag{2.21}$$

leading to

$$\overset{b}{Q}{}^{(0)} = 0, \quad \overset{a}{Q}{}^{(0)} = 0. \tag{2.22}$$

In terms of the classical theory (3.3.3) the conditions (2.20) may be written as

$$T^{(0)}_\alpha = 0, \quad S^{(0)}_{\alpha\beta} = 0 \quad \text{at} \quad \alpha = \alpha_0, \tag{2.23}$$

coinciding with the traditional form. Assuming both plane and anti-plane boundary layers to be zero, we conclude that the original hypothesis of non-deformable normals providing reduction of the 3D problem to a 2D one, is more accurate for symmetric case than for anti-symmetric one.

At $s = 1$ we have

$$\tau^{(1)}_{\alpha\alpha} = \tau^{(1)}_{\alpha\beta} = 0 \quad \text{at} \quad \alpha = \alpha_0 \tag{2.24}$$

or

$$T_\alpha^{(1)} = S_{\alpha\beta}^{(1)} = 0 \quad \text{at} \quad \alpha = \alpha_0$$

and

$$\overset{b}{\sigma}\,\overset{(1)}{_{\alpha\alpha}} = 0, \quad \overset{b}{\sigma}\,\overset{(1)}{_{\alpha\gamma}} = -\sigma_{\alpha\gamma}^{(0)} \; (\alpha = \alpha_0) \quad \text{at} \quad t = 0 \tag{2.25}$$

implying $\overset{b}{Q}{}^{(1)} \neq 0$. Similarly,

$$\overset{a}{\sigma}\,\overset{(1)}{_{\alpha\beta}} = 0 \quad \text{at} \quad t = 0 \quad \Rightarrow \quad \overset{a}{Q}{}^{(1)} = 0. \tag{2.26}$$

Restricting our consideration to the approximations for $s = 0, 1$ and combining (2.23) and (2.24), we obtain the free edge boundary conditions

$$T_\alpha = 0, \quad S_{\alpha\beta} = 0 \quad \text{at} \quad \alpha = \alpha_0. \tag{2.27}$$

We note that the classical boundary conditions are more accurate for symmetric motion than for the antisymmetric one, providing asymptotic accuracy of $O(\varepsilon^2)$ compared to $O(1)$. It may be observed that the anti-plane boundary layer plays a crucial role in bending problem, whereas in case of symmetric problem it is of secondary importance, with the plane boundary layer effect arising only at the next order of $s = 1$. Application of the classical theory may cause misleading results for $\sigma_{\alpha\gamma}, \sigma_{\gamma\gamma}$, since both of these stresses are of order $O(\varepsilon^{-1})$, see (2.25). In general case of loading the conditions (2.19) and (2.27) along with equations (3.3.5)-(3.3.11) provide the asymptotic expansion of the outer solution with accuracy of $\sim O(\varepsilon^2)$.

4.3   Variational Matching

The matching between the outer solution and the boundary layer may also be achieved through variational principles depending on the type of boundary conditions. Let us illustrate this approach by considering the first type of boundary value problem. Let the boundary conditions be imposed as

$$\sigma_{\alpha\alpha} = \bar\sigma_{\alpha\alpha}, \quad \sigma_{\alpha\beta} = \bar\sigma_{\alpha\beta}, \quad \sigma_{\alpha\gamma} = \bar\sigma_{\alpha\gamma} \quad \text{at} \quad \alpha = \alpha_0 \tag{3.1}$$

where the boundary values of stresses $\bar\sigma_{\alpha\alpha}, \bar\sigma_{\alpha\beta}, \bar\sigma_{\alpha\gamma}$ may depend on the small parameter ε. Assume a general form of loading

$$\bar\sigma_{\alpha\alpha} = \varepsilon^{-2+s} f_1^{(s)}(\zeta,\eta), \quad \bar\sigma_{\alpha\beta} = \varepsilon^{-2+s} f_2^{(s)}(\zeta,\eta), \quad \bar\sigma_{\alpha\gamma} = \varepsilon^{-1+s} f_3^{(s)}(\zeta,\eta). \tag{3.2}$$

As earlier, we adapt the summation convention over repeated index s from 0 to S. The solution is given by

$$J = Q^{out} + \varepsilon^\chi \overset{a}{Q} + \varepsilon^\mu \overset{b}{Q}. \tag{3.3}$$

Consider bending problem. Bearing in mind the structure of the outer solution and the boundary layer given by formulae (2.4), (2.7), (5.16), (5.19), (5.41), (5.43), (5.50) of Chapter 3, we deduce

$$\sigma_{\alpha\alpha} = \varepsilon^{-2+s}\left(\zeta\tau_{\alpha\alpha}^{(s)} + \sigma_{\alpha\alpha}^{*(s)}\right)$$

$$+\varepsilon^{\mu+s}\left(\overset{b}{\sigma}{}_{\alpha n}(t,\eta,\zeta)A_n^{(s)} + \overset{b}{\sigma}{}_\alpha^{*(s)}(t,\eta,\zeta)\right) + \varepsilon^{\chi+s}\overset{a}{\sigma}{}_{\alpha\alpha}^{(s)}(t,\eta,\zeta), \quad (\alpha,\beta),$$

$$\sigma_{\alpha\beta} = \varepsilon^{-2+s}\left(\zeta\tau_{\alpha\beta}^{(s)} + \sigma_{\alpha\beta}^{*(s)}\right)$$

$$+\varepsilon^{\chi+s}\left(\overset{a}{\sigma}{}_{\alpha\beta n}(t,\eta,\zeta)B_n^{(s)}(\eta) + \overset{a}{\sigma}{}_{\alpha\beta}^{*(s)}(t,\eta,\zeta)\right) + \varepsilon^{\mu+s}\overset{b}{\sigma}{}_{\alpha\beta}^{(s)}(t,\eta,\zeta),$$

$$\sigma_{\alpha\gamma} = \varepsilon^{-1+s}\left((\zeta^2-1)\tau_{\alpha\gamma}^{(s)} + \frac{1}{2}X_2^{(s)} + \sigma_{\alpha\gamma}^{*(s)}\right) \quad (3.4)$$

$$+\varepsilon^{\mu+s}\left(\overset{b}{\sigma}{}_{\alpha\gamma n}A_n^{(s)} + \overset{b}{\sigma}{}_{\alpha\gamma}^{*(s)}\right) + \varepsilon^{\chi+s}\overset{a}{\sigma}{}_{\alpha\gamma}^{(s)}, \quad (\alpha,\beta;X,Y),$$

$$u_\alpha = \varepsilon^{-2+s}\left(-\zeta a H_\alpha\frac{\partial w^{(s)}}{\partial\alpha} + u_\alpha^{*(s)}\right)$$

$$+a\varepsilon^{\mu+s+1}\left(\overset{b}{u}{}_{\alpha n}A_n^{(s)} + \overset{b}{u}{}_\alpha^{*(s)}\right) + \varepsilon^{\chi+1+s}\overset{a}{u}{}_\alpha^{(s)},$$

$$u_\beta = \varepsilon^{-2+s}\left(-\zeta a I\!I_\beta\frac{\partial w^{(s)}}{\partial\beta} + u_\beta^{*(s)}\right)$$

$$+a\varepsilon^{\chi+1+s}\left(\overset{a}{u}{}_{\beta n}B_n^{(s)} + \overset{a}{u}{}_\beta^{*(s)}\right) + a\varepsilon^{\mu+1+s}\overset{b}{u}{}_\beta^{(s)},$$

$$w = \varepsilon^{-3+s}\left(w^{(s)} + W^{*(s)}\right)$$

$$+a\varepsilon^{\mu+1+s}\left(\overset{b}{w}{}_n(t,\zeta,\eta)A_n^{(s)} + \overset{b}{w}{}^{*(s)}(t,\zeta,\eta)\right) + a\varepsilon^{\chi+1+s}\overset{a}{w}{}^{(s)}.$$

Here $A_n^{(s)}$ and $B_n^{(s)}$ are arbitrary constants (functions of η) of the plane and anti-plane boundary layer solutions, respectively, $\overset{b}{\sigma}{}_{\alpha n},\overset{b}{\sigma}{}_{\alpha\gamma n},\overset{b}{u}{}_{\alpha n},\overset{b}{w}{}_n$ are known quantities, coefficients of A_n in (3.5.19), $\overset{a}{\sigma}{}_{\alpha\beta n},\overset{a}{u}{}_{\beta n}$ are also known, determined from (3.5.41) and (3.5.43) as coefficients in the anti-plane boundary layer. The associated quantities of the boundary layer solution $\overset{a}{\sigma}{}_{\alpha\alpha}^{(s)},\overset{a}{\sigma}{}_{\alpha\gamma}^{(s)},\overset{a}{u}{}_\alpha^{(s)},\overset{a}{w}{}^{(s)}, \overset{b}{\sigma}{}_{\alpha\beta}^{(s)},\overset{b}{u}{}_\beta^{(s)}$ are known provided that the outer solution is obtained along with the boundary layer components of the previous orders (i.e. they do not contain arbitrary constants).

In order to satisfy conditions (3.1) one may apply the Lagrange variational principle. Due to the fact that the representation (3.4) satisfies the equations of 3D elasticity along with the boundary conditions at $\gamma = \pm h$, the variational Lagrange equation may be written as (Leibenzon, 1947)

$$\int_{s_\sigma}\left((\sigma_{\alpha\alpha} - \bar{\sigma}_{\alpha\alpha})\delta u_\alpha + (\sigma_{\alpha\beta} - \bar{\sigma}_{\alpha\beta})\delta u_\beta + (\sigma_{\alpha\gamma} - \bar{\sigma}_{\alpha\gamma})\delta w\right)ds_\sigma = 0. \quad (3.5)$$

Taking into account that

$$\int_{s_\sigma} ds_\sigma = \int_{-h}^{+h} d\gamma \int_\Gamma ds_\beta$$

where ds_σ is elementary surface part, $ds_\beta = \dfrac{1}{H_\beta} d\beta$ is element of the coordinate line $\alpha = \alpha_0$, equation (3.5) may be reformulated as

$$\int_\Gamma \left(\int_{-h}^{+h} \left((\sigma_{\alpha\alpha} - \bar{\sigma}_{\alpha\alpha}) \, \delta u_\alpha + (\sigma_{\alpha\beta} - \bar{\sigma}_{\alpha\beta}) \, \delta u_\beta \right.\right.$$

$$\left.\left. + (\sigma_{\alpha\gamma} - \bar{\sigma}_{\alpha\gamma}) \delta w \right) d\gamma \right) ds_\beta = 0. \tag{3.6}$$

Now substituting (3.4), and variations $\delta u_\alpha, \delta u_\beta, \delta w$ into (3.6), using the Cauchy rule of series multiplication and equating the coefficients with respect to $\delta w^{(s)}$, $\dfrac{\partial \delta w^{(s)}}{\partial \alpha}$, $\delta A_n^{(s)}$, $\delta B_n^{(s)}$ to zero, we arrive at conditions for 2D outer problem and a closed homogeneous algebraic system in terms of the arbitrary constants of the boundary layers. These may be written using the notation of the classical theory as

$$w^{out} = \sum_{s=0}^{S} \left(w_0^{(s)} + W_0^{*(s)} \right) = \varepsilon^{-3+s} \left(w^{(s)} + W^{*(s)} \right),$$

$$u_\alpha^{out} = \varepsilon^{-2+s} u_\alpha^{(s)} = \varepsilon^{-2+s} \left(-\zeta \alpha H_\alpha \frac{\partial w^{(s)}}{\partial \alpha} + u_\alpha^{*(s)} \right)$$

$$= \sum_{s=0}^{S} \left(-\gamma H_\alpha \frac{\partial w_0^{(s)}}{\partial \alpha} + u_{\alpha 0}^{*(s)} \right), \quad (\alpha, \beta),$$

$$M_\alpha = \sum_{s=0}^{S} M_\alpha^{(s)},$$

$$M_\alpha^{(s)} = \int_{-h}^{+h} \gamma \dot{\sigma}_{\alpha\alpha}^{(s)} d\gamma, \quad \dot{\sigma}_{\alpha\alpha}^{(s)} = \varepsilon^{-2+s} \left(\zeta \tau_{\alpha\alpha}^{(s)} + \sigma_{\alpha\alpha}^{*(s)} \right), \quad (\bar{\Sigma}_s), \tag{3.7}$$

$$\bar{M}_\alpha = \sum_{s=0}^{S} \bar{M}_\alpha^{(s)}, \quad \bar{M}_\alpha^{(s)} = \int_{-h}^{+h} \gamma \dot{f}_1^{(s)} d\gamma, \quad \dot{f}_1^{(s)} = \varepsilon^{-2+s} f_1^{(s)}, \quad (\bar{\Sigma}_s),$$

$$N_\alpha = \sum_{s=0}^{S} N_\alpha^{(s)}, \quad (N_\alpha, \bar{N}_\alpha), \quad (N_\alpha^{(s)}, \bar{N}_\alpha^{(s)}) = \int_{-h}^{+h} \left(\dot{\sigma}_{\alpha\gamma}^{(s)}, \dot{f}_3^{(s)} \right) d\gamma,$$

$$\dot{\sigma}_{\alpha\gamma}^{(s)} = \varepsilon^{-1+s} \left((\zeta^2 - 1)\tau_{\alpha\gamma}^{(s)} + \frac{1}{2} X_2^{(s)} + \sigma_{\alpha\gamma}^{*(s)} \right), \quad \dot{f}_3^{(s)} = \varepsilon^{-1+s} f_3^{(s)}, \quad (\bar{\Sigma}_s),$$

$$\left(\overset{b}{M}{}_{\alpha n}^{(s)}, \overset{b}{M}{}_{\alpha}^{*(s)} \right) = \varepsilon^{-2+s} \int_{-h}^{+h} \gamma \left(\overset{b}{\sigma}{}_{\alpha n}, \overset{b}{\sigma}{}_{\alpha}^{*(s)} \right) d\gamma \; (\bar{\Sigma}_s), \quad (a, b),$$

$$\overset{b}{u}{}_{\alpha m}^{(s)} = a\varepsilon^{-1+s} \overset{b}{u}{}_{\alpha m}, \quad \overset{b}{w}{}_{m}^{(s)} = a\varepsilon^{-1+s} \overset{b}{w}{}_{m},$$

$$\overset{b}{\sigma}{}_{\alpha n}^{(s)} = \varepsilon^{-2+s} \overset{b}{\sigma}{}_{\alpha n}, \quad \overset{b}{\sigma}{}_{\alpha\gamma n}^{(s)} = \varepsilon^{-2+s} \overset{b}{\sigma}{}_{\alpha\gamma n}, \quad (a, b),$$

$$\dot{\sigma}^{(s)}_{\alpha\beta} = \varepsilon^{-2+s}\left(\zeta\tau^{(s)}_{\alpha\beta} + \sigma^{*(s)}_{\alpha\beta}\right), \quad \dot{f}^{(s)}_2 = \varepsilon^{-2+s}f^{(s)}_2,$$

$$\overset{\dot{a}}{\sigma}\,{}^{*(s)}_{\alpha\beta} = \varepsilon^{-2+s}\overset{a}{\sigma}\,{}^{*(s)}_{\alpha\beta}, \quad (\bar{\Sigma}_s),$$

$$\overset{\dot{b}}{\sigma}\,{}^{(s)}_{\alpha\beta} = \varepsilon^{-2+s}\overset{b}{\sigma}\,{}^{(s)}_{\alpha\beta}, \quad \overset{\dot{a}}{u}\,{}^{(s)}_{\beta m} = a\varepsilon^{-1+s}\overset{a}{u}\,{}_{\beta m},$$

where the quantities with an over dot have the meaning of usual displacements and stresses, and symbol $\bar{\Sigma}_s$ denotes that there is no summation along s. Equating to zero the coefficients of $\dfrac{\partial\delta w^{(s)}}{\partial\alpha}$ and $\delta w^{(s)}$, we get

$$M^{(s)}_{\alpha} - \bar{M}^{(s)}_{\alpha} + \varepsilon^{\mu+2}\overset{b}{M}\,{}^{*(s)}_{\alpha} + \varepsilon^{\chi+2}\overset{a}{M}\,{}^{(s)}_{\alpha} = 0, \tag{3.8}$$

$$N^{(s)}_{\alpha} - \bar{N}^{(s)}_{\alpha} + \varepsilon^{\mu+2}\overset{b}{N}\,{}^{(s)}_{\alpha} + \varepsilon^{\chi+2}\overset{a}{N}\,{}^{(s)}_{\alpha} = 0 \quad \text{at} \quad \alpha = \alpha_0.$$

The latter are consistent provided $\chi = \mu = -2$, then

$$M^{(s)}_{\alpha} - \bar{M}^{(s)}_{\alpha} + M^{(s)}_{\alpha\,comp} = 0,$$

$$N^{(s)}_{\alpha} - \bar{N}^{(s)}_{\alpha} + N^{(s)}_{\alpha\,comp} = 0 \qquad \text{at} \qquad \alpha = \alpha_0(t=0), \tag{3.9}$$

where

$$M^{(s)}_{\alpha\,comp} = \overset{b}{M}\,{}^{*(s)}_{\alpha} + \overset{a}{M}\,{}^{(s)}_{\alpha}$$

$$= a^2\varepsilon^s\int_{-1}^{+1}\zeta\left(\overset{b}{\sigma}\,{}^{*(s)}_{\alpha} + \overset{a}{\sigma}\,{}^{(s)}_{\alpha}\right)\Bigg|_{t=0}d\zeta = a^2\varepsilon^s$$

$$\times\left(-\int_{-1}^{+1}\zeta\,d\zeta\int_0^{\infty}\left(\overset{b}{R}\,{}^{(s-1)}_{\alpha} + \overset{a}{R}\,{}^{(s-1)}_{\alpha}\right)dt\right.$$

$$\left.+\int_{-1}^{+1}d\zeta\int_0^{\infty}\left(\overset{b}{R}\,{}^{(s-1)}_{\gamma} + \overset{a}{R}\,{}^{(s-1)}_{\gamma}\right)t\,dt\right), \tag{3.10}$$

$$N^{(s)}_{\alpha\,comp} = \overset{b}{N}\,{}^{(s)}_{\alpha} + \overset{a}{N}\,{}^{(s)}_{\alpha} = a\varepsilon^{-1+s}\int_{-1}^{+1}\left(\overset{b}{\sigma}\,{}^{(s)}_{\alpha\gamma} + \overset{a}{\sigma}\,{}^{(s)}_{\alpha\gamma}\right)\Bigg|_{t=0}d\zeta$$

$$= -a\varepsilon^s\int_{-1}^{+1}d\zeta\int_0^{\infty}\left(\overset{b}{R}\,{}^{(s)}_{\gamma} + \overset{a}{R}\,{}^{(s)}_{\gamma}\right)dt.$$

In particular,

$$M^{(0)}_{\alpha\,comp} = 0, \qquad N^{(0)}_{\alpha\,comp} = \frac{\partial}{\partial s_{\beta}}\left(H^{(0)}_{\alpha\beta}(\alpha_0) - \bar{H}^{(0)}_{\alpha\beta}\right) \tag{3.11}$$

then conditions (3.9) can be rewritten as

$$M^{(0)}_{\alpha} = \bar{M}^{(0)}_{\alpha},$$

$$N^{(0)}_{\alpha} + \frac{\partial H^{(0)}_{\alpha\beta}}{\partial s_{\beta}} = \bar{N}^{(0)}_{\alpha} + \frac{\partial\bar{H}^{(0)}_{\alpha\beta}}{\partial s_{\beta}} \qquad \text{at} \qquad \alpha = \alpha_0 \tag{3.12}$$

which coincide with the classical form of boundary conditions for loaded edge.

At $s = 1$, bearing in mind $(3.10)_1$ along with $(3.5.8)_{1,2}$ and (2.15), we obtain

$$M^{(1)}_{\alpha\ comp} = -\sqrt{\frac{a_{44}}{a_{66}}}AhH_{\beta0}\frac{\partial H^{(0)}_{\alpha\beta}(\alpha_0)}{\partial\beta},$$

$$N^{(1)}_{\alpha\ comp} = \left(\frac{\partial H^{(1)}_{\alpha\beta}}{\partial s_\beta} - \frac{\partial \bar{H}^{(1)}_{\alpha\beta}}{\partial s_\beta} - \sqrt{\frac{a_{44}}{a_{66}}}Ah\frac{\partial(k_{\beta0}H^{(0)}_{\alpha\beta})}{\partial s_\beta}\right)_{\alpha=\alpha_0}. \qquad (3.13)$$

Therefore, considering approximations for $s = 0,1$ and representing $M_\alpha = M^{(0)}_\alpha + M^{(1)}_\alpha$, $N_\alpha = N^{(0)}_\alpha + N^{(1)}_\alpha$, $H_{\alpha\beta} = H^{(0)}_{\alpha\beta} + H^{(1)}_{\alpha\beta}$ $(Q,\bar{Q})$, using (3.12) and (3.13), it follows from (3.9) that (up to $O(\varepsilon^2)$)

$$M_\alpha - \sqrt{\frac{a_{44}}{a_{66}}}Ah\frac{\partial H^{(0)}_{\alpha\beta}}{\partial s_\beta} = \bar{M}_\alpha, \quad (3.14)$$

$$N_\alpha + \frac{\partial H_{\alpha\beta}}{\partial s_\beta} - \sqrt{\frac{a_{44}}{a_{66}}}Ah\frac{\partial(k_{\beta0}H^{(0)}_{\alpha\beta})}{\partial s_\beta} = \bar{N}_\alpha + \frac{\partial \bar{H}_{\alpha\beta}}{\partial s_\beta} \qquad \text{at} \qquad \alpha = \alpha_0.$$

In case of free faces the latter coincides with the earlier derived boundary conditions (2.19).

Equating the coefficients of variations $\delta A^{(s)}_m$ and $\delta B^{(s)}_m$ to zero, we arrive at an algebraic system in respect of the unknowns $A^{(s)}_n$ and $B^{(s)}_n$ of the plane and anti-plane boundary layers, respectively

$$a_{mn}A^{(s)}_n + b^{(s)}_m = 0 \qquad\qquad (3.15)$$

$$b_{mn}B^{(s)}_n + c^{(s)}_m = 0 \qquad\qquad (3.16)$$

$$(s = 0, 1, 2, ..., \ m = 1, 2, ..., n)$$

where

$$a_{mn} = \int_{-1}^{+1}\left(\overset{b}{\sigma}_{\alpha n}\overset{b}{u}_{\alpha m} + \overset{b}{\sigma}_{\alpha\gamma n}\overset{b}{w}_m\right)d\zeta,$$

$$b^{(s)}_m = \sum_{k=0}^{s}\left(\int_{-1}^{+1}\left(\left(\dot{\sigma}^{(k)}_{\alpha\alpha} + \overset{\dot{b}}{\sigma}{}^{*(k)}_{\alpha} + \overset{\dot{a}}{\sigma}{}^{(k)}_{\alpha\alpha} - \dot{f}^{(k)}_1\right)\overset{b}{u}_{\alpha m}\right.\right.$$

$$\left.+ \left(\dot{\sigma}^{(k)}_{\alpha\gamma} + \overset{\dot{b}}{\sigma}{}^{*(k)}_{\alpha\gamma} - \dot{f}^{(k)}_3 + \overset{\dot{a}}{\sigma}{}^{(k)}_{\alpha\gamma}\right)\overset{b}{w}_m\right)d\zeta\right)\varepsilon^{2-s} \qquad (3.17)$$

$$+\sum_{k=0}^{s-1}\left(\int_{-1}^{+1}\left(\overset{\dot{b}}{\sigma}{}^{(k)}_{\alpha n}\overset{b}{u}_{\alpha m} + \overset{\dot{b}}{\sigma}{}^{(k)}_{\alpha\gamma n}\overset{b}{w}_m\right)d\zeta\right)\varepsilon^{2-s}A^{(k)}_n,$$

$$b_{mn} = \int_{-1}^{+1}\overset{a}{\sigma}_{\alpha\beta n}\overset{a}{u}_{\beta m}d\zeta,$$

$$c^{(s)}_m = \sum_{k=0}^{s}\left(\int_{-1}^{+1}\left(\dot{\sigma}^{(k)}_{\alpha\beta} + \overset{\dot{b}}{\sigma}{}^{(k)}_{\alpha\beta} + \overset{\dot{a}}{\sigma}{}^{*(k)}_{\alpha\beta} - \dot{f}^{(k)}_2\right)\overset{a}{u}_{\beta m}d\zeta\right)\varepsilon^{2-s}$$

$$+\sum_{k=0}^{s-1}\left(\int_{-1}^{+1}\overset{\dot{a}}{\sigma}{}^{(k)}_{\alpha\beta n}\overset{a}{u}_{\beta m}d\zeta\right)\varepsilon^{2-s}B^{(k)}_n.$$

We note that quantities in the integrand are evaluated at $\alpha = \alpha_0 (t = 0)$. One more useful observation is that the matrices of the systems (3.15), (3.16) do not change with order of approximations, hence their inverse should only be calculated once. If the side surface of the plate is free of loading $\left(f_i^{(s)} \equiv 0 \right)$, then $b_m^{(0)} = 0$, $c_m^{(0)} \neq 0$, hence $A_n^{(0)} = 0$, $B_n^{(0)} \neq 0$. Therefore $\overset{b}{Q}{}^{(0)} = 0$, $\overset{a}{Q}{}^{(0)} \neq 0$, which gives the same result as that obtained in Section 4.2 within difference approach. Indeed, in bending problem for a plate with a free edge the anti-plane boundary layer dominates the plane boundary layer, which could be neglected.

Finally, we note that similar analysis is possible for symmetric problem, and for other types of boundary conditions, requiring other types of variational principles. Here we have just illustrated the possibility of a variational approach to matching the outer solution and the boundary layer, obtained from the asymptotic method.

4.4 Two Types of Mixed Boundary Conditions (a Simply Supported Plate)

Consider mixed boundary conditions of the first type, when

$$\sigma_{\alpha\alpha} = \sigma_{\alpha\beta} = w = 0 \quad \text{at} \quad \alpha = \alpha_0. \tag{4.1}$$

The quantities $\sigma_{\alpha\alpha}$, $\sigma_{\alpha\beta}$ are determined by (2.1), whereas

$$w = \varepsilon^c W^{(s)} + a\varepsilon^{\chi+1+s} \overset{a}{w}{}^{(s)} + a\varepsilon^{\mu+1+s} \overset{b}{w}{}^{(s)} \tag{4.2}$$

with $c = s - 3$ and $c = s - 1$ for the bending and extensional problems, respectively. Substituting the expressions for $\sigma_{\alpha\alpha}$, $\sigma_{\alpha\beta}$, w into (4.1), we obtain consistent results provided $\chi = \mu = -2$. Therefore

$$\sigma_{\alpha\alpha}^{(s)} + \overset{b}{\sigma}{}_{\alpha\alpha}^{(s)} + \overset{a}{\sigma}{}_{\alpha\alpha}^{(s)} = 0, \quad \sigma_{\alpha\beta}^{(s)} + \overset{a}{\sigma}{}_{\alpha\beta}^{(s)} + \overset{b}{\sigma}{}_{\alpha\beta}^{(s)} = 0, \tag{4.3}$$

$$W^{(s)} + a\overset{b}{w}{}^{(d)} + a\overset{a}{w}{}^{(d)} = 0 \quad \text{at} \quad \alpha = \alpha_0 \ (t = 0)$$

where the values of d for the symmetric and anti-symmetric problems are $d = s$ and $d = s - 2$, respectively. It follows from the first condition of (4.3) that $\overset{b}{\sigma}{}_{\alpha\alpha}^{(s)}(t = 0) = -\sigma_{\alpha\alpha}^{(s)}(\alpha_0) - \overset{a}{\sigma}{}_{\alpha\alpha}^{(s)}(t = 0)$. Substituting the latter into the compatibility condition (2.3), and performing similar analysis with that of the free edge boundary conditions, for the bending problem we obtain

$$\tau_{\alpha\alpha}^{(0)}(\alpha = \alpha_0) = 0 \quad \text{or} \quad M_\alpha^{(0)} = 0 \quad \text{at} \quad \alpha = \alpha_0 \tag{4.4}$$

with the third condition of (4.3) implying

$$w^{(0)} = 0, \quad w^{(1)} = 0 \quad \text{at} \quad \alpha = \alpha_0. \tag{4.5}$$

In other words, at $s = 0$ the outer field of the plate in case of the bending problem is governed by the conditions (in classical notation)

$$M_\alpha^{(0)} = 0, \quad w_0^{(0)} = 0 \quad \text{at} \quad \alpha = \alpha_0 \tag{4.6}$$

which coincide with the conditions for a simply supported plate within the classical context. Thus, the leading order approximation of the outer problem corresponds to the Kirchhoff theory for both governing equations and boundary conditions.

Returning back to (4.3) and using (4.4), (4.5) and the fact that $W^{(2)}(\alpha_0) = w^{(2)}(\alpha_0) + \left(\frac{1}{2}a\zeta^2\right)a_{23}\tau_{\beta\beta}^{(0)}(\alpha_0)$ we obtain for the plane boundary layer

$$\overset{b}{\sigma}{}_{\alpha\alpha}^{(0)} = 0, \quad \overset{b}{w}{}^{(0)} = -\left(\frac{1}{a}w^{(2)}(\alpha_0) + \frac{1}{2}\zeta^2 a_{23}\tau_{\beta\beta}^{(0)}(\alpha_0)\right) \quad \text{at} \quad t = 0. \tag{4.7}$$

The right-hand sides of (4.7) should satisfy the conditions of existence of a decaying solution in the plane problem for a semi-infinite strip (the compatibility conditions) in case of mixed edge boundary conditions. The first of these conditions, which was used earlier in derivation of (4.4), is related to self-equilibrated form of the edge load, whereas the second condition takes the form (Aghalovyan and Khachatryan, 1975)

$$(b_{13} + a_{55}) \int_0^1 \zeta^3 f_1(\zeta)d\zeta + 3 \int_0^1 (1 - \zeta^2) f_2(\zeta)d\zeta = 0 \tag{4.8}$$

where f_1, f_2 are the boundary values of $\overset{b}{\sigma}{}_{\alpha\alpha}$, $\overset{b}{w}$, respectively. In our case

$$f_1 = 0, \quad f_2 = -\left(\frac{1}{a}w^{(2)}(\alpha_0) + \frac{1}{2}\zeta^2 a_{23}\tau_{\beta\beta}^{(0)}(\alpha_0)\right).$$

Substituting the latter into (4.8), we get

$$w^{(2)}(\alpha_0, \eta) = -\frac{a}{10}a_{23}\tau_{\beta\beta}^{(0)}(\alpha_0), \tag{4.9}$$

which is one of the conditions for the next order approximation of the outer problem. Taking it into consideration, (4.7) may be reformulated as

$$\overset{b}{\sigma}{}_{\alpha\alpha}^{(0)} = 0, \quad \overset{b}{w}{}^{(0)} = \frac{1}{10}a_{23}(1 - 5\zeta^2)\tau_{\beta\beta}^{(0)}(\alpha_0) \quad \text{at} \quad t = 0. \tag{4.10}$$

The plane boundary layer solution, corresponding to (4.10), is given by

$$\overset{b}{Q}{}^{(0)} = \frac{1}{10}a_{23}\tau_{\beta\beta}^{(0)}(\alpha_0)\,\overset{b}{Q}{}^{[0]} \tag{4.11}$$

where $\overset{b}{Q}{}^{[0]}$ is the plane boundary layer solution (3.5.7) at $s = 0$, satisfying the following relations

$$\overset{b}{\sigma}{}_{\alpha\gamma}^{[0]} = \overset{b}{\sigma}{}_{\gamma\gamma}^{[0]} = 0 \quad \text{at} \quad \zeta = \pm 1, \quad \overset{b}{Q}{}^{[0]} \to 0 \quad \text{at} \quad t \to +\infty$$

$$\overset{b}{\sigma}{}_{\alpha\alpha}^{[0]} = 0, \quad \overset{b}{w}{}^{[0]} = 1 - 5\zeta^2 \quad \text{at} \quad t = 0. \tag{4.12}$$

The solution for $\overset{b}{Q}{}^{[0]}$ is obtained similarly, using the generalized orthogonality property (3.5.39) of the functions $F_n(\zeta)$, namely

$$\overset{b}{\sigma}{}^{[0]}_{\alpha\alpha} = A_k^{[0]} F_k'' \exp\left(-\lambda_k t\right), \qquad \overset{b}{\sigma}{}^{[0]}_{\alpha\gamma} = A_k^{[0]} \lambda_k F_k' \exp\left(-\lambda_k t\right),$$

$$\overset{b}{\sigma}{}^{[0]}_{\gamma\gamma} = A_k^{[0]} \lambda_k^2 F_k \exp\left(-\lambda_k t\right),$$

$$\overset{b}{u}{}^{[0]}_{\alpha} = -A_k^{[0]} \left(\frac{b_{11}}{\lambda_k} F_k'' + b_{13}\lambda_k F_k\right) \exp\left(-\lambda_k t\right), \quad (4.13)$$

$$\overset{b}{w}{}^{[0]} = -A_k^{[0]} \left(\frac{b_{11}}{\lambda_k^2} F_k''' + (b_{13} + a_{55}) F_k'\right) \exp\left(-\lambda_k t\right),$$

$$A_k^{[0]} = \frac{10\lambda_k^2}{b_{11}\Delta_k} \int_0^1 \zeta F_k d\zeta \quad (\bar{\Sigma}_k), \quad \Delta_k = \int_0^1 \left((F_k'')^2 - \frac{b_{33}}{b_{11}}\lambda_k^4 F_k^2\right) d\zeta$$

or

$$\overset{b}{Q}{}^{(0)} = \frac{a_{23}}{b_{11}} \frac{\lambda_k^2}{\Delta_k} \tau_{\beta\beta}^{(0)}(\alpha_0) \cdot \int_0^1 \zeta F_k d\zeta \cdot \overset{b}{Q}{}^{[0]}_k(\zeta,t). \qquad (4.14)$$

Here $\overset{b}{Q}{}^{[0]}_k(\zeta,t)$ are coefficients of $A_k^{[0]}$ for all of the quantities on the left-hand sides of (4.13), $F_k(\zeta)$ are the eigenfunctions of the boundary value problem (3.5.20). We also assume summation along the repeated index k (for all eigenvalues λ_k such that $Re\lambda_k > 0$), unless otherwise stated.

It follows from (4.3) that the boundary condition for the anti-plane boundary layer may be written as

$$\overset{a}{\sigma}{}^{(0)}_{\alpha\beta} = -\zeta\tau_{\alpha\beta}^{(0)}(\alpha_0) \qquad \text{at} \qquad t = 0. \qquad (4.15)$$

Hence, the anti-plane boundary layer is governed by (2.13).

Since $\tau_{\beta\beta}^{(0)}(\alpha_0)$, $\tau_{\alpha\beta}^{(0)}(\alpha_0) \neq 0$, both plane and anti-plane boundary layer solutions are non-zero, therefore the Kirchhoff plate theory does not provide accurate results for any of the stress components in the near-edge vicinity. Indeed, the stress components near the edge are of order $O(\varepsilon^{-2})$ and are compatible with the outer stresses, with the displacements being of order $O(\varepsilon^{-1})$.

At $s = 1$ one of the conditions for the outer problem is given by (4.5), whereas the second follows from the compatibility conditions (2.3)

$$\int_{-1}^{+1} \zeta\sigma_{\alpha\alpha}^{(1)}(\alpha_0)d\zeta = J_1 + J_2 + J_3 + J_4. \qquad (4.16)$$

Bearing in mind (3.5.8), (3.5.10), (4.13), (4.14) and omitting some of the intermediate computations, we obtain

$$J_1 = \int_{-1}^{+1} \zeta\, d\zeta \int_0^\infty \overset{b}{R}{}^{(0)}_\alpha\, dt = ak_{\beta 0} \int_0^\infty dt \int_{-1}^{+1} \zeta\, \overset{b}{\sigma}{}^{(0)}_{\beta\beta}\, d\zeta$$

$$= -ak_{\beta 0}\frac{a_{23}}{a_{22}} \int_0^\infty dt \int_{-1}^{+1} \zeta\, \overset{b}{\sigma}{}^{(0)}_{\gamma\gamma}\, d\zeta = -ak_{\beta 0}B_1 D_1 \cdot \frac{2}{3}\tau^{(0)}_{\beta\beta}(\alpha_0),$$

$$J_2 = \int_{-1}^{+1} \zeta\, d\zeta \int_0^\infty \overset{a}{R}{}^{(0)}_\alpha\, dt$$

$$= -\int_{-1}^{+1} \zeta\, d\zeta \int_0^\infty \left(d_{20}\, \overset{a}{\sigma}{}^{(0)}_{\alpha\beta} + 2ak_{\alpha 0}\, \overset{a}{\sigma}{}^{(0)}_{\alpha\beta} \right) dt$$

$$= \frac{1}{2a} A \sqrt{\frac{a_{44}}{a_{66}}} H_{\alpha 0} H_{\beta 0} \frac{\partial}{\partial\beta} \left(\frac{H^{(0)}_{\alpha\beta}(\alpha_0)}{H_{\alpha 0}} \right), \qquad (4.17)$$

$$J_3 = \int_{-1}^{+1} d\zeta \int_0^\infty \left(\xi_1 d_{11}\, \overset{b}{\sigma}{}^{(0)}_{\alpha\gamma} + ak_{\beta 0}\, \overset{b}{\sigma}{}^{(0)}_{\alpha\gamma} \right) t\, dt = 0,$$

$$J_4 = \int_{-1}^{+1} d\zeta \int_0^\infty \left(-\overset{a}{R}{}^{(0)}_\gamma \right) t\, dt$$

$$= \int_{-1}^{+1} d\zeta \int_0^\infty \left(d_{20}\, \overset{a}{\sigma}{}^{(0)}_{\beta\gamma} + ak_{\alpha 0}\, \overset{a}{\sigma}{}^{(0)}_{\beta\gamma} \right) t\, dt$$

$$= \frac{1}{2a} A \sqrt{\frac{a_{44}}{a_{66}}} \frac{H_{\beta 0}}{H_{\alpha 0}} \frac{\partial}{\partial\beta} \left(H_{\alpha 0}\, H^{(0)}_{\alpha\beta}(\alpha_0) \right),$$

$$B_1 = \frac{a_{23}^2}{a_{11}a_{22} - a_{12}^2} = \frac{E_1 E_2}{E_3^2} \frac{\nu_{32}^2}{1 - \nu_{12}\nu_{21}} = \frac{E_1}{E_3} \frac{\nu_{23}\nu_{32}}{1 - \nu_{12}\nu_{21}},$$

$$D_1 = 3 \sum_{k=1}^\infty \frac{\lambda_k^3 \left(\int_0^1 \zeta F_k\, d\zeta \right)^2}{\Delta_k}.$$

We note that the series for D_1 is convergent since it is obtained by integration of a uniformly convergent series for $\overset{b}{\sigma}{}^{[0]}_{\alpha\gamma}$ with respect to ζ and t. It may be observed that D_1 depends only on the material parameters but not the geometry, therefore it may be computed only once as a material constant using (3.5.25), (3.5.28), (3.5.31) etc. If the function $F_k(\zeta)$ is specified as (3.5.25), then

$$D_1 = \frac{3}{\beta^3} D^*, \qquad D^* = 4 \sum_{n=1}^\infty Re \left(\frac{1}{\mu_n \cos^4 \mu_n} \right) \approx -0.02904 \qquad (4.18)$$

where D^* is the constant value for isotropic material, see also Goldenveizer (1961, 1976).

In case when $F_k(\zeta)$ is defined as (3.5.28), which typically corresponds to orthotropic materials, then from (4.17) we get

$$D_1 = \frac{3}{2} \left(\frac{b_{11}}{b_{33}} \right)^2 (\beta_1 + \beta_2)^5 \bar{D}_1, \qquad \bar{D}_1 = \sum_{n=1}^\infty \frac{1}{z_n^5} \left(2 - z_n \frac{\cos z_n + \cos \omega z_n}{\sin z_n} \right)^2 \qquad (4.19)$$

where z_n are roots of $\omega \sin z_n - \sin \omega z_n = 0$. Table 10 contains the values of D_1 for several materials, studied earlier, see Table 6 in Chapter 3.

Table 10

Elastic characteristics	Materials			
	Unidirectional fiberglass winding	Fiberglass reinforced orthogonally 2:1	Fiberglass STET	Fiberglass ASTT(b) - C_2-O and PN-3
$b_{11} \cdot 10^{11}\text{Pa}^{-1}$	1.75426	2.74743	2.76737	5.60142
$b_{33} \cdot 10^{11}\text{Pa}^{-1}$	6.11647	8.56069	5.09112	22.95888
β_1	1.22365	0.808386	0.646914	0.816764
β_2	3.002848	2.183592	2.096657	2.478728
ω	0.421	0.46	0.528	0.504
$\bar{D}_1$	0.00104	0.00073	0.001	0.00339
D_1	0.173068	0.027042	0.068894	0.117649
$\bar{D}_2$	0.00079	0.00167	0.01004	0.03793
D_2	0.043822	0.020621	0.230564	0.43878
Elastic characteristics	Materials			
	SVAM 5:1	SVAM 10:1	SVAM 15:1	Graphite-epoxy material
$b_{11} \cdot 10^{11}\text{Pa}^{-1}$	3.21966	2.55529	2.1599	12.34118
$b_{33} \cdot 10^{11}\text{Pa}^{-1}$	12.28338	9.85754	8.52854	1.18027
β_1	0.744809	0.652242	0.638488	0.23067
β_2	2.622459	3.0113	3.112207	1.34067
ω	0.558	0.644	0.66	0.706
$\bar{D}_1$	0.00073	0.00036	0.00041	0.00023
D_1	0.032568	0.02395	0.02928	0.361338
$\bar{D}_2$	0.00533	0.00229	0.0039	0.00174
D_2	0.079262	0.050776	0.092836	0.911201

Substituting (4.17) into (4.16), we obtain

$$M_\alpha^{(1)} = -h k_{\beta 0} B_1 D_1 M_\beta^{(0)} + \sqrt{\frac{a_{44}}{a_{66}}} A h \frac{\partial H_{\alpha\beta}^{(0)}}{\partial s_\beta} \quad \text{at} \quad \alpha = \alpha_0 \qquad (4.20)$$

which acts as boundary conditions (together with the second condition of (4.5)) for the next order outer problem.

Considering the approximations for $s = 0, 1$, uniting (4.4), (4.5) and (4.20), we obtain up to $O(\varepsilon^2)$

$$\underline{M_\alpha + hk_{\beta 0}\frac{E_1}{E_3}\frac{\nu_{23}\nu_{32}}{1-\nu_{12}\nu_{21}}D_1 M_\beta^{(0)}} - \sqrt{\frac{G_{12}}{G_{23}}}Ah\frac{\partial H_{\alpha\beta}^{(0)}}{\partial s_\beta} = 0,$$

$$W = 0 \quad \text{at} \quad \alpha = \alpha_0 \qquad (4.21)$$

with the underlined terms providing refinement to the classical Kirchhoff theory. In case of isotropic medium the order of refinements is small $\left(O(\varepsilon^1)\right)$, whereas in case of anisotropic media the contribution could be more significant for materials with relatively small shear stiffness and which are relatively weak for extension/compression in the transverse direction. As may be seen from (4.21), the kinematic condition does not change, hence, the kinematic condition of the classical theory provides more accurate description of the outer problem, than the static one.

After the quantities of the outer problem has been determined for $s = 0, 1$, returning to (4.3), the boundary values $\overset{b}{\sigma}\overset{(1)}{_{\alpha\alpha}}$, $\overset{b}{w}{}^{(1)}$, $\overset{a}{\tau}\overset{(1)}{_{\alpha\beta}}$ may be found, and then the boundary layer solutions may be obtained in a manner similar to that for $s = 0$. Naturally, the asymptotic process may be continued to higher order.

Consider now symmetric problem (extension/compression). It follows from (4.3) and (2.4) that

$$\tau_{\alpha\alpha}^{(0)} = 0, \quad \tau_{\alpha\beta}^{(0)} = 0 \qquad (4.22)$$

or $T_\alpha^{(0)} = 0$, $\quad S_{\alpha\beta}^{(0)} = 0 \quad$ at $\quad \alpha = \alpha_0$.

Returning to (4.3), the following conditions may be written for the plane boundary layer

$$\overset{b}{\sigma}\overset{(0)}{_{\alpha\alpha}} = 0, \quad a\overset{b}{w}{}^{(0)} = -w^{(0)}(\alpha_0) \quad \text{at } t = 0 \qquad (4.23)$$

whereas for the anti-plane boundary layer

$$\overset{a}{\sigma}\overset{(0)}{_{\alpha\beta}} = 0 \quad \text{at } t = 0. \qquad (4.24)$$

Satisfying these, we obtain

$$\overset{a}{Q}{}^{(0)} = 0, \quad \overset{b}{Q}{}^{(0)} = -a_{23}\tau_{\beta\beta}^{(0)}(\alpha_0)\overset{b}{Q}{}^{[0]} \qquad (4.25)$$

where $\overset{b}{Q}{}^{[0]}$ is the symmetric plane boundary layer solution, satisfying the boundary conditions $\overset{b}{\sigma}\overset{(0)}{_{\alpha\alpha}} = 0$, $\overset{b}{w}{}^{(0)} = \zeta$ at $t = 0$. The quantity $\overset{b}{Q}{}^{[0]}$ is defined by (4.13), however, the functions $F_k(\zeta)$ in case of symmetric problem are defined by formulae (5.24), (5.26) and (5.29) of Chapter 3, with

$$A_k^{[0]} = -\frac{1}{b_{11}}\frac{\lambda_k^2}{\Delta_k}\int_0^1 F_k d\zeta \quad (\bar{\Sigma}_k). \qquad (4.26)$$

Here and below notation $\bar{\Sigma}_k$ means that no summation is assumed along the repeated index k.

Thus, in case of a symmetric problem the conditions (4.1) model the free edge. The classical theory is more accurate for symmetric problem, than in case of plate bending, where both boundary layers were present. The results of the classical theory in case of extension/compression in the near-edge vicinity are not accurate for the stress components $\sigma_{\alpha\alpha}$, $\sigma_{\alpha\gamma}$, $\sigma_{\beta\beta}$, $\sigma_{\gamma\gamma}$ since their intensity in the boundary layer solutions is of order $O(\varepsilon^{-2})$, matching these in the outer problem. However, the results of the classical theory are reasonably accurate for the displacement components. At $s = 1$ for the conditions of the outer problem we have

$$2\tau_{\alpha\alpha}^{(1)}(\alpha_0) = -\int_{-1}^{+1} d\zeta \int_0^\infty \left(\xi_1 d_{11} \overset{b}{\sigma}\overset{(0)}{{}_{\alpha\alpha}} + ak_{\beta 0} \left(\overset{b}{\sigma}\overset{(0)}{{}_{\alpha\alpha}} - \overset{b}{\sigma}\overset{(0)}{{}_{\beta\beta}} \right) \right) dt$$

$$= a\frac{a_{23}}{a_{22}} k_{\beta 0} \int_{-1}^{+1} \zeta \overset{b}{\sigma}\overset{(0)}{{}_{\alpha\gamma}}(t = 0) d\zeta = -ak_{\beta 0} B_1 D_2 2\tau_{\beta\beta}^{(0)}(\alpha_0),$$

$$2\tau_{\alpha\beta}^{(1)}(\alpha_0) = J_5 + J_6, \quad (4.27)$$

$$J_5 = -\int_{-1}^{+1} d\zeta \int_0^\infty \left(d_{20} \overset{b}{\sigma}\overset{(0)}{{}_{\beta\beta}} + ak_{\alpha 0} \left(\overset{b}{\sigma}\overset{(0)}{{}_{\beta\beta}} - \overset{b}{\sigma}\overset{(0)}{{}_{\alpha\alpha}} \right) \right) dt$$

$$= a\frac{a_{23}}{a_{22}} H_{\beta 0} \frac{\partial}{\partial\beta} \left(\int_{-1}^{+1} d\zeta \int_0^\infty \overset{b}{\sigma}\overset{(0)}{{}_{\gamma\gamma}} dt \right) = aB_1 D_2 H_{\beta 0} \frac{\partial}{\partial\beta} \left(2\tau_{\beta\beta}^{(0)}(\alpha_0) \right),$$

$$J_6 = -\int_{-1}^{+1} d\zeta \int_0^\infty \left(\xi_1 d_{11} \overset{a}{\sigma}\overset{(0)}{{}_{\alpha\beta}} + 2ak_{\beta 0} \overset{a}{\sigma}\overset{(0)}{{}_{\alpha\beta}} \right) dt = 0,$$

$$D_2 = \sum_{k=1}^\infty \frac{\lambda_k^3}{\Delta_k} \left(\int_0^1 F_k(\zeta) d\zeta \right)^2.$$

These may be reformulated in terms of the classical theory as

$$T_\alpha^{(1)} + hB_1 k_{\beta 0} D_2 T_\beta^{(0)}(\alpha_0) = 0, \quad (4.28)$$

$$S_{\alpha\beta}^{(1)} - hB_1 D_2 \frac{\partial T_\beta^{(0)}(\alpha_0)}{\partial s_\beta} = 0 \qquad \text{at} \qquad \alpha = \alpha_0.$$

Here D_2 depends on the material properties only. Its value may be determined from (3.5.24), (3.5.26) and (3.5.29). In case of isotropic media when $F_n(\zeta)$ is determined from (3.5.24)

$$D_2 = \frac{1}{\beta^3} D_2^*, \quad D_2^* = 4\sum_{n=1}^\infty Re\left(\frac{1}{z_n \cos^4 z_n} \right) \approx -0.19136 \quad (4.29)$$

where z_n is the root of $\sin 2z_n + 2z_n = 0$.

When $F_n(\zeta)$ is defined by (3.5.26)

$$D_2 = \frac{1}{2}\left(\frac{b_{11}}{b_{33}} \right)^2 (\beta_1 + \beta_2)^5 \bar{D}_2, \quad \bar{D}_2 = \sum_{n=1}^\infty \frac{1}{z_n^3} \left(\frac{\cos\omega z_n - \cos z_n}{\sin z_n} \right)^2 \quad (4.30)$$

where z_n is a solution of $\omega \sin z_n + \sin\omega z_n = 0$. The third case, corresponding to (3.5.29), may be treated in a similar way.

Uniting the conditions (4.22) and (4.28), we obtain (up to $O(\varepsilon^2)$)

$$T_\alpha + h\frac{E_1}{E_3}\frac{\nu_{23}\nu_{32}}{1 - \nu_{12}\nu_{21}} k_\beta D_2 T_\beta^{(0)} = 0 \quad (4.31)$$

$$S_{\alpha\beta} - h\frac{E_1}{E_3}\frac{\nu_{23}\nu_{32}}{1 - \nu_{12}\nu_{21}} D_2 \frac{\partial T_\beta^{(0)}}{\partial s_\beta} = 0 \qquad \text{at} \qquad \alpha = \alpha_0$$

where $T_\beta^{(0)}$ is the force calculated within the framework of the Kirchhoff theory. Equations (4.31) govern the outer field up to $\sim O(\varepsilon^2)$. After the outer problem is solved, the boundary layer components at $s = 1$ may be evaluated from (4.3).

Thus, the 3D formulation (4.1) has been reduced to conditions (4.21), (4.31) of the 2D problem. The latter generalize the classical conditions modeling pinned edge in bending problems and generalized free edge in symmetric problem.

Consider now mixed conditions of the second type, when the boundary conditions on $\alpha = \alpha_0$ are given by

$$\sigma_{\alpha\alpha} = 0, \ v = w = 0 \qquad \text{at} \qquad \alpha = \alpha_0. \tag{4.32}$$

We begin with substitution of the asymptotic expansions

$$\sigma_{\alpha\alpha} = \varepsilon^{-2+s}\sigma_{\alpha\alpha}^{(s)} + \varepsilon^{\chi+s}\overset{a}{\sigma}{}_{\alpha\alpha}^{(s)} + \varepsilon^{\mu+s}\overset{b}{\sigma}{}_{\alpha\alpha}^{(s)},$$

$$v = \varepsilon^{-2+s}V^{(s)} + \varepsilon^{\chi+1+s}a\,\overset{a}{u}{}_\beta^{(s)} + a\varepsilon^{\mu+1+s}\overset{b}{u}{}_\beta^{(s)}, \tag{4.33}$$

$$w = \varepsilon^c W^{(s)} + a\varepsilon^{\chi+1+s}\overset{a}{w}{}^{(s)} + a\varepsilon^{\mu+1+s}\overset{b}{w}{}^{(s)}$$

into (4.32), where $c = s - 3$ and $c = s - 1$ for bending and extension/compression, respectively. The consistency of the asymptotic procedure for the outer problem and the boundary layer dictates $\chi = \mu = -2$, resulting in

$$\sigma_{\alpha\alpha}^{(s)} + \overset{a}{\sigma}{}_{\alpha\alpha}^{(s)} + \overset{b}{\sigma}{}_{\alpha\alpha}^{(s)} = 0,$$

$$V^{(s)} + a\,\overset{a}{u}{}_\beta^{(s-1)} + a\,\overset{b}{u}{}_\beta^{(s-1)} = 0 \qquad \text{at} \qquad \alpha = \alpha_0(t = 0), \tag{4.34}$$

$$W^{(s)} + a\,\overset{a}{w}{}^{(d)} + a\,\overset{b}{w}{}^{(d)} = 0.$$

Here $d = s - 2$ for bending problem and $d = s$ for extension/compression problem.

In case of the anti-symmetric problem it follows from the third condition of (4.34) that

$$w^{(0)} = 0, \ \ w^{(1)} = 0 \quad \text{at} \quad \alpha = \alpha_0 \tag{4.35}$$

with the second one implying $V^{(0)}(\alpha_0) = 0$. On the other hand $V^{(0)} = -\zeta a H_\beta \dfrac{\partial w^{(0)}}{\partial \beta}$, therefore provided that the first condition of (4.35) is satisfied, $V^{(0)}(\alpha_0) = 0$, hence, the second condition of (4.34) at $s = 0$ is satisfied automatically. Treating the first condition of (4.34) at $s = 0$ similarly to the previous case, we obtain $\tau_{\alpha\alpha}^{(0)}(\alpha_0) = 0$. Therefore, at $s = 0$ we have

$$\tau_{\alpha\alpha}^{(0)} = 0, \ \ w^{(0)} = 0 \tag{4.36}$$

or

$$M_\alpha^{(0)} = 0, \ \ w_0^{(0)} = 0 \qquad \text{at} \qquad \alpha = \alpha_0.$$

In view of (4.35) $V^{(1)}(\alpha_0) = 0$, therefore, since $\overset{b}{u}{}_\beta^{(0)} \equiv 0$, we deduce from (4.34) $(s = 1)$ that $\overset{a}{u}{}_\beta^{(0)} = 0$ at $t = 0$, leading to

$$\overset{a}{Q}{}^{(0)} \equiv 0. \tag{4.37}$$

It follows from the latter that unlikely to the previously considered pinned edge boundary conditions, the anti-plane boundary layer arises even at the leading order. From physical point of view in the first case of the pinned edge the edge had a torsion tendency, which is no longer the case for the second type of mixed boundary conditions.

The plane boundary layer solution is governed by (4.7), therefore, since the outer solution does not change, the solution is given by (4.11), and (4.13).

Within the classical approach there is no distinction between the conditions (4.1) and (4.32), since both of these correspond to (4.36). However, these describe different stress fields. Indeed, in case of (4.32) the anti-plane boundary layer solution is zero, which is not the case for (4.1). This distinction becomes even more clear from the next order analysis.

At $s = 1$, expressing $\overset{b}{\sigma}{}^{(1)}_{\alpha\alpha}$ through $\sigma^{(1)}_{\alpha\alpha}$, $\overset{a}{\sigma}{}^{(1)}_{\alpha\alpha}$ in (4.34) and substituting the results into the compatibility conditions (3.5.13) or (2.3), we obtain

$$\int_{-1}^{+1} \zeta \sigma^{(1)}_{\alpha\alpha}(\alpha_0)\,d\zeta = J_1 + J_2 + J_3 + J_4 \tag{4.38}$$

where

$$J_1 = \int_{-1}^{+1} \zeta\,d\zeta \int_0^{\infty} \overset{b}{R}{}^{(0)}_{\alpha}\,dt = ak_{\beta0} \int_0^{\infty} dt \int_{-1}^{+1} \zeta \overset{b}{\sigma}{}^{(0)}_{\beta\beta}\,d\zeta$$

$$= -ak_{\beta0}B_1 D_1 \cdot \frac{2}{3}\tau^{(0)}_{\beta\beta}(\alpha_0),$$

$$J_2 = \int_{-1}^{+1} \zeta\,d\zeta \int_0^{\infty} \overset{a}{R}{}^{(0)}_{\alpha}\,dt$$

$$= -\int_0^{\infty} dt \int_{-1}^{+1} \zeta \left(d_{20} \overset{a}{\sigma}{}^{(0)}_{\alpha\beta} + 2ak_{\alpha0} \overset{a}{\sigma}{}^{(0)}_{\alpha\beta} \right) d\zeta = 0, \tag{4.39}$$

$$J_3 = -\int_{-1}^{+1} d\zeta \int_0^{\infty} \overset{b}{R}{}^{(0)}_{\gamma}\,t\,dt$$

$$= \int_{-1}^{+1} d\zeta \int_0^{\infty} \left(\xi_1 d_{11} \overset{b}{\sigma}{}^{(0)}_{\alpha\gamma} + ak_{\beta0} \overset{b}{\sigma}{}^{(0)}_{\alpha\gamma} \right) t\,dt = 0,$$

$$J_4 = -\int_{-1}^{+1} d\zeta \int_0^{\infty} \overset{a}{R}{}^{(0)}_{\gamma}\,t\,dt$$

$$= \int_{-1}^{+1} d\zeta \int_0^{\infty} \left(d_{20} \overset{a}{\sigma}{}^{(0)}_{\beta\gamma} + ak_{\alpha0} \overset{a}{\sigma}{}^{(0)}_{\beta\gamma} \right) t\,dt = 0$$

or

$$\frac{2}{3}\tau^{(1)}_{\alpha\alpha} = -ak_{\beta0}B_1 D_1 \cdot \frac{2}{3}\tau^{(0)}_{\beta\beta}(\alpha_0), \quad w^{(1)} = 0 \quad \text{at} \quad \alpha = \alpha_0. \tag{4.40}$$

Restricting our attention to approximations for $s = 0, 1$, from conditions (4.36) and (4.40), we deduce the following boundary conditions for the outer problem

$$M_\alpha + hk_{\beta0}B_1 D_1 M^{(0)}_\beta = 0, \quad w = 0 \quad \text{at} \quad \alpha = \alpha_0. \tag{4.41}$$

It should be emphasized that there is a principal difference between (4.41) and (4.21). As mentioned earlier, the classical theory does not reveal any difference between (4.1) and (4.32), however, comparison of (4.41) and (4.21) shows that these are indeed different in terms of the effective 2D problem. Therefore, it could be several different 3D boundary value problems, which may be reduced to the same 2D problem within the classical framework. Naturally, the question of accuracy of the Kirchhoff theory should arise only after the direct correspondence between the approximate 2D formulation and the original 3D statement has been established. In our case, if we start from (4.32), the accuracy of the Kirchhoff theory is better than that for (4.1), due to the fact that no anti-plane boundary layer solution arises at leading order, therefore, conditions (4.32) are preferable compared to (4.1). Taking into account the conditions (4.41) and (4.21) are therefore rather important since it underlines the difference in the boundary conditions. When the material parameters of the plate demonstrate small shear stiffness and weak resistance to transverse extension/compression, making use of (4.41) and (4.21) becomes crucial. Since the static but not kinematic boundary conditions have been first affected by reduction from 3D formulation to 2D statement, we again deduce that the kinematic relations of the classical theory are more accurate than the static ones.

In case of the symmetric problem it follows from (4.34) for the leading order approximation of the outer problem $(s = 0)$ that

$$\tau_{\alpha\alpha}^{(0)} = 0, \ v^{(0)} = 0 \quad \text{at} \quad \alpha = \alpha_0, \tag{4.42}$$

whereas for the plane boundary layer solution

$$\overset{b}{\sigma}{}_{\alpha\alpha}^{(0)} = 0, \ \overset{b}{w}{}^{(0)} = -a_{23}\tau_{\beta\beta}^{(0)}(\alpha_0) \quad \text{at} \quad t = 0. \tag{4.43}$$

The solution corresponding to (4.43) is given by formulae (4.25), (4.26). In order to derive the formulation for the anti-plane boundary layer, we express $\overset{a}{u}{}_{\beta}^{(s)}$ through $V^{(s+1)}$ in (4.34), and substitute $\overset{b}{u}{}_{\beta}^{(s)}$ into the compatibility conditions (3.5.15). (Note that these are valid for both indices "a" and "b".) As a result, we get

$$\int_{-1}^{+1} V^{(s+1)}(\alpha_0)d\zeta = a \int_0^\infty dt \int_{-1}^{+1} \left(\overset{a}{R}{}_{\beta}^{(s-1)} + \overset{b}{R}{}_{\beta}^{(s-1)} \right) d\zeta \tag{4.44}$$

$$-aa_{66} \int_0^\infty tdt \int_{-1}^{+1} \left(\overset{a}{R}{}_{\tau}^{(s-1)} + \overset{b}{R}{}_{\tau}^{(s-1)} \right) d\zeta,$$

from which

$$V^{(1)}(\alpha_0) = 0. \tag{4.45}$$

The second condition of (4.34) at $s = 1$ implies $a\overset{a}{u}{}_{\beta}^{(0)} = 0$ at $t = 0$, therefore

$$\overset{a}{Q}{}^{(0)} \equiv 0. \tag{4.46}$$

In other words, at leading order boundary layer solutions of the symmetric problem, corresponding to conditions (4.1) and (4.32), coincide. The conditions (4.1) model the free edge in the outer problem, whereas (4.32) describe the free edge in the longitudinal direction ($T_\alpha^0 = 0$, $V^{(0)} = 0$ at $\alpha = \alpha_0$).

At next order we obtain

$$T_\alpha + hB_1 D_2 k_{\beta 0} T_\beta^{(0)} = 0, \quad V = 0 \quad \text{at} \quad \alpha = \alpha_0. \tag{4.47}$$

Thus, comparing the anti-symmetric and the symmetric cases, we conclude that within the classical theory conditions (4.1) and (4.32) describe the same type of boundary conditions in the bending problem, namely, the pinned edge, however, in the symmetric problem these conditions characterize different boundary value problems: condition (4.1) corresponds in 2D framework to the generalized free edge, whereas (4.32) describes the generalized free edge in the longitudinal direction.

4.5 Boundary Conditions at Clamped Edge

Consider the following boundary conditions

$$u = v = w = 0 \quad \text{at} \quad \alpha = \alpha_0. \tag{5.1}$$

Let us substitute the following expansions

$$u = \varepsilon^{-2+s} U^{(s)} + a\varepsilon^{\chi+1+s}\, \overset{a}{u}_\alpha^{(s)} + a\varepsilon^{\mu+1+s}\, \overset{b}{u}_\alpha^{(s)}, \qquad (u, v; \alpha, \beta), \tag{5.2}$$

$$w = \varepsilon^c W^{(s)} + a\varepsilon^{\chi+1+s}\, \overset{a}{w}^{(s)} + a\varepsilon^{\mu+1+s}\, \overset{b}{w}^{(s)}$$

into the boundary conditions (5.1), where $c = s - 3$ and $c = s - 1$ in case of anti-symmetric and symmetric problems, respectively. Once again, the consistency of the asymptotic procedure is only possible for $\chi = \mu = -2$. In view of (5.2) conditions (5.1) become

$$U^{(s)} + a\,\overset{a}{u}_\alpha^{(s-1)} + a\,\overset{b}{u}_\alpha^{(s-1)} = 0,$$

$$V^{(s)} + a\,\overset{a}{u}_\beta^{(s-1)} + a\,\overset{b}{u}_\beta^{(s-1)} = 0 \quad \text{at} \quad \alpha = \alpha_0, \tag{5.3}$$

$$W^{(s)} + a\,\overset{a}{w}^{(d)} + a\,\overset{b}{w}^{(d)} = 0,$$

where $d = s - 2$ and $d = s$ for anti-symmetric and symmetric problems, respectively. Consider the bending problem first. It follows from the leading order ($s = 0$) of the first and third equations of (5.3) that

$$u^{(0)} = 0, \quad w^{(0)} = 0 \quad \text{at} \quad \alpha = \alpha_0. \tag{5.4}$$

The second equation gives $V^{(0)}(\alpha_0) = 0$, which is satisfied automatically in view of (5.4). At next order ($s = 1$) equations (5.3) imply $w^{(1)}(\alpha_0) = 0$, which also gives $V^{(1)}(\alpha_0) = 0$. Therefore, from the second condition of (5.3) at $s = 1$ we obtain for the anti-plane boundary layer $a\,\overset{a}{u}_\beta^{(0)} = 0$ at $t = 0$, which leads to $\overset{a}{Q}{}^{(0)} \equiv 0$. In order to derive the conditions for the plane boundary layer let us follow the idea suggested by Goldenveizer (1961, 1969, 1976) for isotropic shells. Indeed, it follows from (5.3) that

$$a \overset{b}{u} {}^{(0)}_{\alpha}(t=0) = \zeta a H_\alpha \frac{\partial w^{(1)}(\alpha_0)}{\partial \alpha}, \qquad (5.5)$$

$$a \overset{b}{w} {}^{(0)}(t=0) = -w^{(2)}(\alpha_0) - \frac{1}{2}a\zeta^2 \left(a_{13}\tau^{(0)}_{\alpha\alpha}(\alpha_0) + a_{23}\tau^{(0)}_{\beta\beta}(\alpha_0) \right).$$

The right-hand sides in the latter should enable existence of decaying boundary layer solutions, corresponding to the boundary conditions (5.5). Therefore the functions on the right-hand sides should satisfy some sort of compatibility conditions. In order to derive these conditions let us study a 2D problem for a semi-infinite strip $\Omega = \{(t,\zeta), \quad 0 \le t \le \ell, \quad -1 \le \zeta \le +1\}$ subject to conditions

$$\overset{b}{\sigma}{}_{\alpha\gamma} = \overset{b}{\sigma}{}_{\gamma\gamma} = 0 \qquad \text{at} \qquad \zeta = \pm 1 \qquad (5.6)$$

$$\overset{b}{u}{}_\alpha = \overset{b}{w} = 0 \qquad \text{at} \qquad t = \ell$$

and also conditions (5.5) at $t = 0$. The value of ℓ may be associated with the zone o decay of the plane boundary layer, i.e. $1 + \exp\left(- (\min Re\lambda_n)\,\ell\right) \approx 1$ or the quantity corresponding to the opposite edge. Since the problem is anti-symmetric, a bending moment $M^{(0)}$ and the transverse force $N^{(0)}$ should arise at $t = \ell$. Now if we impose $M^{(0)} = N^{(0)} = 0$, then from the absence of reaction at $t = \ell$ we deduce that the stresses are self-equilibrated at $t = 0$, hence, from the Saint-Venant principle the solution of the problem for a semi-strip decays.

Due to linearity of the problem, the 2D solution satisfying (5.6), may be presented as

$$Q^{(0)} = H_\alpha \frac{\partial w^{(1)}}{\partial \alpha} Q^{[1]} - \frac{1}{a}w^{(2)}(\alpha_0)Q^{[2]} - \frac{1}{2}\left(a_{13}\tau^{(0)}_{\alpha\alpha}(\alpha_0) + a_{23}\tau^{(0)}_{\beta\beta}(\alpha_0)\right) \cdot Q^{[3]} \quad (5.7)$$

where $Q^{[1]},\ Q^{[2]}, Q^{[3]}$ are solutions of the same problem, subject to conditions

$$\overset{b}{u}{}^{[1]}_\alpha = \zeta, \quad \overset{b}{w}{}^{[1]} = 0; \quad \overset{b}{u}{}^{[2]}_\alpha = 0, \quad \overset{b}{w}{}^{[2]} = 1; \quad \overset{b}{u}{}^{[3]}_\alpha = 0, \quad \overset{b}{w}{}^{[3]} = \zeta^2 \qquad \text{at} \qquad t = 0, \qquad (5.8)$$

respectively. These solutions may be obtained through a more general approach presented earlier in Section 10 of Chapter 1. The values of $M^{(0)}, \ N^{(0)}$ are then found as

$$M^{(0)} = H_\alpha \frac{\partial w^{(1)}}{\partial \alpha} M^{[1]} - \frac{1}{a}w^{(2)}(\alpha_0)M^{[2]}$$
$$- \frac{1}{2}\left(a_{13}\tau^{(0)}_{\alpha\alpha}(\alpha_0) + a_{23}\tau^{(0)}_{\beta\beta}\right) \cdot M^{[3]}, \qquad (5.9)$$

$$N^{(0)} = H_\alpha \frac{\partial w^{(1)}}{\partial \alpha} N^{[1]} - \frac{1}{a}w^{(2)}(\alpha_0)N^{[2]}$$
$$- \frac{1}{2}\left(a_{13}\tau^{(0)}_{\alpha\alpha}(\alpha_0) + a_{23}\tau^{(0)}_{\beta\beta}\right) \cdot N^{[3]}.$$

Satisfying the conditions $M^{(0)} = N^{(0)} = 0$ at $t = \ell$, we obtain

$$H_\alpha \frac{\partial w^{(1)}(\alpha_0)}{\partial \alpha} = \frac{1}{2}\left(a_{13}\tau^{(0)}_{\alpha\alpha}(\alpha_0) + a_{23}\tau^{(0)}_{\beta\beta}(\alpha_0)\right)m_1, \qquad (5.10)$$

$$w^{(2)}(\alpha_0) = \frac{a}{2}\left(a_{13}\tau^{(0)}_{\alpha\alpha}(\alpha_0) + a_{23}\tau^{(0)}_{\beta\beta}(\alpha_0)\right)m_2$$

with

$$m_1 = \frac{M^{[3]}N^{[2]} - M^{[2]}N^{[3]}}{M^{[1]}N^{[2]} - M^{[2]}N^{[1]}}, \qquad m_2 = \frac{M^{[3]}N^{[1]} - M^{[1]}N^{[3]}}{M^{[1]}N^{[2]} - M^{[2]}N^{[1]}}. \qquad (5.11)$$

Therefore, the outer field at $s = 1$ is determined by the following conditions at $\alpha = \alpha_0$

$$w^{(1)}(\alpha_0) = 0, \qquad H_\alpha \frac{\partial w^{(1)}}{\partial \alpha} = \frac{1}{2}\left(a_{13}\tau^{(0)}_{\alpha\alpha}(\alpha_0) + a_{23}\tau^{(0)}_{\beta\beta}(\alpha_0)\right)m_1, \qquad (5.12)$$

whereas for the plane boundary layer

$$\overset{b}{u}{}^{(0)}_{\alpha} = \frac{1}{2}\zeta\left(a_{13}\tau^{(0)}_{\alpha\alpha}(\alpha_0) + a_{23}\tau^{(0)}_{\beta\beta}(\alpha_0)\right)m_1, \qquad (5.13)$$

$$\overset{b}{w}{}^{(0)} = -\frac{1}{2}\left(a_{13}\tau^{(0)}_{\alpha\alpha}(\alpha_0) + a_{23}\tau^{(0)}_{\beta\beta}(\alpha_0)\right)\left(m_2 + \zeta^2\right) \qquad \text{at} \quad t = 0.$$

The corresponding solution may be found through approach described in Chapter 1. We note that $\overset{b}{Q}{}^{(0)} \neq 0$, therefore the results of the classical theory for $\sigma_{\alpha\alpha}, \sigma_{\beta\beta}, \sigma_{\alpha\gamma}, \sigma_{\gamma\gamma}$ are not accurate in the near-edge vicinity. We also remark that the anti-plane boundary layer solution plays a secondary role here.

In terms of (3.3.2) the conditions (5.4) and (5.12) may be written as

$$w = 0, \qquad (5.14)$$

$$H_\alpha \frac{\partial w}{\partial \alpha} + \frac{hm_1}{2D_3}\frac{1}{1 - \nu_{13}\nu_{23}}\left(\nu_{13}M^0_\alpha + \nu_{23}M^0_\beta\right) = 0 \qquad \text{at} \quad \alpha = \alpha_0$$

where $D_3 = \dfrac{2}{3}\dfrac{E_3 h^3}{1 - \nu_{13}\nu_{23}}$ is the transverse stiffness, and m_1 and m_2 are material constants which can be computed only once. These computations show that $m_1 \ll 1$, $m_2 \approx -0.2$. Therefore the conditions (5.14) are important for materials with weak stiffness in transverse direction. For plates composed of other materials the classical theory should provide results of reasonable accuracy for the outer field. We note that the contribution of the boundary layer in the near-edge zone cannot be neglected.

In the symmetric case it follows from (5.3) that

$$U^{(0)} = V^{(0)} = 0 \qquad \text{at} \quad \alpha = \alpha_0, \qquad (5.15)$$

which acts as boundary conditions for the outer problem. Hence, $a\overset{a}{u}{}^{(0)}_{\beta}(t = 0) = -V^{(1)}(\alpha_0)$. Substituting the latter into the compatibility conditions (3.5.15) or

(4.4.4), we deduce $V^{(1)}(\alpha_0) = 0$, therefore $\overset{a}{Q}{}^{(0)} \equiv 0$. The conditions for plane boundary layer may also be derived from (5.3) as

$$\overset{b}{u}{}^{(0)}_{\alpha} = -\frac{1}{a}v^{(1)}_{\alpha}(\alpha_0), \tag{5.16}$$

$$\overset{b}{w}{}^{(0)} = -\zeta\left(a_{13}\tau^{(0)}_{\alpha\alpha}(\alpha_0) + a_{23}\tau^{(0)}_{\beta\beta}(\alpha_0)\right) \quad \text{at} \quad t = 0.$$

The right-hand sides of (5.16) should allow decaying solutions. In order to derive the associated conditions, let us once again consider a problem for a semi-strip Ω subject to conditions (5.6) and (5.16). Since the quantities $\overset{b}{u}{}_{\alpha}, \overset{b}{\sigma}{}_{\alpha\alpha}, \overset{b}{\sigma}{}_{\gamma\gamma}$ are even with respect to ζ, and $\overset{b}{w}, \overset{b}{\sigma}{}_{\alpha\gamma}$ are odd in the problem under consideration, the only emerging force at $t = \ell$ is $T^{(0)}_1$. The corresponding compatibility condition, therefore, corresponds to vanishing of this force. Then according to the Saint-Venant principle there exists a decaying solution of the problem subject to (5.6). The leading order solution ($s = 0$) of the boundary value problem (3.5.7), (5.6) and (5.16) is given by

$$\overset{b}{Q}{}^{(0)} = -\frac{1}{a}v^{(1)}_{\alpha}(\alpha_0)Q^{[1]} - \left(a_{13}\tau^{(0)}_{\alpha\alpha}(\alpha_0) + a_{23}\tau^{(0)}_{\beta\beta}(\alpha_0)\right)Q^{[2]} \tag{5.17}$$

where $Q^{[1]}$, $Q^{[2]}$ are solutions of the same problem when the conditions (5.16) are replaced by

$$\overset{b}{u}{}^{[1]}_{\alpha} = 1, \quad \overset{b}{w}{}^{[1]} = 0 \qquad \text{problem [1]} \tag{5.18}$$

$$\overset{b}{u}{}^{[2]}_{\alpha} = 0, \quad \overset{b}{w}{}^{[2]} = \zeta \qquad \text{problem [2]} \quad \text{at} \quad t = 0,$$

leading to

$$T^{(0)}_1 = -\frac{1}{a}v^{(1)}_{\alpha}(\alpha_0)T^{[1]}_1 - \left(a_{13}\tau^{(0)}_{\alpha\alpha}(\alpha_0) + a_{23}\tau^{(0)}_{\beta\beta}(\alpha_0)\right)T^{[2]}_1. \tag{5.19}$$

Since $T^{(0)}_1(t = \ell) = 0$, (5.19) provides one more condition for the outer problem. In view of the above stated, at $s = 1$ we have

$$v^{(1)}_{\alpha} = -am_3\left(a_{13}\tau^{(0)}_{\alpha\alpha}(\alpha_0) + a_{23}\tau^{(0)}_{\beta\beta}(\alpha_0)\right), \quad v^{(1)}_{\beta} = 0 \qquad \text{at} \qquad \alpha = \alpha_0 \tag{5.20}$$

with $m_3 = T^{[2]}_1/T^{[1]}_1$. The condition (5.20) governs the outer solution, with the plane boundary layer solution following from

$$\overset{b}{u}{}^{(0)}_{\alpha} = m_3\left(a_{13}\tau^{(0)}_{\alpha\alpha} + a_{23}\tau^{(0)}_{\beta\beta}\right), \tag{5.21}$$

$$\overset{b}{w}{}^{(0)} = -\zeta\left(a_{13}\tau^{(0)}_{\alpha\alpha} + a_{23}\tau^{(0)}_{\beta\beta}\right) \qquad \text{at} \quad t = 0 \ (\alpha = \alpha_0).$$

Some more details of the solution procedure has been presented in Chapter 1. $\overset{b}{Q}{}^{(0)} \neq 0$, i.e. similarly to bending problem, the main contribution near clamped end is that of the plane boundary layer, the affect of the anti-plane boundary layer may be neglected. Restricting our consideration to $s = 0, 1$, we derive from (5.16) and (5.20) that

$$u - \frac{m_3}{2E_3}\left(\nu_{13}T_\alpha^0 + \nu_{23}T_\beta^0\right) = 0, \quad V = 0 \quad \text{at} \quad \alpha = \alpha_0. \tag{5.22}$$

The obtained conditions are generalizing the classical ones, giving the outer field up to $\sim O(\varepsilon^2)$. In order to obtain results of similar accuracy in the near-end vicinity, the boundary layer solutions should be computed, with the major contribution coming from the plane boundary layer.

In case of the third type of mixed boundary conditions (fixed edge) we have

$$u = \sigma_{\alpha\beta} = w = 0 \quad \text{at} \quad \alpha = \alpha_0. \tag{5.23}$$

As in the previous cases, substituting asymptotic expansions for $u, \sigma_{\alpha\beta}, w$ into (5.23), we deduce that the procedure is consistent provided $\chi = \mu = -2$. Equations (5.23) are transformed to

$$U^{(s)} + a\overset{a}{u}{}_\alpha^{(s-1)} + a\overset{b}{u}{}_\alpha^{(s-1)} = 0,$$

$$\sigma_{\alpha\beta}^{(s)} + \overset{a}{\sigma}{}_{\alpha\beta}^{(s)} + \overset{b}{\sigma}{}_{\alpha\beta}^{(s)} = 0 \quad \text{at} \quad \alpha = \alpha_0 \ (t = 0), \tag{5.24}$$

$$W^{(s)} + a\overset{a}{w}{}^{(d)} + a\overset{b}{w}{}^{(d)} = 0$$

where $d = s - 2$ and $d = s$ for anti-symmetric and symmetric problems, respectively. Let us concentrate on the anti-symmetric problem first. It then follows from (5.24) that

$$w^{(0)} = 0, \quad \frac{\partial w^{(0)}}{\partial s_\alpha} = 0 \quad \text{at} \quad \alpha = \alpha_0 \tag{5.25}$$

which are the conditions for the outer problem. For the anti-plane boundary layer $\overset{a}{\sigma}{}_{\alpha\beta}^{(0)}(t = 0) = -\zeta\tau_{\alpha\beta}^{(0)}(\alpha_0)$, with the corresponding solution given by (2.13).

We note that the 2D conditions (5.25), coinciding with the classical conditions of clamped ends, correspond to two different types of 3D boundary conditions. However, the asymptotic method reveals that these cases are indeed different. Indeed, in case of (5.1) the anti-plane boundary layer vanishes, whereas it is present for conditions (5.23). The plane boundary layer is the same for both (5.1) and (5.23). We also remark that conditions (5.1) are preferable to (5.23) in a sense of increased stiffness of the edge, since in case of (5.23) the edge demonstrates torsion tendency due to the presence of the anti-plane boundary layer.

Similar considerations may be repeated for $s = 1$, leading to (5.12) and consequently (5.14) as conditions of the outer problem. The plane boundary layer is then given by (5.13). Therefore conditions (5.1) and (5.23) are both modeled through the 2D formulation (5.14), giving the same results (up to $\sim O(\varepsilon^2)$) for the outer solution and for the plane boundary layer. It should be mentioned that in this case the conditions (5.1) and (5.23) are closer to each other than conditions (1.2) and (1.3) modeling pinned edge within the bending problem, when the associated 2D conditions coincided only at leading order.

In case of the symmetric problem (5.24) imply

$$u_0^{(0)} = S_{\alpha\beta}^{(0)} = 0 \qquad \text{at} \qquad \alpha = \alpha_0. \tag{5.26}$$

This shows that in 2D formulation (5.23) model free boundary in tangential direction. Since $\tau_{\alpha\beta}^{(0)}(\alpha_0) = 0$ it follows that $\overset{a}{Q}{}^{(0)} \equiv 0$. The plane boundary layer is determined from (5.21), with the meaning of $\tau_{\alpha\alpha}^{(0)}$, $\tau_{\beta\beta}^{(0)}$ being in line with (5.26). Therefore,

$$\overset{b}{Q}{}^{(0)} = \left(a_{13}\tau_{\alpha\alpha}^{(0)}(\alpha_0) + a_{23}\tau_{\beta\beta}^{(0)}(\alpha_0) \right) \overset{b}{Q}{}^{[0]} \tag{5.27}$$

where $\overset{b}{Q}{}^{[0]}$ is the plane boundary layer solution, corresponding to the following edge boundary conditions $\overset{b}{u}{}_\alpha^{[0]} = m_3$, $\overset{b}{w}{}^{[0]} = -\zeta$ at $t = 0$.

At $s = 1$ the first condition for the outer problem is

$$v_\alpha^{(1)} = -a \left(a_{13}\tau_{\alpha\alpha}^{(0)} + a_{23}\tau_{\beta\beta}^{(0)} \right) m_3 \qquad \text{at} \qquad \alpha = \alpha_0 \tag{5.28}$$

whereas the second condition follows from (2.4)

$$2\tau_{\alpha\beta}^{(1)}(\alpha_0) = \int_{-1}^{+1} d\zeta \int_0^\infty \left(\overset{a}{R}{}_\tau^{(0)} + \overset{b}{R}{}_\tau^{(0)} \right) dt. \tag{5.29}$$

In our case $\overset{a}{R}{}_\tau^{(0)} = 0$, $\overset{b}{R}{}_\tau^{(0)} = -d_{20}\overset{b}{\sigma}{}_{\beta\beta}^{(0)} - ak_{\alpha0}\left(\overset{b}{\sigma}{}_{\beta\beta}^{(0)} - \overset{b}{\sigma}{}_{\alpha\alpha}^{(0)} \right)$, therefore

$$\int_{-1}^{+1} d\zeta \int_0^\infty \overset{b}{R}{}_\tau^{(0)} dt$$

$$= \frac{\nu_{32}}{2E_3} m_4 H_{\beta0} \frac{\partial}{\partial\beta} \left(\nu_{13}\left(2a\tau_{\alpha\alpha}^{(0)}(\alpha_0) \right) + \nu_{23}\left(2a\tau_{\beta\beta}^{(0)}(\alpha_0) \right) \right), \tag{5.30}$$

$$m_4 = \int_{-1}^{+1} d\zeta \int_0^\infty \overset{b}{\sigma}{}_{\gamma\gamma}^{[0]} dt = \sum_{(\lambda_n)} \frac{1}{\lambda_n} A_n^{[0]} \int_{-1}^{+1} F_n d\zeta.$$

Uniting the conditions (5.28), (5.29), and (5.26), it is possible to reformulate these in terms of the classical theory as

$$u - \frac{m_3}{2E_3} \left(\nu_{13}T_\alpha^0 + \nu_{23}T_\beta^0 \right) = 0, \tag{5.31}$$

$$S_{\alpha\beta} - h\frac{\nu_{32}}{2E_3} m_4 \frac{\partial}{\partial s_\beta} \left(\nu_{13}T_\alpha^0 + \nu_{23}T_\beta^0 \right) = 0 \qquad \text{at} \qquad \alpha = \alpha_0.$$

We note that conditions (5.1) and (5.23) model clamped edge boundary conditions with high degree of accuracy in case of the antisymmetric problem, whereas in the symmetric case they correspond to another type of boundary conditions. Indeed, comparison of (5.22) and (5.31) reveals that the first case describes the generalized clamped end, whereas the second one corresponds to a generalized free end in the tangential direction.

Analysis of formulae (5.14), (5.22) and (5.31) shows that refinement of the kinematic and static conditions of the classical theory is related to the Poisson effect in the transverse direction, which needs to be taken into account. We also mention that for some materials incorporation of the refined terms into analysis may become crucial, especially for plates composed of materials which are weak on bending and extension/compression in the cross sections perpendicular to the mid-plane.

4.6 Refinement of the Classical Theory for Bending and Extension of Anisotropic Plates

The results obtained in Chapters 3 and 4, lead to suggestions on possible refinement of the classical theory for bending or extension/compression of anisotropic plates.

As shown above, the field of the plate may be decomposed into two parts which are qualitatively different, namely, the outer solution and the boundary layer solution. The boundary layer solution is also composed of two parts including the plane and the anti-plane boundary layers. It has been shown in Chapter 3 that at leading order the formulation of the outer problem is identical to the corresponding classical equations of bending or extension/compression for anisotropic plates. Moreover, in Chapter 4 it has been observed that at leading order the analysis of the outer problem is identical to the classical theory not only in equations, but also in boundary conditions. Analysis of the higher orders of the outer problem combined with the boundary layer effects provide refinement of the classical formulation.

The problem of refining the classical theory for anisotropic plates is now reduced to construction of a number of successive approximations of the outer problem and the boundary layer that provide the desired accuracy both in the near-edge vicinity and also away from the edge of the plate.

Here we present two of the relatively simple methods of refinement relying on the previous results of Chapters 3 and 4. The first method allows to obtain refined results away from the end of the plate, whereas the second approach is oriented to refinement of the results in the near-edge vicinity, which could be practically important, in particular, for problems of stress concentration near holes.

It has been shown in Chapter 3 that the iterative process for the outer problem is given by equations (3.3.5) and (3.3.10). At leading order the equations coincide with that of the classical theory. At next order ($s = 1$) we obtain similar equations, but these are already homogeneous, since $X_1^{(1)} = Y_1^{(1)} = Z_1^{(1)} = 0, (1, 2)$, $Q^{*(1)} = 0$. Therefore, if we restrict our consideration to the two-term asymptotic expansions for $T_\alpha, T_\beta, S_{\alpha\beta}, M_\alpha, H_{\alpha\beta}, M_\beta$, then these quantities satisfy the equations for anisotropic plates, and at the same time they contain terms of order ε. Analysis of the corresponding boundary conditions suggests the following refined approach to classical plate theory.

The problem is reduced to construction of the classical equations, which take into account the moments arising from transforming the surface load to the load acting on the mid-plane. These equations are written below

$$\frac{1}{AB}\frac{\partial BT_\alpha}{\partial\alpha} - \frac{1}{AB}\frac{\partial B}{\partial\alpha}T_\beta + \frac{1}{B}\frac{\partial S_{\alpha\beta}}{\partial\beta}$$

$$+2\frac{1}{AB}\frac{\partial A}{\partial\beta}S_{\alpha\beta} = -\varepsilon^{-1}X_1, \qquad (\alpha,\beta;1,2),$$

$$\frac{1}{AB}\left(\frac{\partial}{\partial\alpha}\left(BN_\alpha\right)+\frac{\partial}{\partial\beta}\left(AN_\beta\right)\right)=-Z_2,\tag{6.1}$$

$$\frac{\partial}{\partial\alpha}\left(BM_\alpha\right)+\frac{\partial}{\partial\beta}\left(AH_{\alpha\beta}\right)+\frac{\partial A}{\partial\beta}H_{\alpha\beta}-\frac{\partial B}{\partial\alpha}M_\beta$$

$$=ABN_\alpha-aABX_2\qquad(\alpha,\beta;X,Y).$$

It should be mentioned here that since the tangential load has been originally specified as $\varepsilon^{-1}X$, the right-hand sides in (6.1) differ slightly from the conventional formulation

$$T_\alpha=C_{11}\varepsilon_1+C_{12}\varepsilon_2,\quad S_{\alpha\beta}=C_{66}\omega,\qquad(\alpha,\beta;1,2),$$

$$M_\alpha=D_{11}\chi_1+D_{12}\chi_2,\quad H_{\alpha\beta}=H_{\beta\alpha}=D_{66}\tau,\tag{6.2}$$

$$C_{ik}=2hB_{ik},\quad D_{ik}=\frac{2}{3}h^3B_{ik}.$$

The corresponding boundary conditions take into consideration some additional terms arising from higher order approximations.

In case of a free edge ($\sigma_{\alpha\alpha}=\sigma_{\alpha\beta}=\sigma_{\alpha\gamma}=0$ at $\alpha=\alpha_0$)

$$T_\alpha=0,\quad S_{\alpha\beta}=0,$$

$$M_\alpha-\sqrt{\frac{G_{12}}{G_{23}}}Ah\frac{\partial H_{\alpha\beta}}{\partial s_\beta}=0\qquad\text{at}\qquad\alpha=\alpha_0,\tag{6.3}$$

$$N_\alpha+\frac{\partial H_{\alpha\beta}}{\partial s_\beta}-\sqrt{\frac{G_{12}}{G_{23}}}Ah\frac{\partial(k_{\beta0}H_{\alpha\beta})}{\partial s_\beta}=0.$$

In case of the pinned end of the first type ($\sigma_{\alpha\alpha}=\sigma_{\alpha\beta}=w=0$)

$$T_\alpha+h\frac{E_1E_2}{E_3^2}\frac{\nu_{32}^2}{1-\nu_{12}\nu_{21}}k_{\beta0}D_2T_\beta=0,$$

$$S_{\alpha\beta}-h\frac{E_1E_2}{E_3^2}\frac{\nu_{32}^2}{1-\nu_{12}\nu_{21}}D_2\frac{\partial T_\beta}{\partial s_\beta}=0,\tag{6.4}$$

$$M_\alpha-\sqrt{\frac{G_{12}}{G_{23}}}Ah\frac{\partial H_{\alpha\beta}}{\partial s_\beta}+hk_{\beta0}\frac{E_1E_2}{E_3^2}\frac{\nu_{32}^2}{1-\nu_{12}\nu_{21}}D_1M_\beta=0\qquad\text{at}\qquad\alpha=\alpha_0,$$

$$W=0.$$

In case of the pinned end of the second type ($\sigma_{\alpha\alpha}=v=w=0$)

$$T_\alpha+h\frac{E_1E_2}{E_3^2}\frac{\nu_{32}^2}{1-\nu_{12}\nu_{21}}k_{\beta0}D_2T_\beta=0,$$

$$V=0,\tag{6.5}$$

$$M_\alpha+hk_{\beta0}\frac{E_1E_2}{E_3^2}\frac{\nu_{32}^2}{1-\nu_{12}\nu_{21}}D_1M_\beta=0\qquad\text{at}\qquad\alpha=\alpha_0,$$

$$W=0.$$

For clamped end of the first type $(u = v = w = 0)$

$$u - \frac{m_3}{2E_3}\left(\nu_{13}T_\alpha + \nu_{23}T_\beta\right) = 0,$$

$$V = 0, \tag{6.6}$$

$$W = 0 \qquad \text{at} \quad \alpha = \alpha_0,$$

$$H_\alpha \frac{\partial W}{\partial \alpha} + \frac{hm_1}{2D_3}\frac{1}{1 - \nu_{13}\nu_{23}}\left(\nu_{13}M_\alpha + \nu_{23}M_\beta\right) = 0$$

for clamping of the second type $(u = \sigma_{\alpha\beta} = w = 0)$

$$u - \frac{m_3}{2E_3}\left(\nu_{13}T_\alpha + \nu_{23}T_\beta\right) = 0,$$

$$S_{\alpha\beta} - h\frac{\nu_{32}}{2E_3}m_4\frac{\partial}{\partial s_\beta}\left(\nu_{13}T_\alpha + \nu_{23}T_\beta\right) = 0,$$

$$W = 0 \qquad \text{at} \quad \alpha = \alpha_0, \tag{6.7}$$

$$H_\alpha \frac{\partial W}{\partial \alpha} + \frac{hm_1}{2D_3}\frac{1}{1 - \nu_{13}\nu_{23}}\left(\nu_{13}M_\alpha + \nu_{23}M_\beta\right) = 0.$$

As before, the underlined terms are refinements to the classical theory. It may be noticed from (6.3)-(6.7), that the contribution from the refined terms becomes more important for plates composed of materials with small shear stiffness, small extension/compression or bending stiffness in the transverse plane. We note that if the above mentioned stiffness parameters are large compared to the corresponding parameters in the mid-plane, then the results of the classical theory are providing a good approximation. For example, in case of (6.3), assuming $G_{23} = +\infty$, we obtain the classical conditions. In other words, the Kirchhoff hypotheses are almost true for plates composed of materials with large stiffness for shear and extension/compression in the transverse direction.

The approach proposed above is accurate up to $\sim O(\varepsilon^2)$. In order to increase the accuracy, higher order terms should be included both into the governing equations and the boundary conditions, i.e. the next order iterative theory should be constructed for the outer solution and the boundary layer. We note that as shown above, the construction of the boundary layer is not the goal itself, but the boundary layer is affecting the outer solution through the boundary conditions, and the refined terms of the outer solution have been affected by the boundary layer. We remark that within the fully 3D approach to the problem the refinement of the classical results is only possible in case of simultaneous refinement of the outer solution and the boundary layer. Therefore, all of the theories which aim only at refinement of the outer solution and do not contain refinement of the boundary layer are inconsistent and mathematically incorrect.

In case of the problems of stress concentration, loading with discontinuities, change of type of boundary conditions etc. the phenomena in the near-boundary domain or close to discontinuities is of vital importance. In terms of the proposed asymptotic method this corresponds to construction of the boundary layer solutions, which becomes a necessity for these problems.

Indeed, as may be seen from (3.2.4) and (3.2.5), the stress and displacement components of the outer problem are of order

$$\sigma_{\alpha\alpha}, \ \sigma_{\alpha\beta}, \ \sigma_{\beta\beta} \sim \varepsilon^{-2}, \quad \sigma_{\alpha\gamma}, \ \sigma_{\beta\gamma} \sim \varepsilon^{-1}, \tag{6.8}$$
$$\sigma_{\gamma\gamma} \sim \varepsilon^{0}, \quad u_{\alpha}, u_{\beta} \sim \varepsilon^{-2}, \quad w \sim \varepsilon^{-3}.$$

At the boundary these quantities should be combined with these of the boundary layer solution. We recall that in the bending problem and in case of the first boundary value problem the major contribution is delivered by the anti-plane boundary layer, therefore, as follows from Section 2 of Chapter 4,

$$\sigma_{\alpha\beta}, \ \sigma_{\beta\gamma} \sim O(\varepsilon^{-2}), \quad \sigma_{\alpha\alpha}, \ \sigma_{\beta\beta}, \sigma_{\alpha\gamma}, \ \sigma_{\gamma\gamma} \sim O(\varepsilon^{-1}), \tag{6.9}$$
$$u_{\beta} \sim O(\varepsilon^{-1}), \quad u_{\alpha}, w \sim O(\varepsilon^{0}).$$

In case of the symmetric problem the stresses of the plane boundary layer are of order $O(\varepsilon^{-1})$, whereas these of anti-plane boundary layer are of order $O(\varepsilon^{0})$, with the corresponding displacements being of order $O(\varepsilon^{0}), O(\varepsilon^{1})$, respectively.

In case of the mixed boundary conditions (1.2) for bending problem the stress components for both plane and anti-plane boundary layer solutions are of order $O(\varepsilon^{-2})$

$$\sigma_{\alpha\alpha}, \sigma_{\alpha\beta}, \ \sigma_{\beta\beta}, \ \sigma_{\alpha\gamma}, \sigma_{\beta\gamma}, \sigma_{\gamma\gamma} \sim O(\varepsilon^{-2}), \tag{6.10}$$
$$u_{\alpha}, \ u_{\beta}, w \sim O(\varepsilon^{-1}).$$

In case of the symmetric problem the major contribution arrives from the plane boundary layer, containing stresses of order $O(\varepsilon^{-2})$ and displacements of order $O(\varepsilon^{-1})$. The corresponding quantities of the anti-plane boundary layer are of order $O(\varepsilon^{-1}), O(\varepsilon^{0})$ respectively.

As shown earlier in Section 4, in case of the mixed boundary conditions (1.3) the plane boundary layer is dominant for both symmetric and anti-symmetric problems. The dominant stress components are of order $O(\varepsilon^{-2})$, whereas the displacement components are of order $\sim O(\varepsilon^{-1})$, i.e.

$$\sigma_{\alpha\alpha}, \ \sigma_{\beta\beta}, \ \sigma_{\alpha\gamma}, \sigma_{\gamma\gamma} \sim O(\varepsilon^{-2}), \quad \sigma_{\alpha\beta}, \sigma_{\beta\gamma} \sim O(\varepsilon^{-1}), \tag{6.11}$$
$$u_{\alpha}, w \sim O(\varepsilon^{-1}), \quad u_{\beta} \sim O(\varepsilon^{0}).$$

In case of the second boundary value problem of elasticity (clamped edge) the major contribution is once again coming from the plane boundary layer. The asymptotic orders of stresses and displacements are the same as in (6.11).

In case of clamping of the second type (1.5) for bending problem both the plane and anti-plane boundary layers contribute to the leading order approximation, with the asymptotic orders being identical to (6.10). In case of extension/compression problem the major contribution arises from the plane boundary layer. The associated stresses are of order $O(\varepsilon^{-2})$, displacements are of order $O(\varepsilon^{-1})$. As for the anti-plane boundary layer, the orders of stresses and displacements are $O(\varepsilon^{-1})$ and $O(\varepsilon^0)$, respectively.

We remark that the classical theory typically neglects the stresses and displacements of the boundary layers (6.9)-(6.11). However, all of the stresses (or some of these) are of the same orders as the corresponding stresses of the outer solution (6.8). Therefore, when analyzing the stress field of the problems involving stress concentrations etc., the boundary layers cannot be neglected and should be constructed even at leading order.

In the simplest case of the leading order approximation, in order to refine the results of the classical theory, the stresses predicted by the latter, should be complemented by leading order approximations of the stresses arising from

a) in case of a free edge - the anti-plane boundary layer solution;

b) in case of pinned edge of the first type - both plane and anti-plane boundary layer;

c) in case of pinned edge of the second type - plane boundary layer;

d) in case of a clamped edge - plane boundary layer;

e) in case of the clamping of the second type - both plane and anti-plane boundary layers.

The formulae for the complementary stresses of the boundary layers have been presented earlier in this Chapter. The accuracy of this approximate theory is of order $O(\varepsilon)$. Further refinement of this theory may be performed through taking into account higher order approximations of the iterative asymptotic procedure.

Comparison of the relations (6.8) and (6.9)-(6.11) reveals that the classical theory provides reasonably accurate results for the displacement components.

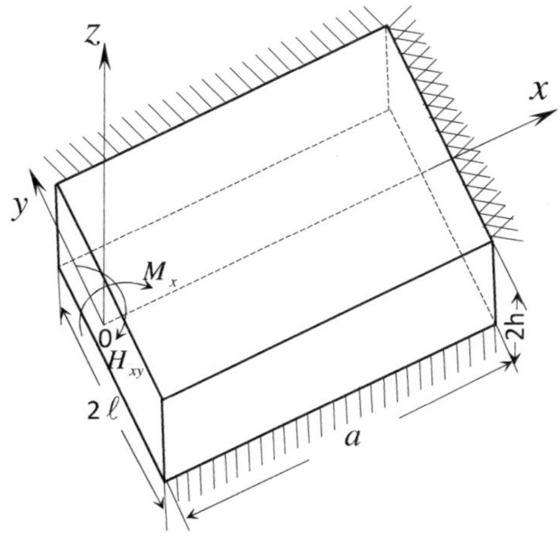

Fig. 4.1

As an illustration, let us consider an example of a bending problem for a transversely isotropic rectangular plate, with two ends pinned, one clamped edge and one edge loaded by bending and twisting moments, see Fig. 4.1. Here we assume that XOY is the plane of isotropy. Consider the 3D problem in elasticity for this plate in the domain $\Omega = \{(x, y, z) : 0 \leq x \leq a, |y| \leq \ell, |z| \leq h, h \ll a, \ell\}$ subject to the following boundary conditions

$$X^\pm = Y^\pm = Z^\pm = 0 \quad \text{at} \quad z = \pm h,$$

$$\sigma_x = \bar{\sigma}_x, \quad \sigma_{xy} = \bar{\sigma}_{xy}, \quad \sigma_{xz} = \bar{\sigma}_{xz} \quad \text{at} \quad x = 0, \tag{6.12}$$

$$u_\alpha = u_\beta = u_\gamma = 0 \quad \text{at} \quad x = a,$$

$$\sigma_y = u_\alpha = u_\gamma = 0 \quad \text{at} \quad y = \pm \ell.$$

Let

$$\bar{\sigma}_x = \frac{M_0}{h^2 \bar{f}_1} f_1(\zeta) \sin \frac{\pi}{\ell} y, \tag{6.13}$$

$$\bar{\sigma}_{xy} = \frac{H_0}{h^2 \bar{f}_2} f_2(\zeta) \cos \frac{\pi}{\ell} y, \quad \bar{\sigma}_{xz} = 0,$$

where

$$\bar{f}_i = \int_{-1}^{+1} f_i(\zeta) \zeta \, d\zeta, \quad (i = 1, 2),$$

and f_1, f_2 are odd functions. The solution of the formulated 3D problem is composed of the outer solution and the boundary layers. According to (6.12) and results presented earlier in Chapters 3 and 4, the outer problem reduces to the following effective equation $\Delta \Delta w = 0$ in the domain $\bar{\Omega} = \{(x, y) : 0 \leq x \leq a, |y| \leq \ell\}$, subject to

$$M_x - \sqrt{\frac{G_{12}}{G_{23}}} Ah \frac{\partial H_{xy}}{\partial y} = \bar{M}_x, \tag{6.14}$$

$$N_x + \frac{\partial H_{xy}}{\partial y} - \sqrt{\frac{G_{12}}{G_{23}}} Ah \frac{\partial (k_y H_{xy})}{\partial y} = \bar{N}_x + \frac{\partial \bar{H}_{xy}}{\partial y} \quad \text{at} \quad x = 0$$

$$M_y + h\bar{B}_1 k_y D_1 M_x = 0, \quad w = 0 \quad \text{at} \quad y = \pm \ell \tag{6.15}$$

$$w = 0, \quad \frac{\partial w}{\partial x} + \frac{h}{2D_3} \frac{m_1}{1 - \nu_{13}\nu_{23}} (\nu_{13} M_x + \nu_{23} M_y) = 0 \quad \text{at} \quad x = a. \tag{6.16}$$

In our case

$$k_x = k_y = 0, \quad G_{12} = G, \quad G_{13} = G_{23} = G', \quad \nu_{13} = \nu_{23} = \nu', \tag{6.17}$$

$$\bar{M}_x = M_0 \sin \alpha \, y, \quad \bar{H}_{xy} = H_0 \cos \alpha \, y, \quad \bar{N}_x = 0 \quad (\alpha = \pi/\ell).$$

The bending and twisting moments along with the transverse forces may be expressed through w as

$$M_x = -D \left(\frac{\partial^2 w}{\partial x^2} + \nu \frac{\partial^2 w}{\partial y^2} \right), \quad (x, y),$$

$$H_{xy} = H_{yx} = -D(1 - \nu) \frac{\partial^2 w}{\partial x \partial y}, \tag{6.18}$$

$$N_x = -D \frac{\partial}{\partial x} \Delta w, \quad (x, y), \quad D = \frac{2}{3} \frac{Eh^3}{1 - \nu^2}.$$

The solution is sought in the form

$$w = w_0(x) \sin \alpha y. \tag{6.19}$$

Then the conditions (6.15) are satisfied automatically, and w_0 is to be determined from the following equation

$$w_0^{IV} - 2\alpha^2 w_0'' + \alpha^4 w_0 = 0. \tag{6.20}$$

Therefore,

$$w = \left((C_1 + C_2 x) e^{-\alpha x} + (C_3 + C_4 x) e^{\alpha x} \right) \sin \alpha y. \tag{6.21}$$

Using (6.18) and satisfying the conditions (6.14), (6.16) the constants C_i yield

$$C_1 = \frac{1}{2\alpha} \left((1 - 2a\alpha + w_2) C_2 + (1 - w_2) C_4 e^{2\alpha a} \right),$$

$$C_3 = -\frac{1}{2\alpha} \left((1 + w_2) C_2 e^{-2\alpha a} + (1 + 2a\alpha - w_2) C_4 \right),$$

$$C_2 = \frac{1}{\alpha \Delta_1 D} \left(M_0 \left(3 + \nu + (1 - \nu)(2a\alpha - w_2) \right) + H_0 \left(3 + \nu + (1 - \nu) \right. \right.$$
$$\left. \times (\alpha w_1 - (1 + \alpha w_1)(2a\alpha - w_2)) \right)$$
$$\left. + e^{2a\alpha} \left(M_0 + H_0 (1 - \alpha w_1) \right) (1 - \nu)(1 - w_2) \right), \tag{6.22}$$

$$C_4 = \frac{1}{\alpha \Delta_1 D} \left(e^{2a\alpha} \left((1 - \nu)(2a\alpha - w_2) (M_0 + H_0 (1 - \alpha w_1)) + H_0 (1 + \alpha w_1) \right. \right.$$
$$\left. \times (3 + \nu) - 2\alpha w_1 (M_0 + 2H_0) \right) - (1 - \nu)(1 + w_2)(M_0 - H_0 (1 + \alpha w_1)) \right),$$

$$\Delta_1 = (1 - \nu)(1 + w_2)(3 + \nu + 2\alpha w_1) + e^{2a\alpha} \left((3 + \nu)^2 + (1 - \nu)^2 \right.$$
$$\left. \times (1 - 4\alpha a w_2 + 4\alpha^2 a^2) - 4\alpha w_1 (1 - \nu)(\alpha a - w_2) \right)$$
$$+ (1 - \nu)(1 - w_2)(3 + \nu - 2\alpha w_1) e^{4\alpha a},$$

$$w_1 = \sqrt{\frac{G}{G'}} Ah, \quad w_2 = \frac{D}{D_3} \frac{h}{2} \frac{m_1 \nu_{13}(1 + \nu)}{1 - \nu_{13}^2}.$$

The results of the classical theory may be easily re-cast from the latter by assuming $w_1 = w_2 = 0$, with the moments and forces defined by (6.18). The outer stresses are then given by

$$\sigma_x = \frac{M_x}{\frac{2}{3} h^2} \zeta, \quad \sigma_{xy} = \frac{H_{xy}}{\frac{2}{3} h^2} \zeta, \quad \sigma_{xz} = \frac{N_x}{\frac{4}{3} h} (1 - \zeta^2), \quad (x, y), \tag{6.23}$$

$$\sigma_z = \frac{1}{4} \left(\frac{\partial N_x}{\partial x} + \frac{\partial N_y}{\partial y} \right) (\zeta^3 - \zeta).$$

Since the dependence on thickness variable of (6.23) differs from that of (6.13), the plane and anti-plane boundary layers arise in the vicinity of $x = 0, a$.

In order to reveal the influence of the refined terms in (6.14), along with that of the boundary layers associated with $x = 0$, let us consider the same problem in the domain $\Omega = \{(x, y, z) : 0 \le x < +\infty, |y| \le \ell, |z| \le h\}$ subject to conditions (6.14), (6.15) and condition of finite solution at $x \to +\infty$. The solution is given by

$$w = (C_1 + C_2 x) e^{-\alpha x} \sin \alpha y \tag{6.24}$$

where

$$C_1 = -\frac{1}{D\alpha\,(1-\nu)\,\Delta_1}\left(M_0\,(1+\nu) - H_0\,(2 - (1-\nu)\alpha\omega_1)\right),$$

$$C_2 = \frac{1}{D\Delta_1}\left(M_0 + H_0\,(1-\alpha\omega_1)\right), \qquad (6.25)$$

$$\Delta_1 = \alpha\,(3+\nu-2\alpha\omega_1).$$

The rest of the sought-for quantities may be determined from (6.18), (6.23).

The effect of the additional term of condition (6.14) on outer solution is characterized by the quantity $\alpha\omega_1$. If $\alpha\omega_1$ is of order unity, then the refinement is of the same order as the classical terms, hence, the classical theory is not applicable. It may be observed that

$$\alpha\omega_1 = \sqrt{G/G'}\,Ah\frac{\pi}{\ell} \approx \sqrt{G/G'}\,1.26\frac{\pi h}{\ell}, \qquad (6.26)$$

therefore for isotropic plates $\alpha\omega_1$ is of order unity if $\ell \approx 4h$. In other words, the classical theory is valid for plates having typical size of the mid-plane $\ell > 4h$, which is usually the case in real-life structures. We remark that for anisotropic plates a similar condition for applicability of the classical theory is $\ell > 4h\sqrt{G/G'}$, hence, once again we deduce that the accuracy of the classical plate theory decreases with decrease of shear stiffness in cross sectional area.

In the considered problem there is no boundary layer arising at $y = \pm\ell$, therefore the classical theory provides results of reasonable accuracy up to these ends of the plate. Both plane and anti-plane boundary layer solutions arise in the vicinity of $x = 0$. At leading order they are governed by

$$\overset{b}{\sigma}_x = \frac{M_0}{h^2}\left(\frac{f_1}{\bar{f}_1} - \frac{3}{2}\zeta\right)\sin\alpha y, \quad \overset{b}{\sigma}_{xz} = 0 \qquad \text{at} \qquad x = 0 \qquad (6.27)$$

$$\overset{a}{\sigma}_{xy} = \frac{1}{h^2}\left(H_0\frac{f_2}{\bar{f}_2} - \frac{3}{2}\zeta H_{xy}(x=0)\right)\cos\alpha y \qquad \text{at} \qquad x = 0. \qquad (6.28)$$

The solution of the latter may be written as

$$\overset{a}{v} = \Phi(t,\zeta,y), \qquad \overset{a}{\sigma}_{xy} = G\frac{\partial\Phi}{\partial t}, \qquad \overset{a}{\sigma}_{yz} = G'\frac{\partial\Phi}{\partial\zeta},$$

$$\Phi = \sum_{n=1}^{\infty} A_n \sin(2n-1)\frac{\pi}{2}\zeta \exp\left(-\sqrt{G'/G}(2n-1)\frac{\pi}{2}\frac{x}{h}\right)\cos\alpha y, \qquad (6.29)$$

$$A_n = -\frac{2}{\pi}\frac{1}{\sqrt{GG'}}\frac{1}{2n-1}$$

$$\times \int_{-1}^{+1}\frac{1}{h^2}\left(H_0\frac{f_2}{\bar{f}_2} - \frac{3}{2}\zeta H_{xy}(x=0)\right)\sin(2n-1)\frac{\pi}{2}\zeta d\zeta,$$

$$H_{xy}(x=0) = -\frac{1}{3+\nu-2\alpha\omega_1}\left(2M_0 - (1+\nu)H_0\right).$$

From the analysis of (6.29) it may be observed that the anti-plane boundary layer decays away from the edge without oscillations as $\exp\left(-\sqrt{G'/G}\pi\frac{x}{2h}\right)$. This result may also connect the refined theory of Ambartsumyan (1967) predicting shear boundary phenomena decaying as

$$\exp\left(-\frac{\pi}{\ell}\sqrt{1+\frac{10}{\pi^2}\frac{G'}{G}\frac{\ell^2}{(2h)^2}}\,x\right).$$

We note that since typically $\dfrac{10}{\pi^2}\dfrac{G'}{G}\dfrac{\ell^2}{(2h)^2} \gg 1$, and $\pi^2 \approx 10$, the velocity of attenuation coincides with that following from the asymptotic method.

The theory of S.A. Ambartsumyan describes not only the decay of shear boundary effects along the mid-plane, but also takes into account its influence on the outer stress field. If we consider a particular case of the considered illustrative example and impose only bending moment at the edge $x = 0$, the values of the constants C_1, C_2 determined by the so-called "general theory", see Ambartsumyan (1967) (p.142) are given by

$$C_1 = -\frac{M_0(1+\nu)}{D\alpha(1-\nu)}\frac{1}{\Delta_2}, \quad C_2 = \frac{M_0}{D\Delta_2} \tag{6.30}$$

where

$$\Delta_2 = \alpha\left(3+\nu-4\frac{\alpha^2}{\delta^2}(\omega-1)\right), \quad \omega = \sqrt{1+\frac{\delta^2}{\alpha^2}}, \quad \delta^2 = \frac{10}{(2h)^2}\frac{G'}{G}. \tag{6.31}$$

Since $\omega \approx \delta/\alpha \gg 1 \Rightarrow \Delta_2 \approx \alpha\left(3+\nu-4\frac{\alpha}{\delta}\right)$. If we assume $1.26\pi \approx 4$, then $\Delta_1 \approx \Delta_2$ hence the formulae (6.25) and (6.30) coincide, i.e. the same outer solutions are predicted by the Ambartsumyan theory and the asymptotic method.

It should be emphasized that within the asymptotic method the effect of the anti-plane boundary layer on the outer stress field is realized as a boundary condition, see (6.14), whereas in case of the theory Ambartsumyan (1967) it is a consequence of two simultaneous equations subject to Poisson type boundary conditions. Thus, the general theory of Ambartsumyan (1967, 1974) is able to take into account one of the boundary layer components, namely the anti-plane one. In this respect the general theory of S.A. Ambartsumyan is preferable to his earlier iterative theories, which were not capable of capturing any of the boundary layer effects.

The plane boundary layer solution is given by

$$\overset{b}{\sigma}_x = \sum_{(\lambda_n)} A_n \frac{F_n''}{\lambda_n^2}\exp\left(-\lambda_n\frac{x}{h}\right)\sin\alpha y,$$

$$\overset{b}{\sigma}_{xz} = \sum_{(\lambda_n)} A_n \frac{F_n'}{\lambda_n}\exp\left(-\lambda_n\frac{x}{h}\right)\sin\alpha y,$$

$$\overset{b}{\sigma}_z = \sum_{(\lambda_n)} A_n F_n \exp\left(-\lambda_n\frac{x}{h}\right)\sin\alpha y, \tag{6.32}$$

$$\overset{b}{\sigma}_y = -\frac{a_{12}}{a_{22}}\overset{b}{\sigma}_x - \frac{a_{23}}{a_{22}}\overset{b}{\sigma}_z = \nu\overset{b}{\sigma}_x + \nu'\overset{b}{\sigma}_z,$$

$$\overset{b}{u} = -\ell \sum_{(\lambda_n)} \frac{A_n}{\lambda_n^3}\left(b_{11}F_n'' + b_{13}\lambda_n^2 F_n\right)\sin\alpha y\exp\left(-\lambda_n\frac{x}{h}\right),$$

$$\overset{b}{w} = -\ell \sum_{\substack{(\lambda_n) \\ n}} \frac{A_n}{\lambda_n^4} \left(b_{11} F_n''' + (b_{13} + a_{55})\lambda_n^2 F_n' \right) \exp\left(-\lambda_n \frac{x}{h}\right) \sin \alpha y,$$

$$b_{11} = \frac{1 - \nu^2}{E}, \quad b_{33} = \frac{1 - (\nu')^2}{E'}, \quad b_{13} = -\frac{\nu'}{E'}(1 + \nu), \quad a_{55} = \frac{1}{G'}.$$

Here λ_n characterized the velocity of decay and is calculated as shown earlier in Section 5 of Chapter 3, with functions F_n following from (3.5.25), (3.5.28), and (3.5.31). In order to determine the values of the constants A_n, corresponding to a particular function f_1, the boundary collocation method could be implemented, for more details see Chapter 1.

We underline that at the moment none of the existing technical theories of plates relying on hypotheses allow incorporation of plane boundary layer effects.

The results presented in Chapters 3 and 4 allow some more remarks and philosophical insights on the history of the problem.

It is well known that one of the first methods of derivation of the plate bending equation was based on expanding all of the quantities as power series in a small parameter γ. This approach was used in particular by Cauchy, who claimed that by retaining enough number of terms the problem might be solved with any desired accuracy. However, the theory of Cauchy was at a stall due to the fact that the basic two-dimensional partial differential equation remained an equation of the fourth-order, of elliptic type, and it was impossible to satisfy the three natural boundary conditions of Cauchy-Poisson. This circumstance delayed any further progress of the plate theory for several further decades. Many researchers, including Saint-Venant, believed that such an expansion did not cover all solutions. In particular, it was assumed that the sought for solution should contain a decaying component. Since the theory of Cauchy did not answer the question of how many and what kind of boundary conditions should be satisfied on a given boundary, the theory of Kirchhoff based on well-known assumptions received a widespread attention in the sequel. Half a century after Cauchy, the Kirchhoff theory resolved the issue of boundary conditions under the assumption of the correctness of this theory. With the advent of refined theories the question of boundary conditions again became important, and the interest in representations of Cauchy grew once again. The asymptotic approach allows the clarification of this implicit dispute.

One must bear in mind that there are at least two possible approaches to solution relying on the two types of small parameters which may be used. Cauchy used variable small parameter γ (the distance from the middle surface at its normal to a given point on the plate). This theory was further developed in the well-known works of Kilchevskii (1939, 1962, 1963), who was using the power series expansion for all of the sought-for quantities with respect to this parameter and obtained the equations of the theory of shells, at the same time suggesting a method of derivation of the corresponding boundary conditions. The feature of this theory is that with increasing number of terms in the expansion, the order of the corresponding governing equations increases. We note that the increase in the order of equations is typical for theories, employing the variable as a parameter. For instance, in the theory by I. N. Vekua, where the displacement and stress fields in question were expanded in Fourier series in terms of the Legendre functions of the argument, with the increase in number of terms in the expansion, the order of governing equations also increases.

If the equations obtained by the power series method of Cauchy are rewritten in dimensionless variables, then the obtained equations will contain a small parameter in the coefficients of the higher order operators. As was proved by Goldenveizer (1968b), these equations may be solved efficiently by the method of direct asymptotic integration, which automatically singles out the rapidly decaying solutions (boundary layer). In view of this, equations obtained by the method of power series lose their own meaning and become auxiliary tools to get the results to which the asymptotic method can be applied immediately. Therefore the equations obtained by the power series method contain the corresponding rapidly decaying integrals. However, it should be noted that at the time of Cauchy such theories were not developed. The theory of singularly perturbed equations, as was stated in Section 1, was developed relatively recently, and so, despite his greatness, Cauchy could not claim that solutions of such equations would contain rapidly decaying terms.

If the small parameter is introduced as a constant $\varepsilon = h/a$, then, as was shown in Chapters 1 and 3, use of expansion method only does not allow solution of the problem. Indeed, in addition to the outer representation one must also have a formulation for the boundary layer. The latter is determined by the decay of the stress field from the boundary or line of the perturbation. Therefore implementation of a constant small parameter does not increase the order of the main operator (classical), but causes increase in the number of expansions. On the other hand, the expansion in small variable parameter increases the order of the main operator, but the problem in this case turns out to be the result of singular perturbation of the small dimensionless parameter and the asymptotic method is again required for its solution.

We note that the area of plates and shells theories is still research-active, with some more contributions of the 1990s including Alfutov (1992); Vasiliev (1992, 1995); Darevskii (1995); Goldenveizer et al. (1993); Goldenveizer (1994); Volokh (1994); Zhilin (1992, 1995). We also mention an important monograph on the subject published by Kaplunov et al. (1998). Some more results of 2000s in respect of plates of more advanced physical properties (including pre-stressed plates) are listed below (Kaplunov et al., 2002, 2005; Pichugin and Rogerson, 2002; Nolde et al., 2004). Finally, we cite a wide variety of recent contributions (Shariyat, 2010; Thai and Choi, 2013).

Offering some familiarity with the publications below, it is the desire of the author that the monograph will help the reader to understand the subject and confidently move to the problems of the theory of thin plates and shells.

Chapter 5

Elastic Plates of General Anisotropy

5.1 Governing Equations

It has been noted in previous chapters that one of the advantages of the asymptotic method is its wide area of applicability to thin walled structures including the cases when other analytic methods run into complications. The case of elastic plates of general anisotropy is seemingly one of such areas. Indeed, most of the contributions on anisotropic plates assume that a plate possesses a plane of symmetry, see e.g. Ambartsumyan (1967); Lekhnitskii (1968). It seems that it was guessed a priori that the Kirchhoff hypothesis is not always valid for plates of general anisotropy. We shall see below in this chapter that these qualms are indeed justified.

There are only a few publications containing analysis of elastic plates of general anisotropy, and all of these use either variational techniques or rely on asymptotic methods. The variational approach was applied by Shoykhet (1973), who have derived the approximate equations for a plate of general anisotropy, corresponding to the Kirchhoff hypothesis. The technique of the cited work relied on minimization of energy functional. In the same work B.A. Shoykhet has investigated proximity (by integral norm) of the obtained approximate solution to the corresponding exact 3D formulation. Some generalizations of these results were obtained later by energy method (Berdichevsky, 1975, 1983).

The asymptotic method has been used by Widera (1969), who has obtained leading order approximations for the outer problem. However, in the cited work there is no analysis of terms which characterize the general anisotropy. To the best of our knowledge there have been no analysis of boundary layer effects in case of plates of general anisotropy. The results of full asymptotic analysis for plates of general anisotropy are presented below in this chapter.

Consider a plate of thickness $2h$ in Cartesian system $Oxyz$. Here we assume that the plate is generally anisotropic and satisfies the generalized Hooke's law. The following boundary conditions are imposed at the faces $z = \pm h$

$$\sigma_z = \pm Z^{\pm}(x,y), \quad \sigma_{xz} = \pm(a/h)X^{\pm}(x,y) \quad (x,y) \qquad (1.1)$$

The boundary conditions at the edges of the plate could be these of the first or second problem of elasticity, or mixed type. We note that large parameter is involved

163

with $X^{\pm}$ introduced in order to balance the terms so that the normal and tangential loading contributions to stress field are of the same asymptotic order.

The governing equations of a generally anisotropic plate are adopted in the form (Lekhnitskii, 1977)

$$\frac{\partial \sigma_x}{\partial x} + \frac{\partial \sigma_{xy}}{\partial y} + \frac{\partial \sigma_{xz}}{\partial z} = 0, \quad (x, y, z),$$

$$\frac{\partial u}{\partial x} = a_{11}\sigma_x + a_{12}\sigma_y + a_{13}\sigma_z + a_{14}\sigma_{yz} + a_{15}\sigma_{xz} + a_{16}\sigma_{xy},$$

$$\frac{\partial v}{\partial y} = a_{12}\sigma_x + a_{22}\sigma_y + a_{23}\sigma_z + a_{24}\sigma_{yz} + a_{25}\sigma_{xz} + a_{26}\sigma_{xy},$$

$$\frac{\partial w}{\partial z} = a_{13}\sigma_x + a_{23}\sigma_y + a_{33}\sigma_z + a_{34}\sigma_{yz} + a_{35}\sigma_{xz} + a_{36}\sigma_{xy}, \qquad (1.2)$$

$$\frac{\partial v}{\partial z} + \frac{\partial w}{\partial y} = a_{14}\sigma_x + a_{24}\sigma_y + a_{34}\sigma_z + a_{44}\sigma_{yz} + a_{45}\sigma_{xz} + a_{46}\sigma_{xy},$$

$$\frac{\partial w}{\partial x} + \frac{\partial u}{\partial z} = a_{15}\sigma_x + a_{25}\sigma_y + a_{35}\sigma_z + a_{45}\sigma_{yz} + a_{55}\sigma_{xz} + a_{56}\sigma_{xy},$$

$$\frac{\partial v}{\partial x} + \frac{\partial u}{\partial y} = a_{16}\sigma_x + a_{26}\sigma_y + a_{36}\sigma_z + a_{46}\sigma_{yz} + a_{56}\sigma_{xz} + a_{66}\sigma_{xy}.$$

5.2   Iterative Procedure for the Outer Solution

Let us transform to dimensionless variables,

$$x = a\xi, \quad y = a\eta, \quad z = h\zeta \qquad (2.1)$$

and therefore introduce the small parameter $\varepsilon = h/a$, where a is a typical size of the plate. The system is singularly perturbed, therefore its solution is composed of two parts, namely, the outer solution and the boundary layer solution. The outer solution Q is constructed following a procedure described in Chapters 1 and 3

$$Q = \varepsilon^{-q} \sum_{s=0}^{S} \varepsilon^s Q^{(s)} \qquad (2.2)$$

where Q is any of the stresses or dimensionless displacements, and $U = u/a$ (u, v, w). The whole number q is chosen in order to obtain consistent asymptotic procedure after substitution of (2.2) into (1.2). It may be shown that $q = 2$ for $\sigma_x, \sigma_{xy}, \sigma_y, u, v$, $q = 1$ for σ_{xz}, σ_{yz}, $q = 0$ for σ_z, and $q = 3$ for w. There are no restrictions yet on the material parameters of anisotropy a_{ik}, these will arise naturally in the analysis of the outer solution and the boundary layer. Substituting (2.2) into (1.2), one results in

$$\frac{\partial \sigma_x^{(s)}}{\partial \xi} + \frac{\partial \sigma_{xy}^{(s)}}{\partial \eta} + \frac{\partial \sigma_{xz}^{(s)}}{\partial \zeta} = 0, \quad (x, y, z; \xi, \eta, \zeta),$$

$$\frac{\partial U^{(s)}}{\partial \xi} = a_{11}\sigma_x^{(s)} + a_{12}\sigma_y^{(s)} + a_{13}\sigma_z^{(s-2)}$$

$$+ a_{14}\sigma_{yz}^{(s-1)} + a_{15}\sigma_{xz}^{(s-1)} + a_{16}\sigma_{xy}^{(s)},$$

$$\frac{\partial V^{(s)}}{\partial \eta} = a_{12}\sigma_x^{(s)} + a_{22}\sigma_y^{(s)} + a_{23}\sigma_z^{(s-2)}$$

$$+a_{24}\sigma_{yz}^{(s-1)} + a_{25}\sigma_{xz}^{(s-1)} + a_{26}\sigma_{xy}^{(s)},$$

$$\frac{\partial W^{(s)}}{\partial \zeta} = a_{13}\sigma_x^{(s-2)} + a_{23}\sigma_y^{(s-2)} + a_{33}\sigma_z^{(s-4)}$$

$$+a_{34}\sigma_{yz}^{(s-3)} + a_{35}\sigma_{xz}^{(s-3)} + a_{36}\sigma_{xy}^{(s-2)}, \qquad (2.3)$$

$$\frac{\partial V^{(s)}}{\partial \zeta} + \frac{\partial W^{(s)}}{\partial \eta} = a_{14}\sigma_x^{(s-1)} + a_{24}\sigma_y^{(s-1)}$$

$$+a_{34}\sigma_z^{(s-3)} + a_{44}\sigma_{yz}^{(s-2)} + a_{45}\sigma_{xz}^{(s-2)} + a_{46}\sigma_{xy}^{(s-1)},$$

$$\frac{\partial W^{(s)}}{\partial \xi} + \frac{\partial U^{(s)}}{\partial \zeta} = a_{15}\sigma_x^{(s-1)} + a_{25}\sigma_y^{(s-1)} + a_{35}\sigma_z^{(s-3)}$$

$$+a_{45}\sigma_{yz}^{(s-2)} + a_{55}\sigma_{xz}^{(s-2)} + a_{56}\sigma_{xy}^{(s-1)},$$

$$\frac{\partial V^{(s)}}{\partial \xi} + \frac{\partial U^{(s)}}{\partial \eta} = a_{16}\sigma_x^{(s)} + a_{26}\sigma_y^{(s)} + a_{36}\sigma_z^{(s-2)}$$

$$+a_{46}\sigma_{yz}^{(s-1)} + a_{56}\sigma_{xz}^{(s-1)} + a_{66}\sigma_{xy}^{(s)}.$$

The system (2.3) may be integrated with respect to ζ. Satisfying the conditions (1.1), we obtain

$$W^{(s)} = w^{(s)}(\xi, \eta) + w^{*(s)}(\xi, \eta, \zeta),$$

$$U^{(s)} = -\zeta \frac{\partial w^{(s)}}{\partial \xi} + u^{(s)}(\xi, \eta) + u^{*(s)}(\xi, \eta, \zeta), \quad (u, v; \xi, \eta),$$

$$\sigma_x^{(s)} = \zeta \tau_{x1}^{(s)} + \tau_{x0}^{(s)} + \sigma_x^{*(s)}, \quad (x, y),$$

$$\sigma_{xy}^{(s)} = \zeta \tau_{xy1}^{(s)} + \tau_{xy0}^{(s)} + \sigma_{xy}^{*(s)}, \qquad (2.4)$$

$$\sigma_{xz}^{(s)} = \frac{1}{2}\zeta^2 \tau_{xz2}^{(s)} + \zeta \tau_{xz1}^{(s)} + \tau_{xz0}^{(s)} + \sigma_{xz}^{*(s)}, \quad (x, y),$$

$$\sigma_z^{(s)} = \frac{1}{6}\zeta^3 \tau_{z3}^{(s)} + \frac{1}{2}\zeta^2 \tau_{z2}^{(s)} + \zeta \tau_{z1}^{(s)} + \tau_{z0}^{(s)} + \sigma_z^{*(s)},$$

where

$$\tau_{x1}^{(s)} = -\left(B_{11}\frac{\partial^2 w^{(s)}}{\partial \xi^2} + B_{12}\frac{\partial^2 w^{(s)}}{\partial \eta^2} + 2B_{16}\frac{\partial^2 w^{(s)}}{\partial \xi \partial \eta} \right), \quad (x, y, ; 1, 2),$$

$$\tau_{xy1}^{(s)} = -\left(B_{16}\frac{\partial^2 w^{(s)}}{\partial \xi^2} + B_{26}\frac{\partial^2 w^{(s)}}{\partial \eta^2} + 2B_{66}\frac{\partial^2 w^{(s)}}{\partial \eta \partial \xi} \right),$$

$$\tau_{xz(k+1)}^{(s)} = -\left(\frac{\partial \tau_{xk}^{(s)}}{\partial \xi} + \frac{\partial \tau_{xyk}^{(s)}}{\partial \eta} \right), \quad (x, y), \quad (k = 0, 1),$$

$$\tau_{z(m+1)}^{(s)} = -\left(\frac{\partial \tau_{xzm}^{(s)}}{\partial \xi} + \frac{\partial \tau_{yzm}^{(s)}}{\partial \eta} \right), \quad (m = 0, 1, 2), \qquad (2.5)$$

$$\tau_{x0}^{(s)} = B_{11}\frac{\partial u^{(s)}}{\partial \xi} + B_{12}\frac{\partial v^{(s)}}{\partial \eta} + B_{16}\left(\frac{\partial u^{(s)}}{\partial \eta} + \frac{\partial v^{(s)}}{\partial \xi} \right), \quad (x, y; 1, 2),$$

$$\tau_{xy0}^{(s)} = B_{16}\frac{\partial u^{(s)}}{\partial \xi} + B_{26}\frac{\partial v^{(s)}}{\partial \eta} + B_{66}\left(\frac{\partial u^{(s)}}{\partial \eta} + \frac{\partial v^{(s)}}{\partial \xi}\right),$$

$$\tau_{xz0}^{(s)} = X_2^{(s)} - \frac{1}{2}\left(\sigma_{xz}^{*(s)}(\zeta = 1) + \sigma_{xz}^{*(s)}(\zeta = -1)\right) - \frac{1}{2}\tau_{xz2}^{(s)}, \quad (x, y),$$

$$\tau_{z0}^{(s)} = Z_1^{(s)} - \frac{1}{2}\left(\frac{\partial p_1^{(s)}}{\partial \xi} + \frac{\partial p_2^{(s)}}{\partial \eta}\right) - \frac{1}{2}\left(\sigma_z^{*(s)}(\zeta = 1) + \sigma_z^{*(s)}(\zeta = -1)\right),$$

$$X_i^{(0)} = X_i, \ (X, Y), \quad Z_i^{(0)} = Z_i, \quad X_i^{(s)} = Y_i^{(s)} = Z_i^{(s)} = 0, \ (s > 0), \quad (i = 1, 2),$$

$$X_1 = \frac{1}{2}\left(X^+ + X^-\right), \quad X_2 = \frac{1}{2}\left(X^+ - X^-\right), \quad (X, Y),$$

$$Z_1 = \frac{1}{2}\left(Z^+ - Z^-\right), \quad Z_2 = \frac{1}{2}\left(Z^+ + Z^-\right).$$

We remark that the quantities with the asterisk may be determined for every approximation s as soon as the previous approximations are constructed, in particular, $Q^{*(0)} \equiv 0$. They are

$$w^{*(s)} = \int_0^\zeta \left(a_{13}\sigma_x^{(s-2)} + a_{23}\sigma_y^{(s-2)} + a_{33}\sigma_z^{(s-4)}\right.$$
$$\left. + a_{34}\sigma_{yz}^{(s-3)} + a_{35}\sigma_{xz}^{(s-3)} + a_{36}\sigma_{xy}^{(s-2)}\right) d\zeta,$$

$$u^{*(s)} = \int_0^\zeta \left(a_{15}\sigma_x^{(s-1)} + a_{25}\sigma_y^{(s-1)} + a_{35}\sigma_z^{(s-3)}\right.$$
$$\left. + a_{45}\sigma_{yz}^{(s-2)} + a_{55}\sigma_{xz}^{(s-2)} + a_{56}\sigma_{xy}^{(s-1)} - \frac{\partial w^{*(s)}}{\partial \xi}\right) d\zeta,$$

$$v^{*(s)} = \int_0^\zeta \left(a_{14}\sigma_x^{(s-1)} + a_{24}\sigma_y^{(s-1)} + a_{34}\sigma_z^{(s-3)}\right.$$
$$\left. + a_{44}\sigma_{yz}^{(s-2)} + a_{45}\sigma_{xz}^{(s-2)} + a_{46}\sigma_{xy}^{(s-1)} - \frac{\partial w^{*(s)}}{\partial \eta}\right) d\zeta,$$

$$\sigma_x^{*(s)} = B_{11}\frac{\partial u^{*(s)}}{\partial \xi} + B_{12}\frac{\partial v^{*(s)}}{\partial \eta} + B_{16}\left(\frac{\partial u^{*(s)}}{\partial \eta} + \frac{\partial v^{*(s)}}{\partial \xi}\right)$$
$$+ a_1\sigma_z^{(s-2)} + a_2\sigma_{yz}^{(s-1)} + a_3\sigma_{xz}^{(s-1)},$$

$$\sigma_y^{*(s)} = B_{12}\frac{\partial u^{*(s)}}{\partial \xi} + B_{22}\frac{\partial v^{*(s)}}{\partial \eta} + B_{26}\left(\frac{\partial u^{*(s)}}{\partial \eta} + \frac{\partial v^{*(s)}}{\partial \xi}\right)$$
$$+ b_1\sigma_z^{(s-2)} + b_2\sigma_{yz}^{(s-1)} + b_3\sigma_{xz}^{(s-1)}, \quad (2.6)$$

$$\sigma_{xy}^{*(s)} = B_{16}\frac{\partial u^{*(s)}}{\partial \xi} + B_{26}\frac{\partial v^{*(s)}}{\partial \eta} + B_{66}\left(\frac{\partial u^{*(s)}}{\partial \eta} + \frac{\partial v^{*(s)}}{\partial \xi}\right)$$
$$+ c_1\sigma_z^{(s-2)} + c_2\sigma_{yz}^{(s-1)} + c_3\sigma_{xz}^{(s-1)},$$

$$\sigma_{xz}^{*(s)} = -\int_0^\zeta \left(\frac{\partial \sigma_x^{*(s)}}{\partial \xi} + \frac{\partial \sigma_{xy}^{*(s)}}{\partial \eta}\right) d\zeta, \quad (x, y; \xi, \eta),$$

$$\sigma_z^{*(s)} = - \int_0^\zeta \left(\frac{\partial \sigma_{xz}^{*(s)}}{\partial \xi} + \frac{\partial \sigma_{yz}^{*(s)}}{\partial \eta} \right) d\zeta.$$

The elastic coefficients B_{ik}, a_i, b_i, c_i are expressed through a_{ik} as

$$B_{11} = \left(a_{22}a_{66} - a_{26}^2\right)/\Omega, \quad B_{22} = \left(a_{11}a_{66} - a_{16}^2\right)/\Omega,$$

$$B_{12} = \left(a_{16}a_{26} - a_{12}a_{66}\right)/\Omega, \quad B_{66} = \left(a_{11}a_{22} - a_{12}^2\right)/\Omega,$$

$$B_{16} = \left(a_{12}a_{26} - a_{22}a_{16}\right)/\Omega, \quad B_{26} = \left(a_{12}a_{16} - a_{11}a_{26}\right)/\Omega, \tag{2.7}$$

$$\Omega = \left(a_{11}a_{22} - a_{12}^2\right)a_{66} + 2a_{12}a_{16}a_{26} - a_{11}a_{26}^2 - a_{22}a_{16}^2,$$

$$a_{i-2} = -\left(a_{1i}B_{11} + a_{2i}B_{12} + a_{i6}B_{16}\right),$$

$$b_{i-2} = -\left(a_{1i}B_{12} + a_{2i}B_{22} + a_{i6}B_{26}\right),$$

$$c_{i-2} = -\left(a_{1i}B_{16} + a_{2i}B_{26} + a_{i6}B_{66}\right) \quad (i = 3, 4, 5).$$

It may be observed from (2.5) that all of the quantities involved in the outer solution (2.4) are expressed through $u^{(s)}, v^{(s)}, w^{(s)}$. The latter may be obtained from the relations following from (1.1), (2.3) and (2.5)

$$\ell_{11}u^{(s)} + \ell_{12}v^{(s)} = p_1^{(s)}, \quad \ell_{12}u^{(s)} + \ell_{22}v^{(s)} = p_2^{(s)} \tag{2.8}$$

$$B_{11}\frac{\partial^4 w^{(s)}}{\partial \xi^4} + 4B_{16}\frac{\partial^4 w^{(s)}}{\partial \xi^3 \partial \eta} + 2\left(B_{12} + 2B_{66}\right)\frac{\partial^4 w^{(s)}}{\partial \xi^2 \partial \eta^2} \tag{2.9}$$

$$+4B_{26}\frac{\partial^4 w^{(s)}}{\partial \xi \partial \eta^3} + B_{22}\frac{\partial^4 w^{(s)}}{\partial \eta^4} = q^{(s)}.$$

Here ℓ_{ik} are differential operators

$$\ell_{11} = B_{11}\frac{\partial^2}{\partial \xi^2} + B_{66}\frac{\partial^2}{\partial \eta^2} + 2B_{16}\frac{\partial^2}{\partial \xi \partial \eta}, \quad (1, 2; \xi, \eta), \tag{2.10}$$

$$\ell_{12} = B_{16}\frac{\partial^2}{\partial \xi^2} + \left(B_{12} + B_{66}\right)\frac{\partial^2}{\partial \xi \partial \eta} + B_{26}\frac{\partial^2}{\partial \eta^2}$$

and $p_1^{(s)}, p_2^{(s)}, q^{(s)}$ are generalized loading functions

$$p_1^{(s)} = -X_1^{(s)} + \frac{1}{2}\left(\sigma_{xz}^{*(s)}(\zeta = 1) - \sigma_{xz}^{*(s)}(\zeta = -1)\right),$$

$$p_2^{(s)} = -Y_1^{(s)} + \frac{1}{2}\left(\sigma_{yz}^{*(s)}(\zeta = 1) - \sigma_{yz}^{*(s)}(\zeta = -1)\right), \tag{2.11}$$

$$q^{(s)} = \frac{3}{2}\left(2Z_2^{(s)} + 2\frac{\partial X_2^{(s)}}{\partial \xi} + 2\frac{\partial Y_2^{(s)}}{\partial \eta} - \frac{\partial \sigma_{xz}^{*(s)}(\zeta = 1)}{\partial \xi} - \frac{\partial \sigma_{xz}^{*(s)}(\zeta = -1)}{\partial \xi}\right.$$

$$\left. - \frac{\partial}{\partial \eta}\left(\sigma_{yz}^{*(s)}(\zeta = 1) + \sigma_{yz}^{*(s)}(\zeta = -1)\right) - \left(\sigma_z^{*(s)}(\zeta = 1) - \sigma_z^{*(s)}(\zeta = -1)\right)\right).$$

The solution of (2.8), (2.9) in conjunction with (2.2), (2.4) provides the sought-for solutions for stresses and displacements of the outer problem. Similarly to the previously considered cases, the analyzed system is of the same form for every s.

5.3   Applicability of the Kirchhoff Hypothesis

Let us reveal the links between the iterative process of Section 2 with the results that could be obtained from adopting the Kirchhoff-type hypotheses. Returning to notation of the classical theory through $w_0^{(s)} = a\varepsilon^{-3+s}w^{(s)}$, $u_0^{(s)} = a\varepsilon^{-2+s}u^{(s)}$ (u,v), we obtain

$$\bar{\ell}_{11}u_0^{(s)} + \bar{\ell}_{12}v_0^{(s)} = \bar{p}_1^{(s)}, \quad \bar{\ell}_{12}u_0^{(s)} + \bar{\ell}_{22}v_0^{(s)} = \bar{p}_2^{(s)} \tag{3.1}$$

$$D_{11}\frac{\partial^4 w_0^{(s)}}{\partial x^4} + 4D_{16}\frac{\partial^4 w_0^{(s)}}{\partial x^3 \partial y} + 2\left(D_{12} + 2D_{66}\right)\frac{\partial^4 w_0^{(s)}}{\partial x^2 \partial y^2}$$
$$+4D_{26}\frac{\partial^4 w_0^{(s)}}{\partial x \partial y^3} + D_{22}\frac{\partial^4 w_0^{(s)}}{\partial y^4} = q_0^{(s)} \tag{3.2}$$

where

$$\bar{\ell}_{11} = C_{11}\frac{\partial^2}{\partial x^2} + C_{66}\frac{\partial^2}{\partial y^2} + 2C_{16}\frac{\partial^2}{\partial x \partial y}, \quad (1,2;x,y),$$
$$\bar{\ell}_{12} = C_{16}\frac{\partial^2}{\partial x^2} + (C_{12} + C_{66})\frac{\partial^2}{\partial x \partial y} + C_{26}\frac{\partial^2}{\partial y^2}, \tag{3.3}$$
$$C_{ik} = 2hB_{ik},$$
$$\bar{p}_1^{(s)} = 2\varepsilon^{-1}p_1^{(s)}, \quad \bar{p}_2^{(s)} = 2\varepsilon^{-1}p_2^{(s)}, \quad q_0^{(s)} = \frac{2}{3}q^{(s)}.$$

It is clear that at $s = 0$ the relations (3.1) coincide with equations of the generalized plane problem, and relations (3.2) - with the classical equation of plate bending in case of plane of symmetry (Lekhnitskii, 1968). These equations may be derived using the Kirchhoff-type hypothesis, in particular, one has to neglect the stress components σ_{yz}, σ_{xz} in addition to σ_z in the first three equations of elasticity. The changes for higher order approximations $s > 0$ are occurring only on the right-hand sides, i.e. loading involving the coefficients of general anisotropy. Thus, the outer solution for a plate of general anisotropy has been reduced to a problem for a plate possessing a plane of symmetry.

The asymptotic method allows clarification of one more non-trivial matter. It is often thought, see e.g. Lekhnitskii (1968), that if the plate is loaded by both normal and tangential loading, then due to combined effect superposition principle is not valid. It may be observed that since

$$p_1^{(0)} = -X_1, \quad p_2^{(0)} = -Y_1, \quad q^{(0)} = 3\left(Z_2 + \frac{\partial X_2}{\partial \xi} + \frac{\partial Y_2}{\partial \eta}\right) \tag{3.4}$$

then at $s = 0$ the symmetric and the antisymmetric problems may be decoupled. It also follows that starting from $s = 1$ due to combined effect the quantities $p_i^{(s)}$ contain terms of bending nature, whereas $q^{(s)}$ involve terms associated with the plane problem, more specifically

$$p_1^{(1)} = \frac{1}{6}\left(a_3\left(\frac{\partial^2 \ell_{11}}{\partial \xi^2} + 2\frac{\partial^2 \ell_{12}}{\partial \xi \partial \eta}\right) + a_2\left(\frac{\partial^2 \ell_{12}}{\partial \xi^2} + 2\frac{\partial^2 \ell_{22}}{\partial \xi \partial \eta}\right) + 2c_2\frac{\partial^2 \ell_{22}}{\partial \eta^2}\right.$$
$$\left. + 2c_3\frac{\partial^2 \ell_{12}}{\partial \eta^2} - b_2\frac{\partial^2 \ell_{12}}{\partial \eta^2} - b_3\frac{\partial^2 \ell_{11}}{\partial \eta^2}\right)w^{(0)}$$
$$- \left(a_2\frac{\partial^2 Y_2}{\partial \xi^2} + a_3\frac{\partial^2 X_2}{\partial \xi^2} + c_2\frac{\partial^2 Y_2}{\partial \eta^2} + c_3\frac{\partial^2 X_2}{\partial \eta^2}\right),$$

$$p_2^{(1)} = \frac{1}{6}\left(b_3\left(\frac{\partial^2 \ell_{22}}{\partial \eta^2} + 2\frac{\partial^2 \ell_{12}}{\partial \xi \partial \eta}\right) + b_2\left(\frac{\partial^2 \ell_{12}}{\partial \eta^2} + 2\frac{\partial^2 \ell_{11}}{\partial \xi \partial \eta}\right) + 2c_2\frac{\partial^2 \ell_{11}}{\partial \xi^2}\right. \quad (3.5)$$
$$\left. + 2c_3\frac{\partial^2 \ell_{12}}{\partial \xi^2} - a_3\frac{\partial^2 \ell_{22}}{\partial \xi^2} - a_2\frac{\partial^2 \ell_{12}}{\partial \xi^2}\right)w^{(0)}$$
$$- \left(b_2\frac{\partial^2 Y_2}{\partial \eta^2} + b_3\frac{\partial^2 X_2}{\partial \eta^2} + c_2\frac{\partial^2 Y_2}{\partial \xi^2} + c_3\frac{\partial^2 X_2}{\partial \xi^2}\right),$$

$$q^{(1)} = 3\left(\left(a_3\left(\frac{\partial^2 \ell_{11}}{\partial \xi^2} + \frac{\partial^2 \ell_{12}}{\partial \xi \partial \eta}\right) + a_2\left(\frac{\partial^2 \ell_{12}}{\partial \xi^2} + \frac{\partial^2 \ell_{22}}{\partial \xi \partial \eta}\right) + c_3\left(\frac{\partial^2 \ell_{11}}{\partial \xi \partial \eta} + \frac{\partial^2 \ell_{12}}{\partial \eta^2}\right)\right.\right.$$
$$\left. + c_2\left(\frac{\partial^2 \ell_{12}}{\partial \xi \partial \eta} + \frac{\partial^2 \ell_{22}}{\partial \eta^2}\right)\right)u^{(0)} + \left(b_3\left(\frac{\partial^2 \ell_{11}}{\partial \xi \partial \eta} + \frac{\partial^2 \ell_{12}}{\partial \eta^2}\right) + b_2\left(\frac{\partial^2 \ell_{22}}{\partial \eta^2} + \frac{\partial^2 \ell_{12}}{\partial \xi \partial \eta}\right)\right.$$
$$\left. + c_3\left(\frac{\partial^2 \ell_{11}}{\partial \xi^2} + \frac{\partial^2 \ell_{12}}{\partial \xi \partial \eta}\right) + c_2\left(\frac{\partial^2 \ell_{12}}{\partial \xi^2} + \frac{\partial^2 \ell_{22}}{\partial \xi \partial \eta}\right)\right)v^{(0)} + a_2\frac{\partial^2 Y_1}{\partial \xi^2} + a_3\frac{\partial^2 X_1}{\partial \xi^2}$$
$$\left. + b_2\frac{\partial^2 Y_1}{\partial \eta^2} + b_3\frac{\partial^2 X_1}{\partial \eta^2} + 2c_2\frac{\partial^2 Y_1}{\partial \xi \partial \eta} + 2c_3\frac{\partial^2 X_1}{\partial \xi \partial \eta}\right).$$

Hence, the error corresponding to application of superposition principle and treatment of symmetric and antisymmetric problems independently is of order $O(\varepsilon^1)$. We note that in case of isotropic and orthotropic plates separation for symmetric and antisymmetric problems is fully justified within the framework of linear elasticity (clearly, when the amplitude of loading is large or properties of material are nonlinear, decoupling is no longer valid).

Since $p_1^{(1)} \neq 0$, $p_2^{(1)} \neq 0$, $q^{(1)} \neq 0$, the error arising from the Kirchhoff hypothesis, is of order $O(\varepsilon^1)$. We note that in case of isotropic plate this error is of order $\sim O(\varepsilon^2)$, see Goldenveizer (1962). It may be shown that typically for orthotropic plates having a plane of symmetry, the error is also of order $\sim O(\varepsilon^2)$, provided that there is no strong anisotropy. However, if the material parameters, say, a_1, b_1, c_1 are large compared to the others, i.e. $(a_1, b_1, c_1) \sim O\left(\varepsilon^{-1}(a_i, b_i, c_i)\right)$ $(i = 2, 3)$, then the error is of order $\sim O(\varepsilon^1)$. Thus, as could be expected, in case plates of general anisotropy the error arising from the Kirchhoff hypotheses is more than for isotropic plates. It follows from (2.3) and (3.5) that in case of general anisotropy the leading order correction should follow from taking into account coupled effects, i.e. taking into account shear in cross-section area caused by extension in transverse direction, and shear caused by tangential stresses. The impact of these factors may be lowered to $O(\varepsilon^2)$ if the coupling coefficients are small, i.e. $|\eta_{\alpha,\beta 3}|, |\mu_{\alpha 3,\alpha \beta}| \sim O(\varepsilon^1)$,

$(\alpha, \beta = 1, 2; \alpha \neq \beta)$; (here $(\eta_{i,j}, \mu_{i,j}$ are technical constants). Neglecting the effect of transverse shear associated with G_{13}, G_{23} leads to an error of order $\sim O(\varepsilon^2)$. This error decreases with decrease of G_{13}, G_{23} or could increase to $O(\varepsilon^1)$ if any of the ratios G_{12}/G_{13}, G_{12}/G_{23}, E_1/G_{13}, $E_1/G_{23} \sim O(\varepsilon^{-1})$. We remark that the Kirchhoff hypotheses become not applicable if these ratios are of order $O(\varepsilon^{-2})$. In case of elastic plates of general anisotropy the validity of Kirchhoff assumption could be violated even earlier if

$$|(a_{14}, a_{15})/a_{11}| = |\eta_{23,1}\eta_{13,1}|, \quad |(a_{24}, a_{25})/a_{22}| = |\eta_{23,2}, \eta_{13,2}|,$$
$$|(a_{46}, a_{56})/a_{66}| = |\mu_{23,12}, \mu_{13,12}| \sim O(\varepsilon^{-1}).$$

To conclude this section we note that in practical problems these problematic cases may be excluded from consideration by varying the size of the plate.

5.4   Special Features of the Boundary Layer

The system of equations governing the outer problem is of order 8, therefore it is not possible to satisfy the three spatial boundary conditions for plate edges. Similarly to previous cases, the mathematical problem degenerates, hence there is a necessity in having additional solution, i.e. the boundary layer solution.

    The boundary layer may be constructed by the method described in Chapters 1 and 3. A new variable $t = \xi/\varepsilon$ is introduced in (1.2), and the solution of the transformed equations is sought in the form

$$R_b = \sum_{s=0}^{N} \varepsilon^{\chi_b + s} R_b^{(x)}(\eta, \zeta) \exp(-\lambda t). \tag{4.1}$$

Here $Re\lambda > 0$, $\chi_{\sigma_i} = \chi$, $\chi_{u_i} = \chi + 1$, with the whole number χ determined from the edge boundary conditions similarly to the previously considered case of orthotropic plates. The stress components σ_{xzb}, σ_{yzb}, σ_{zb} should vanish at $\zeta = \pm 1$, and the boundary solution should decay away from the edge $\xi = 0$.

    The sought-for quantities $R_b^{(s)}$— may be expressed through $\sigma_{yzb}^{(s)}$ and $\sigma_{zb}^{(s)}$ as

$$\sigma_{xb}^{(s)} = \frac{1}{\lambda^2}\frac{\partial^2 \sigma_{zb}^{(s)}}{\partial\zeta^2} + R_x^{(s-1)}, \quad \sigma_{xyb}^{(s)} = \frac{1}{\lambda}\frac{\partial \sigma_{yzb}^{(s)}}{\partial\zeta} + R_{xy}^{(s-1)},$$

$$\sigma_{xzb}^{(s)} = \frac{1}{\lambda}\frac{\partial \sigma_{zb}^{(s)}}{\partial\zeta} + R_{xz}^{(s-1)}, \tag{4.2}$$

$$\sigma_{yb}^{(s)} = -\frac{1}{a_{22}}\left(\frac{a_{12}}{\lambda^2}\frac{\partial^2 \sigma_{zb}^{(s)}}{\partial\zeta^2} + \frac{a_{25}}{\lambda}\frac{\partial \sigma_{zb}^{(s)}}{\partial\zeta} \right.$$

$$\left. + \frac{a_{26}}{\lambda}\frac{\partial \sigma_{yzb}^{(s)}}{\partial\zeta} + a_{23}\sigma_{zb}^{(s)} + a_{24}\sigma_{yzb}^{(s)} \right) + R_y^{(s-1)},$$

$$u_b^{(s)} = -\frac{A_{11}}{\lambda^3}\frac{\partial^2 \sigma_{zb}^{(s)}}{\partial \zeta^2} - \frac{A_{15}}{\lambda^2}\frac{\partial \sigma_{zb}^{(s)}}{\partial \zeta} - \frac{A_{16}}{\lambda^2}\frac{\partial \sigma_{yzb}^{(s)}}{\partial \zeta} - \frac{A_{13}}{\lambda}\sigma_{zb}^{(s)} - \frac{A_{14}}{\lambda}\sigma_{yzb}^{(s)} + R_u^{(s-1)},$$

$$v_b^{(s)} = -\frac{A_{16}}{\lambda^3}\frac{\partial^2 \sigma_{zb}^{(s)}}{\partial \zeta^2} - \frac{A_{25}}{\lambda^2}\frac{\partial \sigma_{zb}^{(s)}}{\partial \zeta} - \frac{A_{26}}{\lambda^2}\frac{\partial \sigma_{yzb}^{(s)}}{\partial \zeta} - \frac{A_{23}}{\lambda}\sigma_{zb}^{(s)} - \frac{A_{24}}{\lambda}\sigma_{yzb}^{(s)} + R_v^{(s-1)},$$

$$w_b^{(s)} = -\frac{A_{11}}{\lambda^4}\frac{\partial^3 \sigma_{zb}^{(s)}}{\partial \zeta^3} - \frac{A_{16}}{\lambda^3}\frac{\partial^2 \sigma_{yzb}^{(s)}}{\partial \zeta^2} - \frac{(A_{13}+A_{35})}{\lambda^2}\frac{\partial \sigma_{zb}^{(s)}}{\partial \zeta} - \frac{2A_{15}}{\lambda^3}\frac{\partial^2 \sigma_{zb}^{(s)}}{\partial \zeta^2}$$

$$-\frac{(A_{14}+A_{25})}{\lambda^2}\frac{\partial \sigma_{yzb}^{(s)}}{\partial \zeta} - \frac{A_{33}}{\lambda}\sigma_{zb}^{(s)} - \frac{A_{34}}{\lambda}\sigma_{yzb}^{(s)} + R_w^{(s-1)},$$

where

$$A_{k1} = \left(a_{22}a_{kk} - a_{k2}^2\right)a_{22}^{-1}, \quad (k=1,4),$$

$$A_{1k} = (a_{1k}a_{22} - a_{12}a_{2k})/a_{22},$$

$$A_{2k} = (a_{6k}a_{22} - a_{2k}a_{26})/a_{22}, \quad (k=3,4,5,6),$$

$$A_{3k} = (a_{k5}a_{22} - a_{2k}a_{25})a_{22}^{-1}, \quad (k=3,4,5),$$

$$A_{4k} = (a_{3k}a_{22} - a_{23}a_{2k})a_{22}^{-1}, \quad (k=3,4),$$

$$R_x^{(s-1)} = \frac{1}{\lambda^2}\frac{\partial^2 \sigma_{yzb}^{(s-1)}}{\partial \eta \partial \zeta} + \frac{1}{\lambda}\frac{\partial \sigma_{xyb}^{(s-1)}}{\partial \eta},$$

$$R_{xy}^{(s-1)} = \frac{1}{\lambda}\frac{\partial \sigma_{yb}^{(s-1)}}{\partial \eta}, \quad R_{xz}^{(s-1)} = \frac{1}{\lambda}\frac{\partial \sigma_{yzb}^{(s-1)}}{\partial \eta}, \quad (4.3)$$

$$R_y^{(s-1)} = \frac{1}{a_{22}}\left(a_{12}R_x^{(s-1)} + a_{25}R_{xz}^{(s-1)} + a_{26}R_{xy}^{(s-1)} - \frac{\partial v_b^{(s-1)}}{\partial \eta}\right),$$

$$R_u^{(s-1)} = \frac{1}{\lambda}\left(a_{11}R_x^{(s-1)} + a_{12}R_y^{(s-1)} + a_{15}R_{xz}^{(s-1)} + a_{16}R_{xy}^{(s-1)}\right),$$

$$R_v^{(s-1)} = \frac{1}{\lambda}\left(a_{16}R_x^{(s-1)} + a_{26}R_y^{(s-1)} + a_{56}R_{xz}^{(s-1)} + a_{66}R_{xy}^{(s-1)} - \frac{\partial u_b^{(s-1)}}{\partial \eta}\right),$$

$$R_w^{(s-1)} = \frac{1}{\lambda}\left(a_{15}R_x^{(s-1)} + a_{25}R_y^{(s-1)} + a_{55}R_{xz}^{(s-1)} + a_{56}R_{xy}^{(s-1)} + \frac{\partial R_u^{(s-1)}}{\partial \zeta}\right).$$

The quantities $\sigma_{zb}^{(s)}$ and $\sigma_{yzb}^{(s)}$ yield

$$L_1 \sigma_{zb}^{(s)} + L_2 \sigma_{yzb}^{(s)} = R_1^{(s-1)},$$

$$L_2 \sigma_{zb}^{(s)} + L_3 \sigma_{yzb}^{(s)} = R_2^{(s-1)}, \qquad (4.4)$$

where

$$L_1 = A_{11}\frac{\partial^4}{\partial \zeta^4} + 2A_{15}\lambda\frac{\partial^3}{\partial \zeta^3} + (2A_{13}+A_{35})\lambda^2\frac{\partial^2}{\partial \zeta^2} + 2A_{45}\lambda^3\frac{\partial}{\partial \zeta} + A_{43}\lambda^4,$$

$$L_2 = A_{16}\lambda\frac{\partial^3}{\partial \zeta^3} + (A_{14}+A_{25})\lambda^2\frac{\partial^2}{\partial \zeta^2} + (A_{23}+A_{34})\lambda^3\frac{\partial}{\partial \zeta} + A_{44}\lambda^4, \quad (4.5)$$

$$L_3 = A_{26}\lambda^2\frac{\partial^2}{\partial \zeta^2} + 2A_{24}\lambda^3\frac{\partial}{\partial \zeta} + A_{41}\lambda^4,$$

$$R_1^{(s-1)} = -\left(\frac{\partial R_w^{(s-1)}}{\partial \zeta} + a_{13}R_x^{(s-1)}\right.$$

$$\left. +a_{23}R_y^{(s-1)} + a_{35}R_{xz}^{(s-1)} + a_{36}R_{xy}^{(s-1)}\right)\lambda^4,$$

$$R_2^{(s-1)} = -\left(\frac{\partial R_v^{(s-1)}}{\partial \zeta} - \frac{\partial w_b^{(s-1)}}{\partial \eta} + a_{14}R_x^{(s-1)}\right.$$

$$\left. +a_{24}R_y^{(s-1)} + a_{45}R_{xz}^{(s-1)} + a_{46}R_{xy}^{(s-1)}\right)\lambda^4.$$

We note that the quantities $R_i^{(s)}$ are known for $s \geq 0$, whereas $R_i^{(0)} \equiv 0$ for $s < 0$.

In particular case of orthotropic (isotropic) plates the only non-zero coefficients out of a set of A_{ik} are A_{11}, A_{13}, A_{26}, A_{35}, A_{41}, A_{43} ($A_{ik} \neq A_{ki}$), hence the operator L_2 cancels out, and the system is decoupled into plane and anti-plane boundary layers. We emphasize that in case of a plate of general anisotropy the boundary layer solution cannot be decomposed into plane and anti-plane components, see (4.4). Thus, if the coupling effect could be neglected at leading order of the outer problem, it is no longer the case for boundary layer solution.

At leading order $s = 0$ the solution of (4.4) is given by

$$\sigma_{yzb}^{(0)} = L_1\Phi, \quad \sigma_{zb}^{(0)} = -L_2\Phi \tag{4.6}$$

where $\Phi(\eta, \zeta)$ is a solution of the following boundary value problem

$$\left(L_1 L_3 - L_2^2\right)\Phi = 0 \tag{4.7}$$

$$L_1\Phi(\pm 1) = 0, \quad L_2\Phi(\pm 1) = 0, \quad L_2\Phi'(\pm 1) = 0. \tag{4.8}$$

The conditions (4.8) ensure that $\sigma_{xzb}^{(0)} = \sigma_{yzb}^{(0)} = \sigma_{zb}^{(0)} = 0$ at $\zeta = \pm 1$. At higher orders $s > 0$ solution is combined from the general solution of the homogeneous problem (4.8), (4.9) and particular solution of the non-homogeneous problem (4.4). Using (4.5), equation (4.7) may be transformed to

$$B_0\frac{\partial^6\Phi}{\partial\zeta^6} + B_1\lambda\frac{\partial^5\Phi}{\partial\zeta^5} + B_2\lambda^2\frac{\partial^4\Phi}{\partial\zeta^4} + B_3\lambda^3\frac{\partial^3\Phi}{\partial\zeta^3} \tag{4.9}$$

$$+B_4\lambda^4\frac{\partial^2\Phi}{\partial\zeta^2} + B_5\lambda^5\frac{\partial\Phi}{\partial\zeta} + B_6\lambda^6\Phi = 0,$$

where

$$B_0 = A_{11}A_{26} - A_{16}^2,$$

$$B_1 = 2(A_{11}A_{24} + A_{15}A_{26} - A_{14}A_{16} - A_{16}A_{25}),$$

$$B_2 = 4A_{15}A_{24} + A_{11}A_{41} + A_{26}(2A_{13} + A_{35})$$

$$-2A_{16}(A_{23} + A_{34}) - (A_{14} + A_{25})^2,$$

$$B_3 = 2A_{24}(2A_{13} + A_{35}) + A_{15}A_{41} + A_{26}A_{33}$$

$$-A_{16}A_{14} - (A_{14} + A_{25})(A_{23} + A_{34}), \tag{4.10}$$

$$B_4 = 4A_{24}A_{33} + A_{26}A_{43} + A_{41}(2A_{13} + A_{35})$$

$$-2A_{44}(A_{14} + A_{25}) - (A_{23} + A_{34})^2,$$

$$B_5 = 2(A_{24}A_{43} + A_{33}A_{41} - A_{44}A_{23}), \quad B_6 = A_{43}A_{41} - A_{44}^2.$$

The boundary value problem (4.8), (4.9) is a generalized eigenvalue problem for operator bundle. Similar problems have been investigated by Tamarkin (1917); Palant (1961); Markus (1962); Vizitei and Markus (1965); Orazov (1976). It follows from the results of Palant (1961); Orazov (1976) that the system of eigenfunctions of (4.9) is six times complete, with the system of eigenfunctions corresponding to $Re\lambda > 0$ is triply complete. Therefore the three given functions specified at the edge, possessing appropriate properties, may be expanded along the eigenfunctions, which guarantees that the combination of the outer solution and the boundary layer satisfies the spatial edge boundary conditions. We note that this may be realized through one of the methods described in Chapter 4. In order to obtain the eigenvalues λ_n, solutions of (4.9) have to be substituted into (4.8), hence, the determinant of the system should vanish as a condition of existence of non-trivial solutions, which gives a transcendental equation $\Delta = 0$ for both plane and anti-plane boundary layer solutions. The eigenfunctions may then be found through (4.6).

5.5 Matching of the Outer Solution with the Boundary Layer

Let us now present the results of matching the outer solution and the boundary layer, with the edge boundary conditions specified in the form of the first boundary value problem and mixed form. Other types of boundary conditions may be treated similarly.

Consider the edge $x = 0$, with the boundary conditions written as

$$\sigma_x = \sigma_x^0(y, z), \quad \sigma_{xy} = \sigma_{xy}^0(y, z), \quad \sigma_{xz} = \sigma_{xz}^0(y, z) \tag{5.1}$$

where σ_x^0, σ_{xy}^0, σ_{xz}^0 are given functions.

The general integral of the problem is given by

$$J = Q + R_b \tag{5.2}$$

where Q is the outer solution, governed by (2.2), (2.4), and R_b is the boundary layer solution following from (4.1), (4.2). Then, bearing in mind (5.2) the quantities of (5.1) take the form

$$\sigma_x = \varepsilon^{-2} \sum_{s=0}^{S} \varepsilon^s \sigma_x^{(s)} + \varepsilon^\chi \sum_{s=0}^{N} \varepsilon^s \sum_{(\lambda)} \sigma_{xb}^{(s)}(\eta, \zeta, \lambda) \exp(-\lambda t),$$

$$\sigma_{xy} = \varepsilon^{-2} \sum_{s=0}^{S} \varepsilon^s \sigma_{xy}^{(s)} + \varepsilon^\chi \sum_{s=0}^{N} \varepsilon^s \sum_{(\lambda)} \sigma_{xyb}^{(s)}(\eta, \zeta, \lambda) \exp(-\lambda t), \tag{5.3}$$

$$\sigma_{xz} = \varepsilon^{-1} \sum_{s=0}^{S} \varepsilon^s \sigma_{xz}^{(s)} + \varepsilon^\chi \sum_{s=0}^{N} \varepsilon^s \sum_{(\lambda)} \sigma_{xzb}^{(s)}(\eta, \zeta, \lambda) \exp(-\lambda t).$$

We note that summation over λ in (5.3) is performed for those satisfying $Re\lambda > 0$.

Substituting the expressions (5.3) into (5.1) at $x = 0$ $(t = 0)$, we deduce the only value $\chi = -1$ allowing consistent asymptotic procedure for boundary conditions of the outer solution and the boundary layer solution. Therefore

$$\sigma_x^{(s)}(\xi = 0, \eta, \zeta) + \sum_{(\lambda)} \sigma_{xb}^{(s-1)}(\eta, \zeta, \lambda) = \sigma_x^{0(s-2)}(a\eta, h\zeta),$$

$$\sigma_{xy}^{(s)}(\xi = 0, \eta, \zeta) + \sum_{(\lambda)} \sigma_{xyb}^{(s-1)}(\eta, \zeta, \lambda) = \sigma_{xy}^{0(s-2)}(a\eta, h\zeta), \qquad (5.4)$$

$$\sigma_{xz}^{(s)}(\xi = 0, \eta, \zeta) + \sum_{(\lambda)} \sigma_{xzb}^{(s)}(\eta, \zeta, \lambda) = \sigma_{xz}^{0(s-1)}(a\eta, h\zeta),$$

where

$$\sigma_x^{0(0)} = \sigma_x^0(a\eta, h\zeta), \qquad \sigma_{xy}^{0(0)} = \sigma_{xy}^0, \qquad \sigma_{xz}^{0(0)} = \sigma_{xz}^0,$$
$$\sigma_x^{0(s)} = \sigma_{xy}^{0(s)} = \sigma_{xz}^{0(s)} = 0 \qquad \text{at} \qquad s \neq 0.$$

In further analysis we shall use the following properties of the boundary layer solution. Due to $\sigma_{zb} = \sigma_{xzb} = \sigma_{yzb} = 0$ at $\zeta = \pm 1$, equations (4.2) imply

$$\int_{-1}^{+1} \sigma_{xb}^{(s)} d\zeta = \int_{-1}^{+1} R_x^{(s-1)} d\zeta,$$

$$\int_{-1}^{+1} \zeta \sigma_{xb}^{(s)} d\zeta = \int_{-1}^{+1} \zeta R_x^{(s-1)} d\zeta, \qquad (5.5)$$

$$\int_{-1}^{+1} \sigma_{xyb}^{(s)} d\zeta = \int_{-1}^{+1} R_{xy}^{(s-1)} d\zeta,$$

$$\int_{-1}^{+1} \sigma_{xzb}^{(s)} d\zeta = \int_{-1}^{+1} R_{xz}^{(s-1)} d\zeta.$$

At leading order $s = 0$ the properties (5.5) mean simply that the corresponding stresses are self-equilibrated.

It may be deduced from (5.4) that

$$\sigma_x^{(0)}(0, \eta, \zeta) = 0, \qquad \sigma_{xy}^{(0)}(0, \eta, \zeta) = 0. \qquad (5.6)$$

Taking into account (2.4), it follows from (5.6) that

$$\tau_{x0}^{(0)} = 0, \qquad \tau_{x1}^{(0)} = 0, \qquad \tau_{xy0}^{(1)} = 0 \qquad \text{at} \qquad \xi = 0. \qquad (5.7)$$

On the other hand, from the last equation of (5.4) we have

$$\sum_{(\lambda)} \sigma_{xzb}^{(s)}(\eta, \zeta, \lambda) = \sigma_{xz}^{0(s-1)} - \sigma_{xz}^{(s)}(0, \eta, \zeta). \qquad (5.8)$$

Substituting (5.8) into (5.5), we arrive at

$$\int_{-1}^{+1} \sigma_{xz}^{(s)}(0, \eta, \zeta) d\zeta = \int_{-1}^{+1} \left(\sigma_{xz}^{0(s-1)}(a\eta, h\zeta) - \sum_{(\lambda)} R_{xz}^{(s-1)}(\zeta, \lambda) \right) d\zeta \qquad (5.9)$$

leading to $\int_{-1}^{+1} \sigma_{xz}^{(0)}(0,\eta,\zeta)d\zeta = 0$ or

$$\frac{1}{3}\tau_{xz2}^{(0)} + 2\tau_{xz0}^{(0)} = 0 \qquad \text{at} \qquad \xi = 0. \tag{5.10}$$

Relations (5.7) and (5.10) may act as boundary conditions for the outer solution at $s = 0$. Rewritten in classical notation at $s = 0$, they give

$$T^{(0)} = 0, \quad S_{12}^{(0)} = 0,$$
$$M^{(0)} = 0, \quad N^{(0)} = 0 \quad \text{at} \quad x = 0. \tag{5.11}$$

Now, once the outer solution is determined at $s = 0$, returning back to (5.4), we have

$$\sum_{(\lambda)} \sigma_{xzb}^{(0)}(\eta,\zeta,\lambda) = -\sigma_{xz}^{(0)}(0,\eta,\zeta) \tag{5.12}$$

which is a condition for boundary layer (since the right-hand side is already known).

It follows from (5.4) that

$$\sum_{(\lambda)} \sigma_{xb}^{(0)}(\eta,\zeta,\lambda) = \sigma_x^{0(s-1)}(a\eta,h\zeta) - \sigma_x^{(s+1)}(\xi=0,\eta,\zeta) \tag{5.13}$$

$$\sum_{(\lambda)} \sigma_{xyb}^{(s)}(\eta,\zeta,\lambda) = \sigma_{xy}^{0(s-1)} - \sigma_{xy}^{(s+1)}(0,\eta,\zeta).$$

Substituting the latter into properties (5.5), we obtain

$$\int_{-1}^{+1} \sigma_x^{(s+1)}(0,\eta,\zeta)d\zeta$$
$$= \int_{-1}^{+1} \left(\sigma_x^{0(s-1)} - \sum_{(\lambda)} R_x^{(s-1)}(\eta,\zeta,\lambda) \right) d\zeta,$$

$$\int_{-1}^{+1} \zeta\sigma_x^{(s+1)}(0,\eta,\zeta)d\zeta$$
$$= \int_{-1}^{+1} \zeta \left(\sigma_x^{0(s-1)}(a\eta,h\zeta) - \sum_{(\lambda)} R_x^{(s-1)}(\eta,\zeta,\lambda) \right) d\zeta, \tag{5.14}$$

$$\int_{-1}^{+1} \sigma_{xy}^{(s+1)}(0,\eta,\zeta)d\zeta$$
$$= \int_{-1}^{+1} \left(\sigma_{xy}^{0(s-1)}(a\eta,h\zeta) - \sum_{(\lambda)} R_{xy}^{(s-1)}(\eta,\zeta,\lambda) \right) d\zeta$$

which imply the conditions for the outer solution at next order $s = 1$

$$\int_{-1}^{+1} \sigma_x^{(1)}(0,\eta,\zeta)d\zeta = 0, \qquad \int_{-1}^{+1} \zeta\sigma_x^{(1)}(0,\eta,\zeta)d\zeta = 0, \qquad \int_{-1}^{+1} \sigma_{xy}^{(1)}(0,\eta,\zeta)d\zeta = 0$$

or

$$2\tau_{x0}^{(1)} = -\int_{-1}^{+1} \sigma_x^{*(1)}(0,\eta,\zeta)d\zeta, \qquad \frac{2}{3}\tau_{x1}^{(1)} = -\int_{-1}^{+1} \zeta\sigma_x^{*(1)}(0,\eta,\zeta)d\zeta, \qquad (5.15)$$

$$2\tau_{xy0}^{(1)} = -\int_{-1}^{+1} \sigma_{xy}^{*(1)}(0,\eta,\zeta)d\zeta \qquad \text{at} \quad \xi = 0.$$

Therefore, in view of (2.4) the right-hand sides of (5.13) are known, so the boundary layer at $s = 0$ is governed by

$$\sum_{(\lambda)} \sigma_{xb}^{(0)}(\eta,\zeta,\lambda) = -\sigma_x^{(1)}(\xi = 0,\eta,\zeta),$$

$$\sum_{(\lambda)} \sigma_{xyb}^{(0)}(\eta,\zeta,\lambda) = -\sigma_{xy}^{(1)}(0,\eta,\zeta), \qquad (5.16)$$

$$\sum_{(\lambda)} \sigma_{xzb}^{(0)}(\eta,\zeta,\lambda) = -\sigma_{xz}^{(0)}(0,\eta,\zeta).$$

Since the system of eigenfunctions of the boundary layer solution corresponding to $Re\lambda > 0$ is triply complete, we conclude that three conditions (5.16) are sufficient to fully determine the boundary layer solution.

The fourth condition for the outer solution follows from (5.9) at $s = 1$. It becomes

$$\int_{-1}^{+1} \sigma_{xz}^{(1)}(0,\eta,\zeta)d\zeta = \int_{-1}^{+1} \left(\sigma_{xz}^0(a\eta,h\zeta) - \sum_{(\lambda)} R_{xz}^{(0)}(\eta,\zeta,\lambda) \right) d\zeta \qquad (5.17)$$

or

$$\frac{1}{3}\tau_{xz2}^{(1)} + 2\tau_{xz0}^{(1)} = \int_{-1}^{+1} \left(\sigma_{xz}^0(a\eta,h\zeta) - \sum_{(\lambda)} R_{xz}^{(0)}(\eta,\zeta,\lambda) \right) d\zeta.$$

The next order of the outer solution (at $s = 1$) is now fully determined from (5.16) and (5.17). It is worth mentioning that the boundary layer is affecting the outer stress field through the boundary condition (5.17), therefore, higher order approximations (starting from $s = 1$) should be coupled with the boundary layer solution. This clearly indicates that in case of the *ad hoc* approach the hypotheses should take into account the boundary layer phenomena. The presented asymptotic procedure may be continued for higher orders $s \geq 2$.

As follows from (5.17), the influence of edge loading on the outer field occurs starting from $s = 1$, therefore its contribution is of order $O(\varepsilon)$ compared to the contribution of load of the same intensity acting on the faces of the plate. In other words, the influence of edge loading would be comparable to that of surface loading provided that the intensity of edge loading is $O(\varepsilon^{-1})$.

Consider now the case of mixed edge boundary conditions at $x = 0$

$$\sigma_x = 0, \qquad \sigma_{xy} = 0, \qquad w = 0 \qquad \text{at} \qquad x = 0 \ (t = 0), \qquad (5.18)$$

modeling pinned edge. The stress components σ_x, σ_{xy} are defined by (5.3), whereas for w we have

$$w/a = \varepsilon^{-3} \sum_{s=0}^{S} W^{(s)} + \varepsilon^{\chi+1} \sum_{s=0}^{N} \varepsilon^s \sum_{(\lambda)} w_b^{(s)}(\eta, \zeta, \lambda) \exp(-\lambda t). \qquad (5.19)$$

Satisfying (5.18) we obtain $\chi = -2$ from consistency of the asymptotic procedure. Then (5.18) takes the form

$$\sigma_x^{(s)} + \sum_{(\lambda)} \sigma_{xb}^{(0)}(\eta, \zeta, \lambda) = 0,$$

$$\sigma_{xy}^{(s)} + \sum_{(\lambda)} \sigma_{xyb}^{(0)}(\eta, \zeta, \lambda) = 0 \qquad \text{at} \quad x = 0 \ (t = 0), \qquad (5.20)$$

$$W^{(s)} + \sum_{(\lambda)} w_b^{(s-2)}(\eta, \zeta, \lambda) = 0.$$

Expressing the stresses of the boundary layer solution through $\sigma_x^{(s)}$, $\sigma_{xy}^{(s)}$ and making use of (5.5), we obtain

$$\int_{-1}^{+1} \sigma_x^{(s)}(\xi = 0)d\zeta = -\int_{-1}^{+1} \sum_{(\lambda)} R_x^{(s-1)}(\eta, \zeta, \lambda)d\zeta,$$

$$\int_{-1}^{+1} \zeta\sigma_x^{(s)}(\xi = 0)d\zeta = -\int_{-1}^{+1} \zeta \sum_{(\lambda)} R_x^{(s-1)}(\eta, \zeta, \lambda)d\zeta, \qquad (5.21)$$

$$\int_{-1}^{+1} \sigma_{xy}^{(s)}(\xi = 0)d\zeta = -\int_{-1}^{+1} \sum_{(\lambda)} R_{xy}^{(s-1)}(\eta, \zeta, \lambda)d\zeta.$$

Now at $s = 0$ it may be deduced that

$$\int_{-1}^{+1} \sigma_x^{(0)}(\xi = 0)d\zeta = 0, \qquad \int_{-1}^{+1} \zeta\sigma_x^{(0)}(\xi = 0)d\zeta = 0, \qquad \int_{-1}^{+1} \sigma_{xy}^{(0)}(\xi = 0)d\zeta = 0$$

or

$$2\tau_{x0}^{(0)} = 0, \qquad \frac{2}{3}\tau_{x1}^{(0)} = 0, \qquad 2\tau_{xy0}^{(0)} = 0 \qquad \text{at} \quad x = 0. \qquad (5.22)$$

Rewriting the latter in terms of classical notation, we have

$$T^{(0)} = 0, \qquad M^{(0)} = 0, \qquad S_{12}^{(0)} = 0 \qquad \text{at} \quad x = 0. \qquad (5.23)$$

It also follows from the last equation of (5.20) that

$$W^{(0)} = 0, \qquad W^{(1)} = 0 \qquad \text{at} \quad x = 0. \qquad (5.24)$$

Thus, at leading order $s = 0$ the conditions for the outer solution are

$$T^{(0)} = S_{12}^{(0)} = M^{(0)} = W^{(0)} = 0 \qquad \text{at} \quad x = 0, \qquad (5.25)$$

which coincide with the well-known conditions for pinned end within the classical plate theory.

After the outer solution is obtained for $s = 0$, the quantities $\sigma_x^{(0)}$, $\sigma_{xy}^{(0)}$ are known. In addition, it follows from (2.4) and (5.20) that $\sigma_x^{(0)}(x = 0) = 0$, $\sigma_{xy}^{(0)}(x = 0) = \zeta\tau_{xy1}^{(0)}(\xi = 0)$. Moreover, it may be deduced from (5.20) that $w_b^{(0)}$ is expressed through $W^{(2)}$, which is small. Therefore, returning back to (5.20) the conditions for the boundary layer are formulated as

$$\sum_{(\lambda)} \sigma_{xb}^{(0)}(\eta, \zeta, \lambda) = 0,$$

$$\sum_{(\lambda)} \sigma_{xyb}^{(0)}(\eta, \zeta, \lambda) = -\zeta\tau_{xy1}^{(0)}(\xi = 0), \tag{5.26}$$

$$\sum_{(\lambda)} w_b^{(0)}(\eta, \zeta, \lambda) \approx 0.$$

Since $\tau_{xy1}^{(0)}(\xi = 0) \neq 0$, the stress components of the boundary layer solution are of order $O(\varepsilon^{-2})$, being comparable to the leading order stress field of the outer solution.

At next order $s = 1$ the relations (5.21), (5.24) as well as (2.4) imply the following conditions for the outer problem

$$2\tau_{x0}^{(1)} = -\int_{-1}^{+1} \left(\sigma_x^{*(1)}(\xi = 0) + \sum_{(\lambda)} R_x^{(0)}(\eta, \zeta, \lambda) \right) d\zeta,$$

$$\frac{2}{3}\tau_{x1}^{(1)} = -\int_{-1}^{+1} \zeta \left(\sigma_x^{*(1)}(\xi = 0) + \sum_{(\lambda)} R_x^{(0)}(\eta, \zeta, \lambda) \right) d\zeta \quad \text{at} \quad x = 0, \tag{5.27}$$

$$2\tau_{xy0}^{(1)} = -\int_{-1}^{+1} \left(\sigma_{xy}^{*(1)}(\xi = 0) + \sum_{(\lambda)} R_{xy}^{(0)}(\eta, \zeta, \lambda) \right) d\zeta,$$

$$W^{(1)} = 0.$$

Thus, once again, it may be observed that the boundary layer influences the outer solution through boundary conditions. The quantities $R_x^{(0)}$, $R_{xy}^{(0)}$ appearing in (5.27) may be determined from (4.3). After the outer solution is obtained, the next order approximation of the boundary layer ($s = 1$) may be found from (5.20). The iterative asymptotic procedure may be continued as long as necessary. However, for practical computations it is usually sufficient to consider two-term asymptotic expansions.

Chapter 6

Non-Classical Boundary Value Problems for Anisotropic Plates

6.1 Formulations of the Non-Classical Boundary Value Problems

The classical theory for plates and shells and existing refined theories are originating from the case, when the face boundary conditions are formulated in terms of the corresponding stress components (the conditions of the first boundary value problem of elasticity). A natural question then arises whether it is possible to extend the assumptions of classical theory for plates and shells for other types of face boundary conditions, including the second boundary value problem or mixed problems. It should be noted that the discussed problems are not just of theoretical interest but also possess a number of practical applications, e.g. in calculations of interaction of soft thin walled structures with more rigid, in building foundations, airstrips, etc.

These problems have been considered earlier in simplest cases within the framework of traditional elasticity, mostly for infinite isotropic rods and plates. The cases of the second boundary value problem and mixed face boundary conditions for infinite isotropic strip has been investigated by Uflyand (1967) through Fourier integral transforms. Mixed problems for isotropic layer were considered by Kupradze et al. (1976). The approach of Vorovich et al. (1974) also relied on integral transforms.

In this chapter we present asymptotic analysis for such spatial boundary value problems for anisotropic plates, when the face boundary conditions are formulated in terms of displacements or taken in mixed form. Here and in what follows we refer to these types of boundary value problems as non-classical ones. We emphasize a difference between, say, mixed problems in a sense of mixed boundary conditions imposed on the edges of a plate or shell and these non-classical, when the main emphasis is put on the face boundary conditions. Another fundamental difference between these problems is that as will be shown later, these non-classical boundary value problems cannot be solved through using classical assumptions of plates and shells theory. On the other hand, if one considers exact formulation in 3D, the discussed problems are classical boundary value problems of elasticity.

Consider anisotropic plate $\Omega = \{(\alpha, \beta, \gamma) : \alpha, \beta \in \Omega_0, -h \leq \gamma \leq h\}$ of thickness $2h$. Let us associate the mid-plane Ω_0 to the curvilinear coordinate system α, β, with γ axis directed along the normal to the mid-plane (Fig. 6.1).

We assume that the anisotropic properties are characterized by 21 different material constants, and that the coordinate directions are elastically equivalent at arbitrary point.

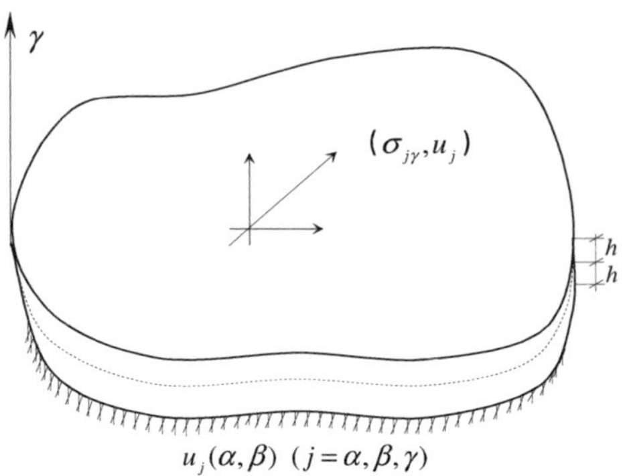

Fig. 6.1

The governing equations of elasticity in curvilinear coordinates α, β, γ over the domain Ω are written as follows:
the equations of equilibrium

$$\frac{1}{A}\frac{\partial \sigma_{\alpha\alpha}}{\partial \alpha} + \frac{1}{B}\frac{\partial \sigma_{\alpha\beta}}{\partial \beta} + \frac{\partial \sigma_{\alpha\gamma}}{\partial \gamma} + k_\beta(\sigma_{\alpha\alpha} - \sigma_{\beta\beta}) + 2k_\alpha\sigma_{\alpha\beta} = 0, \quad (\alpha, \beta; A, B),$$

$$\frac{1}{A}\frac{\partial \sigma_{\alpha\gamma}}{\partial \alpha} + \frac{1}{B}\frac{\partial \sigma_{\beta\gamma}}{\partial \beta} + \frac{\partial \sigma_{\gamma\gamma}}{\partial \gamma} + k_\beta\sigma_{\alpha\gamma} + k_\alpha\sigma_{\beta\gamma} = 0 \quad (1.1)$$

the constitutive relations (the Hooke's law)

$$\frac{1}{A}\frac{\partial u_\alpha}{\partial \alpha} + k_\alpha u_\beta = a_{11}\sigma_{\alpha\alpha} + a_{12}\sigma_{\beta\beta} + a_{13}\sigma_{\gamma\gamma} + a_{14}\sigma_{\beta\gamma} + a_{15}\sigma_{\alpha\gamma} + a_{16}\sigma_{\alpha\beta},$$

$$\frac{1}{B}\frac{\partial u_\beta}{\partial \beta} + k_\beta u_\alpha = a_{12}\sigma_{\alpha\alpha} + a_{22}\sigma_{\beta\beta} + a_{23}\sigma_{\gamma\gamma} + a_{24}\sigma_{\beta\gamma} + a_{25}\sigma_{\alpha\gamma} + a_{26}\sigma_{\alpha\beta},$$

$$\frac{\partial u_\gamma}{\partial \gamma} = a_{13}\sigma_{\alpha\alpha} + a_{23}\sigma_{\beta\beta} + a_{33}\sigma_{\gamma\gamma} + a_{34}\sigma_{\beta\gamma} + a_{35}\sigma_{\alpha\gamma} + a_{36}\sigma_{\alpha\beta}, \quad (1.2)$$

$$\frac{1}{B}\frac{\partial u_\alpha}{\partial \beta} + \frac{1}{A}\frac{\partial u_\beta}{\partial \alpha} - k_\alpha u_\alpha - k_\beta u_\beta$$
$$= a_{16}\sigma_{\alpha\alpha} + a_{26}\sigma_{\beta\beta} + a_{36}\sigma_{\gamma\gamma} + a_{46}\sigma_{\beta\gamma} + a_{56}\sigma_{\alpha\gamma} + a_{66}\sigma_{\alpha\beta},$$

$$\frac{1}{A}\frac{\partial u_\gamma}{\partial \alpha} + \frac{\partial u_\alpha}{\partial \gamma} = a_{15}\sigma_{\alpha\alpha} + a_{25}\sigma_{\beta\beta} + a_{35}\sigma_{\gamma\gamma} + a_{45}\sigma_{\beta\gamma} + a_{55}\sigma_{\alpha\gamma} + a_{56}\sigma_{\alpha\beta},$$

$$\frac{1}{B}\frac{\partial u_\gamma}{\partial \beta} + \frac{\partial u_\beta}{\partial \gamma} = a_{14}\sigma_{\alpha\alpha} + a_{24}\sigma_{\beta\beta} + a_{34}\sigma_{\gamma\gamma} + a_{44}\sigma_{\beta\gamma} + a_{45}\sigma_{\alpha\gamma} + a_{46}\sigma_{\alpha\beta},$$

where A, B are coefficients of the first quadratic form of the coordinate surface, $k_\alpha = \dfrac{1}{AB}\dfrac{\partial A}{\partial \beta}$ $(\alpha, \beta; A, B)$ are geometric curvatures of the α and β coordinate lines, and a_{ik} are material constants.

Consider the problem (1.1), (1.2) over the domain Ω subject to the boundary conditions specified in terms of displacements on the face $\gamma = -h$ (in particular, the fixed face boundary condition $u_\alpha(-h) = u_\beta(-h) = u_\gamma(-h) = 0$)

$$u_\alpha(-h) = u^-(\alpha, \beta), \quad u_\beta(-h) = v^-(\alpha, \beta), \quad u_\gamma(-h) = w^-(\alpha, \beta) \qquad (1.3)$$

whereas at the other face $\gamma = h$ several types of boundary conditions may be considered including:

the first boundary value problem

$$\sigma_{\alpha\gamma}(h) = \varepsilon^{-1}\sigma_{\alpha\gamma}^+(\alpha, \beta), \quad \sigma_{\beta\gamma}(h) = \varepsilon^{-1}\sigma_{\beta\gamma}^+(\alpha, \beta), \quad \sigma_{\gamma\gamma}(h) = \varepsilon^{-1}\sigma_{\gamma\gamma}^+(\alpha, \beta) \quad (1.4)$$

or the second boundary value problem

$$u_\alpha(h) = u^+(\alpha, \beta), \quad u_\beta(h) = v^+(\alpha, \beta), \quad u_\gamma(h) = w^+(\alpha, \beta) \qquad (1.5)$$

or mixed type boundary conditions

$$a) \quad u_\alpha(h) = u^+(\alpha, \beta), \quad u_\beta(h) = v^+(\alpha, \beta), \quad \sigma_{\gamma\gamma}(h) = \varepsilon^{-1}\sigma_{\gamma\gamma}^+(\alpha, \beta) \qquad (1.6)$$

$$b) \quad \sigma_{\alpha\gamma}(h) = \varepsilon^{-1}\sigma_{\alpha\gamma}^+(\alpha, \beta), \quad \sigma_{\beta\gamma}(h) = \varepsilon^{-1}\sigma_{\beta\gamma}^+(\alpha, \beta), \quad u_\gamma(h) = w^+(\alpha, \beta). \quad (1.7)$$

The boundary conditions on the edge of the plate $\partial\Omega$ are not specified here, since their form does not influence the analysis, which depends on the type of prescribed face boundary conditions. We remark that the physical meaning of a small parameter ε is essentially the same as that of the previous chapters.

6.2 The Outer Solution

In order to solve the posed boundary value problems, we again proceed to the dimensionless coordinates ξ, $\eta\,\zeta$, and introduce the dimensionless displacements through the formulae

$$\alpha = a\xi, \quad \beta = a\eta, \quad \gamma = h\zeta = a\varepsilon\zeta \qquad (2.1)$$

$$u_\alpha = au, \quad u_\beta = av, \quad u_\gamma = aw$$

where $\varepsilon = h/a$ is a small parameter and a is a typical size of the mid-plane of the plate.

The system of equations (1.1), (2.1) referred to the dimensionless coordinates, is singularly perturbed with respect to the parameter ε of the system. Its solution, in accordance with the theory of such equations, consists of the solutions of the outer problem and the boundary layer. The outer solution is again sought-for as in the asymptotic expansion

$$Q = \sum_{s=0}^{S} \varepsilon^{\chi_Q + s} Q^{(s)} \qquad (2.2)$$

where Q is any of the stress and the non-dimensional displacement components with $Q^{(s)} \equiv 0$ for $s < 0$. The integers χ_μ, χ_σ, characterizing the intensity of the components of the displacement and the stress, should be chosen so that after the substitution (2.2) into the system of equations (1.1), (2.1) in the referred dimensionless coordinates and comparing the same powers of the small parameter ε on both sides of each equation, the obtained recurrent system with respect to $Q^{(s)}$ must be consistent. To this end, $Q^{(s)}$ should not only satisfy, successively, equations (1.1), (1.2), but also the boundary conditions (1.3)-(1.7). Such values of χ_u and χ_σ are uniquely defined. Let us note that the correct definition of the values χ_u, χ_σ is the most difficult and crucial point of the asymptotic theory and a particular case of the more general problems of the natural sciences. The fact is that, in any physical problem (and not only the physical ones) there exist quantities which play the dominant role, and quantities which only characterize adjunct effects. Establishing the asymptotically dominant quantities (formulation of the corresponding laws) is the most important and difficult problem. It is by no means a coincidence that many authors consider the choice of the correct asymptotic behavior (the order of magnitude) as an art (Barancev, 1976; Babich and Buldyrev, 1977). In our case the goal is only achieved when

$$\chi_u = 0, \ \chi_\sigma = -1. \tag{2.3}$$

Asymptotic expansions of the unknown quantities given by (2.2), (2.3) is fundamentally different from that of the same quantities within the classical theory for isotropic and anisotropic plates (3.2.4), (3.2.5). In contrast to the classical theory, all stresses as well as displacements here are equivalent. Therefore the hypothesis of the classical theory of plates is not applicable to formulations of the problems. Such particular structure of the solution is reflected in the final results. We see below, for instance, that the solution of the outer problem is completely determined by the conditions on the outer surface, which is not encountered in the classical theory. Furthermore, there is no need to supply any further hypothesis to determine the outer stress-strain field.

The asymptotic formula (2.2) indicates that the same order of stress-strain field, induced on the plate separately by the displacements and the external load, is only maintained if the intensity of the loading is greater by ε^{-1} times than the intensity of the imparted displacements. Therefore in order to ensure the same order of contribution from the external loads and displacements to general field, the parameter ε^{-1} is kept in the boundary conditions (1.4)-(1.7). Otherwise the intensity of stress in the plate, caused by external loads, will be less than the intensity of the stress-strain field arising from the displacements imposed on the outer surface of the plate. This indicates that the functions given by the boundary conditions (1.3)-(1.7) may be chosen in more general form

$$\sigma_{j\gamma}(h) = \varepsilon^{-1+s}\sigma_{j\gamma}^{+(s)}, \quad u_j(h) = \varepsilon^s u_j^{+(s)}, \quad s = \overline{0, S} \tag{2.4}$$

where notation $s = \overline{0, S}$ indicates summation along index s varying from zero to any chosen integer S. Usually, part of the known quantities $\sigma_{j\gamma}^{+(s)}$, $u_j^{+(s)}$ may be taken zero. Thus, the representation (2.4) may significantly extend the spectrum of the external loads.

Substituting (2.2) into the transformed equations (1.1) and (1.2), and taking into consideration (2.3), one arrives at a system in respect of the unknown coefficients $Q^{(s)}$. Its solution may be written in recurrent form for the sought-for displacements and stresses of approximation of order s (Aghalovyan and Gevorkyan, 1984)

$$\sigma_{\alpha\gamma}^{(s)} = \sigma_{\alpha\gamma0}^{(s)}(\xi,\eta) + \sigma_{\alpha\gamma*}^{(s)}(\xi,\eta,\zeta), \qquad (\alpha,\beta),$$

$$\sigma_{\gamma\gamma}^{(s)} = \sigma_{\gamma\gamma0}^{(s)}(\xi,\eta) + \sigma_{\gamma\gamma*}^{(s)}(\xi,\eta,\zeta),$$

$$\sigma_{\alpha\alpha}^{(s)} = A_{13}\sigma_{\gamma\gamma0}^{(s)} + A_{14}\sigma_{\beta\gamma0}^{(s)} + A_{15}\sigma_{\alpha\gamma0}^{(s)} + \sigma_{\alpha\alpha*}^{(s)}(\xi,\eta,\zeta),$$

$$\sigma_{\beta\beta}^{(s)} = A_{23}\sigma_{\gamma\gamma0}^{(s)} + A_{24}\sigma_{\beta\gamma0}^{(s)} + A_{25}\sigma_{\alpha\gamma0}^{(s)} + \sigma_{\beta\beta*}^{(s)}(\xi,\eta,\zeta),$$

$$\sigma_{\alpha\beta}^{(s)} = A_{63}\sigma_{\gamma\gamma0}^{(s)} + A_{64}\sigma_{\beta\gamma0}^{(s)} + A_{65}\sigma_{\alpha\gamma0}^{(s)} + \sigma_{\alpha\beta*}^{(s)}(\xi,\eta,\zeta),$$

$$u^{(s)} = \zeta\left(A_{53}\sigma_{\gamma\gamma0}^{(s)} + A_{54}\sigma_{\beta\gamma0}^{(s)} + A_{55}\sigma_{\alpha\gamma0}^{(s)}\right) + u_0^{(s)}(\xi,\eta) + u_*^{(s)}(\xi,\eta,\zeta), \qquad (2.5)$$

$$v^{(s)} = \zeta\left(A_{43}\sigma_{\gamma\gamma0}^{(s)} + A_{44}\sigma_{\beta\gamma0}^{(s)} + A_{45}\sigma_{\alpha\gamma0}^{(s)}\right) + v_0^{(s)}(\xi,\eta) + v_*^{(s)}(\xi,\eta,\zeta),$$

$$w^{(s)} = \zeta\left(A_{33}\sigma_{\gamma\gamma0}^{(s)} + A_{34}\sigma_{\beta\gamma0}^{(s)} + A_{35}\sigma_{\alpha\gamma0}^{(s)}\right) + w_0^{(s)}(\xi,\eta) + w_*^{(s)}(\xi,\eta,\zeta),$$

where

$$\sigma_{\alpha\gamma*}^{(s)} = -\int_0^\zeta \left(\frac{1}{A}\frac{\partial\sigma_{\alpha\alpha}^{(s-1)}}{\partial\xi} + \frac{1}{B}\frac{\partial\sigma_{\alpha\beta}^{(s-1)}}{\partial\eta}\right.$$

$$\left. +ak_\beta\left(\sigma_{\alpha\alpha}^{(s-1)} - \sigma_{\beta\beta}^{(s-1)}\right) + 2ak_\alpha\sigma_{\alpha\beta}^{(s-1)}\right)d\zeta$$

$$(\alpha,\beta;\xi,\eta;A,B)$$

$$\sigma_{\gamma\gamma*}^{(s)} = -\int_0^\zeta \left(\frac{1}{A}\frac{\partial\sigma_{\alpha\gamma}^{(s-1)}}{\partial\xi} + \frac{1}{B}\frac{\partial\sigma_{\beta\gamma}^{(s-1)}}{\partial\eta} + ak_\beta\sigma_{\alpha\gamma}^{(s-1)} + ak_\alpha\sigma_{\beta\gamma}^{(s-1)}\right)d\zeta,$$

$$\sigma_{\alpha\alpha*}^{(s)} = B_{11}R_1^{(s)} + B_{12}R_2^{(s)} + B_{16}R_3^{(s)},$$

$$\sigma_{\beta\beta*}^{(s)} = B_{12}R_1^{(s)} + B_{22}R_2^{(s)} + B_{26}R_3^{(s)},$$

$$\sigma_{\alpha\beta*}^{(s)} = B_{16}R_1^{(s)} + B_{26}R_2^{(s)} + B_{66}R_3^{(s)},$$

$$u_*^{(s)} = \int_0^\zeta \left(a_{15}\sigma_{\alpha\alpha*}^{(s)} + a_{25}\sigma_{\beta\beta*}^{(s)} + \dots + a_{65}\sigma_{\alpha\beta*}^{(s)} - \frac{1}{A}\frac{\partial w^{(s-1)}}{\partial\xi}\right)d\zeta,$$

$$v_*^{(s)} = \int_0^\zeta \left(a_{14}\sigma_{\alpha\alpha*}^{(s)} + a_{24}\sigma_{\beta\beta*}^{(s)} + \dots + a_{64}\sigma_{\alpha\beta*}^{(s)} - \frac{1}{B}\frac{\partial w^{(s-1)}}{\partial\eta}\right)d\zeta, \qquad (2.6)$$

$$w_*^{(s)} = \int_0^\zeta \left(a_{13}\sigma_{\alpha\alpha*}^{(s)} + a_{23}\sigma_{\beta\beta*}^{(s)} + \dots + a_{63}\sigma_{\alpha\beta*}^{(s)}\right)d\zeta,$$

$$R_1^{(s)} = \frac{1}{A}\frac{\partial u^{(s-1)}}{\partial\xi} + ak_\alpha v^{(s-1)} - a_{13}\sigma_{\gamma\gamma*}^{(s)} - a_{14}\sigma_{\beta\gamma*}^{(s)} - a_{15}\sigma_{\alpha\gamma*}^{(s)},$$

$$R_2^{(s)} = \frac{1}{B}\frac{\partial v^{(s-1)}}{\partial\eta} + ak_\beta u^{(s-1)} - a_{23}\sigma_{\gamma\gamma*}^{(s)} - a_{24}\sigma_{\beta\gamma*}^{(s)} - a_{25}\sigma_{\alpha\gamma*}^{(s)},$$

$$R_3^{(s)} = \frac{1}{B}\frac{\partial u^{(s-1)}}{\partial \eta} + \frac{1}{A}\frac{\partial v^{(s-1)}}{\partial \xi} - ak_\alpha u^{(s-1)}$$

$$-ak_\beta v^{(s-1)} - a_{36}\sigma_{\gamma\gamma *}^{(s)} - a_{46}\sigma_{\beta\gamma *}^{(s)} - a_{56}\sigma_{\alpha\gamma *}^{(s)},$$

$$B_{ij} = (a_{ik}a_{jk} - a_{ij}a_{kk})/\Delta, \quad (i \neq j \neq k \neq i),$$

$$B_{kk} = (a_{ii}a_{jj} - a_{ij}^2)/\Delta, \quad B_{ij} = B_{ji}, \qquad i,j,k = 1,2,6$$

$$A_{kl} = -a_{1l}B_{k1} - a_{2l}B_{k2} - a_{6l}B_{k6}, \qquad l,m = 3,4,5$$

$$A_{ml} = a_{m1}A_{1l} + a_{m2}A_{2l} + a_{m6}A_{6l} + a_{ml}, \quad A_{ml} \neq A_{lm},$$

$$\Delta = a_{11}a_{22}a_{66} + 2a_{12}a_{26}a_{16} - a_{11}a_{26}^2 - a_{22}a_{16}^2 - a_{66}a_{12}^2.$$

Here $\sigma_{\alpha\gamma 0}^{(s)}(\xi,\eta)$, $\sigma_{\beta\gamma 0}^{(s)}(\xi,\eta)$, $\sigma_{\gamma\gamma 0}^{(s)}(\xi,\eta)$, $u_0^{(s)}(\xi,\eta)$, $v_0^{(s)}(\xi,\eta)$, $w_0^{(s)}(\xi,\eta)$ are so far unknown functions depending on ξ, η, which are to be determined from the face boundary conditions at $\gamma = \pm h$. Satisfying conditions at $\gamma = -h$, which are common for all of the considered boundary value problems, it is possible to express the quantities $u_0^{(s)}$, $v_0^{(s)}$, $w_0^{(s)}$ through $\sigma_{\alpha\gamma 0}^{(s)}$, $\sigma_{\beta\gamma 0}^{(s)}$, $\sigma_{\gamma\gamma 0}^{(s)}$ as

$$u_0^{(s)} = A_{53}\sigma_{\gamma\gamma 0}^{(s)} + A_{54}\,\sigma_{\beta\gamma 0}^{(s)} + A_{55}\,\sigma_{\alpha\gamma 0}^{(s)} - u_*^{(s)}(\zeta = -1) + u^{-(s)},$$

$$v_0^{(s)} = A_{43}\sigma_{\gamma\gamma 0}^{(s)} + A_{44}\,\sigma_{\beta\gamma 0}^{(s)} + A_{45}\,\sigma_{\alpha\gamma 0}^{(s)} - v_*^{(s)}(\zeta = -1) + v^{-(s)}, \qquad (2.7)$$

$$w_0^{(s)} = A_{33}\sigma_{\gamma\gamma 0}^{(s)} + A_{34}\,\sigma_{\beta\gamma 0}^{(s)} + A_{35}\,\sigma_{\alpha\gamma 0}^{(s)} - w_*^{(s)}(\zeta = -1) + w^{-(s)},$$

where

$$u^{-(0)} = u^-/a, \; v^{-(0)} = v^-/a, \; w^{-(0)} = w^-$$

$$u^{-(s)} = v^{-(s)} = w^{-(s)} = 0 \qquad \text{at} \qquad s \neq 0.$$

Substituting (2.7) into (2.6), we obtain

$$u^{(s)}(\xi,\eta,\zeta) = (1 + \zeta)(A_{53}\sigma_{\gamma\gamma 0}^{(s)} + A_{54}\,\sigma_{\beta\gamma 0}^{(s)} + A_{55}\,\sigma_{\alpha\gamma 0}^{(s)})$$

$$+u^{-(s)} - u_*^{(s)}(\zeta = -1) + u_*^{(s)}(\xi,\eta,\zeta),$$

$$v^{(s)}(\xi,\eta,\zeta) = (1 + \zeta)(A_{43}\sigma_{\gamma\gamma 0}^{(s)} + A_{44}\,\sigma_{\beta\gamma 0}^{(s)} + A_{45}\,\sigma_{\alpha\gamma 0}^{(s)}) \qquad (2.8)$$

$$+v^{-(s)} - v_*^{(s)}(\zeta = -1) + v_*^{(s)}(\xi,\eta,\zeta),$$

$$w^{(s)}(\xi,\eta,\zeta) = (1 + \zeta)(A_{33}\sigma_{\gamma\gamma 0}^{(s)} + A_{34}\,\sigma_{\beta\gamma 0}^{(s)} + A_{35}\,\sigma_{\alpha\gamma 0}^{(s)})$$

$$+w^{-(s)} - w_*^{(s)}(\zeta = -1) + w_*^{(s)}(\xi,\eta,\zeta).$$

The unknown functions $\sigma_{\alpha\gamma 0}^{(s)}$, $\sigma_{\beta\gamma 0}^{(s)}$, $\sigma_{\gamma\gamma 0}^{(s)}$ are determined from the conditions on the opposite face $\gamma = h$. The solutions for all of the considered types of boundary value problems are described below.

6.3 First Boundary Value Problem

Let us begin with the first boundary value problem at $\gamma = h$, i.e. conditions (1.4). Using the first three relations of (2.6), we obtain from (1.4)

$$\sigma_{\alpha\gamma 0}^{(s)}(\xi,\eta) = \sigma_{\alpha\gamma}^{+(s)} - \sigma_{\alpha\gamma*}^{(s)}(\zeta = 1),$$
$$\sigma_{\beta\gamma 0}^{(s)}(\xi,\eta) = \sigma_{\beta\gamma}^{+(s)} - \sigma_{\beta\gamma*}^{(s)}(\zeta = 1), \qquad (3.1)$$
$$\sigma_{\gamma\gamma 0}^{(s)}(\xi,\eta) = \sigma_{\gamma\gamma}^{+(s)} - \sigma_{\gamma\gamma*}^{(s)}(\zeta = 1),$$

where

$$\sigma_{\alpha\gamma}^{+(0)} = \sigma_{\alpha\gamma}^{+}, \ \sigma_{\beta\gamma}^{+(0)} = \sigma_{\beta\gamma}^{+}, \ \sigma_{\gamma\gamma}^{+(0)} = \sigma_{\gamma\gamma}^{+}, \ \sigma_{\alpha\gamma}^{+(s)} = \sigma_{\beta\gamma}^{+(s)} = \sigma_{\gamma\gamma}^{+(s)} = 0 \text{ for } s \neq 0.$$

If the surface loading is specified in the form (2.4), i.e. $\sigma_{\alpha\gamma}^{+(s)}, \sigma_{\beta\gamma}^{+(s)}, \sigma_{\gamma\gamma}^{+(s)} \neq 0$, then these quantities enter (3.1) as $Q^{+(s)}$ at higher orders $s > 0$. The solution of (1.1)-(1.4) is then given by (2.2)-(2.3) with

$$\sigma_{\alpha\gamma}^{(s)} = \sigma_{\alpha\gamma}^{+(s)} - \sigma_{\alpha\gamma*}^{(s)}(\zeta = 1) + \sigma_{\alpha\gamma*}^{(s)}(\xi,\eta,\zeta), \qquad (\alpha,\beta),$$

$$\sigma_{\gamma\gamma}^{(s)} = \sigma_{\gamma\gamma}^{+(s)} - \sigma_{\gamma\gamma*}^{(s)}(\zeta = 1) + \sigma_{\gamma\gamma*}^{(s)}(\xi,\eta,\zeta),$$

$$\sigma_{\alpha\alpha}^{(s)} = A_{13}(\sigma_{\gamma\gamma}^{+(s)} - \sigma_{\gamma\gamma*}^{(s)}(\zeta = 1)) + A_{14}(\sigma_{\beta\gamma}^{+(s)} - \sigma_{\beta\gamma*}^{(s)}(\zeta = 1))$$
$$+ A_{15}(\sigma_{\alpha\gamma}^{+(s)} - \sigma_{\alpha\gamma*}^{(s)}(\zeta = 1)) + \sigma_{\alpha\alpha*}^{(s)}(\xi,\eta,\zeta),$$

$$\sigma_{\beta\beta}^{(s)} = A_{23}(\sigma_{\gamma\gamma}^{+(s)} - \sigma_{\gamma\gamma*}^{(s)}(\zeta = 1)) + A_{24}(\sigma_{\beta\gamma}^{+(s)} - \sigma_{\beta\gamma*}^{(s)}(\zeta = 1))$$
$$+ A_{25}(\sigma_{\alpha\gamma}^{+(s)} - \sigma_{\alpha\gamma*}^{(s)}(\zeta = 1)) + \sigma_{\beta\beta*}^{(s)}(\xi,\eta,\zeta),$$

$$\sigma_{\alpha\beta}^{(s)} = A_{63}(\sigma_{\gamma\gamma}^{+(s)} - \sigma_{\gamma\gamma*}^{(s)}(\zeta = 1)) + A_{64}(\sigma_{\beta\gamma}^{+(s)} - \sigma_{\beta\gamma*}^{(s)}(\zeta = 1)) \qquad (3.2)$$
$$+ A_{65}(\sigma_{\alpha\gamma}^{+(s)} - \sigma_{\alpha\gamma*}^{(s)}(\zeta = 1)) + \sigma_{\alpha\beta*}^{(s)}(\xi,\eta,\zeta),$$

$$u^{(s)}(\xi,\eta,\zeta) = (1+\zeta)(A_{53}(\sigma_{\gamma\gamma}^{+(s)} - \sigma_{\gamma\gamma*}^{(s)}(\zeta = 1))$$
$$+ A_{54}(\sigma_{\beta\gamma}^{+(s)} - \sigma_{\beta\gamma*}^{(s)}(\zeta = 1))$$
$$+ A_{55}(\sigma_{\alpha\gamma}^{+(s)} - \sigma_{\alpha\gamma*}^{(s)}(\zeta = 1))) + u^{-(s)} - u_{*}^{(s)}(\zeta = -1) + u_{*}^{(s)}(\xi,\eta,\zeta),$$

$$v^{(s)}(\xi,\eta,\zeta) = (1+\zeta)(A_{43}(\sigma_{\gamma\gamma}^{+(s)} - \sigma_{\gamma\gamma*}^{(s)}(\zeta = 1))$$
$$+ A_{44}(\sigma_{\beta\gamma}^{+(s)} - \sigma_{\beta\gamma*}^{(s)}(\zeta = 1))$$
$$+ A_{45}(\sigma_{\alpha\gamma}^{+(s)} - \sigma_{\alpha\gamma*}^{(s)}(\zeta = 1))) + v^{-(s)} - v_{*}^{(s)}(\zeta = -1) + v_{*}^{(s)}(\xi,\eta,\zeta),$$

$$w^{(s)}(\xi,\eta,\zeta) = (1+\zeta)(A_{33}(\sigma_{\gamma\gamma}^{+(s)} - \sigma_{\gamma\gamma*}^{(s)}(\zeta = 1))$$
$$+ A_{34}(\sigma_{\beta\gamma}^{+(s)} - \sigma_{\beta\gamma*}^{(s)}(\zeta = 1))$$
$$+ A_{35}(\sigma_{\alpha\gamma}^{+(s)} - \sigma_{\alpha\gamma*}^{(s)}(\zeta = 1))) + w^{-(s)} - w_{*}^{(s)}(\zeta = -1) + w_{*}^{(s)}(\xi,\eta,\zeta),$$

where $\sigma_{\alpha\gamma}^{+(0)} = \sigma_{\alpha\gamma}^{+}$, $\sigma_{\beta\gamma}^{+(0)} = \sigma_{\beta\gamma}^{+}$, $\sigma_{\gamma\gamma}^{+(0)} = \sigma_{\gamma\gamma}^{+}$, $u^{-(0)} = u^{-}/a$ (u,v,w). In case of $s \neq 0$ the quantities $\sigma_{\alpha\gamma}^{+(s)} = \sigma_{\beta\gamma}^{+(s)} = \sigma_{\gamma\gamma}^{+(s)} = u^{-(s)} = v^{-(s)} = w^{-(s)} = 0$. If the boundary values are imposed in the form (2.4), then $Q^{+(s)} \neq 0$ and coincides with the coefficient of (2.4). The quantities with asterisks are determined from (2.6).

For rectangular plates $A = B = 1$, $k_\alpha = k_\beta = 0$. It is worth noting that in case of rectangular anisotropy and the boundary functions of (1.3)-(1.7) specified as algebraic polynomials of α, β, the iterative process ends after finite number of steps (exceeding by one the highest order of the polynomials), and the result is an exact solution of a spatial problem for anisotropic layer.

Consider now several particular cases.

a) Let one of the face surfaces ($\gamma = -h$) of anisotropic plates be fixed, whereas on the other face ($\gamma = h$) the normal and tangential components of the loading have constant intensity

$$u_\alpha(-h) = u_\beta(-h) = u_\gamma(-h) = 0,$$

$$\sigma_{\alpha\gamma}(h) = \sigma_{\alpha\gamma}^+ = \text{const}, \quad \sigma_{\beta\gamma}(h) = \sigma_{\beta\gamma}^+ = \text{const}, \tag{3.3}$$

$$\sigma_{\gamma\gamma}(h) = \sigma_{\gamma\gamma}^+ = \text{const}.$$

Then the iterative process stops at leading order. Using (2.2), (2.6), and (3.2) we result in

$$\begin{aligned}
\sigma_{\alpha\alpha} &= A_{13}\sigma_{\gamma\gamma}^+ + A_{14}\sigma_{\beta\gamma}^+ + A_{15}\sigma_{\alpha\gamma}^+, \\
\sigma_{\beta\beta} &= A_{23}\sigma_{\gamma\gamma}^+ + A_{24}\sigma_{\beta\gamma}^+ + A_{25}\sigma_{\alpha\gamma}^+, \\
\sigma_{\alpha\beta} &= A_{63}\sigma_{\gamma\gamma}^+ + A_{64}\sigma_{\beta\gamma}^+ + A_{65}\sigma_{\alpha\gamma}^+, \\
\sigma_{\alpha\gamma} &= \sigma_{\alpha\gamma}^+, \quad \sigma_{\beta\gamma} = \sigma_{\beta\gamma}^+, \quad \sigma_{\gamma\gamma} = \sigma_{\gamma\gamma}^+, \\
u_\alpha &= (\gamma + h)(A_{53}\sigma_{\gamma\gamma}^+ + A_{54}\sigma_{\beta\gamma}^+ + A_{55}\sigma_{\alpha\gamma}^+), \\
u_\beta &= (\gamma + h)(A_{43}\sigma_{\gamma\gamma}^+ + A_{44}\sigma_{\beta\gamma}^+ + A_{45}\sigma_{\alpha\gamma}^+), \\
u_\gamma &= (\gamma + h)(A_{33}\sigma_{\gamma\gamma}^+ + A_{34}\sigma_{\beta\gamma}^+ + A_{35}\sigma_{\alpha\gamma}^+).
\end{aligned} \tag{3.4}$$

For orthotropic plates the solution takes simpler form

$$\begin{aligned}
\sigma_{\alpha\alpha} &= A_{13}\sigma_{\gamma\gamma}^+, \quad \sigma_{\beta\beta} = A_{23}\sigma_{\gamma\gamma}^+, \quad \sigma_{\alpha\beta} = 0, \\
\sigma_{\alpha\gamma} &= \sigma_{\alpha\gamma}^+, \quad \sigma_{\beta\gamma} = \sigma_{\beta\gamma}^+, \quad \sigma_{\gamma\gamma} = \sigma_{\gamma\gamma}^+, \\
u_\alpha &= (\gamma + h)A_{55}\sigma_{\alpha\gamma}^+, \quad u_\beta = (\gamma + h)A_{44}\sigma_{\beta\gamma}^+, \\
u_\gamma &= (\gamma + h)A_{33}\sigma_{\gamma\gamma}^+.
\end{aligned} \tag{3.5}$$

b) The face surface $\gamma = -h$ is fixed, and the opposite face $\gamma = h$ is loaded by normal component depending linearly on α and β

$$u_\alpha(-h) = u_\beta(-h) = u_\gamma(-h) = 0, \tag{3.6}$$

$$\sigma_{\alpha\gamma}(h) = \sigma_{\beta\gamma}(h) = 0, \quad \sigma_{\gamma\gamma}(h) = b\alpha + c\beta.$$

For the sake of brevity here we consider orthotropic plate, however, general anisotropy may also be considered. Due to linearity of loading the iterative procedure breaks after two steps, leading to the resulting solution in the form

$$\sigma_{\alpha\alpha} = A_{13}(b\alpha + c\beta), \quad \sigma_{\beta\beta} = A_{23}(b\alpha + c\beta), \quad \sigma_{\alpha\beta} = 0,$$

$$\sigma_{\gamma\gamma} = b\alpha + c\beta, \quad \sigma_{\beta\gamma} = A_{23}c(h - \gamma),$$

$$\sigma_{\alpha\gamma} = A_{13}b(h - \gamma), \tag{3.7}$$

$$u_\alpha = \frac{1}{2}A_{13}a_{55}b(3h^2 + 2\gamma h - \gamma^2) - \frac{1}{2}A_{33}b(\gamma + h)^2,$$

$$u_\beta = \frac{1}{2}A_{23}a_{44}c(3h^2 + 2\gamma h - \gamma^2) - \frac{1}{2}A_{33}c(\gamma + h)^2,$$

$$u_\gamma = A_{33}(\gamma + h)(b\alpha + c\beta).$$

c) The face $\gamma = -h$ is fixed, with quadratic (in α and β) normal loading prescribed on the opposite face $\gamma = h$

$$u_\alpha(-h) = u_\beta(-h) = u_\gamma(-h) = 0, \tag{3.8}$$

$$\sigma_{\alpha\gamma}(h) = \sigma_{\beta\gamma}(h) = 0, \quad \sigma_{\gamma\gamma}(h) = b\alpha^2 + c\beta^2 + 2d\alpha\beta.$$

The solution for orthotropic plate is given by

$$\sigma_{\alpha\gamma} = 2A_{13}(h - \gamma)(b\alpha + d\beta), \quad \sigma_{\beta\gamma} = 2A_{23}(h - \gamma)(c\beta + d\alpha),$$

$$\sigma_{\gamma\gamma} = b\alpha^2 + c\beta^2 + 2d\alpha\beta + (h - \gamma)^2(A_{13}b + A_{23}c),$$

$$\sigma_{\alpha\alpha} = A_{13}(b\alpha^2 + c\beta^2 + 2d\alpha\beta) + A_{13}(h - \gamma)^2(A_{13}b + A_{23}c)$$

$$+((ba_{22}a_{55}A_{13} - ca_{12}a_{44}A_{23})(3h^2 + 2\gamma h - \gamma^2)$$

$$+A_{33}(ca_{12} - ba_{22})(\gamma + h)^2)/(a_{11}a_{22} - a_{12}^2),$$

$$\sigma_{\beta\beta} = A_{23}(b\alpha^2 + c\beta^2 + 2d\alpha\beta) + A_{23}(h - \gamma)^2(A_{13}b + A_{23}c) \tag{3.9}$$

$$+((3h^2 + 2\gamma h - \gamma^2)(ca_{11}a_{44}A_{23} - ba_{12}a_{55}A_{13})$$

$$+A_{33}(\gamma + h)^2(ba_{12} - ca_{11}))/(a_{11}a_{22} - a_{12}^2),$$

$$\sigma_{\alpha\beta} = d\left((a_{55}A_{13} + a_{44}A_{23})(3h^2 + 2\gamma h - \gamma^2) - A_{33}(\gamma + h)^2\right)/a_{66},$$

$$u_\alpha = (b\alpha + d\beta)\left(A_{13}a_{55}(3h^2 + 2\gamma h - \gamma^2) - A_{33}(\gamma + h)^2\right),$$

$$u_\beta = (d\alpha + c\beta)\left(A_{23}a_{44}(3h^2 + 2\gamma h - \gamma^2) - A_{33}(\gamma + h)^2\right),$$

$$u_\gamma = A_{33}(\gamma + h)(b\alpha^2 + c\beta^2 + 2d\alpha\beta) + \frac{2}{3}A_{33}(A_{13}b + A_{23}c)$$

$$\times(\gamma^3 + 3h^2\gamma + 4h^3) + \frac{1}{3}\left(ba_{55}A_{13}^2 + ca_{44}A_{23}^2\right)(\gamma^3 - 3h\gamma^2 - 9h^2\gamma - 5h^3).$$

The cases of other polynomials may be considered in a similar manner. Therefore, in view of the Weierstrass theorem, an arbitrary loading function may be approximated by appropriate polynomials, leading to an exact solution of the corresponding spatial problem.

Comparison of (3.5), (3.7), and (3.9) reveals the influence of the variation of the loading function on the solution, causing additional terms in both stresses and displacements. It may be observed that the major components are those arising at leading order approximation, with the others being of secondary importance. For example, in case of (3.7) the main are the stress components $\sigma_{\gamma\gamma}, \sigma_{\alpha\alpha}, \sigma_{\beta\beta}$ with the components $\sigma_{\alpha\gamma}, \sigma_{\beta\gamma}$ of the next asymptotic order. In (3.9) in expression for $\sigma_{\gamma\gamma}$ the first three terms are main, and the last term caused by variation of loading function is less by two orders, which could be readily confirmed by rewriting the results in dimensionless variables $\xi = \alpha/a, \ \eta = \beta/a, \ \zeta = \gamma/h$, namely

$$\sigma_{\gamma\gamma} = \sigma_{\gamma\gamma}^{(0)} + \varepsilon\sigma_{\gamma\gamma}^{(1)} + \varepsilon^2\sigma_{\gamma\gamma}^{(2)},$$

$$\sigma_{\gamma\gamma}^{(0)} = a^2(b\xi^2 + c\eta^2 + 2d\xi\eta), \tag{3.10}$$

$$\sigma_{\gamma\gamma}^{(1)} = 0, \quad \sigma_{\gamma\gamma}^{(2)} = a^2(1-\zeta)^2(A_{13}b + A_{23}c).$$

The expression for the normal displacement u_γ is given by

$$u_\gamma = a\varepsilon w^{(0)} + a\varepsilon^2 w^{(1)} + a\varepsilon^3 w^{(2)},$$

$$w^{(0)} = a^2 A_{33}(1+\zeta)(b\xi^2 + c\eta^2 + 2d\xi\eta), \quad w^{(1)} = 0, \tag{3.11}$$

$$w^{(2)} = \frac{2}{3}a^2 A_{33}(A_{13}b + A_{23}c)(\zeta^3 + 3\zeta + 4)$$

$$+\frac{1}{3}a^2(a_{55}A_{13}^2 b + a_{44}A_{23}^2 c)(\zeta^3 - 3\zeta^2 - 9\zeta - 5),$$

which confirms the above-stated.

It may be deduced from the presented exact solutions, that the normal stress $\sigma_{\gamma\gamma}$ is usually of the same order as the tangential stresses $\sigma_{\alpha\alpha}, \sigma_{\beta\beta}$. It confirms once again the asymptotic equivalence of stress components for the considered boundary value problems, following from (2.2), (2.3). Note that in classical problems for plates the stress $\sigma_{\gamma\gamma}$ is less by two orders than the tangential stresses.

6.4　Second Boundary Value Problem

Consider now the face boundary conditions (1.3), (1.5) at $\gamma = \pm h$. Using (2.8) in order to satisfy the latter, we have

$$\sigma_{\gamma\gamma 0}^{(s)} = \left(B_{35}^* V_\alpha^{(s)} + B_{34}^* V_\beta^{(s)} + B_{33}^* V_\gamma^{(s)} \right)/\Delta_*,$$

$$\sigma_{\beta\gamma 0}^{(s)} = \left(B_{45}^* V_\alpha^{(s)} + B_{44}^* V_\beta^{(s)} + B_{43}^* V_\gamma^{(s)} \right)/\Delta_*, \tag{4.1}$$

$$\sigma_{\alpha\gamma 0}^{(s)} = \left(B_{55}^* V_\alpha^{(s)} + B_{54}^* V_\beta^{(s)} + B_{53}^* V_\gamma^{(s)} \right)/\Delta_*,$$

where

$$V_\alpha^{(s)}(\xi, \eta) = \frac{1}{2}\left(u_*^{(s)}(\zeta = -1) - u_*^{(s)}(\zeta = +1) + u^{+(s)} - u^{-(s)} \right),$$

$$(\alpha, \beta, \gamma; u, v, w)$$

$$u^{\pm(0)} = u^\pm/a, \quad u^{\pm(s)} = 0, \quad s \neq 0, \quad (u, v, w),$$

$$B_{ij}^* = A_{kk}A_{ij} - A_{ik}A_{kj}, \quad B_{ij}^* \neq B_{ji}^*, \tag{4.2}$$

$$B_{kk}^* = A_{ij}A_{ji} - A_{ii}A_{jj}, \quad (i,j,k = 3,4,5; i \neq j \neq k \neq i),$$

$$\Delta_* = A_{33}A_{45}A_{54} + A_{44}A_{35}A_{53} + A_{55}A_{34}A_{43}$$

$$-A_{33}A_{44}A_{55} - A_{34}A_{45}A_{53} - A_{35}A_{54}A_{43}.$$

The solution of the boundary value problem (1.1)-(1.3), (1.5) is given by expressions (2.2), (2.3), (2.5), (2.8), and (4.1), with the coefficients of (2.2) written as

$$\sigma_{\gamma\gamma}^{(s)} = \left(B_{35}^* V_\alpha^{(s)} + B_{34}^* V_\beta^{(s)} + B_{33}^* V_\gamma^{(s)}\right)/\Delta_* + \sigma_{\gamma\gamma*}^{(s)}(\xi,\eta,\zeta),$$

$$\sigma_{\beta\gamma}^{(s)} = \left(B_{45}^* V_\alpha^{(s)} + B_{44}^* V_\beta^{(s)} + B_{43}^* V_\gamma^{(s)}\right)/\Delta_* + \sigma_{\beta\gamma*}^{(s)}(\xi,\eta,\zeta),$$

$$\sigma_{\alpha\gamma}^{(s)} = \left(B_{55}^* V_\alpha^{(s)} + B_{54}^* V_\beta^{(s)} + B_{53}^* V_\gamma^{(s)}\right)/\Delta_* + \sigma_{\alpha\gamma*}^{(s)}(\xi,\eta,\zeta),$$

$$\sigma_{\alpha\alpha}^{(s)} = A_{15}^* V_\alpha^{(s)} + A_{14}^* V_\beta^{(s)} + A_{13}^* V_\gamma^{(s)} + \sigma_{\alpha\alpha*}^{(s)}(\xi,\eta,\zeta), \qquad (4.3)$$

$$\sigma_{\beta\beta}^{(s)} = A_{25}^* V_\alpha^{(s)} + A_{24}^* V_\beta^{(s)} + A_{23}^* V_\gamma^{(s)} + \sigma_{\beta\beta*}^{(s)}(\xi,\eta,\zeta),$$

$$\sigma_{\alpha\beta}^{(s)} = A_{65}^* V_\alpha^{(s)} + A_{64}^* V_\beta^{(s)} + A_{63}^* V_\gamma^{(s)} + \sigma_{\alpha\beta*}^{(s)}(\xi,\eta,\zeta),$$

$$u^{(s)} = (1+\zeta)V_\alpha^{(s)} + u^{-(s)} - u_*^{(s)}(\zeta=-1) + u_*^{(s)}(\xi,\eta,\zeta),$$

$$v^{(s)} = (1+\zeta)V_\beta^{(s)} + v^{-(s)} - v_*^{(s)}(\zeta=-1) + v_*^{(s)}(\xi,\eta,\zeta),$$

$$w^{(s)} = (1+\zeta)V_\gamma^{(s)} + w^{-(s)} - w_*^{(s)}(\zeta=-1) + w_*^{(s)}(\xi,\eta,\zeta),$$

$$A_{ik}^* = (A_{i3}B_{3k}^* + A_{i4}B_{4k}^* + A_{i5}B_{5k}^*)/\Delta_*, \quad (i=1,2,6; k=3,4,5)$$

with $\sigma_{\alpha\alpha*}^{(s)}, \sigma_{\alpha\gamma*}^{(s)}, \sigma_{\alpha\beta*}^{(s)}, u_*^{(s)}$ $(\alpha,\beta,\gamma; u,v,w)$ defined through formulae (2.6).

As in the previous section, in case of rectangular anisotropy provided that the loading functions are polynomials of α, β, the iterative procedure involves a finite number of steps giving the exact solution of the spatial problem.

Similarly to the previous section, let us consider several particular cases.

a) Let the face surface $\gamma = -h$ be fixed, whereas a constant displacement is prescribed on the face $(\gamma = h)$

$$u_\alpha(-h) = u_\beta(-h) = u_\gamma(-h) = 0,$$

$$u_\alpha(h) = u^+ = const, \; u_\beta(h) = v^+ = const, \qquad (4.4)$$

$$u_\gamma(h) = w^+ = const.$$

The iteration then stops at leading order, giving

$$\sigma_{\gamma\gamma} = (B_{35}^* u^+ + B_{34}^* v^+ + B_{33}^* w^+)/(2h\Delta_*),$$

$$\sigma_{\beta\gamma} = (B_{45}^* u^+ + B_{44}^* v^+ + B_{43}^* w^+)/(2h\Delta_*),$$

$$\sigma_{\alpha\gamma} = (B_{55}^* u^+ + B_{54}^* v^+ + B_{53}^* w^+)/(2h\Delta_*), \quad (4.5)$$

$$\sigma_{\alpha\alpha} = (A_{15}^* u^+ + A_{14}^* v^+ + A_{13}^* w^+)/(2h\Delta_*),$$

$$\sigma_{\beta\beta} = (A_{25}^* u^+ + A_{24}^* v^+ + A_{23}^* w^+)/(2h\Delta_*),$$

$$u_\alpha = (h+\gamma)u^+/(2h), \; u_\beta = (h+\gamma)v^+/(2h), \; u_\gamma = (h+\gamma)w^+/(2h).$$

In case of orthotropic plate the solution may be simplified to the form

$$\sigma_{\gamma\gamma} = \frac{w^+}{2hA_{33}}, \; \sigma_{\beta\gamma} = \frac{v^+}{2ha_{44}}, \; \sigma_{\alpha\gamma} = \frac{u^+}{2ha_{55}},$$

$$\sigma_{\alpha\alpha} = \frac{A_{13}w^+}{2hA_{33}}, \; \sigma_{\beta\beta} = \frac{A_{23}w^+}{2hA_{33}}, \; \sigma_{\alpha\beta} = 0, \quad (4.6)$$

$$u_\alpha = (h+\gamma)u^+/(2h), \; u_\beta = (h+\gamma)v^+/(2h), \; u_\gamma = (h+\gamma)w^+/(2h).$$

b) In case of a fixed face $\gamma = -h$, with the normal displacement proportional to α and β imposed on the opposite face $\gamma = h$

$$u_\alpha(\pm h) = 0, \quad u_\beta(\pm h) = 0, \tag{4.7}$$
$$u_\gamma(-h) = 0, \quad u_\gamma(h) = b\alpha + c\beta$$

the iterative procedure involves two steps only, leading to the following solution for orthotropic plates

$$\sigma_{\gamma\gamma} = \frac{1}{2hA_{33}}(b\alpha + c\beta),$$
$$\sigma_{\beta\gamma} = \frac{c}{2ha_{44}A_{33}}(A_{33}h - a_{44}A_{23}\gamma),$$
$$\sigma_{\alpha\gamma} = \frac{b}{2ha_{55}A_{33}}(A_{33}h - a_{55}A_{13}\gamma), \quad \sigma_{\alpha\beta} = 0,$$
$$\sigma_{\alpha\alpha} = \frac{A_{13}}{2hA_{33}}(b\alpha + c\beta), \quad \sigma_{\beta\beta} = \frac{A_{23}}{2hA_{33}}(b\alpha + c\beta), \tag{4.8}$$
$$u_\alpha = b(h^2 - \gamma^2)(A_{55}A_{13} + A_{33})/(4hA_{33}),$$
$$u_\beta = c(h^2 - \gamma^2)(A_{44}A_{23} + A_{33})/(4hA_{33}),$$
$$u_\gamma = (h + \gamma)(b\alpha + c\beta)/(2h).$$

6.5   Mixed Boundary Value Problems

Consider now the boundary conditions on the face $\gamma = -h$ prescribed in the form (1.3), with that at the opposite face $\gamma = h$ assumed in the form (1.6). Using (2.5), and (2.8), from satisfying the face boundary conditions, we obtain

$$\sigma_{\gamma\gamma 0}^{(s)} = \sigma_{\gamma\gamma}^{+(s)} - \sigma_{\gamma\gamma*}^{(s)}(\zeta = 1),$$
$$\sigma_{\beta\gamma 0}^{(s)} = (B_{43}^*(\sigma_{\gamma\gamma}^{+(s)} - \sigma_{\gamma\gamma*}^{(s)}(\zeta = 1)) + A_{45}V_\alpha^{(s)} - A_{55}V_\beta^{(s)})/B_{33}^*, \tag{5.1}$$
$$\sigma_{\alpha\gamma 0}^{(s)} = (B_{53}^*(\sigma_{\gamma\gamma}^{+(s)} - \sigma_{\gamma\gamma*}^{(s)}(\zeta = 1)) + A_{54}V_\beta^{(s)} - A_{44}V_\alpha^{(s)})/B_{33}^*,$$

where $\sigma_{\gamma\gamma}^{+(0)} = \sigma_{\gamma\gamma}^+$, $\sigma_{\gamma\gamma}^{+(s)} = 0$ at $s \neq 0$, with the rest of the quantities involved in (5.1) determined by (4.2). Solution of the boundary value problem (1.1)-(1.3), (1.6) is given by expressions (2.2), (2.3), (2.5), (2.8), and (5.1), with the appropriate coefficients of (2.2) taking the form

$$\sigma_{\alpha\gamma}^{(s)} = (B_{53}^*(\sigma_{\gamma\gamma}^{+(s)} - \sigma_{\gamma\gamma*}^{(s)}(\zeta = 1)) - A_{44}V_\alpha^{(s)} + A_{54}V_\beta^{(s)})/B_{33}^* + \sigma_{\alpha\gamma*}^{(s)}(\xi, \eta, \zeta),$$
$$\sigma_{\beta\gamma}^{(s)} = (B_{43}^*(\sigma_{\gamma\gamma}^{+(s)} - \sigma_{\gamma\gamma*}^{(s)}(\zeta = 1)) + A_{45}V_\alpha^{(s)} - A_{55}V_\beta^{(s)})/B_{33}^* + \sigma_{\beta\gamma*}^{(s)}(\xi, \eta, \zeta),$$
$$\sigma_{\gamma\gamma}^{(s)} = \sigma_{\gamma\gamma}^{+(s)} - \sigma_{\gamma\gamma*}^{(s)}(\zeta = 1) + \sigma_{\gamma\gamma*}^{(s)}(\xi, \eta, \zeta),$$
$$\sigma_{\alpha\alpha}^{(s)} = (A_{13}^*(\sigma_{\gamma\gamma}^{+(s)} - \sigma_{\gamma\gamma*}^{(s)}(\zeta = 1)) - B_{15}^*V_\alpha^{(s)} - B_{14}^*V_\beta^{(s)})/B_{33}^* + \sigma_{\alpha\alpha*}^{(s)}(\xi, \eta, \zeta),$$
$$\sigma_{\beta\beta}^{(s)} = (A_{23}^*(\sigma_{\gamma\gamma}^{+(s)} - \sigma_{\gamma\gamma*}^{(s)}(\zeta = 1)) - B_{25}^*V_\alpha^{(s)} - B_{24}^*V_\beta^{(s)})/B_{33}^* + \sigma_{\beta\beta*}^{(s)}(\xi, \eta, \zeta),$$
$$\sigma_{\alpha\beta}^{(s)} = (A_{63}^*(\sigma_{\gamma\gamma}^{+(s)} - \sigma_{\gamma\gamma*}^{(s)}(\zeta = 1)) - B_{65}^*V_\alpha^{(s)} - B_{64}^*V_\beta^{(s)})/B_{33}^* + \sigma_{\alpha\beta*}^{(s)}(\xi, \eta, \zeta),$$

$$u^{(s)} = (1+\zeta)V_\alpha^{(s)} + u^{-(s)} - u_*^{(s)}(\zeta = -1) + u_*^{(s)}(\xi, \eta, \zeta),$$

$$v^{(s)} = (1+\zeta)V_\alpha^{(s)} + v^{-(s)} - v_*^{(s)}(\zeta = -1) + v_*^{(s)}(\xi, \eta, \zeta), \qquad (5.2)$$

$$w^{(s)} = (1+\zeta)(A_{33}^*(\sigma_{\gamma\gamma}^{+(s)} - \sigma_{\gamma\gamma*}^{(s)}(\zeta = 1)) - B_{35}^* V_\alpha^{(s)} - B_{34}^* V_\beta^{(s)})/B_{33}^*$$

$$+ w^{-(s)} - w_*^{(s)}(\zeta = -1) + w_*^{(s)}(\xi, \eta, \zeta),$$

$$A_{i3}^* = A_{i3}B_{33}^* + A_{i4}B_{43}^* + A_{i5}B_{53}^*, \qquad (i = 1, 2, ..., 6).$$

Similarly to the previous two sections, the asymptotic procedure involves finite number of approximations in case of polynomial loading. Let us present the solutions of several particular problems.

a) the face $\gamma = -h$ is fixed, and the opposite face $\gamma = h$ is loaded by normal force of constant intensity

$$u_\alpha(\pm h) = u_\beta(\pm h) = u_\gamma(-h) = 0, \qquad (5.3)$$

$$\sigma_{\gamma\gamma}(h) = \sigma_{\gamma\gamma}^+ = const.$$

The solution is given by

$$\sigma_{\alpha\gamma} = B_{53}^*\sigma_{\gamma\gamma}^+, \qquad \sigma_{\beta\gamma} = B_{43}^*\sigma_{\gamma\gamma}^+, \qquad \sigma_{\gamma\gamma} = \sigma_{\gamma\gamma}^+,$$

$$\sigma_{\alpha\alpha} = \frac{A_{13}^*}{B_{33}^*}\sigma_{\gamma\gamma}^+, \qquad \sigma_{\beta\beta} = \frac{A_{23}^*}{B_{33}^*}\sigma_{\gamma\gamma}^+, \qquad \sigma_{\alpha\beta} = \frac{A_{63}^*}{B_{33}^*}\sigma_{\gamma\gamma}^+, \qquad (5.4)$$

$$u_\alpha = u_\beta = 0, \qquad u_\gamma = \frac{A_{33}^*}{B_{33}^*}(h+\gamma)\sigma_{\gamma\gamma}^+.$$

In case of orthotropic layer the latter reduces to

$$\sigma_{\alpha\gamma} = \sigma_{\beta\gamma} = 0, \qquad \sigma_{\gamma\gamma} = \sigma_{\gamma\gamma}^+,$$

$$\sigma_{\alpha\alpha} = A_{13}\sigma_{\gamma\gamma}^+, \qquad \sigma_{\beta\beta} = A_{23}\sigma_{\gamma\gamma}^+, \qquad \sigma_{\alpha\beta} = 0, \qquad (5.5)$$

$$u_\alpha = u_\beta = 0, \qquad u_\gamma = A_{33}(h+\gamma)\sigma_{\gamma\gamma}^+.$$

b) consider fixed face boundary conditions at $\gamma = -h$, with normal loading at $\gamma = h$ proportional to α and β

$$u_\alpha(\pm h) = u_\beta(\pm h) = u_\gamma(-h) = 0, \qquad (5.6)$$

$$\sigma_{\gamma\gamma}(h) = b\alpha + c\beta.$$

The iterative procedure contains two steps

$$\sigma_{\alpha\alpha} = A_{13}(b\alpha + c\beta), \qquad \sigma_{\beta\beta} = A_{23}(b\alpha + c\beta),$$

$$\sigma_{\alpha\beta} = 0, \qquad \sigma_{\alpha\gamma} = b\left(\frac{A_{33}}{a_{55}}h - \gamma A_{13}\right),$$

$$\sigma_{\beta\gamma} = c\left(\frac{A_{33}}{a_{44}}h - \gamma A_{23}\right), \qquad \sigma_{\gamma\gamma} = b\alpha + c\beta, \qquad (5.7)$$

$$u_\alpha = \frac{1}{2}b(h^2 - \gamma^2)(A_{33} + a_{55}A_{13}),$$

$$u_\beta = \frac{1}{2}c(h^2 - \gamma^2)(A_{33} + a_{44}A_{23}),$$

$$u_\gamma = A_{33}(h+\gamma)(b\alpha + c\beta).$$

Comparison of (5.5) and (5.7) reveals that the tangential displacements may arise due to variation of normal loading. We remarks that these displacements may be significant in plates with small shear stiffness in transverse direction.

Consider now one more mixed problem, which possesses important practical applications. Let us impose conditions (1.3) at $\gamma = -h$, along with conditions (1.7) at $\gamma = h$, modeling action of a rigid stamp of a given profile, with friction forces taken into account. Satisfying these boundary conditions and using (2.5), and (2.8), we obtain

$$\sigma_{\alpha\gamma 0}^{(s)} = \sigma_{\alpha\gamma}^{+(s)} - \sigma_{\alpha\gamma *}^{(s)}(\zeta = 1), \quad (\alpha, \beta),$$

$$\sigma_{\gamma\gamma 0}^{(s)} = \frac{1}{A_{33}}(V_\gamma^{(s)} - A_{35}(\sigma_{\alpha\gamma}^{+(s)} - \sigma_{\alpha\gamma *}^{(s)}(\zeta = 1))$$

$$- A_{34}(\sigma_{\beta\gamma}^{+(s)} - \sigma_{\beta\gamma *}^{(s)}(\zeta = 1))), \tag{5.8}$$

$$V_\gamma^{(s)} = \frac{1}{2}(w^{+(s)} - w^{-(s)} + w_*^{(s)}(\zeta = -1) - w_*^{(s)}(\zeta = 1)),$$

$$\sigma_{\alpha\gamma}^{+(0)} = \sigma_{\alpha\gamma}^+, \quad \sigma_{\beta\gamma}^{+(0)} = \sigma_{\beta\gamma}^+, \quad w^{\pm(0)} = w^\pm/a,$$

$$\sigma_{\alpha\gamma}^{+(s)} = \sigma_{\beta\gamma}^{+(s)} = w^{\pm(s)} = 0, \quad s \neq 0.$$

Solution of the mixed boundary value problem (1.1)-(1.3), (1.7) is then given by the formulae (2.2), (2.5), (2.8), with the functions $\sigma_{\alpha\gamma 0}^{(s)}$, $\sigma_{\beta\gamma 0}^{+(s)}$, $\sigma_{\gamma\gamma 0}^{(s)}$ computed through (5.8). Consider now several particular cases.

a) Let the face $\gamma = -h$ be rigidly fixed, with the opposite face $\gamma = h$ free from tangential load (including friction), with a constant normal loading imposed, namely

$$u_\alpha(-h) = u_\beta(-h) = u_\gamma(-h) = 0, \tag{5.9}$$

$$\sigma_{\alpha\gamma}(h) = \sigma_{\beta\gamma}(h) = 0, \quad u_\gamma(h) = w^+ = const.$$

The stress-strain field of such anisotropic plate is then governed by

$$\sigma_{\alpha\gamma} = \sigma_{\beta\gamma} = 0, \quad \sigma_{\gamma\gamma} = w^+/(2hA_{33}),$$

$$\sigma_{\alpha\alpha} = \frac{A_{13}}{A_{33}} \cdot \frac{w^+}{2h}, \quad \sigma_{\beta\beta} = \frac{A_{23}}{A_{33}} \cdot \frac{w^+}{2h}, \tag{5.10}$$

$$\sigma_{\alpha\beta} = \frac{A_{63}}{A_{33}} \cdot \frac{w^+}{2h}, \quad u_\alpha = (h + \gamma)\frac{A_{53}}{A_{33}} \cdot \frac{w^+}{2h},$$

$$u_\beta = (h + \gamma)A_{43}w^+/(2hA_{33}), \quad u_\gamma = (h + \gamma)w^+/(2h).$$

In case of orthotropic plate the latter may be written in simplified form as

$$\sigma_{\alpha\gamma} = \sigma_{\beta\gamma} = 0, \quad \sigma_{\gamma\gamma} = w^+/(2hA_{33}),$$

$$\sigma_{\alpha\alpha} = A_{13}w^+/(2hA_{33}), \quad \sigma_{\beta\beta} = A_{23}w^+/(2hA_{33}), \tag{5.11}$$

$$\sigma_{\alpha\beta} = u_\alpha = u_\beta = 0, \quad u_\gamma = (h + \gamma)w^+/(2h).$$

We note that the solution (5.11) corresponding to the boundary conditions (5.9) coincides with the solution (4.6) of the problem (4.4), if one assumes $u^+ = v^+ = 0$. Thus, the same outer solution may correspond to different boundary value problems;

b) Let the face $\gamma = -h$ be fixed, and the normal loading proportional to the coordinates α and β is prescribed on the opposite face $\gamma = h$ in the absence of tangential loading

$$u_\alpha(-h) = u_\beta(-h) = u_\gamma(-h) = 0, \tag{5.12}$$
$$\sigma_{\alpha\gamma}(h) = \sigma_{\beta\gamma}(h) = 0, \quad u_\gamma(h) = b\alpha + c\beta.$$

The corresponding solution of the spatial problem for orthotropic layer is

$$\sigma_{\alpha\gamma} = bA_{13}(h - \gamma)/(2hA_{33}), \quad \sigma_{\beta\gamma} = cA_{23}(h - \gamma)/(2hA_{33}),$$
$$\sigma_{\gamma\gamma} = (b\alpha + c\beta)/(2hA_{33}), \quad \sigma_{\alpha\alpha} = (b\alpha + c\beta)A_{13}/(2hA_{33}),$$
$$\sigma_{\beta\beta} = (b\alpha + c\beta)A_{23}/(2hA_{33}), \quad \sigma_{\alpha\beta} = 0, \tag{5.13}$$
$$u_\alpha = (3h^2 + 2h\gamma - \gamma^2)ba_{55}A_{13}/(4hA_{33}) - (h + \gamma)^2 b/(4h),$$
$$u_\beta = (3h^2 + 2h\gamma - \gamma^2)ca_{44}A_{23}/(4hA_{33}) - (h + \gamma)^2 c/(4h),$$
$$u_\gamma = (h + \gamma)(b\alpha + c\beta)/(2h).$$

c) Consider fixed face $\gamma = -h$, with the normal load at $\gamma = h$ being quadratic in α and β in the absence of friction, i.e.

$$u_\alpha(-h) = u_\beta(-h) = u_\gamma(-h) = 0, \tag{5.14}$$
$$\sigma_{\alpha\gamma}(h) = \sigma_{\beta\gamma}(h) = 0, \quad u_\gamma(h) = b\alpha^2 + c\beta^2 + 2d\alpha\beta.$$

The iterative asymptotic process contains three steps only, for orthotropic plate resulting in

$$\sigma_{\alpha\gamma} = \frac{A_{13}}{hA_{33}}(h - \gamma)(b\alpha + d\beta), \quad \sigma_{\beta\gamma} = \frac{A_{23}}{hA_{33}}(h - \gamma)(c\beta + \alpha d),$$

$$\sigma_{\gamma\gamma} = \frac{1}{2hA_{33}}(b\alpha^2 + c\beta^2 + 2\alpha\beta d) + \frac{4h}{3A_{33}^2}(A_{55}A_{13}^2 b + A_{44}A_{23}^2 c)$$
$$+ \frac{1}{6hA_{33}}(bA_{13} + cA_{23})(3\gamma^2 - 6\gamma h - 5h^2),$$

$$\sigma_{\alpha\alpha} = \frac{A_{13}}{2hA_{33}}(b\alpha^2 + c\beta^2 + 2\alpha\beta d) + \frac{4hA_{13}}{3A_{33}^2}(A_{13}^2 A_{55}b + A_{23}^2 A_{44}c)$$
$$- \frac{5hA_{13}}{6A_{33}}(A_{13}b + A_{23}c) - \frac{1}{2h}(\gamma + h)^2(B_{11}b + B_{12}c) + \frac{1}{2hA_{33}}(h + \gamma)$$
$$\times (3h - \gamma)(A_{13}A_{55}B_{11}b + A_{23}A_{44}B_{12}c) + \frac{A_{13}}{2hA_{33}}\gamma(\gamma - 2h)(A_{13}b + A_{23}c),$$

$$\sigma_{\beta\beta} = \frac{A_{23}}{2hA_{33}}(b\alpha^2 + c\beta^2 + 2\alpha\beta d) + \frac{4hA_{23}}{3A_{33}^2}(A_{13}^2 A_{55}b + A_{23}^2 A_{44}c)$$
$$- \frac{5hA_{23}}{6A_{33}}(A_{13}b + A_{23}c) - \frac{1}{2h}(\gamma + h)^2(B_{12}b + B_{22}c)$$
$$+ \frac{1}{2h}\frac{A_{23}}{A_{33}}\gamma(\gamma - h)(A_{13}b + A_{23}c),$$

$$\sigma_{\alpha\beta} = \frac{B_{66}d}{2hA_{33}}(h+\gamma)(3h-\gamma)(A_{13}A_{55}+A_{23}A_{44}) - B_{66}(\gamma+h)^2\frac{d}{h},$$

$$u_\alpha = \frac{1}{2hA_{33}}(\gamma+h)((3h-\gamma)A_{13}A_{55} - (\gamma+h)A_{33})(b\alpha+\beta d), \qquad (5.15)$$

$$u_\beta = \frac{1}{2hA_{33}}(\gamma+h)((3h-\gamma)A_{23}A_{44} - (\gamma+h)A_{33})(c\beta+\alpha d),$$

$$u_\gamma = \frac{1}{2h}(h+\gamma)(b\alpha^2+c\beta^2+2\alpha\beta d) + \frac{\gamma}{3h}(\gamma^2-h^2)(A_{13}b+cA_{23})$$

$$+\frac{1}{6hA_{33}}(\gamma^2-h^2)(\gamma-3h)(A_{13}^2A_{55}b+cA_{23}^2A_{44}).$$

6.6 Boundary Layer for Non-Classical Boundary Value Problems

The outer solution has been found above, however, the associated stresses and displacements may or may not match the edge boundary conditions. If the conditions are satisfied then the outer solution coincides with the final one, and no boundary layer solution arises. But this happens relatively seldom, with the boundary layer solution typically arising since the outer solution cannot satisfy the edge boundary conditions.

Following the procedure, presented several times already in the previous chapters, we should obtain the boundary layer solution and then perform matching of it with the outer solution, leading to the solution of the spatial problem with any specified asymptotic tolerance, equally valid in the near-edge vicinity as well as in the interior zone, see Aghalovyan (1984); Gevorkyan (1984).

In order to construct the boundary layer solution, we perform the following variable transformations in the governing equations of anisotropic elasticity (1.1), (1.2)

$$\alpha - \alpha_0 = a\varepsilon t, \quad \beta = a\eta, \quad \gamma = a\varepsilon\zeta = h\zeta. \qquad (6.1)$$

The stress components σ_{ij} along with the dimensionless displacements $u = u_\alpha/a$, $v = u_\beta/a$, $w = u_\gamma/a$; are quantities to be determined, $\alpha = \alpha_0$ is the edge surface in the vicinity of which the boundary layer solution is analyzed. The Lamé coefficients H_α, H_β and the geodesic curvatures k_α, k_β are expanded as Taylor series in variable $\xi = \varepsilon t$ around the point $\xi_0 = \varepsilon t_0 = 0$.

The transformed governing equations should then be solved subject to the following homogeneous boundary conditions at $\gamma = -h$

$$u_\alpha(-h) = u_\beta(-h) = u_\gamma(-h) = 0 \qquad (6.2)$$

and the homogeneous conditions imposed at the face $\gamma = h$ of the first boundary value problem

$$\sigma_{\alpha\gamma}(h) = \sigma_{\beta\gamma}(h) = \sigma_{\gamma\gamma}(h) = 0, \qquad (6.3)$$

second boundary value problem

$$u_\alpha(h) = u_\beta(h) = u_\gamma(h) = 0, \qquad (6.4)$$

or mixed boundary value problem

$$u_\alpha(h) = u_\beta(h) = 0, \ \sigma_{\gamma\gamma}(h) = 0 \tag{6.5}$$

or

$$\sigma_{\alpha\gamma}(h) = \sigma_{\beta\gamma}(h) = 0, \ \ u_\gamma(h) = 0. \tag{6.6}$$

We note that the formulated boundary conditions (6.2)-(6.6) are homogeneous due to the linearity of the problem and the fact that the appropriate non-homogeneous conditions have been treated in analysis of the outer problem.

The solution of the governing equations transformed by means of (6.1) are now sought in the form

$$Q = \sum_{s=0}^{S} \varepsilon^{\chi_Q + s} Q^{(s)} \tag{6.7}$$

where Q is any of the unknowns. We infer from the required consistency of the asymptotic procedure in respect of $Q^{(s)}$ that $\chi_\sigma = -1$ for stresses and $\chi_u = 0$ for the displacement components. Following the usual procedure, after some transformations the following system may be obtained

$$H_{\alpha 0} \frac{\partial \sigma_{\alpha\alpha}^{(s)}}{\partial t} + \frac{\partial \sigma_{\alpha\gamma}^{(s)}}{\partial \zeta} = R_u^{(s-1)},$$

$$H_{\alpha 0} \frac{\partial \sigma_{\alpha\gamma}^{(s)}}{\partial t} + \frac{\partial \sigma_{\gamma\gamma}^{(s)}}{\partial \zeta} = R_w^{(s-1)},$$

$$H_{\alpha 0} \frac{\partial u^{(s)}}{\partial t} = a_{11} \sigma_{\alpha\alpha}^{(s)} + a_{12} \sigma_{\beta\beta}^{(s)} + a_{13} \sigma_{\gamma\gamma}^{(s)} + a_{14} \sigma_{\beta\gamma}^{(s)}$$

$$+ a_{15} \sigma_{\alpha\gamma}^{(s)} + a_{16} \sigma_{\alpha\beta}^{(s)} + R_{\alpha\alpha}^{(s-1)}, \tag{6.8}$$

$$\frac{\partial w^{(s)}}{\partial \zeta} = a_{13} \sigma_{\alpha\alpha}^{(s)} + a_{23} \sigma_{\beta\beta}^{(s)} + a_{33} \sigma_{\gamma\gamma}^{(s)} + \cdots + a_{36} \sigma_{\alpha\beta}^{(s)},$$

$$a_{12} \sigma_{\alpha\alpha}^{(s)} + a_{22} \sigma_{\beta\beta}^{(s)} + a_{23} \sigma_{\gamma\gamma}^{(s)} + a_{24} \sigma_{\beta\gamma}^{(s)} + a_{25} \sigma_{\alpha\gamma}^{(s)} + a_{26} \sigma_{\alpha\beta}^{(s)} = R_{\beta\beta}^{(s-1)},$$

$$H_{\alpha 0} \frac{\partial w^{(s)}}{\partial t} + \frac{\partial u^{(s)}}{\partial \zeta} = a_{15} \sigma_{\alpha\alpha}^{(s)} + a_{25} \sigma_{\beta\beta}^{(s)}$$

$$+ a_{35} \sigma_{\gamma\gamma}^{(s)} + \cdots + a_{56} \sigma_{\alpha\beta}^{(s)} + R_{\alpha\gamma}^{(s-1)},$$

and

$$H_{\alpha 0} \frac{\partial \sigma_{\alpha\beta}^{(s)}}{\partial t} + \frac{\partial \sigma_{\beta\gamma}^{(s)}}{\partial \zeta} = R_v^{(s-1)},$$

$$\frac{\partial v^{(s)}}{\partial \zeta} = a_{14} \sigma_{\alpha\alpha}^{(s)} + a_{24} \sigma_{\beta\beta}^{(s)} + a_{34} \sigma_{\gamma\gamma}^{(s)} + \cdots + a_{46} \sigma_{\alpha\beta}^{(s)} + R_{\beta\gamma}^{(s-1)}, \tag{6.9}$$

$$H_{\alpha 0} \frac{\partial v^{(s)}}{\partial t} = a_{16} \sigma_{\alpha\alpha}^{(s)} + a_{26} \sigma_{\beta\beta}^{(s)} + a_{36} \sigma_{\gamma\gamma}^{(s)} + \cdots + a_{66} \sigma_{\alpha\beta}^{(s)} + R_{\alpha\beta}^{(s-1)},$$

where $H_{\alpha s}$ is the general Taylor series coefficient for the Lamé coefficient, and $R_u^{(s-1)}$, $R_v^{(s-1)}$, $R_w^{(s-1)}$, $R_{\alpha\alpha}^{(s-1)}$, $R_{\beta\beta}^{(s-1)}$, $R_{\alpha\gamma}^{(s-1)}$, $R_{\beta\gamma}^{(s-1)}$, $R_{\alpha\beta}^{(s-1)}$ are known quantities which are not equal to zero for $s \geq 1$ only. These are determined from the following recurrent relations

$$R_u^{(s-1)} = -H_{\alpha(1+m)}t^{1+m}\frac{\partial\sigma_{\alpha\alpha}^{(s-1-m)}}{\partial t} - H_{\beta\,m}t^m\frac{\partial\sigma_{\alpha\beta}^{(s-1-m)}}{\partial\eta}$$

$$-ak_{\beta m}t^m\left(\sigma_{\alpha\alpha}^{(s-1-m)} - \sigma_{\beta\beta}^{(s-1-m)}\right) - 2ak_{\alpha\,m}t^m\sigma_{\alpha\beta}^{(s-1-m)},$$

$$R_v^{(s-1)} = -H_{\alpha(1+m)}t^{1+m}\frac{\partial\sigma_{\alpha\beta}^{(s-1-m)}}{\partial t} - H_{\beta\,m}t^m\frac{\partial\sigma_{\beta\beta}^{(s-1-m)}}{\partial\eta}$$

$$-ak_{\alpha\,m}t^m\left(\sigma_{\beta\beta}^{(s-1-m)} - \sigma_{\alpha\alpha}^{(s-1-m)}\right) - 2ak_{\beta\,m}t^m\sigma_{\alpha\beta}^{(s-1-m)},$$

$$R_w^{(s-1)} = -H_{\alpha(1+m)}t^{1+m}\frac{\partial\sigma_{\alpha\gamma}^{(s-1-m)}}{\partial t} - H_{\beta\,m}t^m\frac{\partial\sigma_{\beta\beta}^{(s-1-m)}}{\partial\eta} \qquad (6.10)$$

$$-ak_{\alpha\,m}t^m\sigma_{\alpha\gamma}^{(s-1-m)} - ak_{\beta\,m}t^m\sigma_{\beta\gamma}^{(s-1-m)},$$

$$R_{\alpha\alpha}^{(s-1)} = -H_{\alpha(1+m)}t^{1+m}\frac{\partial u^{(s-1-m)}}{\partial t} - ak_{\alpha\,m}t^m v^{(s-1-m)},$$

$$R_{\alpha\beta}^{(s-1)} = -H_{\alpha(1+m)}t^{1+m}\frac{\partial v^{(s-1-m)}}{\partial t} - H_{\beta\,m}t^m\frac{\partial u^{(s-1-m)}}{\partial\eta}$$

$$+ak_{\alpha\,m}t^m u^{(s-1-m)} + ak_{\beta\,m}t^m v^{(s-1-m)},$$

$$R_{\alpha\gamma}^{(s-1)} = -H_{\alpha(1+m)}t^{1+m}\frac{\partial w^{(s-1-m)}}{\partial t},$$

$$R_{\beta\gamma}^{(s-1)} = -H_{\beta\,m}t^m w^{(s-1-m)},$$

$$R_{\beta\beta}^{(s-1)} = H_{\beta\,m}t^m\frac{\partial v^{(s-1-m)}}{\partial t} + k_{\beta\,m}u^{(s-1-m)} \qquad m = \overline{0, s-1}.$$

We remark that summation over repeated index m is assumed in (6.10) in the limits $m = 0, s-1$ $(s \geq 1)$.

In case of general anisotropy involving 21 elastic constants due to coupling effect the systems (6.8) and (6.9) are not separated and should be solved simultaneously. Both systems are homogeneous at $s = 0$ and non-homogeneous at higher orders $s \geq 1$. The general solution is then composed of the appropriate general solution of the homogeneous and particular solutions of the non-homogeneous systems.

At $s = 0$ the boundary layer type solution is determined from (6.8) and (6.9). Using the boundary conditions (6.2)-(6.6), it is possible to determine the velocity of attenuation (order of exponential decay) for each boundary value problem. The structure of the homogeneous solutions is the same for all s, therefore, the conditions (6.2)-(6.6) may only be satisfied approximately. This difference will be treated later during the matching process and appropriate changes of the outer solution. Therefore, we restrict ourselves to solution for arbitrary s at the moment.

The solution of (6.8), (6.9) satisfying the boundary conditions (6.2)-(6.6) is sought according to Aghalovyan (1972a) as

$$\sigma_{ik}^{(s)} = \sum_{(\omega)} \sigma_{jl}^{(s)}(\zeta)\exp(-\omega x),$$

$$x = \frac{t}{H_{\alpha 0}} = \frac{1}{\varepsilon H_{\alpha 0}}\xi, \quad (i,k = \alpha,\beta,\gamma; j,l = 1,2,3), \tag{6.11}$$

$$u_i^{(s)} = \sum_{(\omega)} u_j^{(s)}(\zeta)\exp(-\omega x), \quad Re\,\omega > 0,$$

$$(u_i^{(s)} = u^{(s)}, v^{(s)}, w^{(s)}; j = 1,2,3).$$

Substituting (6.11) into (6.8) and (6.9) and expressing all of the unknown quantities through $\sigma_{33}^{(s)}$ and $\sigma_{23}^{(s)}$, we obtain

$$\sigma_{11}^{(s)} = \frac{1}{\omega^2}\sigma''^{\,(s)}_{33}, \quad \sigma_{13}^{(s)} = \frac{1}{\omega}\sigma'^{\,(s)}_{33}, \quad \sigma_{12}^{(s)} = \frac{1}{\omega}\sigma'^{\,(s)}_{23},$$

$$\sigma_{22}^{(s)} = -\frac{1}{a_{22}}\left(\frac{a_{12}}{\omega^2}\sigma''^{\,(s)}_{33} + \frac{a_{25}}{\omega}\sigma'^{\,(s)}_{33} + a_{23}\sigma_{33}^{(s)} + \frac{a_{26}}{\omega}\sigma'^{\,(s)}_{23} + a_{24}\sigma_{23}^{(s)}\right),$$

$$u_1^{(s)} = -\frac{1}{\omega}\left(\frac{A_{11}}{\omega^2}\sigma''^{\,(s)}_{33} + \frac{A_{15}}{\omega}\sigma'^{\,(s)}_{33} + A_{13}\sigma_{33}^{(s)} + \frac{A_{16}}{\omega}\sigma'^{\,(s)}_{23} + A_{14}\sigma_{23}^{(s)}\right),$$

$$u_2^{(s)} = -\frac{1}{\omega}\left(\frac{A_{16}}{\omega^2}\sigma''^{\,(s)}_{33} + \frac{A_{56}}{\omega}\sigma'^{\,(s)}_{33} + A_{36}\sigma_{33}^{(s)} + \frac{A_{66}}{\omega}\sigma'^{\,(s)}_{23} + A_{46}\sigma_{23}^{(s)}\right), \tag{6.12}$$

$$u_3^{(s)} = -\frac{1}{\omega}\left(\frac{A_{11}}{\omega^3}\sigma'''^{\,(s)}_{33} + \frac{2A_{15}}{\omega^2}\sigma''^{\,(s)}_{33} + (A_{13} + A_{55})\frac{1}{\omega}\sigma'^{\,(s)}_{33} + A_{35}\sigma_{33}^{(s)}\right.$$

$$\left. + \frac{A_{16}}{\omega^2}\sigma''^{\,(s)}_{23} + \frac{A_{14} + A_{56}}{\omega}\sigma'^{\,(s)}_{23} + A_{45}\sigma_{23}^{(s)}\right),$$

$$A_{ij} = (a_{22}a_{ij} - a_{2i}a_{2j})/a_{22}, \quad A_{ij} = A_{ji}, \quad (i,j = 1,3,4,5,6).$$

The quantities $\sigma_{33}^{(s)}, \sigma_{23}^{(s)}$ are determined from the system

$$L_{11}\sigma_{33}^{(s)} + L_{12}\sigma_{23}^{(s)} = 0, \tag{6.13}$$

$$L_{12}\sigma_{33}^{(s)} + L_{22}\sigma_{23}^{(s)} = 0$$

where L_{11}, L_{12}, L_{22} are the differential operators

$$L_{11} = A_{11}\frac{d^4}{d\zeta^4} + 2A_{15}\omega\frac{d^3}{d\zeta^3}$$

$$+ (A_{55} + 2A_{13})\omega^2\frac{d^2}{d\zeta^2} + 2A_{35}\omega^3\frac{d}{d\zeta} + A_{33}\omega^4,$$

$$L_{12} = A_{16}\omega\frac{d^3}{d\zeta^3} + (A_{14} + A_{56})\omega^2\frac{d^2}{d\zeta^2} + (A_{36} + A_{45})\omega^3\frac{d}{d\zeta} + A_{34}\omega^4, \tag{6.14}$$

$$L_{22} = A_{66}\omega^2\frac{d^2}{d\zeta^2} + 2A_{46}\omega^3\frac{d}{d\zeta} + A_{44}\omega^4.$$

Thus, the system (6.13) should be solved subject to boundary conditions (6.2)-(6.6). Let us introduce a function of stress $\Phi^{(s)}$ through

$$\sigma_{33}^{(s)} = L_{22}\Phi^{(s)}, \quad \sigma_{23}^{(s)} = -L_{12}\Phi^{(s)} \tag{6.15}$$

then the system (6.13) may be reduced to an ordinary differential equation of the sixth order

$$\left(L_{11}L_{22} - L_{12}^2\right)\Phi^{(s)} = 0. \tag{6.16}$$

Let $\Phi^{(s)} = \sum_{k=1}^{6} C_k^{(s)}\psi_k(\zeta,\omega)$ be the solution of (6.16). Then, using (6.12) in order to determine the stresses and displacements for $s = 0$, and satisfying the boundary conditions (6.2)-(6.6), for each boundary value problem we result in an algebraic system of order 6 in respect of the unknown coefficients $C_k^{(0)}$. The transcendental equation in respect of ω is then inferred as $\Delta = 0$, the condition of existence of non-trivial solution of an appropriate algebraic system. In order to ensure decay of the solution along the α coordinate line, one has to focus his attention on the roots of the transcendental equation satisfying $Re\,\omega > 0$. After ω is defined, the sought-for displacement and stress components may be obtained from (6.7), (6.11), and (6.12).

In view of (6.16), since the structure of the homogeneous solution is the same for all values of s, for higher order approximations $s \geq 1$ the appropriate particular solutions of the systems (6.8), (6.9) should be added to the homogeneous solution (with new constants $C_k^{(s)}$). It may be observed that the stresses and displacements corresponding to homogeneous component of the solution, will automatically satisfy the conditions (6.2)-(6.6). However, in view of the particular solutions of the non-homogeneous systems (6.8), (6.9) these conditions are not satisfied. The additional terms (all of these are known functions) arising at $\gamma = \pm h$ may be eliminated by modification of the higher order solutions of the outer problem ($s \geq 1$), in which the values of the appropriate boundary functions at $\zeta = \pm 1$ should be subtracted.

We remark that this approach is fully valid. Even though some of the boundary functions may possess higher variability, as follows from the outer solution (2.5), (3.1) the functions $\sigma_{j\gamma}^{+(s)}$ are involved in the solution, but not their derivatives. The appropriate solution of the outer problem, involving the terms decaying away from the edge, combined with the boundary layer solution, will provide the correct solution and remove the difference at $\zeta = \pm 1$. The influence of variability of the boundary functions may arise at higher orders, which are usually neglected due to required tolerance of the practical computations.

In case of orthotropic plates, when the principal directions of anisotropy coincide with the directions of the coordinate lines, the governing system for boundary layer splits into two separate systems (6.8) and (6.9), which describe the plane and anti-plane boundary layer solutions. Since $L_{12} \equiv 0$, it follows from (6.13) that

$$L_{11}\sigma_{33}^{(s)} = 0, \quad L_{22}\sigma_{23}^{(s)} = 0. \tag{6.17}$$

The characteristic equations of the differential operators (6.17) are given by

$$A_{11}r^4 + (A_{55} + 2A_{23})r^2 + A_{33} = 0, \tag{6.18}$$

$$A_{66}\lambda^2 + A_{44} = 0.$$

In case of real orthotropic materials, the roots of the first equation of (6.18) could either be different imaginary or repeated purely imaginary or complex conjugate, whereas the roots of the second equation could be only purely imaginary.

Solving the system (6.17), (6.8) and (6.9) and satisfying the boundary conditions at $\gamma = \pm h$, the general boundary layer solution may be obtained. Below we present these solutions for each of the boundary value problems.

6.7 Boundary Layer for the First Boundary Value Problem

Let us present the solutions for the plane and anti-plane boundary layers for an orthotropic plate, with one face fixed, and the other subjected to the first boundary value problem

$$u_\alpha(-h) = u_\beta(-h) = u_\gamma(-h) = 0, \tag{7.1}$$
$$\sigma_{\alpha\gamma}(h) = \sigma_{\beta\gamma}(h) = \sigma_{\gamma\gamma}(h) = 0.$$

Let the roots of the first characteristic equation of (6.18) corresponding to the plane boundary layer solution, are distinct purely imaginary ones, which is often the case for real orthotropic materials

$$r_{1,2} = \pm i q_1, \quad r_{3,4} = \pm i q_2 \tag{7.2}$$

where

$$q_{1,2}^2 = \frac{A_{55} + 2A_{13} \mp \sqrt{D}}{2A_{11}}, \quad D = (A_{55} + 2A_{13})^2 - 4A_{11}A_{33} > 0$$

then at leading order $s = 0$ the solution is

$$\sigma_{33}^{(0)} = C_1^{(0)} \cos q_1 \omega \zeta + C_2^{(0)} \sin q_1 \omega \zeta + C_3^{(0)} \cos q_2 \omega \zeta + C_4^{(0)} \sin q_2 \omega \zeta \tag{7.3}$$

with $C_1^{(0)}, C_2^{(0)}, C_3^{(0)}, C_4^{(0)}$ denoting the arbitrary constants of integration. Then, substituting (7.3) into (6.11) and (6.12), we have for the plane boundary layer solution (with superscript "b")

$$\overset{b}{\sigma}{}_{\gamma\gamma}^{(0)} = \sum_\omega C_\omega^{(0)} \left(p_2 q_2 \left(\sin(1+\zeta)q_1\omega - \sin 2q_1\omega \cos(1-\zeta)q_2\omega \right) \right.$$

$$+ p_1 q_1 \left(\sin(1+\zeta)q_2\omega - \sin 2q_2\omega \cos(1-\zeta)q_1\omega \right)$$

$$+ p_1 q_2 \cos 2q_2\omega \sin(1-\zeta)q_1\omega$$

$$\left. + q_1 p_2 \cos 2q_1\omega \sin(1-\zeta)q_2\omega \right) \exp(-\omega x),$$

$$\overset{b}{\sigma}{}_{\alpha\gamma}^{(0)} = \sum_\omega C_\omega^{(0)} \left(p_2 q_2 \left(q_1 \cos(1+\zeta)q_1\omega - q_2 \sin 2q_1\omega \sin(1-\zeta)q_2\omega \right) \right.$$

$$+ p_1 q_1 \left(q_2 \cos(1+\zeta)q_2\omega - q_1 \sin 2q_2\omega \sin(1-\zeta)q_1\omega \right)$$

$$- p_1 q_1 q_2 \cos 2q_2\omega \cos(1-\zeta)q_1\omega \tag{7.4}$$

$$\left. - q_1 q_2 p_2 \cos 2q_1\omega \cos(1-\zeta)q_2\omega \right) \exp(-\omega x),$$

$$\overset{b}{\sigma}\,\overset{(0)}{_{\alpha\alpha}} = \sum_{\omega} C_{\omega}^{(0)} \left(p_2 q_2 \left(q_2^2 \sin 2q_1\omega \cos(1-\zeta)q_2\omega - q_1^2 \sin(1+\zeta)q_1\omega \right) \right.$$

$$+ p_1 q_1 \left(q_1^2 \sin 2q_2\omega \cos(1-\zeta)q_1\omega - q_2^2 \sin(1+\zeta)q_2\omega \right)$$

$$- q_1 q_2^2 p_2 \cos 2q_1\omega \sin(1-\zeta)q_2\omega$$

$$\left. - q_1^2 q_2 p_1 \cos 2q_2\omega \sin(1-\zeta)q_1\omega \right) \exp(-\omega x),$$

$$\overset{b}{\sigma}\,\overset{(0)}{_{\beta\beta}} = \frac{1}{a_{22}} \sum_{\omega} C_{\omega}^{(0)} \left(p_2 q_2 \left((a_{12}q_1^2 - a_{23}) \sin(1+\zeta)q_1\omega \right. \right.$$

$$\left. - (a_{12}q_2^2 - a_{23}) \sin 2q_1\omega \cos(1-\zeta)q_2\omega \right)$$

$$+ p_1 q_1 \left((a_{12}q_2^2 - a_{23}) \sin(1+\zeta)q_2\omega \right.$$

$$\left. - (a_{12}q_1^2 - a_{23}) \sin 2q_2\omega \cos(1-\zeta)q_1\omega \right)$$

$$+ p_1 q_2 (a_{12}q_1^2 - a_{23}) \cos 2q_2\omega \sin(1-\zeta)q_1\omega$$

$$\left. + q_1 p_2 (a_{12}q_2^2 - a_{23}) \cos 2q_1\omega \sin(1-\zeta)q_2\omega \right) \exp(-\omega x),$$

$$\overset{b}{u}\,\overset{(0)}{} = \sum_{\omega} \frac{1}{2\omega} C_{\omega}^{(0)} \left(q_1 p_1^2 \sin(1+\zeta)q_2\omega + q_2 p_2^2 \sin(1+\zeta)q_1\omega \right.$$

$$+ p_1 q_1 p_2 \left(\cos 2q_1\omega \sin(1-\zeta)q_2\omega - \sin 2q_2\omega \cos(1-\zeta)q_1\omega \right)$$

$$+ q_2 p_1 p_2 \left(\cos 2q_2\omega \sin(1-\zeta)q_1\omega - \sin 2q_1\omega \cos(1-\zeta)q_2\omega \right) \exp(-\omega x),$$

$$\overset{b}{w}\,\overset{(0)}{} = -\sum_{\omega} \frac{1}{2\omega} \left(q_1 q_2 p_1 p_2 \left(\cos(1+\zeta)q_1\omega + \cos(1+\zeta)q_2\omega \right) \right.$$

$$- q_1^2 p_1^2 \sin 2q_2\omega \sin(1-\zeta)q_1\omega - q_2^2 p_2^2 \sin 2q_1\omega \sin(1-\zeta)q_2\omega$$

$$\left. - q_1 q_2 p_1^2 \cos 2q_2\omega \cos(1-\zeta)q_1\omega - q_1 q_2 p_2^2 \cos 2q_1\omega \cos(1-\zeta)q_2\omega \right) \exp(-\omega x),$$

$$p_1 = A_{55} + \sqrt{D}, \quad p_2 = A_{55} - \sqrt{D},$$

where ω satisfying $\mathrm{Re}\,\omega > 0$ is a root of the following transcendental equation

$$\cos 2(q_1 - q_2)\omega - m_1 \cos 2(q_1 + q_2)\omega = m_2,$$

$$m_2 = \frac{4(A_{55}^2 - D)\sqrt{A_{11}A_{33}}}{A_{55}(A_{55}^2 - D) + 2(A_{55}^2 + D)(A_{13} + \sqrt{A_{11}A_{33}})}, \tag{7.5}$$

$$m_1 = \frac{A_{55}(A_{55}^2 - D) + 2(A_{55}^2 + D)(A_{13} - \sqrt{A_{11}A_{33}})}{A_{55}(A_{55}^2 - D) + 2(A_{55}^2 + D)(A_{13} + \sqrt{A_{11}A_{33}})}.$$

If ω is the root of equation (7.5), it may be inferred from the system (6.9) and boundary conditions (7.1) that

$$\overset{b}{\sigma}\,\overset{(0)}{_{\beta\gamma}} = \overset{b}{\sigma}\,\overset{(0)}{_{\alpha\beta}} = \overset{b}{v}\,\overset{(0)}{} \equiv 0 \tag{7.6}$$

According to the Pickard theorem the roots of the transcendental equation (7.5) forms a countable set. The roots may either be real or complex conjugates. We note that summation in (7.4) and below is performed along the roots ω of equation (7.5) such that $\mathrm{Re}\,\omega > 0$. Since every root ω has a corresponding conjugate root $\bar{\omega}$, the solution (7.4) is real expressed in terms of two groups of arbitrary real constants which should be determined.

In case of repeated imaginary roots of the first characteristic equation (in particular for isotropic plate)

$$r_{1,2} = \pm iq_3, \quad r_{3,4} = \pm iq_3,$$
$$q_3^2 = \frac{A_{55} + 2A_{13}}{2A_{11}} = \sqrt{\frac{A_{33}}{A_{11}}},$$
$$D = (A_{55} + 2A_{13})^2 - 4A_{11}A_{33} = 0, \tag{7.7}$$

the plane boundary layer solution takes the form

$$\overset{b}{\sigma}{}^{(0)}_{\gamma\gamma} = \sum_{\omega} B^{(0)}_{\omega} \left(((A_{55} + 4A_{13}) \sin 2q_3\omega \right.$$
$$- (1 + \zeta) A_{55} q_3\omega \cos 2q_3\omega) \sin(1 - \zeta)q_3\omega$$
$$- q_3\omega(1 - \zeta) ((A_{55} + 4A_{13}) \sin(1 + \zeta)q_3\omega$$
$$\left. - 2A_{55}q_3\omega \cos(1 + \zeta)q_3\omega)) \exp(-\omega x),$$

$$\overset{b}{\sigma}{}^{(0)}_{\alpha\gamma} = \sum_{\omega} B^{(0)}_{\omega} q_3 \left(((A_{55} + 4A_{13}) \sin(1 + \zeta)q_3\omega \right.$$
$$- 2A_{55}q_3\omega \cos(1 + \zeta)q_3\omega) - A_{55} \cos 2q_3\omega \sin(1 - \zeta)q_3\omega$$
$$- q_3\omega(1 - \zeta)((A_{55} + 4A_{13}) \cos(1 + \zeta)q_3\omega$$
$$+ 2A_{55}q_3\omega \sin(1 + \zeta)q_3\omega) - (A_{55} + 4A_{13}) \sin 2q_3\omega$$
$$\left. - (1 + \zeta) A_{55} q_3\omega \cos 2q_3\omega) \cos(1 - \zeta)q_3\omega) \exp(-\omega x),$$

$$\overset{b}{\sigma}{}^{(0)}_{\alpha\alpha} = \sum_{\omega} B^{(0)}_{\omega} q_3^2 \left(2((A_{55} + 4A_{13}) \cos(1 + \zeta)q_3\omega \right.$$
$$+ 2A_{55}q_3\omega \sin(1 + \zeta)q_3\omega) + 2A_{55} \cos 2q_3\omega \cos(1 - \zeta)q_3\omega$$
$$+ q_3\omega(1 - \zeta)((A_{55} + 4A_{13}) \sin(1 + \zeta)q_3\omega$$
$$- 2A_{55}q_3\omega \cos(1 + \zeta)q_3\omega) - ((A_{55} + 4A_{13}) \sin 2q_3\omega -$$
$$\left. - (1 + \zeta) A_{55} q_3\omega \cos 2q_3\omega) \sin(1 - \zeta)q_3\omega) \exp(-\omega x),$$

$$\overset{b}{\sigma}{}^{(0)}_{\beta\beta} = \frac{1}{a_{22}} \sum_{\omega} B^{(0)}_{\omega} \left((a_{12}^2 q_3^2 - a_{23}) ((A_{55} + 4A_{13}) \sin 2q_3\omega \right. \tag{7.8}$$
$$- (1 - \zeta) A_{55} q_3\omega \cos q_3\omega \sin(1 - \zeta)q_3\omega - (a_{12}^2 q_3^2 - a_{23})q_3\omega(1 - \zeta)$$
$$\times ((A_{55} + 4A_{13}) \sin(1 + \zeta)q_3\omega - 2A_{55}q_3\omega \cos(1 + \zeta)q_3\omega) - 2A_{55}a_{12}q_3^2$$
$$\times \cos 2q_3\omega \cos(1 - \zeta)q_3\omega + 2a_{13}q_3^2 ((A_{55} + 4A_{13}) \cos(1 + \zeta)q_3\omega$$
$$\left. + 2A_{55}q_3\omega \sin(1 + \zeta)q_3\omega)) \exp(-\omega x),$$

$$\overset{b}{u}{}^{(0)} = - \sum_{\omega} \frac{1}{2\omega} B^{(0)}_{\omega} \left(2A_{55}(A_{55} + 2A_{13}) \cos 2q_3\omega \cos(1 - \zeta)q_3\omega \right.$$
$$+ 2(A_{55} + 2A_{13}) ((A_{55} + 4A_{13}) \cos(1 + \zeta)q_3\omega + 2A_{55}q_3\omega \sin(1 + \zeta)q_3\omega)$$

$$+A_{55}q_3\omega(1-\zeta)\left((A_{55}+4A_{13})\sin(1+\zeta)q_3\omega \; 2A_{55}q_3\omega\cos(1+\zeta)q_3\omega\right)$$
$$-A_{55}((A_{55}+4A_{13})\sin 2q_3\omega$$
$$-(1+\zeta)A_{55}q_3\omega\cos 2q_3\omega)\sin(1-\zeta)q_3\omega)\exp(-\omega x),$$

$$\overset{b}{w}{}^{(0)}=\sum_\omega\frac{1}{2\omega}(A_{55}q_3\omega(1-\zeta)((A_{55}+4A_{13})\cos(1+\zeta)q_3\omega$$

$$+2A_{55}q_3\omega\sin(1+\zeta)q_3\omega)$$
$$-A_{55}(A_{55}+4A_{13})\cos 2q_3\omega\sin(1-\zeta)q_3\omega)$$
$$+A_{55}((A_{55}+4A_{13})\sin 2q_3\omega-(1+\zeta)A_{55}q_3\omega\cos 2q_3\omega)q_3\omega$$
$$+(A_{55}+4A_{13})((A_{55}+4A_{13})\sin(1+\zeta)q_3\omega$$
$$-2A_{55}q_3\omega\cos(1+\zeta)q_3\omega))\exp(-\omega x),$$

$$\overset{b}{\sigma}{}^{(0)}_{\beta\gamma}=\overset{b}{\sigma}{}^{(0)}_{\alpha\beta}=\overset{b}{v}{}^{(0)}=0,$$

where ω are the complex conjugate roots with positive real part $(Re\,\omega>0)$ of the following transcendental equation

$$\cos 4q_3\omega=m_1(4q_3\omega)^2-m_2,$$

$$m_1=\frac{A_{55}}{2(3A_{55}+8A_{13})},\quad m_2=\frac{A_{55}^2+(3A_{55}+8A_{13})^2}{2A_{55}(3A_{55}+8A_{13})}.\qquad(7.9)$$

In case when the roots of the first characteristic equation of (6.18) are complex conjugates

$$r_{1,2}=p\pm iq,\;\; r_{3,4}=-p\pm iq,$$
$$p^2=(2\sqrt{A_{11}A_{33}}+A_{55}+2A_{13})/(4A_{11}),\qquad(7.10)$$
$$q^2=(2\sqrt{A_{11}A_{33}}-A_{55}-2A_{13})/(4A_{11})$$

the plane boundary layer solution may be obtained in a similar manner, with ω defined through

$$\sin 4q\omega=m\mathrm{sh}4p\omega,$$

$$m=\frac{q(A_{55}^2+D)+2pA_{55}\sqrt{-D}}{p(A_{55}^2+D)-2qA_{55}\sqrt{-D}},\quad D=(A_{55}+2A_{13})^2-4A_{11}A_{33}<0.\quad(7.11)$$

As mentioned earlier, in case of orthotropic plates the systems (6.8) and (6.9) decouple at $s=0$. Consider now (6.9) at $s=0$. Using (6.11), the second equation of (6.18) and conditions (7.1), we arrive at a new solution, with the leading order quantities involving $\sigma^{(0)}_{\alpha\beta},\sigma^{(0)}_{\beta\gamma},v^{(0)}$, where ω is given by

$$\omega_n=\frac{\pi}{4}\sqrt{\frac{a_{66}}{a_{44}}}(2n+1)\quad n\in N.\qquad(7.12)$$

Therefore, the anti-plane boundary layer solution (denoted by "a" superscript), following from (6.8) subject to conditions (7.1) at $s = 0$ may be written as

$$\overset{a}{\sigma}{}^{(0)}_{\beta\gamma} = \sum_{n=0}^{\infty} D_n^{(0)} \cos \frac{\pi}{4}(2n+1)(1+\zeta) \exp\left(-\frac{\pi}{4}\sqrt{\frac{a_{66}}{a_{44}}}(2n+1)x\right),$$

$$\overset{a}{\sigma}{}^{(0)}_{\alpha\beta} = -\sqrt{\frac{a_{44}}{a_{66}}} \sum_{n=0}^{\infty} D_n^{(0)} \sin \frac{\pi}{4}(2n+1)(1+\zeta)$$

$$\times \exp\left(-\frac{\pi}{4}\sqrt{\frac{a_{66}}{a_{44}}}(2n+1)x\right), \qquad (7.13)$$

$$\overset{a}{v}{}^{(0)} = \frac{4}{\pi}a_{44} \sum_{n=0}^{\infty} D_n^{(0)} \frac{1}{2n+1} \sin \frac{\pi}{4}(2n+1)(1+\zeta)$$

$$\times \exp\left(-\frac{\pi}{4}\sqrt{\frac{a_{66}}{a_{44}}}(2n+1)x\right),$$

$$\overset{a}{\sigma}{}^{(0)}_{\alpha\alpha} = \overset{a}{\sigma}{}^{(0)}_{\beta\beta} = \overset{a}{\sigma}{}^{(0)}_{\gamma\gamma} = \overset{a}{\sigma}{}^{(0)}_{\alpha\gamma} = \overset{a}{u}{}^{(0)} = \overset{a}{w}{}^{(0)} = 0.$$

At higher orders ($s \geq 1$) the systems (6.8) and (6.9) are non-homogeneous, so two cases should be considered. The first case corresponds to ω being a root of the transcendental equation (7.5), (7.9), or (7.11) of the plane boundary layer, whereas the second case is that of the anti-plane boundary layer, with ω defined by (7.12). The associated quantities will be denoted by superscripts "b" and "a", respectively. The solutions of (6.8) and (6.9) are then written as

$$\overset{b}{Q}{}^{(s)} = \overset{b}{Q}{}^{(s)}_0 + \overset{b}{Q}{}^{(s)}_* \qquad (b, a) \qquad (7.14)$$

where $Q_0^{(s)}$ is a general solution of the homogeneous system, and $Q_*^{(s)}$ is a particular solution of the non-homogeneous one. We note that since the leading order solutions satisfy (7.1), the solutions of the homogeneous systems for both plane and anti-plane boundary layer solutions satisfy conditions (7.1) due to similarity of the structure of the left-hand sides of the systems for any approximation s. In case of the plane boundary layer the system (6.8) should be considered, with ω defined through (7.12). The solution for any s will again be of the form (7.3), with ω known from (7.12), hence the four constants $C_1^{(s)}, C_2^{(s)}, C_3^{(s)}, C_4^{(s)}$ would be sufficient to satisfy the conditions (7.1) in respect of the quantities $\sigma_{\alpha\gamma}, \sigma_{\gamma\gamma}, u_\alpha, u_\gamma$. Similarly, the homogeneous solution of the system (6.9) with superscript "a" satisfies conditions (7.1) in respect of $\sigma_{\beta\gamma}, u_\beta$. On the other hand, the conditions (7.1) may be satisfied in a straightforward manner for arbitrary s by use of the corresponding plane boundary layer (with superscript "b"), bearing in mind (6.11) and (6.18), since ω is now the solution of (7.5), (7.9) or (7.11), but not described by (7.12). Thus, in case of an orthotropic plate the final solution of (6.8), (6.9) for arbitrary s is written as

$$Q^{(s)} = \overset{b}{Q}{}^{(s)}_0 + \overset{b}{Q}{}^{(s)}_* + \overset{a}{Q}{}^{(s)}_0 + \overset{a}{Q}{}^{(s)}_* \qquad (7.15)$$

where Q is any of the unknowns associated with the boundary layer. It has been shown earlier that $\overset{b}{\sigma}{}^{(0)}_{\beta\gamma} = \overset{b}{\sigma}{}^{(0)}_{\alpha\beta} = \overset{b}{u}{}^{(0)}_{\beta} = \overset{a}{\sigma}{}^{(0)}_{\alpha\alpha} = \overset{a}{\sigma}{}^{(0)}_{\beta\beta} =$

$$\overset{a}{\sigma}{}^{(0)}_{\alpha\gamma} = \overset{a}{\sigma}{}^{(0)}_{\gamma\gamma} = \overset{a}{u}{}^{(0)}_{\alpha} = \overset{a}{u}{}^{(0)}_{\gamma} = \overset{b}{Q}{}^{(0)}_{*} = \overset{a}{Q}{}^{(0)}_{*} = 0. \qquad \text{At } s \geq 1$$

the quantities $\overset{b}{\sigma}{}^{(s)}_{\alpha\gamma0}, \overset{b}{\sigma}{}^{(s)}_{\gamma\gamma0}, \overset{b}{u}{}^{(s)}_{\alpha0}, \overset{b}{u}{}^{(s)}_{\gamma0}, \overset{a}{\sigma}{}^{(s)}_{\beta\gamma0}, \overset{a}{u}{}^{(s)}_{\beta0}$ satisfy conditions (7.1), whereas the solution (7.15) satisfies conditions (7.1) by means of the appropriate choice of arbitrary constants of the associated boundary layer solutions for $\overset{a}{\sigma}{}^{(s)}_{\alpha\gamma0}, \overset{a}{\sigma}{}^{(s)}_{\gamma\gamma0}, \overset{a}{u}{}^{(s)}_{\alpha0}, \overset{a}{u}{}^{(s)}_{\gamma0}, \overset{b}{\sigma}{}^{(s)}_{\beta\gamma0}, \overset{b}{u}{}^{(s)}_{\beta0}$.

In case of general anisotropy, when the systems (6.8) and (6.9) have to be solved simultaneously, the solution is presented in the general form $Q^{(s)} = Q^{(s)}_0 + Q^{(s)}_*$, whereas previously, $Q^{(s)}_0$ and $Q^{(s)}_*$ denote the general and particular solutions of the homogeneous and non-homogeneous systems, respectively. Then, in order to satisfy conditions (7.1) at each order $s \geq 1$, as mentioned in Section 6, some amendments are required in the formulation of the outer problem. These amendments essentially cause the quantities $\sigma^{+(s)}_{\alpha\gamma}, \sigma^{+(s)}_{\beta\gamma}, \sigma^{+(s)}_{\gamma\gamma}$ $(s \geq 1)$ of the solution (3.2) to become non-zero, being opposite by sign and identical by absolute value to the corresponding quantities of the boundary layer. Therefore, we may deduce that in case of general anisotropy the boundary layer affects the outer field through boundary conditions at $s \geq 1$, i.e. there is mutual influence between the outer solution and the boundary layer. However, in case of orthotropic plates the boundary layer is only removing the boundary effect on the edge of the plate, so the outer solution affects the boundary layer but the boundary layer solution has no influence on the outer solution. Thus, this phenomena should be emphasized as different from the one observed for the so-called classical problems of plate theory. Indeed, as has been demonstrated earlier, the boundary layer always affects the outer solution through the boundary conditions regardless of the type of anisotropy.

6.8   Boundary Layer for the Second Boundary Value Problem

Consider the boundary layer solutions for anisotropic plates having homogeneous boundary conditions specified on the faces $\gamma = \pm h$ in terms of displacements as

$$u_\alpha(\pm h) = u_\beta(\pm h) = u_\gamma(\pm h) = 0. \tag{8.1}$$

For the sake of brevity here we assume that the plate is orthotropic, with the plane and anti-plane boundary layer solutions arising. In case of the purely imaginary roots of (7.2) at leading order $s = 0$ the plane boundary layer solution is given by

$$\overset{b}{\sigma}{}^{(0)}_{\gamma\gamma} = \sum_{(\omega)} A^{(0)}_\omega \left((p_1 \cos q_2\omega \cos q_1\omega\zeta - p_2 \cos q_1\omega \cos q_2\omega\zeta)F_1(\omega) \right.$$
$$\left. - (p_1 \sin q_2\omega \sin q_1\omega\zeta - p_2 \sin q_1\omega \sin q_2\omega\zeta)F_2(\omega) \right) \exp(-\omega x),$$

$$\overset{b}{\sigma}{}^{(0)}_{\alpha\gamma} = -\sum_{(\omega)} A^{(0)}_\omega \left((p_1 q_1 \cos q_2\omega \sin q_1\omega\zeta - p_2 q_2 \cos q_1\omega \sin q_2\omega\zeta)F_1(\omega) \right.$$
$$\left. + (p_1 q_1 \sin q_2\omega \cos q_1\omega\zeta - p_2 q_2 \sin q_1\omega \cos q_2\omega\zeta)F_2(\omega) \right) \exp(-\omega x),$$

$$\overset{b}{\sigma}\overset{(0)}{_{\alpha\alpha}} = -\sum_{(\omega)} A_\omega^{(0)} \left((p_1 q_1^2 \cos q_2\omega \cos q_1\omega\zeta - p_2 q_2^2 \cos q_1\omega \cos q_2\omega\zeta) F_1(\omega) \right.$$

$$\left. -(p_1 q_1^2 \sin q_2\omega \sin q_1\omega\zeta - p_2 q_2^2 \sin q_1\omega \sin q_2\omega\zeta) F_2(\omega) \right) \exp(-\omega x),$$

$$\overset{b}{\sigma}\overset{(0)}{_{\beta\beta}} = -\sum_{(\omega)} \frac{1}{a_{22}} A_\omega^{(0)} \left((p_1(a_{12}q_1^2 - a_{23}) \cos q_2\omega \cos q_1\omega\zeta \right.$$

$$-p_2(a_{12}q_2^2 - a_{23}) \cos q_1\omega \cos q_2\omega\zeta) F_1(\omega) \qquad (8.2)$$

$$-(p_1(a_{12}q_1^2 - a_{23}) \sin q_2\omega \sin q_1\omega\zeta$$

$$\left. - p_2(a_{12}q_2^2 - a_{23}) \sin q_1\omega \sin q_2\omega\zeta) F_2(\omega) \right) \exp(-\omega x),$$

$$\overset{b}{u}\overset{(0)}{} = \frac{1}{2} p_1 p_2 \sum_{(\omega)} \frac{1}{\omega} A_\omega^{(0)} \left((\cos q_2\omega \cos q_1\omega\zeta - \cos q_1\omega \cos q_2\omega\zeta) F_1(\omega) \right.$$

$$\left. -(\sin q_2\omega \sin q_1\omega\zeta - \sin q_1\omega \sin q_2\omega\zeta) F_2(\omega) \right) \exp(-\omega x),$$

$$\overset{b}{w}\overset{(0)}{} = \sum_{(\omega)} \frac{1}{2\omega} A_\omega^{(0)} \left((q_1 p_1^2 \cos q_2\omega \sin q_1\omega\zeta - q_2 p_2^2 \cos q_1\omega \sin q_2\omega\zeta) F_1(\omega) \right.$$

$$\left. +(q_1 p_1^2 \sin q_2\omega \cos q_1\omega\zeta - q_2 p_2^2 \sin q_1\omega \cos q_2\omega\zeta) F_2(\omega) \right) \exp(-\omega x),$$

$$\overset{b}{\sigma}\overset{(0)}{_{\alpha\beta}} = \overset{b}{\sigma}\overset{(0)}{_{\beta\gamma}} = \overset{b}{v}\overset{(0)}{} = 0,$$

$$F_{1,2}(\omega) = \frac{1}{2}(q_1 p_1^2 - q_2 p_2^2) \sin(q_1 + q_2)\omega \mp \frac{1}{2}(q_1 p_1^2 + q_2 p_2^2) \sin(q_1 - q_2)\omega.$$

We note that summation is assumed in (8.2) along the roots ω of the transcendental equation

$$F_1(\omega) \cdot F_2(\omega) = 0 \qquad (8.3)$$

satisfying $\operatorname{Re}\omega > 0$.

In case of repeated purely imaginary roots (7.7) the solution takes the form

$$\overset{b}{\sigma}\overset{(0)}{_{\gamma\gamma}} = \sum_{(\omega)} B_\omega^{(0)} \left(A_{55} q_3\omega(3A_{55} + 8A_{13})(1 + \zeta) \sin q_3\omega \sin(1 - \zeta) q_3\omega \right.$$

$$+2A_{55}^2 q_3^2\omega^2(1 - \zeta) \sin(1 + \zeta)q_3\omega - 4A_{55}(A_{55} + 2A_{13})q_3\omega \cos(1 + \zeta)q_3\omega$$

$$\left. -2(A_{55} + 2A_{13})(3A_{55} + 8A_{13}) \sin 2q_3\omega \cos(1 - \zeta)q_3\omega) \exp(-\omega x),$$

$$\overset{b}{\sigma}\overset{(0)}{_{\alpha\alpha}} = \sum_{(\omega)} q_3^2 B_\omega^{(0)} \left(8q_3\omega A_{55} A_{13} \cos(1 + \zeta)q_3\omega \right.$$

$$+4A_{13}(3A_{55} + 8A_{13}) \sin 2q_3\omega \cos(1 - \zeta)q_3\omega$$

$$-2A_{55}^2 q_3^2\omega^2(1 - \zeta) \sin(1 + \zeta)q_3\omega$$

$$\left. -A_{55}(3A_{55} + 8A_{13})q_3\omega(1 + \zeta) \sin 2q_3\omega \sin(1 - \zeta)q_3\omega) \exp(-\omega x),$$

$$\overset{b}{\sigma}\overset{(0)}{_{\alpha\gamma}} = \sum_{(\omega)} B_\omega^{(0)} q_3 \left(2A_{55}^2 q_3^2\omega^2(1 - \zeta) \cos(1 + \zeta)q_3\omega \right. \qquad (8.4)$$

$$-A_{55}(3A_{55} + 8A_{13})q_3\omega(1 + \zeta) \sin 2q_3\omega \cos(1 - \zeta)q_3\omega$$

$$-(A_{55} + 4A_{13})(3A_{55} + 8A_{13}) \sin 2q_3\omega \sin(1 - \zeta)q_3\omega$$

$$\left. +2q_3\omega A_{55}(A_{55} + 4A_{13}) \sin(1 + \zeta)q_3\omega) \exp(-\omega x),$$

$$\overset{b}{\sigma}{}^{(0)}_{\beta\beta} = \frac{1}{a_{22}} \sum_{(\omega)} B^{(0)}_{\omega}(A_{55}(3A_{55} + 8A_{13})(a_{12}q_3^2 - a_{23})$$

$$\times q_3\omega(1+\zeta)\sin 2q_3\omega \sin(1-\zeta)q_3\omega$$

$$+2A_{55}^2(a_{12}q_3^2 - a_{23})q_3^2\omega^2(1-\zeta)\sin(1+\zeta)q_3\omega$$

$$+(3A_{55} + 8A_{13})\left(2a_{23}(A_{55} + 2A_{13}) - 4q_3^2 a_{12}A_{13}\right)\sin 2q_3\omega \cos(1-\zeta)q_3\omega$$

$$+ 4q_3\omega A_{55}\left(a_{23}(A_{55} + 2A_{13}) - 2a_{12}A_{13}q_3^2\right)\cos(1+\zeta)q_3\omega\Big)\exp(-\omega x),$$

$$\overset{b}{u}{}^{(0)} = \frac{1}{2}q_3 \sum_{(\omega)} B^{(0)}_{\omega} A_{55}^2(3A_{55} + 8A_{13})(1+\zeta)\sin 2q_3\omega \sin(1-\zeta)q_3\omega$$

$$-2A_{55}q_3\omega(1-\zeta)\sin(1+\zeta)q_3\omega\Big)\exp(-\omega x),$$

$$\overset{b}{w}{}^{(0)} = \frac{1}{2}q_3 \sum_{(\omega)} \frac{1}{\omega} B^{(0)}_{\omega} A_{55}(A_{55}(3A_{55} + 8A_{13})$$

$$\times q_3\omega(1+\zeta)\sin 2q_3\omega \cos(1-\zeta)q_3\omega$$

$$-2A_{55}^2 q_3^2\omega^2(1-\zeta)\cos(1+\zeta)q_3\omega + (3A_{55} + 8A_{13})^2 \sin 2q_3\omega \sin(1-\zeta)q_3\omega$$

$$-2A_{55}(3A_{55} + 8A_{13})q_3\omega \sin(1+\zeta)q_3\omega\Big)\exp(-\omega x),$$

$$\overset{b}{\sigma}{}^{(0)}_{\alpha\beta} = \overset{b}{\sigma}{}^{(0)}_{\beta\gamma} = \overset{b}{v}{}^{(0)} = 0,$$

with ω denoting a root of the transcendental equation

$$\sin 2q_3\omega = \pm \frac{2A_{55}q_3\omega}{3A_{55} + 8A_{13}}. \tag{8.5}$$

The plane boundary layer solution associated with the complex conjugate roots of (7.10) may be obtained analogously, with the equation in respect of ω written as

$$\sin 2q\omega = \pm m \sinh 2p\omega, \tag{8.6}$$

where the values of p, q, m are determined by (7.10), (7.11).

In view of (6.8), (6.9), (6.18) and (8.1) the leading order (at $s = 0$) anti-plane boundary layer solution is given by

$$\overset{a}{\sigma}{}^{(0)}_{\gamma\gamma} = \overset{a}{\sigma}{}^{(0)}_{\alpha\alpha} = \overset{a}{\sigma}{}^{(0)}_{\alpha\gamma} = \overset{a}{\sigma}{}^{(0)}_{\beta\beta} = \overset{a}{u}{}^{(0)} = \overset{a}{w}{}^{(0)} = 0,$$

$$\overset{a}{\sigma}{}^{(0)}_{\beta\gamma} = \sum_{n=1}^{\infty} D^{(0)}_n \cos\frac{\pi n}{2}(1+\zeta)\exp(-\omega_n x),$$

$$\overset{a}{\sigma}{}^{(0)}_{\alpha\beta} = -\sqrt{\frac{a_{44}}{a_{66}}} \sum_{n=1}^{\infty} D^{(0)}_n \sin\frac{\pi n}{2}(1+\zeta)\exp(-\omega_n x), \tag{8.7}$$

$$\overset{a}{v}{}^{(0)} = \frac{2}{\pi}a_{44} \sum_{n=1}^{\infty} \frac{1}{n} D^{(0)}_n \sin\frac{\pi n}{2}(1+\zeta)\exp(-\omega_n x),$$

$$\omega_n = \sqrt{\frac{a_{66}}{a_{44}}}\frac{\pi n}{2}.$$

We remark that the arbitrary constants $A^{(0)}_{\omega}$, $B^{(0)}_{\omega}$, $D^{(0)}_n$ involved in solutions (8.2), (8.4), and (8.7), have to be determined from the edge boundary conditions, i.e. the matching procedure of the outer solution and the boundary layer.

Similarly to the previously considered case of the first boundary value problem (here and below in this chapter it is meant that the terminology of first, second or mixed problems corresponds to the boundary conditions on the face $\gamma = h$) the systems (6.8), (6.9) should be analyzed for $s \geq 1$ for two cases in respect of ω corresponding to the plane and anti-plane boundary layer solutions, denoted as before by $\overset{b}{Q}$ and $\overset{a}{Q}$, given by (7.14) and (7.15). We note that the main properties of the boundary layer solutions for $s \geq 1$ are essentially those observed in Section 6.7 for the first boundary value problem.

The boundary layer functions decay as $\exp(-\omega x)$. Since the real parts $Re\,\omega$ corresponding to equations (8.3), (8.5) and (8.6) are increasing rapidly, the associated boundary layer functions decay away from the edge of the plate. This reduces the number of boundary layer functions required for reasonable accuracy in practical results. As an example of that, we present several of the first values of $Re\,\omega$ for some typical materials.

a) Fiberglass ASTT(b) -C_2-O and PN-3

$$E_1 = 17.6 \cdot 10^3\,\text{MPa}, \quad E_2 = 12.9 \cdot 10^3\,\text{MPa}, \; E_3 = 4.2 \cdot 10^3\,\text{MPa},$$
$$G_{12} = 2.7 \cdot 10^3\,\text{MPa}, \quad G_{23} = 2.4 \cdot 10^3\,\text{MPa}, \; G_{13} = 2.4 \cdot 10^3\,\text{MPa},$$
$$\nu_{12} = 0.15, \quad \nu_{23} = 0.31, \quad \nu_{31} = 0.08 : \qquad (8.8)$$
$$Re\,\omega = 0.5653; \; 1,9001; \; 3.2265; \; 3.8569; \; 4.4288; \; 7.0246;$$
$$7.7142; \quad 8.1634,$$

b) Graphite-epoxy

$$E_1 = 7.3 \cdot 10^3\,\text{MPa}, \quad E_2 = 7.3 \cdot 10^3\,\text{MPa}, \; E_3 = 84.7 \cdot 10^3\,\text{MPa},$$
$$G_{12} = 2.7 \cdot 10^3\,\text{MPa}, \quad G_{23} = 4.2 \cdot 10^3\,\text{MPa}, \; G_{13} = 4.2 \cdot 10^3\,\text{MPa},$$
$$\nu_{12} = 0.323, \quad \nu_{23} = 0.026, \quad \nu_{31} = 0.30 : \qquad (8.9)$$
$$Re\,\omega = 1.1304; \; 3.4991; \; 5.8461; \; 12.9179; \; 13.7722; \; 15.1748;$$
$$17.537; \quad 19.8835,$$

c) Fiberglass (2:1)

$$E_1 = 36 \cdot 10^3\,\text{MPa}, \quad E_2 = 26.3 \cdot 10^3\,\text{MPa}, \; E_3 = 10.8 \cdot 10^3\,\text{MPa},$$
$$G_{12} = 4.9 \cdot 10^3\,\text{MPa}, \quad G_{23} = 4 \cdot 10^3\,\text{MPa}, \; G_{13} = 4.4 \cdot 10^3\,\text{MPa},$$
$$\nu_{12} = 0.105, \quad \nu_{23} = 0.431, \quad \nu_{31} = 0.405 : \qquad (8.10)$$
$$Re\,\omega = 0.3496; \; 2.1848; \; 3.8532; \; 7.8252; \; 11.039; \; 11.8449;$$
$$15.4033; \quad 16.3699.$$

6.9 Boundary Layer for the Mixed Boundary Value Problem

In this section we present the boundary layer solutions for the mixed boundary value problem, when the face plane $\gamma = -h$ of the plate is rigidly fixed, whereas on the opposite face $\gamma = h$ the boundary conditions are prescribed in the form

$$u_\alpha(-h) = u_\beta(-h) = u_\gamma(-h) = 0, \tag{9.1}$$
$$u_\alpha(h) = u_\beta(h) = 0, \ \sigma_{\gamma\gamma}(h) = 0.$$

In case of the purely imaginary roots of the first characteristic equation of (6.18) given by (7.2), the solution is written as

$$\overset{b}{\sigma}\,\overset{(0)}{\gamma\gamma} = \sum_{(\omega)} A_\omega^{(0)}(p_1 \sin 2q_2\omega \sin(1 - \zeta)q_1\omega$$

$$-p_2 \sin 2q_1\omega \sin(1 - \zeta)q_2\omega)\exp(-\omega x),$$

$$\overset{b}{\sigma}\,\overset{(0)}{\alpha\gamma} = \sum_{(\omega)} A_\omega^{(0)}(p_2 q_2 \sin 2q_1\omega \cos(1 - \zeta)q_2\omega$$

$$-p_1 q_1 \sin 2q_2\omega \cos(1 - \zeta)q_1\omega)\exp(-\omega x),$$

$$\overset{b}{\sigma}\,\overset{(0)}{\alpha\alpha} = \sum_{(\omega)} A_\omega^{(0)}(p_2 q_2^2 \sin 2q_1\omega \sin(1 - \zeta)q_2\omega$$

$$-p_1 q_1^2 \sin 2q_2\omega \sin(1 - \zeta)q_1\omega)\exp(-\omega x),$$

$$\overset{b}{\sigma}\,\overset{(0)}{\beta\beta} = \frac{1}{a_{22}} \sum_{(\omega)} A_\omega^{(0)}(p_1(a_{12}q_1^2 - a_{23}) \sin 2q_2\omega \sin(1 - \zeta)q_1\omega$$

$$-p_2(a_{12}q_2^2 - a_{23}) \sin 2q_1\omega \sin(1 - \zeta)q_2\omega)\exp(-\omega x), \tag{9.2}$$

$$\overset{b}{u}\,\overset{(0)}{} = \sum_{(\omega)} \frac{1}{2\omega} A_\omega^{(0)} p_1 p_2 (\sin 2q_2\omega \sin(1 - \zeta)q_1\omega$$

$$- \sin 2q_1\omega \sin(1 - \zeta)q_2\omega)\exp(-\omega x),$$

$$\overset{b}{w}\,\overset{(0)}{} = \sum_{(\omega)} \frac{1}{2\omega} A_\omega^{(0)}(q_1 p_1^2 \sin 2q_2\omega \cos(1 - \zeta)q_1\omega$$

$$-q_2 p_2^2 \sin 2q_1\omega \cos(1 - \zeta)q_2\omega)\exp(-\omega x),$$

$$\overset{b}{\sigma}\,\overset{(0)}{\alpha\beta} = \overset{b}{\sigma}\,\overset{(0)}{\beta\gamma} = \overset{b}{v}\,\overset{(0)}{} = 0,$$

where ω, $(Re\,\omega > 0)$ is the solution of the following equation

$$(q_1 p_1^2 - q_2 p_2^2) \sin 2(q_1 + q_2)\omega - (q_1 p_1^2 + q_2 p_2^2) \sin 2(q_1 - q_2)\omega = 0. \tag{9.3}$$

In the cases of the repeated roots (7.7) and complex conjugate roots (7.10) we obtain the transcendental equations

$$\sin 4q_3\omega + \frac{A_{55}}{3A_{55} + 8A_{13}}(4q_3\omega) = 0 \tag{9.4}$$

and

$$\sin 4q\omega + m \sinh 4p\omega = 0, \tag{9.5}$$

respectively.

In case of the mixed boundary conditions

$$u_\alpha(-h) = u_\beta(-h) = u_\gamma(-h) = 0, \tag{9.6}$$
$$\sigma_{\alpha\gamma}(h) = \sigma_{\beta\gamma}(h) = 0, \ u_\gamma(h) = 0$$

for (7.2) the plane boundary layer solution takes the form

$$\overset{b}{\sigma}{}^{(0)}_{\gamma\gamma} = \sum_{(\omega)} B^{(0)}_{\omega}(p_1 \cos 2q_2\omega \cos(1-\zeta)q_1\omega$$

$$-p_2 \cos 2q_1\omega \cos(1-\zeta)q_2\omega) \exp(-\omega x),$$

$$\overset{b}{\sigma}{}^{(0)}_{\alpha\gamma} = \sum_{(\omega)} B^{(0)}_{\omega}(p_1 q_1 \cos 2q_2\omega \sin(1-\zeta)q_1\omega$$

$$-p_2 q_2 \cos 2q_1\omega \sin(1-\zeta)q_2\omega) \exp(-\omega x),$$

$$\overset{b}{\sigma}{}^{(0)}_{\alpha\alpha} = \sum_{(\omega)} B^{(0)}_{\omega}(p_2 q_2^2 \cos 2q_1\omega \cos(1-\zeta)q_2\omega$$

$$-p_1 q_1^2 \cos 2q_2\omega \cos(1-\zeta)q_1\omega) \exp(-\omega x),$$

$$\overset{b}{\sigma}{}^{(0)}_{\beta\beta} = \frac{1}{a_{22}} \sum_{(\omega)} B^{(0)}_{\omega}(p_1(a_{12}q_1^2 - a_{23}) \cos 2q_2\omega \cos(1-\zeta)q_1\omega$$

$$-p_2(a_{12}q_2^2 - a_{23}) \cos 2q_1\omega \cos(1-\zeta)q_1\omega) \exp(-\omega x), \qquad (9.7)$$

$$\overset{b}{u}{}^{(0)} = \sum_{(\omega)} \frac{1}{2\omega} B^{(0)}_{\omega} p_1 p_2 (\cos 2q_2\omega \cos(1-\zeta)q_1\omega$$

$$- \cos 2q_1\omega \cos(1-\zeta)q_2\omega) \exp(-\omega x),$$

$$\overset{b}{w}{}^{(0)} = \sum_{(\omega)} \frac{1}{2\omega} B^{(0)}_{\omega}(q_2 p_2^2 \cos 2q_1\omega \sin(1-\zeta)q_2\omega$$

$$-q_1 p_1^2 \cos 2q_2\omega \sin(1-\zeta)q_1\omega) \exp(-\omega x),$$

$$\overset{b}{\sigma}{}^{(0)}_{\alpha\beta} = \overset{b}{\sigma}{}^{(0)}_{\beta\gamma} = \overset{b}{v}{}^{(0)} = 0,$$

with ω being the root of the following secular equation

$$\left(q_1 p_1^2 - q_2 p_2^2\right) \sin 2(q_1 + q_2)\omega + \left(q_1 p_1^2 + q_2 p_2^2\right) \sin 2(q_1 - q_2)\omega = 0. \qquad (9.8)$$

The solutions for repeated imaginary roots and complex conjugate roots (7.2) and (7.10) may be written similarly, with the general integral analogous to (9.7). However, for the sake of brevity, these expressions are not presented here. We note only the forms of the associated transcendental equations for the exponential orders ω describing the attenuation of the boundary layer. These equations are given below

$$\sin 4q_3\omega = \frac{A_{55}}{3A_{55} + 8A_{13}}(4q_3\omega) \qquad (9.9)$$

and

$$\sin 4q\omega = m \sinh 4p\omega. \qquad (9.10)$$

A comparison of the transcendental equations (8.3), (8.5) and (8.6) with (9.3)-(9.5) and (9.8)-(9.10) shows that in case of the plane boundary layer solution of the second boundary value problem (with zero displacements assumed on the faces) the order of exponential decay is twice that in case of the plate with one of the faces free in tangential or normal direction. On the other hand, the comparison of

(7.11) and (9.10) reveals the same speed of attenuation of the plane boundary layer in case of one of the faces free or free in tangential direction in absence of normal displacements.

We also note that the anti-plane boundary layer of the boundary value problem (9.1) coincides at $s = 0$ with the corresponding anti-plane boundary layer of the second boundary value problem (8.1), with the solution expressed as (8.7). Similarly, the anti-plane boundary layer of the problem (9.6) coincides with the anti-plane boundary layer solution of the first boundary value problem (7.1), see (7.13).

At higher orders $s \geq 1$ the structure of the boundary layer is generally written in the form of (7.15).

6.10 Matching of the Outer Solution with the Boundary Layer

Since the solution of the spatial problem of elasticity for the anisotropic plate contains the outer solution and the boundary layer solution, which in turn is combined of the plane and anti-plane components, the general solution of the 3D boundary value problem for an orthotropic plate may be written as

$$J = Q^{out} + \overset{b}{Q} + \overset{a}{Q}, \tag{10.1}$$

where $Q^{out}, \overset{b}{Q}, \overset{a}{Q}$ are the outer solution and the plane and anti-plane boundary layer solutions, respectively. According to the previous results of this Chapter, the outer solution does not involve and arbitrary constants which have to be determined, the plane boundary layer solution contains two groups of unknown constants (functions of η) in view of (7.14) and (7.15), with one more set of constants (functions of η) to be determined for the anti-plane boundary layer solution.

It has been demonstrated in Section 6.7 that the boundary layer solutions do not involve arbitrary constants, and the structure of the homogeneous component of the solution does not depend on the order of approximation s. Hence, in view of the conjugate values $\bar{\omega}$, all of the quantities $\overset{b}{Q}\overset{(s)}{_0}$ of the plane boundary layer may be represented as

$$\overset{b}{Q}\overset{(s)}{_0} = \sum_{(\omega)} \left(E_\omega^{(s)}(\eta) Re\,\Omega_Q(x,\varsigma) + F_\omega^{(s)}(\eta) Jm\,\Omega_Q(x,\varsigma) \right). \tag{10.2}$$

Here for all of the displacement and stress components Ω_Q are functions associated with the complex constants $C_\omega^{(0)}, B_\omega^{(0)} A_\omega^{(0)}$, see (7.4), (7.8), (8.2), (8.4), (9.2), and (9.7). The expression for $\overset{a}{Q}\overset{(s)}{_0}$ has a structure of (7.13) or (8.7), and contains one set of real functions $D_n^{(s)}(\eta)$.

Thus, the general solution (10.1) contains three sets of real functions $E_\omega^{(s)}(\eta), F_\omega^{(s)}(\eta), D_n^{(s)}(\eta)$, forming a countable set, that need to be determined from the edge boundary conditions. For a considered class of non-classical problems all of these unknown functions belong to the boundary layer, which is a clear distinction from the classical problems of the plate theory.

Let us describe the procedure for evaluating the unknown functions from the edge boundary conditions. Let the conditions at the edge surface $\alpha = \alpha_0$ be specified in the form of

the first boundary value problem

$$\sigma_{\alpha\alpha}(\alpha_0) = \varepsilon^{-1}\sigma_{\alpha\alpha}^0(\eta,\zeta), \quad \sigma_{\alpha\beta}(\alpha_0) = \varepsilon^{-1}\sigma_{\alpha\beta}^0(\eta,\zeta) \tag{10.3}$$
$$\sigma_{\alpha\gamma}(\alpha_0) = \varepsilon^{-1}\sigma_{\alpha\gamma}^0(\eta,\zeta);$$

second boundary value problem

$$u_\alpha(\alpha_0) = u^0(\eta,\zeta), \quad u_\beta(\alpha_0) = v^0(\eta,\zeta) \tag{10.4}$$
$$u_\gamma(\alpha_0) = w^0(\eta,\zeta);$$

mixed boundary value problem

$$\sigma_{\alpha\alpha}(\alpha_0) = \varepsilon^{-1}\sigma_{\alpha\alpha}^0(\eta,\zeta), \quad u_\gamma(\alpha_0) = w^0(\eta,\zeta) \tag{10.5}$$
$$u_\beta(\alpha_0) = v^0(\eta,\zeta),$$

or

$$u_\alpha(\alpha_0) = u^0(\eta,\zeta), \quad \sigma_{\alpha\beta}(\alpha_0) = \varepsilon^{-1}\sigma_{\alpha\beta}^0(\eta,\zeta) \tag{10.6}$$
$$\sigma_{\alpha\gamma}(\alpha_0) = \varepsilon^{-1}\sigma_{\alpha\gamma}^0(\eta,\zeta).$$

According to (10.1), (6.7), and (7.15) we have

$$\sigma_{\alpha\alpha}^{(s)} = \overset{out}{\sigma}{}_{\alpha\alpha}^{(s)} + \overset{b}{\sigma}{}_{\alpha\alpha}^{(s)} + \overset{b}{\sigma}{}_{\alpha\alpha}^{*(s)} + \overset{a}{\sigma}{}_{\alpha\alpha}^{(s)} + \overset{a}{\sigma}{}_{\alpha\alpha}^{*(s)}, \qquad (\alpha,\beta),$$
$$\sigma_{\alpha\beta}^{(s)} = \overset{out}{\sigma}{}_{\alpha\beta}^{(s)} + \overset{b}{\sigma}{}_{\alpha\beta}^{(s)} + \overset{b}{\sigma}{}_{\alpha\beta}^{*(s)} + \overset{a}{\sigma}{}_{\alpha\beta}^{(s)} + \overset{a}{\sigma}{}_{\alpha\beta}^{*(s)}, \qquad (\alpha,\gamma),$$
$$\sigma_{\alpha\gamma}^{(s)} = \overset{out}{\sigma}{}_{\alpha\gamma}^{(s)} + \overset{b}{\sigma}{}_{\alpha\gamma}^{(s)} + \overset{b}{\sigma}{}_{\alpha\gamma}^{*(s)} + \overset{a}{\sigma}{}_{\alpha\gamma}^{(s)} + \overset{a}{\sigma}{}_{\alpha\gamma}^{*(s)},$$
$$\sigma_{\gamma\gamma}^{(s)} = \overset{out}{\sigma}{}_{\gamma\gamma}^{(s)} + \overset{b}{\sigma}{}_{\gamma\gamma}^{(s)} + \overset{b}{\sigma}{}_{\gamma\gamma}^{*(s)} + \overset{a}{\sigma}{}_{\gamma\gamma}^{(s)} + \overset{a}{\sigma}{}_{\gamma\gamma}^{*(s)}, \tag{10.7}$$
$$u_{\alpha}^{(s)} = \overset{out}{u}{}_{\alpha}^{(s)} + \overset{b}{u}{}_{\alpha}^{(s)} + \overset{b}{u}{}_{\alpha}^{*(s)} + \overset{a}{u}{}_{\alpha}^{(s)} + \overset{a}{u}{}_{\alpha}^{*(s)},$$
$$u_{\gamma}^{(s)} = \overset{out}{u}{}_{\gamma}^{(s)} + \overset{b}{u}{}_{\gamma}^{(s)} + \overset{b}{u}{}_{\gamma}^{*(s)} + \overset{a}{u}{}_{\gamma}^{(s)} + \overset{a}{u}{}_{\gamma}^{*(s)},$$
$$u_{\beta}^{(s)} = \overset{out}{u}{}_{\beta}^{(s)} + \overset{b}{u}{}_{\beta}^{(s)} + \overset{b}{u}{}_{\beta}^{*(s)} + \overset{a}{u}{}_{\beta}^{(s)} + \overset{a}{u}{}_{\beta}^{*(s)}.$$

We note that all the quantities of the outer solution in (10.7) are known, $\overset{b}{Q}{}^{*(s)}$, $\overset{a}{Q}{}^{*(s)}$ are particular solutions of the non-homogeneous systems (6.8), (6.9), and $\overset{a}{\sigma}{}_{\alpha\alpha}^{(s)}$, $\overset{a}{\sigma}{}_{\alpha\gamma}^{(s)}$, $\overset{a}{u}{}_{\alpha}^{(s)}$, $\overset{a}{u}{}_{\gamma}^{(s)}$, $\overset{a}{\sigma}{}_{\gamma\gamma}^{(s)}$, $\overset{b}{\sigma}{}_{\alpha\beta}^{(s)}$, $\overset{b}{u}{}_{\beta}^{(s)}$ are associated with the corresponding boundary layers. $\overset{b}{\sigma}{}_{\alpha\alpha}^{(s)}$, $\overset{b}{\sigma}{}_{\alpha\gamma}^{(s)}$, $\overset{b}{u}{}_{\alpha}^{(s)}$, $\overset{b}{u}{}_{\gamma}^{(s)}$ contain the unknown functions $E_\omega^{(s)}(\eta)$, $F_\omega^{(s)}(\eta)$, whereas $\overset{a}{\sigma}{}_{\alpha\beta}^{(s)}$, $\overset{a}{u}{}_{\beta}^{(s)}$ involve the unknowns $D_n^{(s)}(\eta)$. Using

(2.2), (6.7), (10.2), (10.7) and satisfying conditions (10.3), we get

$$\sum_{(\omega)} \left(E_\omega^{(s)} Re\,\Omega_{\sigma_{\alpha\alpha}}(0,\zeta) + F_\omega^{(s)} J_m\,\Omega_{\sigma_{\alpha\alpha}}(0,\zeta) \right)$$

$$= \bar{\sigma}_{\alpha\alpha}^{(s)} - \overset{out}{\sigma}{}_{\alpha\alpha}^{(s)}(\alpha_0) - \overset{a}{\sigma}{}_{\alpha\alpha}^{(s)}(x=0) - \overset{a}{\sigma}{}_{\alpha\alpha}^{*(s)}(x=0) - \overset{b}{\sigma}{}_{\alpha\alpha}^{*(s)}(x=0),$$

$$\sum_{(\omega)} \left(E_\omega^{(s)} Re\,\Omega_{\sigma_{\alpha\gamma}}(0,\zeta) + F_\omega^{(s)} J_m\,\Omega_{\sigma_{\alpha\gamma}}(0,\zeta) \right) \qquad (10.8)$$

$$= \bar{\sigma}_{\alpha\gamma}^{(s)} - \overset{out}{\sigma}{}_{\alpha\gamma}^{(s)}(\alpha_0) - \overset{a}{\sigma}{}_{\alpha\gamma}^{(s)}(x=0) - \overset{a}{\sigma}{}_{\alpha\gamma}^{*(s)}(x=0) - \overset{b}{\sigma}{}_{\alpha\gamma}^{*(s)}(x=0),$$

$$\bar{\sigma}_{\alpha\alpha}^{(0)} = \sigma_{\alpha\alpha}^0, \quad \bar{\sigma}_{\alpha\gamma}^{(0)} = \sigma_{\alpha\gamma}^0, \quad \bar{\sigma}_{\alpha\alpha}^{(s)} = \bar{\sigma}_{\alpha\gamma}^{(s)} = 0 \quad s > 0$$

along with

$$-\sqrt{\frac{a_{44}}{a_{66}}} \sum_{n=1}^{\infty} D_n^{(s)} \sin\frac{\pi}{4}(2n+1)(1+\zeta)$$

$$= \bar{\sigma}_{\alpha\beta}^{(s)} - \overset{out}{\sigma}{}_{\alpha\beta}^{(s)}(\alpha_0) - \overset{b}{\sigma}{}_{\alpha\beta}^{(s)}(x=0) - \overset{a}{\sigma}{}_{\alpha\beta}^{*(s)}(x=0) - \overset{b}{\sigma}{}_{\alpha\beta}^{*(s)}(x=0), \qquad (10.9)$$

$$\bar{\sigma}_{\alpha\beta}^{(0)} = \sigma_{\alpha\beta}^0, \quad \bar{\sigma}_{\alpha\beta}^{(s)} = 0 \quad (s>0),$$

if the boundary layer satisfies

$$\sigma_{\alpha\gamma}(h) = \sigma_{\beta\gamma}(h) = 0; \quad \sigma_{\gamma\gamma}(h) = 0 \quad \text{or} \quad u_\gamma(h) = 0 \qquad (10.10)$$

or

$$-\sqrt{\frac{a_{44}}{a_{66}}} \sum_{n=1}^{\infty} D_n^{(s)} \sin\frac{\pi n}{2}(1+\zeta)$$

$$= \bar{\sigma}_{\alpha\beta}^{(s)} - \overset{out}{\sigma}{}_{\alpha\beta}^{(s)}(\alpha_0) - \overset{b}{\sigma}{}_{\alpha\beta}^{(s)}(x=0) - \overset{a}{\sigma}{}_{\alpha\beta}^{*(s)}(x=0) - \overset{b}{\sigma}{}_{\alpha\beta}^{*(s)}(x=0) \qquad (10.11)$$

if the boundary layer satisfies the following condition

$$u_\alpha(h) = u_\beta(h) = 0; \quad u_\gamma(h) = 0 \quad \text{or} \quad \sigma_{\gamma\gamma}(h) = 0. \qquad (10.12)$$

We remark that the plane boundary layer solution and the outer solution in (10.8) should correspond to the conditions (10.10) and (10.12).

The unknown coefficients $E_\omega^{(s)}$, $F_\omega^{(s)}$ of the plane boundary layer solution may be found approximately from the system (10.8) with required accuracy. As in the previously considered 2D problems, one may use the boundary collocation method, method of least squares. Another approach could be realized through the Fourier expansion along a system of orthonormal functions, leading to an infinite algebraic system for the unknown coefficients. The coefficients $D_n^{(s)}$ may be obtained exactly from the Fourier method. As follows from (10.9)

$$D_n^{(s)} = -\sqrt{\frac{a_{66}}{a_{44}}} \int_{-1}^{1} \left(\bar{\sigma}_{\alpha\beta}^{(s)} - \overset{out}{\sigma}{}_{\alpha\beta}^{(s)}(\alpha_0) - \overset{b}{\sigma}{}_{\alpha\beta}^{(s)}(x=0) \right.$$

$$\left. - \overset{a}{\sigma}{}_{\alpha\beta}^{*(s)}(x=0) - \overset{b}{\sigma}{}_{\alpha\beta}^{*(s)}(x=0) \right) \sin\frac{\pi}{4}(2n+1)(1+\zeta)d\zeta \qquad (10.13)$$

with conditions (10.1) implying

$$D_n^{(s)} = -\sqrt{\frac{a_{66}}{a_{44}}} \int_{-1}^{1} \left(\bar{\sigma}_{\alpha\beta}^{(s)} - \overset{out}{\sigma}{}_{\alpha\beta}^{(s)}(\alpha_0) - \overset{b}{\sigma}{}_{\alpha\beta}^{(s)}(x=0) \right.$$

$$\left. - \overset{a}{\sigma}{}_{\alpha\beta}^{*(s)}(x=0) - \overset{b}{\sigma}{}_{\alpha\beta}^{*(s)}(x=0) \right) \sin \frac{\pi n}{4}(1+\zeta)d\zeta. \tag{10.14}$$

Satisfying the conditions (10.4), one obtains

$$\sum_{(\omega)} \left(E_\omega^{(s)} Re\,\Omega_{u_\alpha}(0,\zeta) + F_\omega^{(s)} Jm\,\Omega_{u_\alpha}(0,\zeta) \right)$$

$$= \bar{u}_\alpha^{(s)} - \overset{out}{u}{}_\alpha^{(s)}(\alpha_0) - \overset{a}{u}{}_\alpha^{(s)}(x=0) - \overset{a}{u}{}_\alpha^{*(s)}(x=0) - \overset{b}{u}{}_\alpha^{*(s)}(x=0),$$

$$\sum_{(\omega)} \left(E_\omega^{(s)} Re\,\Omega_{u_\gamma}(0,\zeta) + F_\omega^{(s)} Jm\,\Omega_{u_\gamma}(0,\zeta) \right) \tag{10.15}$$

$$= \bar{u}_\gamma^{(s)} - \overset{out}{u}{}_\gamma^{(s)}(\alpha_0) - \overset{a}{u}{}_\gamma^{(s)}(x=0) - \overset{a}{u}{}_\gamma^{*(s)}(x=0) - \overset{b}{u}{}_\gamma^{*(s)}(x=0),$$

$$\bar{u}_\alpha^{(0)} = u^0, \quad \bar{u}_\gamma^{(0)} = w^0, \quad \bar{u}_\alpha^{(s)} = \bar{u}_\gamma^{(s)} = 0 \quad s > 0$$

and

$$\frac{4}{\pi}a_{44} \sum_{n=1}^{\infty} \frac{1}{2n+1} D_n^{(s)} \sin \frac{\pi}{4}(2n+1)(1+\zeta) = \bar{u}_\beta^{(s)}$$

$$- \overset{out}{u}{}_\beta^{(s)}(\alpha=\alpha_0) - \overset{b}{u}{}_\beta^{(s)}(x=0) - \overset{b}{u}{}_\beta^{*(s)}(x=0) - \overset{a}{u}{}_\beta^{*(s)}(x=0), \tag{10.16}$$

$$\bar{u}_\beta^{(0)} = v^{(0)}, \quad \bar{u}_\beta^{(s)} = 0 \quad (s>0)$$

when the boundary layer satisfies conditions (10.10), and

$$\frac{2}{\pi}a_{44} \sum_{n=1}^{\infty} \frac{1}{n} D_n^{(s)} \sin \frac{\pi n}{2}(1+\zeta) = \bar{u}_\beta^{(s)} - \overset{out}{u}{}_\beta^{(s)}(\alpha=\alpha_0) \tag{10.17}$$

$$- \overset{b}{u}{}_\beta^{(s)}(x=0) - \overset{b}{u}{}_\beta^{*(s)}(x=0) - \overset{a}{u}{}_\beta^{*(s)}(x=0)$$

in case of the boundary layer satisfying the relations (10.12).

It follows from (10.16) that

$$D_n^{(s)} = \frac{\pi}{4a_{44}}(2n+1) \int_{-1}^{1} \left(\bar{u}_\beta^{(s)} - \overset{out}{u}{}_\beta^{(s)}(\alpha=\alpha_0) - \overset{b}{u}{}_\beta^{(s)}(x=0) \right.$$

$$\left. - \overset{b}{u}{}_\beta^{*(s)}(x=0) - \overset{a}{u}{}_\beta^{*(s)}(x=0) \right) \sin \frac{\pi}{4}(2n+1)(1+\zeta)d\zeta \tag{10.18}$$

whereas conditions (10.17) give

$$D_n^{(s)} = \frac{\pi n}{2a_{44}} \int_{-1}^{1} \left(\bar{u}_\beta^{(s)} - \overset{out}{u}{}_\beta^{(s)}(\alpha=\alpha_0) - \overset{b}{u}{}_\beta^{(s)}(x=0) \right.$$

$$\left. - \overset{b}{u}{}_\beta^{*(s)}(x=0) - \overset{a}{u}{}_\beta^{*(s)}(x=0) \right) \sin \frac{\pi n}{2}(1+\zeta)d\zeta. \tag{10.19}$$

We note that the coefficients $E_\omega^{(s)}$, $F_\omega^{(s)}$ may be determined through a previously described method. The boundary layer solutions corresponding to the boundary conditions (10.5) and (10.6), may be treated similarly.

Chapter 7

Two-Layer Anisotropic Plates. The Modulus of a Layered Foundation

7.1 Formulation of Boundary Value Problems. Solution of the Outer Problem

The asymptotic method may also be applied to evaluation of stress-strain fields of layered thermo-elastic plates. In this chapter we shall illustrate this for two-layered plates. The generalization of the presented approach to more layers is straightforward, and is not presented here due to lengthy expressions. The results for two-layered plates are also related to the model of a compressible layer on elastic foundation. Within the asymptotic method it is possible to obtain relatively simple relations for all of the required quantities for both foundation and the layer without using any *ad hoc* hypotheses, which are typical for technical theories of foundations, see Vlasov and Leont'ev (1960); Gorbunov-Posadov et al. (1984); Wang et al. (2005).

Consider a 3D problem of thermo-elasticity for a two-layered anisotropic plate occupying the domain $\Omega = \{\alpha, \beta, \gamma : \alpha, \beta \in \Omega_0, -h_2 \leq \gamma \leq h_1\}$, where Ω_0 is the interface plane between the layers, and h_1, h_2 are thicknesses of the layers, see Fig. 7.1.

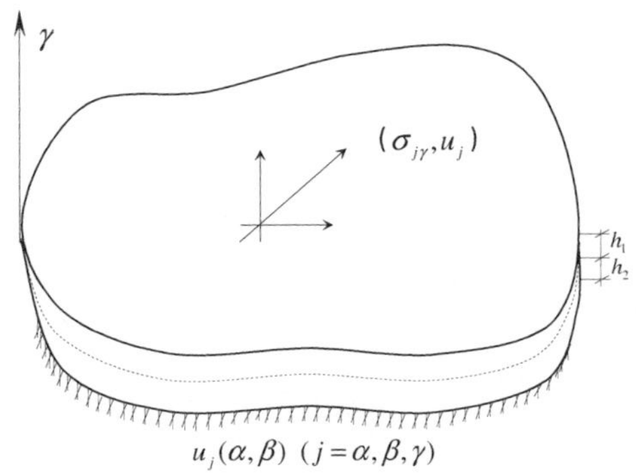

Fig. 7.1

Let us assume the action of given body forces $F_\alpha^{(i)}(\alpha, \beta, \gamma)$, (α, β, γ), in partic-

215

ular, taking into account the weight of the layers and temperature effects. Here we assume that the temperature effects are described by the thermo-elastic theory using the Duhamel-Neumann law, see e.g. Novatsky (1970). The quantities of the upper and lower layers are denoted with indices (1) and (2), respectively.

Let the boundary conditions at the face $\gamma = -h_2$ of the lower layer be formulated in terms of displacements (in particular, the lower face could be fixed)

$$u_\alpha^{(2)}(-h_2) = u^-(\alpha,\beta), \quad u_\beta^{(2)}(-h_2) = v^-(\alpha,\beta), \quad u_\gamma^{(2)}(-h) = w^-(\alpha,\beta) \qquad (1.1)$$

whereas at the upper face $\gamma = h_1$ we consider one of the following types of boundary conditions corresponding to

a) the first boundary value problem

$$\sigma_{\alpha\gamma}^{(1)}(h_1) = \varepsilon^{-1}\sigma_{\alpha\gamma}^+(\alpha,\beta), \quad \sigma_{\beta\gamma}^{(1)}(h_1) = \varepsilon^{-1}\sigma_{\beta\gamma}^+(\alpha,\beta),$$
$$\sigma_{\gamma\gamma}^{(1)}(h_1) = \varepsilon^{-1}\sigma_{\gamma\gamma}^+(\alpha,\beta) \qquad (1.2)$$

b) the second boundary value problem

$$u_\alpha^{(1)}(h_1) = u^+(\alpha,\beta), \quad u_\beta^{(1)}(h_1) = v^+(\alpha,\beta), \quad u_\gamma^{(1)}(h_1) = w^+(\alpha,\beta) \qquad (1.3)$$

c) mixed boundary value problem

$$\sigma_{\alpha\gamma}^{(1)}(h_1) = \varepsilon^{-1}\sigma_{\alpha\gamma}^+(\alpha,\beta), \quad \sigma_{\beta\gamma}^{(1)}(h_1) = \varepsilon^{-1}\sigma_{\beta\gamma}^+(\alpha,\beta),$$
$$u_\gamma^{(1)}(h_1) = w^+(\alpha,\beta) \qquad (1.4)$$

or

$$u_\alpha^{(1)}(h_1) = u^+(\alpha,\beta), \quad u_\beta^{(1)}(h_1) = v^+(\alpha,\beta), \quad \sigma_{\gamma\gamma}^{(1)}(h_1) = \varepsilon^{-1}\sigma_{\gamma\gamma}^+(\alpha,\beta). \qquad (1.5)$$

We also assume general anisotropy for each of the layers, characterized by 21 material parameters. The sought-for solution should also satisfy the boundary conditions imposed on the edge S_Ω, which will be specified later. We note that the edge boundary conditions do not affect the outer solution, but influence significantly on the form of the boundary layer solution. Similarly to the previous chapter, the interaction between the outer solution and the boundary layer solution will be observed at higher orders $s \geq 1$, whereas the leading order solutions $s = 0$ are independent of each other.

We also remark that according to the foundation model (Vlasov and Leont'ev, 1960; Gorbunov-Posadov et al., 1984) the system (1.1), (1.2) describes an elastic layer on a foundation. Consider now the equilibrium equations in 3D problem of elasticity expressed in curvilinear coordinates along with the constitutive relations taking into account temperature deformations in accordance with the Duhamel-Neumann's law. Let us introduce the dimensionless coordinates $\xi = \dfrac{\alpha}{a}$, $\eta = \dfrac{\beta}{a}$ and $\zeta = \dfrac{\gamma}{h_2} = \varepsilon^{-1}\dfrac{\gamma}{a}$, $\varepsilon = \dfrac{h_2}{a}$, $h_1 + h_2 \ll a$ along with the dimensionless displacements $u^{(i)} = \dfrac{u_\alpha^{(i)}}{a}$, $v^{(i)} = \dfrac{u_\beta^{(i)}}{a}$, $w^{(i)} = \dfrac{u_\gamma^{(i)}}{a}$. In case of $h_1 > h_2$ it is useful to introduce

$\zeta = \dfrac{\gamma}{h_1}$, $\varepsilon = \dfrac{h_1}{a}$, or alternatively, $\zeta = \dfrac{\gamma}{h}$, $\varepsilon = \dfrac{h}{a}$, where $h = \max(h_1, h_2)$, and a is typical size of a plate.

The governing system of equations for spatial problem of thermo-elasticity is written as

$$\frac{1}{A}\frac{\partial \sigma_{\alpha\alpha}^{(i)}}{\partial \xi} + \frac{1}{B}\frac{\partial \sigma_{\alpha\beta}^{(i)}}{\partial \eta} + \varepsilon^{-1}\frac{\partial \sigma_{\alpha\gamma}^{(i)}}{\partial \zeta} + a(\sigma_{\alpha\alpha}^{(i)} - \sigma_{\beta\beta}^{(i)})k_\beta$$

$$+ 2ak_\alpha \sigma_{\alpha\beta}^{(i)} + aF_\alpha^{(i)} = 0, \quad (\alpha, \beta; \xi, \eta),$$

$$\frac{1}{A}\frac{\partial \sigma_{\alpha\gamma}^{(i)}}{\partial \xi} + \frac{1}{B}\frac{\partial \sigma_{\beta\gamma}^{(i)}}{\partial \eta} + \varepsilon^{-1}\frac{\partial \sigma_{\gamma\gamma}^{(i)}}{\partial \zeta} + ak_\beta \sigma_{\alpha\gamma}^{(i)} + ak_\alpha \sigma_{\beta\gamma}^{(i)} + aF_\gamma^{(i)} = 0,$$

$$\frac{1}{A}\frac{\partial u^{(i)}}{\partial \xi} + ak_\alpha v^{(i)} = a_{11}^{(i)}\sigma_{\alpha\alpha}^{(i)} + a_{12}^{(i)}\sigma_{\beta\beta}^{(i)}$$

$$+ a_{13}^{(i)}\sigma_{\gamma\gamma}^{(i)} + \ldots + a_{16}^{(i)}\sigma_{\alpha\beta}^{(i)} + \alpha_{11}^{(i)}\theta^{(i)}, \quad (1,2; \alpha, \beta; \xi, \eta; u, v),$$

$$\varepsilon^{-1}\frac{\partial w^{(i)}}{\partial \zeta} = a_{13}^{(i)}\sigma_{\alpha\alpha}^{(i)} + a_{23}^{(i)}\sigma_{\beta\beta}^{(i)} + \ldots + a_{36}^{(i)}\sigma_{\alpha\beta}^{(i)} + \alpha_{33}^{(i)}\theta^{(i)}, \qquad (1.6)$$

$$\frac{1}{B}\frac{\partial u^{(i)}}{\partial \eta} + \frac{1}{A}\frac{\partial v^{(i)}}{\partial \xi} - a(k_\alpha u^{(i)} - k_\beta v^{(i)}) = a_{16}^{(i)}\sigma_{\alpha\alpha}^{(i)}$$

$$+ a_{26}^{(i)}\sigma_{\beta\beta}^{(i)} + \ldots + a_{66}^{(i)}\sigma_{\alpha\beta}^{(i)} + \alpha_{12}^{(i)}\theta^{(i)},$$

$$\frac{1}{A}\frac{\partial w^{(i)}}{\partial \xi} + \varepsilon^{-1}\frac{\partial u^{(i)}}{\partial \zeta} = a_{15}^{(i)}\sigma_{\alpha\alpha}^{(i)} + a_{25}^{(i)}\sigma_{\beta\beta}^{(i)} + \ldots + a_{56}^{(i)}\sigma_{\alpha\beta}^{(i)} + \alpha_{13}^{(i)}\theta^{(i)},$$

$$\frac{1}{B}\frac{\partial w^{(i)}}{\partial \eta} + \varepsilon^{-1}\frac{\partial v^{(i)}}{\partial \zeta} = a_{14}^{(i)}\sigma_{\alpha\alpha}^{(i)} + a_{24}^{(i)}\sigma_{\beta\beta}^{(i)} + \ldots + a_{46}^{(i)}\sigma_{\alpha\beta}^{(i)} + \alpha_{23}^{(i)}\theta^{(i)},$$

where A, B are coefficients of the first quadratic form, k_α, k_β are geodesic curvatures, $a_{jk}^{(i)}$ are the pliability elastic coefficients, $\alpha_{jk}^{(i)}$ are coefficients of thermal expansion, $\theta^{(i)} = T^{(i)} - T_0^{(i)}$ is temperature variation, with $i = 1$ corresponding to the upper layer $0 \leq \gamma \leq h_1$, and $i = 2$ associated with the lower layer $-h_2 \leq \gamma \leq 0$. We remark that variability along coordinate lines may also be taken into account in equations (1.6), see Goldenveizer (1961, 1976).

Since the system (1.6) is singularly perturbed by a small parameter ε, its solution may be once again represented in terms of the outer solution and a boundary layer.

The outer solution is sought in the form (Aghalovyan and Gevorkyan, 1986)

$$Q^{(i)} = \sum_{s=0}^{S} \varepsilon^{\chi_Q + s} Q^{(i,s)} \quad (i = 1, 2) \qquad (1.7)$$

where $Q^{(i)}$ is any of the sought-for displacement or stress components. Substitution of (1.7) into (1.6) reveals the values of $\chi_Q = -1$ for stresses, and $\chi_Q = 0$ for displacements in order to construct consistent asymptotic procedure to determine $Q^{(i,s)}$.

We assume that the influence of the volume forces and temperature effects are of the same asymptotic order as that of the external loading, i.e. temperature caused terms and volume forces arise at leading order. This is possible when

$$F_\alpha^{(i)} = \sum_{s=0}^{S} \varepsilon^{-2+s} F_{\alpha s}^{(i)}(\xi, \eta, \zeta), \quad (\alpha, \beta, \gamma), \tag{1.8}$$

$$\Theta^{(i)} = \sum_{s=0}^{S} \varepsilon^{-1+s} \Theta_s^{(i)}(\xi, \eta, \zeta).$$

It may be observed from (1.7), (1.8) that the order of the stress-strain field caused by the volume forces, external load and temperature field will be of the same order with the field induced by the displacements of the faces provided the intensity of the volume forces exceeds by ε^{-2}, and the intensity of surface loads and temperature fields exceeds by ε^{-1} the intensity of the displacements imposed on the faces of the plate. In the opposite case the associated terms arise at higher orders only.

Following the usual procedure described in more details in the previous chapters, substituting (1.7) into (1.6) and making note of (1.8), one arrives at a system with respect to $Q^{(i,s)}$, which should be resolved subject to conditions (1.1) along with the continuity conditions at the interface $\gamma = 0$ given by

$$u_\alpha^{(1)} = u_\alpha^{(2)}, \quad u_\beta^{(1)} = u_\beta^{(2)}, \quad u_\gamma^{(1)} = u_\gamma^{(2)}, \tag{1.9}$$

$$\sigma_{\alpha\gamma}^{(1)} = \sigma_{\alpha\gamma}^{(2)}, \quad \sigma_{\beta\gamma}^{(1)} = \sigma_{\beta\gamma}^{(2)}, \quad \sigma_{\gamma\gamma}^{(1)} = \sigma_{\gamma\gamma}^{(2)}.$$

As a result, a system of recurrent relations for $Q^{(i,s)}$ is obtained, namely

$$\sigma_{\alpha\alpha}^{(i,s)} = A_{13}^{(i)} \sigma_{\gamma\gamma 0}^{(s)} + A_{14}^{(i)} \sigma_{\beta\gamma 0}^{(s)} + A_{15}^{(i)} \sigma_{\alpha\gamma 0}^{(s)} + \sigma_{\alpha\alpha *}^{(i,s)}(\xi, \eta, \zeta), \quad (\alpha, \beta; 1, 2),$$

$$\sigma_{\alpha\beta}^{(i,s)} = A_{63}^{(i)} \sigma_{\gamma\gamma 0}^{(s)} + A_{64}^{(i)} \sigma_{\beta\gamma 0}^{(s)} + A_{65}^{(i)} \sigma_{\alpha\gamma 0}^{(s)} + \sigma_{\alpha\beta *}^{(i,s)}(\xi, \eta, \zeta),$$

$$\sigma_{\alpha\gamma}^{(i,s)} = \sigma_{\alpha\gamma 0}^{(s)}(\xi, \eta) + \sigma_{\alpha\gamma *}^{(i,s)}(\xi, \eta, \zeta), \quad (\alpha, \beta),$$

$$\sigma_{\gamma\gamma}^{(i,s)} = \sigma_{\gamma\gamma 0}^{(s)}(\xi, \eta) + \sigma_{\gamma\gamma *}^{(i,s)}(\xi, \eta, \zeta),$$

$$u^{(i,s)} = D_{53}^{(i)} \sigma_{\gamma\gamma 0}^{(s)} + D_{54}^{(i)} \sigma_{\beta\gamma 0}^{(s)} + D_{55}^{(i)} \sigma_{\alpha\gamma 0}^{(s)} \tag{1.10}$$

$$+ u^{-(s)} - u_*^{(2,s)}(\zeta = -1) + u_*^{(i,s)}(\xi, \eta, \zeta),$$

$$v^{(i,s)} = D_{43}^{(i)} \sigma_{\gamma\gamma 0}^{(s)} + D_{44}^{(i)} \sigma_{\beta\gamma 0}^{(s)} + D_{45}^{(i)} \sigma_{\alpha\gamma 0}^{(s)}$$

$$+ v^{-(s)} - v_*^{(2,s)}(\zeta = -1) + v_*^{(i,s)}(\xi, \eta, \zeta),$$

$$w^{(i,s)} = D_{33}^{(i)} \sigma_{\gamma\gamma 0}^{(s)} + D_{34}^{(i)} \sigma_{\beta\gamma 0}^{(s)} + D_{35}^{(i)} \sigma_{\alpha\gamma 0}^{(s)}$$

$$+ w^{-(s)} - w_*^{(2,s)}(\zeta = -1) + w_*^{(i,s)}(\xi, \eta, \zeta), \quad (i = 1, 2),$$

where

$$u^{-(0)} = \frac{u^-}{a}, \quad u^{-(s)} = 0, \quad s > 0, \quad (u, v, w),$$

$$A_{kl}^{(i)} = -a_{1l}^{(i)} B_{k1}^{(i)} - a_{2l}^{(i)} B_{k2}^{(i)} - a_{6l}^{(i)} B_{k6}^{(i)},$$

$$A_{ml}^{(i)} = a_{m1}^{(i)} A_{1l}^{(i)} + a_{m2}^{(i)} A_{2l}^{(i)} + a_{m6}^{(i)} A_{6l}^{(i)} + a_{ml}^{(i)},$$

$$A_{ml}^{(i)} \neq A_{lm}^{(i)}, \quad (m, l = 3, 4, 5),$$

$$B_{nj}^{(i)} = \frac{(a_{nk}^{(i)} a_{jk}^{(i)} - a_{nj}^{(i)} a_{kk}^{(i)})}{\Delta^{(i)}}, \quad (n \neq j \neq k \neq n),$$

$$B_{kk}^{(i)} = \frac{(a_{nn}^{(i)} a_{jj}^{(i)} - (a_{nj}^{(i)})^2)}{\Delta^{(i)}}, \quad B_{nj}^{(i)} = B_{jn}^{(i)},$$

$$(n, j, k = 1, 2, 6),$$

$$D_{ml}^{(i)} = \zeta A_{ml}^{(i)} + A_{ml}^{(2)},$$

$$\Delta^{(i)} = a_{11}^{(i)} a_{22}^{(i)} a_{66}^{(i)} + 2 a_{12}^{(i)} a_{26}^{(i)} a_{16}^{(i)} \tag{1.11}$$

$$- a_{11}^{(i)} (a_{26}^{(i)})^2 - a_{22}^{(i)} (a_{16}^{(i)})^2 - a_{66}^{(i)} (a_{12}^{(i)})^2,$$

$$\sigma_{\alpha\gamma*}^{(i,s)} = -\int_0^\zeta \left(\frac{1}{A} \frac{\partial \sigma_{\alpha\alpha}^{(i,s-1)}}{\partial \xi} + \frac{1}{B} \frac{\partial \sigma_{\alpha\beta}^{(i,s-1)}}{\partial \eta} \right.$$

$$+ ak_\beta \left(\sigma_{\alpha\alpha}^{(i,s-1)} - \sigma_{\beta\beta}^{(i,s-1)} \right)$$

$$\left. + 2ak_\alpha \sigma_{\alpha\beta}^{(i,s-1)} + aF_{\alpha s}^{(i)} \right) d\zeta, \quad (\alpha, \beta; \xi, \eta; A, B)$$

and

$$\sigma_{\gamma\gamma*}^{(i,s)} = -\int_0^\zeta \left(\frac{1}{A} \frac{\partial \sigma_{\alpha\gamma}^{(i,s-1)}}{\partial \xi} + \frac{1}{B} \frac{\partial \sigma_{\beta\gamma}^{(i,s-1)}}{\partial \eta} \right.$$

$$\left. + ak_\beta \sigma_{\alpha\gamma}^{(i,s-1)} + ak_\alpha \sigma_{\beta\gamma}^{(i,s-1)} + aF_{\gamma s}^{(i)} \right) d\zeta,$$

$$\sigma_{\alpha\alpha*}^{(i,s)} = B_{11}^{(i)} R_1^{(i,s)} + B_{12}^{(i)} R_2^{(i,s)} + B_{16}^{(i)} R_3^{(i,s)},$$

$$\sigma_{\beta\beta*}^{(i,s)} = B_{12}^{(i)} R_1^{(i,s)} + B_{22}^{(i)} R_2^{(i,s)} + B_{26}^{(i)} R_3^{(i,s)}, \tag{1.12}$$

$$\sigma_{\alpha\beta*}^{(i,s)} = B_{16}^{(i)} R_1^{(i,s)} + B_{26}^{(i)} R_2^{(i,s)} + B_{66}^{(i)} R_3^{(i,s)},$$

$$u_*^{(i,s)} = \int_0^\zeta \left(a_{15}^{(i)} \sigma_{\alpha\alpha*}^{(i,s)} + a_{25}^{(i)} \sigma_{\beta\beta*}^{(i,s)} + \ldots + a_{56}^{(i)} \sigma_{\alpha\beta*}^{(i,s)} \right.$$

$$\left. - \frac{1}{A} \frac{\partial w^{(i,s-1)}}{\partial \xi} + \alpha_{13}^{(i)} \theta_s^{(i)} \right) d\zeta,$$

$$v_*^{(i,s)} = \int_0^\zeta \left(a_{14}^{(i)} \sigma_{\alpha\alpha*}^{(i,s)} + a_{24}^{(i)} \sigma_{\beta\beta*}^{(i,s)} + \ldots + a_{46}^{(i)} \sigma_{\alpha\beta*}^{(i,s)} \right.$$

$$\left. - \frac{1}{B} \frac{\partial w^{(i,s-1)}}{\partial \eta} + \alpha_{23}^{(i)} \theta_s^{(i)} \right) d\zeta$$

$$w_*^{(i,s)} = \int_0^\zeta \left(a_{13}^{(i)} \sigma_{\alpha\alpha*}^{(i,s)} + a_{23}^{(i)} \sigma_{\beta\beta*}^{(i,s)} + \ldots + a_{36}^{(i)} \sigma_{\alpha\beta*}^{(i,s)} + \alpha_{33}^{(i)} \theta_s^{(i)} \right) d\zeta,$$

$$R_1^{(i,s)} = \frac{1}{A} \frac{\partial u^{(i,s-1)}}{\partial \xi} + ak_\alpha v^{(i,s-1)} - a_{13}^{(i)} \sigma_{\gamma\gamma*}^{(i,s)}$$

$$-a_{14}^{(i)} \sigma_{\beta\gamma*}^{(i,s)} - a_{15} \sigma_{\alpha\gamma*}^{(i,s)} - \alpha_{11}^{(i)} \theta_s^{(i)}, \quad (1,2;\xi,\eta;u,v),$$

$$R_3^{(i,s)} = \frac{1}{B} \frac{\partial u^{(i,s-1)}}{\partial \eta} + \frac{1}{A} \frac{\partial v^{(i,s-1)}}{\partial \xi} - a(k_\alpha u^{(i,s-1)} + k_\beta v^{(i,s-1)})$$

$$-a_{36}^{(i)} \sigma_{\gamma\gamma*}^{(i,s-1)} - a_{46}^{(i)} \sigma_{\beta\gamma*}^{(i,s)} - a_{56}^{(i)} \sigma_{\alpha\beta*}^{(i,s)} - \alpha_{12}^{(i)} \theta_3^{(i)}.$$

The obtained solution (1.7), (1.10) contains unknown functions $\sigma_{\gamma\gamma0}^{(s)}(\xi,\eta)$, $\sigma_{\beta\gamma0}^{(s)}(\xi,\eta)$, $\sigma_{\alpha\gamma0}^{(s)}(\xi,\eta)$ which are independent of the number of the layer (i). Therefore, the proposed approach to the outer solution may be realized for plates containing arbitrary number of layers, since the functions are uniquely defined by the face boundary conditions on $\gamma = h_1$.

7.2   Analysis of the First Boundary Value Problem. Modulus of an Orthotropic Foundation

Let the conditions of the first boundary value problem (1.2) be specified on the face $\gamma = h_1 \left(\zeta_0 = \frac{h_1}{h_2} \right)$ of a two-layered anisotropic plate. Using (1.10) and satisfying these boundary conditions, we obtain

$$\sigma_{\alpha\gamma0}^{(s)}(\xi,\eta) = \sigma_{\alpha\gamma}^{+(s)} - \sigma_{\alpha\gamma*}^{(1,s)}(\zeta_0), \quad (\alpha,\beta),$$

$$\sigma_{\gamma\gamma0}^{(s)}(\xi,\eta) = \sigma_{\gamma\gamma}^{+(s)} - \sigma_{\gamma\gamma*}^{(1,s)}(\zeta_0), \tag{2.1}$$

$$\sigma_{\alpha\gamma}^{+(0)} = \sigma_{\alpha\gamma}^+, \quad \sigma_{\beta\gamma}^{+(0)} = \sigma_{\beta\gamma}^+, \quad \sigma_{\gamma\gamma}^{+(0)} = \sigma_{\gamma\gamma}^+,$$

$$\sigma_{\alpha\gamma}^{+(s)} = \sigma_{\beta\gamma}^{+(s)} = \sigma_{\gamma\gamma}^{+(s)} = 0 \quad s \neq 0.$$

Then, substitution of expressions (2.1) into the relations (1.7), (1.10) gives the required solutions for all of the sought-for quantities. Let us illustrate the above-said by several examples.

a) Let the lower face $\gamma = -h_2$ be fixed, whereas the upper face $\gamma = h_1$ be subject to a load of constant intensity

$$u_\alpha(-h_2) = u_\beta(-h_2) = u_\gamma(-h_2) = 0,$$

$$\sigma_{\alpha\gamma}(h_1) = \sigma_{\alpha\gamma}^+, \quad \sigma_{\beta\gamma}(h_1) = \sigma_{\beta\gamma}^+, \quad \sigma_{\gamma\gamma}(h_1) = \sigma_{\gamma\gamma}^+, \tag{2.2}$$

$$\sigma_{j\gamma}^+ = \text{const}, \quad j = \alpha, \beta, \gamma.$$

Using Eqs. (1.7), (1.10), and (2.1), we arrive at

$$\sigma_{\alpha\alpha}^{(i)} = A_{13}^{(i)}\sigma_{\gamma\gamma}^+ + A_{14}^{(i)}\,\sigma_{\beta\gamma}^+ + A_{15}^{(i)}\,\sigma_{\alpha\gamma}^+,$$

$$\sigma_{\beta\beta}^{(i)} = A_{23}^{(i)}\sigma_{\gamma\gamma}^+ + A_{24}^{(i)}\,\sigma_{\beta\gamma}^+ + A_{25}^{(i)}\,\sigma_{\alpha\gamma}^+,$$

$$\sigma_{\alpha\beta}^{(i)} = A_{63}^{(i)}\sigma_{\gamma\gamma}^+ + A_{64}^{(i)}\,\sigma_{\beta\gamma}^+ + A_{65}^{(i)}\,\sigma_{\alpha\gamma}^+,$$

$$\sigma_{\alpha\gamma}^{(i)} = \sigma_{\alpha\gamma}^+, \quad \sigma_{\beta\gamma}^{(i)} = \sigma_{\beta\gamma}^+, \quad \sigma_{\gamma\gamma}^{(i)} = \sigma_{\gamma\gamma}^+, \tag{2.3}$$

$$u_\alpha^{(i)} = h_2\left(D_{53}^{(i)}\sigma_{\gamma\gamma}^+ + D_{54}^{(i)}\,\sigma_{\beta\gamma}^+ + D_{55}^{(i)}\,\sigma_{\alpha\gamma}^+\right),$$

$$u_\beta^{(i)} = h_2\left(D_{43}^{(i)}\sigma_{\gamma\gamma}^+ + D_{44}^{(i)}\,\sigma_{\beta\gamma}^+ + D_{45}^{(i)}\,\sigma_{\alpha\gamma}^+\right),$$

$$u_\gamma^{(i)} = h_2\left(D_{33}^{(i)}\sigma_{\gamma\gamma}^+ + D_{34}^{(i)}\,\sigma_{\beta\gamma}^+ + D_{35}^{(i)}\,\sigma_{\alpha\gamma}^+\right),$$

$$D_{kj}^{(i)} = \zeta A_{kj}^{(i)} + A_{kj}^{(2)}, \quad i = 1,2;\ \zeta = \frac{\gamma}{h_2}.$$

Since the solution (2.3) is exact, it is natural to propose further analysis of relations between the stresses and displacements on the interface $\gamma = 0$ in order to investigate the range of validity of the Fuss-Zimmermann-Winkler hypotheses. These relations are written as

$$u_\alpha^{(c)} = h_2\left(A_{53}^{(2)}\sigma_{\gamma\gamma}^{(c)} + A_{54}^{(2)}\,\sigma_{\beta\gamma}^{(c)} + A_{55}^{(2)}\,\sigma_{\alpha\gamma}^{(c)}\right),$$

$$u_\beta^{(c)} = h_2\left(A_{43}^{(2)}\sigma_{\gamma\gamma}^{(c)} + A_{44}^{(2)}\,\sigma_{\beta\gamma}^{(c)} + A_{45}^{(c)}\,\sigma_{\alpha\gamma}^{(c)}\right), \tag{2.4}$$

$$u_\gamma^{(c)} = h_2\left(A_{33}^{(2)}\sigma_{\gamma\gamma}^{(c)} + A_{34}^{(2)}\,\sigma_{\beta\gamma}^{(c)} + A_{35}^{(2)}\,\sigma_{\alpha\gamma}^{(c)}\right),$$

with index (c) denoting that the quantity is evaluated at the contact plane. As follows from (2.4), even in case of normal load $\sigma_{\gamma\gamma}^+$ there are tangential displacements $u_\alpha^{(c)}, u_\beta^{(c)}$ arising in addition to the normal displacement $u_\gamma^{(c)}$. Moreover, study of the coefficients reveals that at $\sigma_{\alpha\gamma}^+ = \sigma_{\beta\gamma}^+ = 0$

$$u_\alpha^{(c)} = h_2 A_{53}^{(2)}\sigma_{\gamma\gamma}^{(c)}, \quad u_\beta^{(c)} = h_2 A_{43}^{(2)}\sigma_{\gamma\gamma}^{(c)}, \tag{2.5}$$

$$u_\gamma^{(c)} = h_2 A_{33}^{(2)}\sigma_{\gamma\gamma}^{(c)}.$$

This means that in case of a foundation of general anisotropy the Winkler-Fuss assumption is violated, with the error growing along with the increase of the coefficients $A_{53}^{(2)}$, $A_{43}^{(2)}$. This conclusion may also be interpreted from physical point of view.

In case of orthotropic (and, therefore, isotropic) foundations we have $A_{53}^{(2)}$, $A_{43}^{(2)} = 0$, hence, the Winkler-Fuss hypothesis is true, since $u_\alpha^{(c)} = u_\beta^{(c)} = 0$, we infer that

$$\sigma_{\gamma\gamma}^{(c)} = K_{33}u_\gamma^{(c)}, \quad K_{33} = \frac{1}{h_2 A_{33}^{(2)}}. \tag{2.6}$$

The coefficient K_{33}, which is also referred to as modulus of orthotropic foundation, may be presented in the form

$$K_{33} = \frac{(1 - \nu_{\alpha\beta}\nu_{\beta\alpha})E_\gamma}{h_2(1 - \nu_{\beta\gamma}\nu_{\gamma\beta} - \nu_{\alpha\gamma}\nu_{\gamma\alpha} - \nu_{\alpha\beta}\nu_{\beta\alpha} - 2\nu_{\alpha\beta}\nu_{\beta\gamma}\nu_{\gamma\alpha})}, \tag{2.7}$$

where ν is the Poisson ratio, and E is the Young modulus. In case of isotropic foundation the expression for K_{33} reduces to a form, well-known since 1930s, see Gersevanov and Polshin (1948); Gorbunov-Posadov et al. (1984)

$$K = \frac{(1-\nu)E}{h_2(1+\nu)(1-2\nu)}. \tag{2.8}$$

b) Consider the normal load acting on the surface $\gamma = h_1$, depending linearly on α, β, with the lower face $\gamma = -h_2$ being rigidly fixed

$$\sigma_{\gamma\gamma}(h_1) = b_1\alpha + b_2\beta, \qquad \sigma_{\alpha\gamma}(h_1) = \sigma_{\beta\gamma}(h_1) = 0, \tag{2.9}$$

$$u_\alpha(-h_2) = u_\beta(-h_2) = u_\gamma(-h_2) = 0.$$

In this case the first two approximations provide non-trivial solutions, thus

$$\sigma_{\alpha\alpha}^{(i)} = A_{13}^{(i)}(b_1\alpha + b_2\beta), \quad \sigma_{\beta\beta}^{(i)} = A_{23}^{(i)}(b_1\alpha + b_2\beta),$$

$$\sigma_{\alpha\beta}^{(i)} = 0, \quad \sigma_{\gamma\gamma}^{(i)} = b_1\alpha + b_2\beta,$$

$$\sigma_{\alpha\gamma}^{(i)} = b_1(h_1 A_{13}^{(1)} - \gamma A_{13}^{(i)}), \quad \sigma_{\beta\gamma}^{(i)} = b_2(h_1 A_{23}^{(1)} - \gamma A_{23}^{(i)}),$$

$$u_\alpha^{(i)} = b_1 h_1 A_{13}^{(1)}(h_2 A_{55}^{(2)} + \gamma A_{55}^{(i)}) + \frac{1}{2}b_1(h_2^2 A_{55}^{(2)} A_{13}^{(2)} - \gamma^2 A_{55}^{(i)} A_{13}^{(i)}) \tag{2.10}$$

$$- \frac{1}{2}b_1(h_2^2 A_{33}^{(2)} + 2\gamma h_2 A_{33}^{(2)} + \gamma^2 A_{33}^{(i)}),$$

$$u_\beta^{(i)} = b_2 h_1 A_{23}^{(1)}(h_2 A_{44}^{(2)} + \gamma A_{44}^{(i)}) + \frac{1}{2}b_2(h_2^2 A_{44}^{(2)} A_{23}^{(2)} - \gamma^2 A_{44}^{(i)} A_{23}^{(i)})$$

$$- \frac{1}{2}b_2(h_2^2 A_{33}^{(2)} + 2\gamma h_2 A_{33}^{(2)} + \gamma^2 A_{33}^{(i)}),$$

$$u_\gamma^{(i)} = (h_2 A_{33}^{(2)} + \gamma A_{33}^{(i)})(b_1\alpha + b_2\beta), \quad (i = 1, 2).$$

It may be shown that at the contact plane $\gamma = 0$

$$\sigma_{\gamma\gamma}^{(c)} = b_1\alpha + b_2\beta, \quad \sigma_{\alpha\gamma}^{(c)} = A_{13}^{(1)}b_1 h_1, \quad \sigma_{\beta\gamma}^{(c)} = A_{23}^{(1)}b_2 h_1,$$

$$u_\alpha^{(c)} = \frac{1}{2}b_1 h_2(A_{55}^{(2)}(2h_1 A_{13}^{(1)} + h_2 A_{13}^{(2)}) - h_2 A_{33}^{(2)}), \tag{2.11}$$

$$u_\beta^{(c)} = \frac{1}{2}b_2 h_2(A_{44}^{(2)}(2h_1 A_{23}^{(1)} + h_2 A_{23}^{(2)}) - h_2 A_{33}^{(2)}),$$

$$u_\gamma^{(c)} = h_2 A_{33}^{(2)}(b_1\alpha + b_2\beta).$$

In other words, the non-uniform character of normal load distribution leads to non-zero tangential stress components at the interface. Accordingly, in case of general normal loading $\sigma_{\gamma\gamma}^+(\alpha, \beta)$ the asymptotic approximations up to $O(\varepsilon^2)$ give for the quantities at the interface

$$\sigma_{\alpha\gamma}^{(c)} = A_{13}^{(1)}h_1\frac{\partial\sigma_{\gamma\gamma}^+}{\partial\alpha}, \quad \sigma_{\beta\gamma}^{(c)} = A_{23}^{(1)}h_1\frac{\partial\sigma_{\gamma\gamma}^+}{\partial\beta}, \quad \sigma_{\gamma\gamma}^{(c)} = \sigma_{\gamma\gamma}^+,$$

$$u_\alpha^{(c)} = \frac{1}{2}h_2(A_{55}^{(2)}(2h_1 A_{13}^{(1)} + h_2 A_{13}^{(2)}) - h_2 A_{33}^{(2)})\frac{\partial\sigma_{\gamma\gamma}^+}{\partial\alpha}, \tag{2.12}$$

$$u_\beta^{(c)} = \frac{1}{2}h_2(A_{44}^{(2)}(2h_1 A_{23}^{(1)} + h_2 A_{23}^{(2)}) - h_2 A_{33}^{(2)})\frac{\partial\sigma_{\gamma\gamma}^+}{\partial\beta},$$

$$u_\gamma^{(c)} = h_2 A_{33}^{(2)}\sigma_{\gamma\gamma}^{(c)}.$$

It follows from the latter that the Winkler-Fuss hypothesis of proportional dependence between the reactive force and deflection is valid (up to $O(\varepsilon^2)$) for a general form of $\sigma_{\gamma\gamma}^+(\alpha, \beta)$, however, in addition to the Winkler-Fuss model the tangential displacements $u_\alpha^{(c)}, u_\beta^{(c)}$ should be taken into consideration. These displacements may depend substantially on the elastic parameters of the layers and the variabilities $\dfrac{\partial \sigma_{\gamma\gamma}^+}{\partial \alpha}, \dfrac{\partial \sigma_{\gamma\gamma}^+}{\partial \beta}$ of the normal load along the coordinate lines α and β. In case of large variability this could lead to considerable errors of the Winkler-Fuss model.

c) let us calculate the thermal stress components of a two-layered orthotropic plate, having one face free, and the other rigidly fixed. We also assume constant temperatures within the layers in the absence of body forces. The boundary conditions are then given by

$$u_\alpha(-h_2) = u_\beta(-h_2) = u_\gamma(-h_2) = 0,$$
$$\sigma_{\gamma\gamma}(h_1) = \sigma_{\alpha\gamma}(h_1) = \sigma_{\beta\gamma}(h_1) = 0, \tag{2.13}$$
$$F_\alpha = F_\beta = F_\gamma = 0, \quad \theta^{(1)} \approx \text{const}, \quad \theta^{(2)} \approx \text{const}.$$

The iterative process is stopped at leading order, giving

$$\sigma_{\alpha\alpha}^{(i)} = C_{11}^{(i)}\theta^{(i)}, \quad \sigma_{\beta\beta}^{(i)} = C_{22}^{(i)}\theta^{(i)}, \quad \sigma_{\alpha\beta}^{(i)} = -\frac{1}{a_{66}^{(i)}}\alpha_{12}^{(i)}\theta^{(i)},$$

$$\sigma_{\alpha\gamma}^{(i)} = \sigma_{\beta\gamma}^{(i)} = \sigma_{\gamma\gamma}^{(i)} = 0,$$

$$u_\alpha^{(i)} = \alpha_{13}^{(2)}\theta^{(2)}h_2 + \alpha_{13}^{(i)}\theta^{(i)}\gamma \quad (\alpha, \beta; 13, 23), \tag{2.14}$$

$$u_\gamma^{(i)} = C_{33}^{(2)}\theta^{(2)}h_2 + C_{33}^{(i)}\theta^{(i)}\gamma,$$

$$C_{11}^{(i)} = \frac{a_{12}^{(i)}\alpha_{22}^{(i)} - a_{22}^{(i)}\alpha_{11}^{(i)}}{a_{11}^{(i)}a_{22}^{(i)} - (a_{12}^{(i)})^2},$$

$$C_{33}^{(i)} = a_{13}^{(i)}C_{11}^{(i)} + a_{23}^{(i)}C_{22}^{(i)} + \alpha_{33}^{(i)}, \quad (i = 1, 2).$$

If the temperature variation is linear along thickness, i.e. $\theta^{(i)} = b_i\gamma + d$, then the iterative procedure involves two steps, with the resulting solution written as

$$\sigma_{\alpha\alpha}^{(i)} = C_{11}^{(i)}(b_i\gamma + d), \quad \sigma_{\beta\beta}^{(i)} = C_{22}^{(i)}(b_i\gamma + d),$$

$$\sigma_{\alpha\beta}^{(i)} = -\frac{\alpha_{12}^{(i)}}{a_{66}^{(i)}}(b_i\gamma + d), \quad \sigma_{\alpha\gamma}^{(i)} = \sigma_{\beta\gamma}^{(i)} = \sigma_{\gamma\gamma}^{(i)} = 0, \tag{2.15}$$

$$u_\alpha^{(i)} = \alpha_{13}^{(2)}h_2(d - \frac{1}{2}b_2h_2) + \alpha_{13}^{(i)}(\frac{1}{2}b_i\gamma^2 + \gamma d), \quad (\alpha, \beta; 13, 23),$$

$$u_\gamma^{(i)} = C_{33}^{(2)}h_2(d - \frac{1}{2}b_2h_2) + C_{33}^{(i)}(\frac{1}{2}b_i\gamma^2 + \gamma d).$$

7.3 Second Boundary Value Problem

Let us write down the outer solution corresponding to conditions of the second boundary value problem imposed at $\gamma = h_1 \left(\zeta_0 = \frac{h_1}{h_2}\right)$. Satisfying the relations

(1.10), the unknown functions $\sigma_{\alpha\gamma 0}^{(s)}(\xi, \eta)$, $\sigma_{\beta\gamma 0}^{(s)}(\xi, \eta)$, $\sigma_{\gamma\gamma 0}^{(s)}(\xi, \eta)$ may be presented in the form

$$
\begin{aligned}
\sigma_{\alpha\gamma 0}^{(s)}(\xi, \eta) &= C_{55} V_\alpha^{(s)} + C_{54} V_\beta^{(s)} + C_{53} V_\gamma^{(s)}, \\
\sigma_{\beta\gamma 0}^{(s)}(\xi, \eta) &= C_{45} V_\alpha^{(s)} + C_{44} V_\beta^{(s)} + C_{43} V_\gamma^{(s)}, \\
\sigma_{\gamma\gamma 0}^{(s)}(\xi, \eta) &= C_{35} V_\alpha^{(s)} + C_{34} V_\beta^{(s)} + C_{33} V_\gamma^{(s)},
\end{aligned}
\tag{3.1}
$$

where

$$
V_\alpha^{(s)} = u^{+(s)} - u^{-(s)} + u_*^{(2,s)}(\zeta = -1) - u_*^{(1,s)}(\zeta_0),
$$

$$
u^{+(0)} = \frac{u^+}{a}, \quad u^{-(0)} = \frac{u^-}{a}, \quad u^{+(s)} = u^{-(s)} = 0, \quad s \neq 0,
$$

$$
(\alpha, \beta, \gamma; u, v, w),
$$

$$
C_{jk} = \frac{A_{jk} A_{ll} - A_{jl} A_{lk}}{\Delta_1},
\tag{3.2}
$$

$$
C_{ll} = \frac{A_{jk} A_{ll} - A_{jj} A_{kk}}{\Delta_1}, \quad (j \neq k \neq l \neq j; \; j, k, l = 3, 4, 5),
$$

$$
\Delta_1 = A_{53}(A_{44} A_{35} - A_{34} A_{45}) + A_{54}(A_{45} A_{33} - A_{43} A_{35})
$$

$$
+ A_{55}(A_{34} A_{43} - A_{33} A_{44}),
$$

$$
A_{mn} = \zeta_0 A_{mn}^{(1)} + A_{mn}^{(2)} \quad (m, n = 3, 4, 5) \; \zeta_0 = \frac{h_1}{h_2}.
$$

Therefore, in view of (1.10) and (3.1), the solution of the boundary value problem (1.1), (1.3) is given by

$$
\begin{aligned}
\sigma_{\alpha\gamma}^{(i,s)} &= C_{55} V_\alpha^{(s)} + C_{54} V_\beta^{(s)} + C_{53} V_\gamma^{(s)} + \sigma_{\alpha\gamma *}^{(i,s)}(\xi, \eta, \zeta), \\
\sigma_{\beta\gamma}^{(i,s)} &= C_{45} V_\alpha^{(s)} + C_{44} V_\beta^{(s)} + C_{43} V_\gamma^{(s)} + \sigma_{\beta\gamma *}^{(i,s)}(\xi, \eta, \zeta), \\
\sigma_{\gamma\gamma}^{(i,s)} &= C_{35} V_\alpha^{(s)} + C_{34} V_\beta^{(s)} + C_{33} V_\gamma^{(s)} + \sigma_{\gamma\gamma *}^{(i,s)}(\xi, \eta, \zeta), \\
\sigma_{\alpha\alpha}^{(i,s)} &= C_{15}^{(i)} V_\alpha^{(s)} + C_{14}^{(i)} V_\beta^{(s)} + C_{13}^{(i)} V_\gamma^{(s)} + \sigma_{\alpha\alpha *}^{(i,s)}(\xi, \eta, \zeta), \\
\sigma_{\beta\beta}^{(i,s)} &= C_{25}^{(i)} V_\alpha^{(s)} + C_{24}^{(i)} V_\beta^{(s)} + C_{23}^{(i)} V_\gamma^{(s)} + \sigma_{\beta\beta *}^{(i,s)}(\xi, \eta, \zeta), \\
\sigma_{\alpha\beta}^{(i,s)} &= C_{65}^{(i)} V_\alpha^{(s)} + C_{64}^{(i)} V_\beta^{(s)} + C_{63}^{(i)} V_\gamma^{(s)} + \sigma_{\alpha\beta *}^{(i,s)}(\xi, \eta, \zeta),
\end{aligned}
$$

$$
\begin{aligned}
u^{(i,s)} &= E_{55}^{(i)} V_\alpha^{(s)} + E_{54}^{(i)} V_\beta^{(s)} + E_{53}^{(i)} V_\gamma^{(s)} \\
&\quad + u^{-(s)} - u_*^{(2,s)}(\zeta = -1) + u_*^{(i,s)}(\xi, \eta, \zeta),
\end{aligned}
\tag{3.3}
$$

$$
\begin{aligned}
v^{(i,s)} &= E_{45}^{(i)} V_\alpha^{(s)} + E_{44}^{(i)} V_\beta^{(s)} + E_{43}^{(i)} V_\gamma^{(s)} \\
&\quad + v^{-(s)} - v_*^{(2,s)}(\zeta = -1) + v_*^{(i,s)}(\xi, \eta, \zeta), \\
w^{(i,s)} &= E_{35}^{(i)} V_\alpha^{(s)} + E_{34}^{(i)} V_\beta^{(s)} + E_{33}^{(i)} V_\gamma^{(s)} \\
&\quad + w^{-(s)} - w_*^{(2,s)}(\zeta = -1) + w_*^{(i,s)}(\xi, \eta, \zeta),
\end{aligned}
$$

$$
C_{jk}^{(i)} = A_{j3}^{(i)} C_{3k} + A_{j4}^{(i)} C_{4k} + A_{j5}^{(i)} C_{5k}, \quad C_{jk}^{(i)} \neq C_{kj}^{(i)},
$$

$$
E_{lk}^{(i)} = C_{3k} D_{l3}^{(i)} + C_{4k} D_{l4}^{(i)} + C_{5k} D_{l5}^{(i)}, \quad E_{lk}^{(i)} \neq E_{kl}^{(i)},
$$

$$
D_{kl}^{(i)} = \zeta A_{kl}^{(i)} + A_{kl}^{(2)}, \quad (j = 1, 2, 6; l, k = 3, 4, 5).
$$

The solutions for all the quantities $Q_{jk}^{(i,s)}$ expressed in (3.3) should then be substituted into (1.7), thus all the unknown quantities will be determined.

Let us illustrate this case by considering several particular problems.

a) let the normal displacement w^+ is imposed at the face $\gamma = h_1$, with the lower face $\gamma = -h_2$ rigidly fixed

$$u_\alpha(h_1) = u_\beta(h_1) = 0, \ \ u_\gamma(h_1) = w^+ = \text{const}, \tag{3.4}$$
$$u_\alpha(-h_2) = u_\beta(-h_2) = u_\gamma(-h_2) = 0.$$

Here we assume absence of body forces $F_\alpha^{(i)} = F_\beta^{(i)} = F_\gamma^{(i)} = 0$ and constant temperature field $\theta^{(i)} = 0$. The iterative process is then reduced to leading order only, implying

$$\sigma_{\alpha\alpha}^{(i)} = \frac{1}{h_2} C_{13}^{(i)} w^+, \ \ \sigma_{\beta\beta}^{(i)} = \frac{1}{h_2} C_{23}^{(i)} w^+, \ \ \sigma_{\alpha\beta}^{(i)} = \frac{1}{h_2} C_{63}^{(i)} w^+,$$
$$\sigma_{\alpha\gamma}^{(i)} = \frac{1}{h_2} C_{53} w^+, \ \ \sigma_{\beta\gamma}^{(i)} = \frac{1}{h_2} C_{43} w^+, \ \ \sigma_{\gamma\gamma}^{(i)} = \frac{1}{h_2} C_{33} w^+,$$
$$u_\alpha^{(i)} = (C_{33} D_{53}^{(i)} + C_{43} D_{54}^{(i)} + C_{53} D_{55}^{(i)}) w^+, \tag{3.5}$$
$$u_\beta^{(i)} = (C_{33} D_{43}^{(i)} + C_{43} D_{44}^{(i)} + C_{53} D_{45}^{(i)}) w^+,$$
$$u_\gamma^{(i)} = (C_{33} D_{33}^{(i)} + C_{43} D_{34}^{(i)} + C_{53} D_{35}^{(i)}) w^+.$$

The relations between stress and displacement components at the interface $\gamma = 0$ are given by

$$u_\alpha^{(c)} = h_2 \left(A_{53}^{(2)} \sigma_{\gamma\gamma}^{(c)} + A_{54}^{(2)} \sigma_{\beta\gamma}^{(c)} + A_{55}^{(2)} \sigma_{\alpha\gamma}^{(c)} \right),$$
$$u_\beta^{(c)} = h_2 \left(A_{43}^{(2)} \sigma_{\gamma\gamma}^{(c)} + A_{44}^{(2)} \sigma_{\beta\gamma}^{(c)} + A_{45}^{(2)} \sigma_{\alpha\gamma}^{(c)} \right), \tag{3.6}$$
$$u_\gamma^{(c)} = h_2 \left(A_{33}^{(2)} \sigma_{\gamma\gamma}^{(c)} + A_{34}^{(2)} \sigma_{\beta\gamma}^{(c)} + A_{35}^{(2)} \sigma_{\alpha\gamma}^{(c)} \right).$$

We remark that in case of general anisotropy, then due to (3.5) $u_\alpha^{(c)} \neq 0$, $u_\beta^{(c)} \neq 0$, i.e. under the action of a normal stamp the points of the interface will also obtain motions in the tangential direction. Let us express the contact stresses through displacements as

$$\sigma_{\gamma\gamma}^{(c)} = \frac{1}{h_2 \Delta_2} \left(B_{35} u_\alpha^{(c)} + B_{34} u_\beta^{(c)} + B_{33} u_\gamma^{(c)} \right),$$
$$\sigma_{\beta\gamma}^{(c)} = \frac{1}{h_2 \Delta_2} \left(B_{45} u_\alpha^{(c)} + B_{44} u_\beta^{(c)} + B_{43} u_\gamma^{(c)} \right),$$
$$\sigma_{\alpha\gamma}^{(c)} = \frac{1}{h_2 \Delta_2} \left(B_{55} u_\alpha^{(c)} + B_{54} u_\beta^{(c)} + B_{53} u_\gamma^{(c)} \right), \tag{3.7}$$
$$\Delta_2 = A_{53}^{(2)} B_{35} + A_{54}^{(2)} B_{45} + A_{55}^{(2)} B_{55},$$
$$B_{jk} = A_{jk}^{(2)} A_{ll}^{(2)} - A_{jl}^{(2)} A_{lk}^{(2)}, \ \ B_{ll} = A_{jk}^{(2)} A_{kj}^{(2)} - A_{jj}^{(2)} A_{kk}^{(2)},$$
$$B_{jk} \neq B_{kj}, \ \ (j \neq k \neq l \neq j; \ j,k,l = 3,4,5).$$

As follows from the latter, the Winkler-Fuss model is not applicable in case of foundation of general anisotropy. In case of orthotropic foundation

$$\sigma_{\gamma\gamma}^{(c)} = \frac{1}{h_2 A_{33}^{(2)}} u_{\gamma}^{(c)}, \quad \sigma_{\beta\gamma}^{(c)} = \sigma_{\alpha\gamma}^{(c)} = u_{\alpha}^{(c)} = u_{\beta}^{(c)} = 0 \tag{3.8}$$

hence, the Winker-Fuss hypothesis is perfectly valid.

b) Let the upper and lower faces of a two-layer plate be both fixed, the temperature function $\theta^{(i)}$ has arbitrary variation along thickness, in the absence of volume forces, see Fig. 7.2.

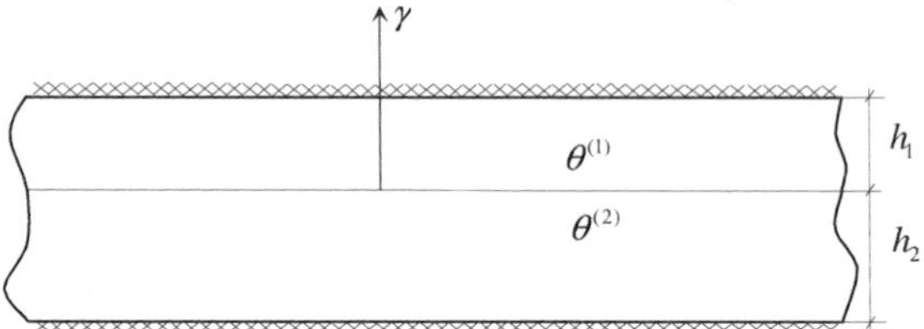

Fig. 7.2

$$u_{\alpha}(h_1) = u_{\beta}(h_1) = 0, \quad u_{\gamma}(h_1) = 0,$$
$$u_{\alpha}(-h_2) = u_{\beta}(-h_2) = u_{\gamma}(-h_2) = 0, \tag{3.9}$$
$$\theta^{(i)} = \theta^{(i)}(\gamma), \quad F_{\alpha} = F_{\beta} = F_{\gamma} = 0.$$

Using (1.8), (1.10) and (3.3), we deduce for the coefficients of the expansion of $Q_{jk}^{(i,s)}$ that

$$\sigma_{\alpha\gamma}^{(i,s)} = \frac{h_2}{h_1 A_{55}^{(1)} + h_2 A_{55}^{(2)}} \left(\alpha_{13}^{(2)} \int_0^{-1} \theta_s^{(2)} d\zeta - \alpha_{13}^{(1)} \int_0^{\zeta_0} \theta_s^{(1)} d\zeta \right),$$

$$u^{(i,s)} = (\zeta A_{55}^{(i)} + A_{55}^{(2)})\sigma_{\alpha\gamma}^{(i,s)} - \alpha_{13}^{(2)} \int_0^{-1} \theta_s^{(2)} d\zeta + \alpha_{13}^{(i)} \int_0^{\zeta_0} \theta_s^{(i)} d\zeta,$$

$$(\alpha, \beta; u, v; 13, 23; 5, 4), \quad \zeta_0 = \frac{h_1}{h_2},$$

$$\sigma_{\gamma\gamma}^{(i,s)} = \frac{h_2}{h_1 A_{33}^{(1)} + h_2 A_{33}^{(2)}} \left(C_{33}^{(2)} \int_0^{-1} \theta_s^{(2)} d\zeta - C_{33}^{(1)} \int_0^{\zeta_0} \theta_s^{(1)} d\zeta \right), \tag{3.10}$$

$$\sigma_{\alpha\alpha}^{(i,s)} = A_{13}^{(i)}\sigma_{\gamma\gamma}^{(i,s)} + C_{11}^{(i)}\theta_s^{(i)}, \quad (\alpha, \beta; 1, 2),$$

$$\sigma_{\alpha\beta}^{(i,s)} = -\frac{\alpha_{12}^{(i)}\theta_s^{(i)}}{a_{66}^{(i)}},$$

$$w^{(i,s)} = (\zeta A_{33}^{(i)} + A_{33}^{(2)})\sigma_{\gamma\gamma}^{(i,s)} - C_{33}^{(2)} \int_0^{-1} \theta_s^{(2)} d\zeta + C_{33}^{(i)} \int_0^{\zeta_0} \theta_s^{(i)} d\zeta.$$

Substitution of (3.10) into (1.7) allows evaluation of all of the sought-for quantities. In particular, in case of linear temperature dependence along the thickness variable

$\theta^{(i)} = b_i\gamma + d$, the solution is given by

$$\sigma_{\alpha\gamma}^{(i)} = -\frac{\alpha_{13}^{(1)}h_1(\frac{1}{2}b_1h_1 + d) - \alpha_{13}^{(2)}h_2(\frac{1}{2}b_2h_2 - d)}{h_1 A_{55}^{(1)} + h_2 A_{55}^{(2)}},$$

$$u_\alpha^{(i)} = (\gamma A_{55}^{(i)} + h_2 A_{55}^{(2)})\sigma_{\alpha\gamma}^{(i)} - \alpha_{13}^{(2)}h_2(\frac{1}{2}b_2h_2 - d)$$

$$+\alpha_{13}^{(i)}\gamma(\frac{1}{2}b_i\gamma + d), \qquad (\alpha,\beta; u, \mathrm{v}; 13, 23; 55, 44),$$

$$\sigma_{\gamma\gamma}^{(i)} = -\frac{C_{33}^{(1)}h_1(\frac{1}{2}b_1h_1 + d) - C_{33}^{(2)}h_2(\frac{1}{2}b_2h_2 - d)}{h_1 A_{33}^{(1)} + h_2 A_{33}^{(2)}}, \qquad (3.11)$$

$$\sigma_{\alpha\alpha}^{(i)} = A_{13}^{(i)}\sigma_{\gamma\gamma}^{(i)} + C_{11}^{(i)}(b_i\gamma + d), \qquad (\alpha,\beta; 1, 2),$$

$$\sigma_{\alpha\beta}^{(i)} = -\frac{\alpha_{12}^{(i)}(b_i\gamma + d)}{a_{66}^{(i)}},$$

$$u_\gamma^{(i)} = (\gamma A_{33}^{(i)} + h_2 A_{33}^{(2)})\sigma_{\gamma\gamma}^{(i)} - C_{33}^{(2)}h_2(\frac{1}{2}b_2h_2 - d)$$

$$+C_{33}^{(i)}\gamma(\frac{1}{2}b_i\gamma + d),$$

$$C_{11}^{(i)} = \frac{a_{12}^{(i)}\alpha_{22}^{(i)} - a_{22}^{(i)}\alpha_{11}^{(i)}}{a_{11}^{(i)}a_{22}^{(i)} - \left(a_{12}^{(i)}\right)^2}, \qquad (1, 2),$$

$$C_{33}^{(i)} = a_{13}^{(i)}C_{11}^{(i)} + a_{23}^{(i)}C_{22}^{(i)} + \alpha_{33}^{(i)}.$$

As follows from (3.11), there is no direct proportion between the normal stress $\sigma_{\gamma\gamma}^{(c)}$ and normal displacement $u_\gamma^{(c)}$ at the contact area $\gamma = 0$, in other words, strictly speaking, the Winkler-Fuss hypothesis is not valid in case of general thermo-elastic foundations.

7.4 Mixed Boundary Value Problems

Consider now the case of the mixed boundary conditions assigned at the upper face $\gamma = h_1$ of a two-layered plate. Satisfying these by means of the conditions (1.10), the unknown functions $\sigma_{\alpha\gamma0}^{(s)}$, $\sigma_{\beta\gamma0}^{(s)}$, $\sigma_{\gamma\gamma0}^{(s)}$ may be found as

$$\sigma_{\alpha\gamma0}^{(s)} = \sigma_{\alpha\gamma}^{+(s)} - \sigma_{\alpha\gamma*}^{(1,s)}(\zeta_0), \qquad (\alpha,\beta), \quad \zeta_0 = \frac{h_1}{h_2},$$

$$\sigma_{\gamma\gamma0}^{(s)} = \frac{1}{A_{33}}(V_\gamma^{(s)} - A_{34}(\sigma_{\beta\gamma}^{+(s)} - \sigma_{\beta\gamma*}^{(1,s)}(\zeta_0)) - A_{35}(\sigma_{\alpha\gamma}^{+(s)} - \sigma_{\alpha\gamma*}^{(1,s)}(\zeta_0))),$$

$$V_\gamma^{(s)} = w^{+(s)} - w^{-(s)} + w_*^{(2,s)}(\zeta = -1) - w_*^{(1,s)}(\zeta_0), \qquad (4.1)$$

$$w^{\pm(0)} = \frac{w^\pm}{a}, \quad \sigma_{\alpha\gamma}^{+(0)} = \sigma_{\alpha\gamma}^+, \quad \sigma_{\beta\gamma}^{+(0)} = \sigma_{\beta\gamma}^+,$$

$$w^{\pm(s)} = 0, \quad \sigma_{\alpha\gamma}^{+(s)} = \sigma_{\beta\gamma}^{+(s)} = 0, \qquad s \neq 0, \quad A_{jk} = \zeta_0 A_{jk}^{(1)} + A_{jk}^{(2)}.$$

The coefficients of expansions (1.10) are then given by

$$\sigma_{\alpha\alpha}^{(i,s)} = \frac{1}{A_{33}}(A_{13}^{(i)}V_\gamma^{(s)} + (A_{14}^{(i)}A_{33} - A_{13}^{(i)}A_{34})(\sigma_{\beta\gamma}^{+(s)} - \sigma_{\beta\gamma*}^{(1,s)}(\zeta_0))$$

$$+ (A_{15}^{(i)}A_{33} - A_{13}^{(i)}A_{35})(\sigma_{\alpha\gamma}^{+(s)} - \sigma_{\alpha\gamma*}^{(1,s)}(\zeta_0))) + \sigma_{\alpha\alpha*}^{(i,s)}(\xi,\eta,\zeta),$$

$$(\alpha\alpha, \beta\beta, \alpha\beta; 1,2,6),$$

$$\sigma_{\alpha\gamma}^{(i,s)} = \sigma_{\alpha\gamma}^{+(s)} - \sigma_{\alpha\gamma*}^{(1,s)}(\zeta_0) + \sigma_{\alpha\gamma*}^{(i,s)}(\xi,\eta,\zeta), \quad (\alpha,\beta),$$

$$\sigma_{\gamma\gamma}^{(i,s)} = (V_\gamma^{(s)} - A_{34}(\sigma_{\beta\gamma}^{+(s)} - \sigma_{\beta\gamma*}^{(1,s)}(\zeta_0))$$

$$-A_{35}(\sigma_{\alpha\gamma}^{+(s)} - \sigma_{\alpha\gamma*}^{(1,s)}(\zeta_0)))\frac{1}{A_{33}} + \sigma_{\gamma\gamma*}^{(i,s)}(\xi,\eta,\zeta), \qquad (4.2)$$

$$u^{(i,s)} = \frac{1}{A_{33}}(D_{53}^{(i)}V_\gamma^{(s)} + (D_{54}^{(i)}A_{33} - D_{53}^{(i)}A_{34})(\sigma_{\beta\gamma}^{+(s)} - \sigma_{\beta\gamma*}^{(1,s)}(\zeta_0))$$

$$+(D_{55}^{(i)}A_{33} - D_{53}^{(i)}A_{35})(\sigma_{\alpha\gamma}^{+(s)} - \sigma_{\alpha\gamma*}^{(1,s)}(\zeta_0))) + u^{-(s)}$$

$$-u_*^{(2,s)}(\zeta = -1) + u_*^{(i,s)}(\xi,\eta,\zeta),$$

$$u^{-(0)} = \frac{u^-}{a}, u^{-(s)} = 0, \ s \neq 0, \ D_{kl}^{(i)} = \zeta A_{kl}^{(i)} + A_{kl}^{(2)},$$

$$(u,v,w; D_{5k}, D_{4k}, D_{3k}), \quad i = 1,2.$$

Substitution of the results (4.2) and (1.12) into (1.7) leads to the sought-for solution.

In case of the mixed boundary conditions (1.5) imposed on the face $\gamma = h_1$, similarly, we obtain

$$\sigma_{\alpha\gamma0}^{(s)} = (C_{53}(\sigma_{\gamma\gamma}^{+(s)} - \sigma_{\gamma\gamma*}^{(1,s)}(\zeta_0)) + A_{54}V_\beta^{(s)} - A_{44}V_\alpha^{(s)})\frac{1}{C_{33}},$$

$$\sigma_{\beta\gamma0}^{(s)} = (C_{43}(\sigma_{\gamma\gamma}^{+(s)} - \sigma_{\gamma\gamma*}^{(1,s)}(\zeta_0)) - A_{55}V_\beta^{(s)} - A_{45}V_\alpha^{(s)})\frac{1}{C_{33}},$$

$$\sigma_{\gamma\gamma0}^{(s)} = \sigma_{\gamma\gamma}^{+(s)} - \sigma_{\gamma\gamma*}^{(1,s)}(\zeta_0), \qquad (4.3)$$

$$V_\alpha^{(s)} = u^{+(s)} - u^{-(s)} + u^{(2,s)}(\zeta = -1) - u_*^{(1,s)}(\zeta_0),$$

$$u^{\pm(0)} = \frac{u^\pm}{a}, \ \sigma_{\gamma\gamma}^{+(0)} = \sigma_{\gamma\gamma}^+, \ u^{\pm(s)} = 0, \ \sigma_{\gamma\gamma}^{+(s)} = 0 \ s \neq 0, \ (\alpha,\beta; u,v),$$

$$C_{33} = A_{45}A_{54} - A_{44}A_{55}, \ C_{53} = A_{53}A_{44} - A_{54}A_{43},$$

$$C_{43} = A_{43}A_{55} - A_{45}A_{53}.$$

The coefficients of (1.10) take the form

$$\sigma_{\alpha\alpha}^{(i,s)} = \frac{1}{C_{33}}((A_{13}^{(i)}C_{33} + A_{14}^{(i)}C_{43} + A_{15}^{(i)}C_{53})(\sigma_{\gamma\gamma}^{+(s)} - \sigma_{\gamma\gamma*}^{(1,s)}(\zeta_0))$$

$$+(A_{15}^{(i)}A_{54} - A_{14}^{(i)}A_{55})V_\beta^{(s)} + (A_{14}^{(i)}A_{45} - A_{15}^{(i)}A_{44})V_\alpha^{(s)}) + \sigma_{\alpha\alpha*}^{(i,s)}(\xi,\eta,\zeta),$$

$$(\alpha\alpha, \beta\beta, \alpha\beta; 1,2,6),$$

$$\sigma_{\alpha\gamma}^{(i,s)} = \frac{1}{C_{33}}(C_{53}(\sigma_{\gamma\gamma}^{+(s)} - \sigma_{\gamma\gamma*}^{(1,s)}(\zeta_0)) + A_{54}V_\beta^{(s)} - A_{44}V_\alpha^{(s)})$$

$$+\sigma_{\alpha\gamma*}^{(i,s)}(\xi,\eta,\zeta), \quad (\alpha,\beta;5,4),$$

$$\sigma_{\gamma\gamma}^{(i,s)} = \sigma_{\gamma\gamma}^{+(s)} - \sigma_{\gamma\gamma*}^{(1,s)}(\zeta_0) + \sigma_{\gamma\gamma*}^{(i,s)}(\xi,\eta,\zeta), \qquad (4.4)$$

$$u^{(i,s)} = \frac{1}{C_{33}}((C_{33}D_{53}^{(i)} + C_{43}D_{54}^{(i)} + C_{53}D_{55}^{(i)})(\sigma_{\gamma\gamma}^{+(s)} - \sigma_{\gamma\gamma*}^{(1,s)}(\zeta_0))$$

$$+(D_{55}^{(i)}A_{54} - D_{54}^{(i)}A_{55})V_\beta^{(s)} + (D_{54}^{(i)}A_{45} - D_{55}^{(i)}A_{44})V_\alpha^{(s)}) + u^{-(s)}$$

$$-u_*^{(2,s)}(\zeta = -1) + u_*^{(i,s)}(\xi,\eta,\zeta),$$

$$(u,v,w; D_{5k}, D_{4k}, D_{3k}), \quad i = 1,2,$$

with $\sigma_{jk*}^{(i,s)}(\xi,\eta,\zeta)$, $u_*^{(i,s)}$, $v_*^{(i,s)}$, $w_*^{(i,s)}$ evaluated from the recurrent formulae (1.12). Substitution of (4.4) into (1.7) provides the solutions for all the sought-for quantities.

Let us present several examples of closed (exact) outer solutions.

a) Let the face $\gamma = -h_2$ of a two-layered anisotropic plate be rigidly fixed, whereas the face $\gamma = h_1$ is subject to action of a rigid frictionless stamp, causing a constant normal displacement w^+

$$u_\alpha(-h_2) = u_\beta(-h_2) = u_\gamma(-h_2) = 0, \qquad (4.5)$$

$$\sigma_{\alpha\gamma}(h_1) = \sigma_{\beta\gamma}(h_1) = 0, \quad u_\gamma(h_1) = w^+ = \text{const.}$$

The iterative procedure ends at leading order, giving the following exact solution

$$\sigma_{\alpha\alpha}^{(i)} = \frac{A_{13}^{(i)}}{\Delta}w^+, \quad \sigma_{\beta\beta}^{(i)} = \frac{A_{23}^{(i)}}{\Delta}w^+, \quad \sigma_{\alpha\beta}^{(i)} = \frac{A_{63}^{(i)}}{\Delta}w^+,$$

$$\sigma_{\gamma\gamma}^{(i)} = \frac{w^+}{\Delta}, \quad \sigma_{\alpha\gamma}^{(i)} = \sigma_{\beta\gamma}^{(i)} = 0, \qquad (4.6)$$

$$u_\gamma^{(i)} = (\gamma A_{33}^{(i)} + h_2 A_{33}^{(2)})\frac{w^+}{\Delta}, \quad u_\alpha^{(i)} = (\gamma A_{53}^{(i)} + h_2 A_{53}^{(2)})\frac{w^+}{\Delta},$$

$$u_\beta^{(i)} = (\gamma A_{43}^{(i)} + h_2 A_{43}^{(2)})\frac{w^+}{\Delta}, \quad \Delta = h_1 A_{33}^{(1)} + h_2 A_{33}^{(2)},$$

whereas at the contact area $\gamma = 0$ we have

$$\sigma_{\alpha\gamma}^{(c)} = \sigma_{\beta\gamma}^{(c)} = 0, \quad \sigma_{\gamma\gamma}^{(c)} = \frac{w^+}{\Delta}, \quad u_\alpha^{(c)} = h_2 A_{53}^{(2)}\sigma_{\gamma\gamma}^{(c)}, \qquad (4.7)$$

$$u_\beta^{(c)} = h_2 A_{43}^{(2)}\sigma_{\gamma\gamma}^{(c)}, \quad u_\gamma^{(c)} = h_2 A_{33}^{(2)}\sigma_{\gamma\gamma}^{(c)},$$

which take simpler form for the orthotropic plate

$$\sigma_{\alpha\gamma}^{(c)} = \sigma_{\beta\gamma}^{(c)} = 0, \quad u_\alpha^{(c)} = u_\beta^{(c)} = 0, \qquad (4.8)$$

$$\sigma_{\gamma\gamma}^{(c)} = \frac{w^+}{\Delta}, \quad u_\gamma^{(c)} = h_2 A_{33}^{(2)}\sigma_{\gamma\gamma}^{(c)}.$$

It follows from (4.7), (4.8) that

$$\sigma_{\gamma\gamma}^{(c)} = K_{33}u_\gamma^{(c)}, \quad K_{33} = \frac{1}{h_2 A_{33}^{(2)}}. \qquad (4.9)$$

However, in case of a plate of general anisotropy $u_\alpha^{(c)} \neq 0$, $u_\beta^{(c)} \neq 0$, which could be compatible to $u_\gamma^{(c)}$, provided that the coefficients $A_{43}^{(2)} A_{53}^{(2)}$ are of the asymptotic order as $A_{33}^{(2)}$. This may lead to the case, when even though the relation (4.9) is formally satisfied, the Winkler-Fuss model is nevertheless non-applicable. We note that in case of orthotropic plate the Winkler-Fuss hypothesis is perfectly valid due to (4.8).

b) Let the lower face $\gamma = -h_2$ be rigidly fixed, with normal load of constant intensity acting on the upper surface $\gamma = h_1$ in the absence of tangential displacements, namely

$$\sigma_{\gamma\gamma}(h_1) = \sigma_{\gamma\gamma}^+ = \text{const}, \quad u_\alpha(h_1) = u_\beta(h_1) = 0, \tag{4.10}$$
$$u_\alpha(-h_2) = u_\beta(-h_2) = u_\gamma(-h_2) = 0.$$

The stress-strain field is then given by

$$\sigma_{\alpha\alpha}^{(i)} = A_{13}^{(i)}\sigma_{\gamma\gamma}^+, \quad \sigma_{\beta\beta}^{(i)} = A_{23}^{(i)}\sigma_{\gamma\gamma}^+, \quad \sigma_{\alpha\beta}^{(i)} = 0,$$
$$\sigma_{\gamma\gamma}^{(i)} = \sigma_{\gamma\gamma}^+, \quad \sigma_{\alpha\gamma}^{(i)} = \sigma_{\beta\gamma}^{(i)} = 0, \tag{4.11}$$
$$u_\gamma^{(i)} = (\gamma A_{33}^{(i)} + h_2 A_{33}^{(2)})\sigma_{\gamma\gamma}^+, \quad u_\alpha^{(i)} = u_\beta^{(i)} = 0,$$

with the following relations on the contact area $\gamma = 0$

$$\sigma_{\alpha\gamma}^{(c)} = \sigma_{\beta\gamma}^{(c)} = 0, \quad u_\alpha^{(c)} = u_\beta^{(c)} = 0, \tag{4.12}$$
$$u_\gamma^{(c)} = h_2 A_{33}^{(2)}\sigma_{\gamma\gamma}^{(c)}.$$

Thus, the validity of the Winkler-Fuss model is established once again.

There are some more examples illustrating the efficiency and wide area of applicability of the asymptotic method. We emphasize once again that in case of polynomial form of the boundary functions, volume forces and temperature functions depending on the coordinates, the asymptotic method gives the exact solution of the associated 3D problem. Indeed, the iterative process contains $n + 1$ steps, where n is the highest degree of a polynomial. In case of general dependence of boundary and temperature functions, the latter may be approximated by appropriate polynomials, for which an exact solution of a spatial problem for a plate could be constructed.

In case of finite dimensions of the plate, the presented solutions of the outer problem are no longer valid in the vicinity of the side surface S_Ω, governed by the arising boundary layer solutions arising, decaying exponentially away from the edge. We note that the boundary layer emerges also near the area of change of type of the boundary conditions. The boundary layer solution of the two-layered plate may be constructed similarly to the previously considered cases of single-layered plate and two-layered strip. The interaction of the boundary layer solution with the outer solution may also be investigated in an analogous manner.

7.5   Modules of Non-Homogeneous Layered Foundations

The solutions of spatial problems obtained in this chapter allows some further remarks on the modules of layered foundations containing n arbitrary layers or even a more general case of a non-homogeneous foundation. In what follows we rely on solution (2.3).

Evaluation of the displacement vector on the surface $\gamma = h_1$ reveals that even in case of action of the normal load $\sigma_{\gamma\gamma}^+$ only, the points on the surface receive tangential displacements $\sigma_{\gamma\gamma}^+$ growing with increase of the coefficients $D_{53}^{(1)}$, $D_{43}^{(1)}$ in addition to the expected normal displacement. Therefore, in case of general anisotropy the concept of having a module of a foundation seems misleading, since the points under the acting normal load exhibit tangential motions along with the normal displacement. The module of a foundation may still be a subject of discussion in case of orthotropic foundation, when the coordinate lines coincide with the principal directions of anisotropy. Then, due to $\sigma_{\beta\gamma}^+ = \sigma_{\alpha\gamma}^+ = 0$, $\sigma_{\gamma\gamma}^+ \neq 0$ we infer from (1.11) and (2.3) that

$$u_\alpha^{(1)}(h_1) = u_\beta^{(1)}(h_1) = 0, \tag{5.1}$$

$$u_\gamma^{(1)}(h_1) = h_2 D_{33}^{(1)}(h_1)\sigma_{\gamma\gamma}^+$$

or

$$\sigma_{\gamma\gamma}^{(1)}(h_1) = \frac{1}{h_1 A_{33}^{(1)} + h_2 A_{33}^{(2)}} u_\gamma^{(1)}(h_1). \tag{5.2}$$

Thus, the normal stress is proportional to the resulting normal displacement with the coefficient of the foundation

$$K_2 = \frac{1}{h_1 A_{33}^{(1)} + h_2 A_{33}^{(2)}}. \tag{5.3}$$

Comparing this result with the corresponding result (2.6) for single-layered foundation, a rather straightforward pattern may be deduced. The formula (5.3) may therefore be generalized to a case of arbitrary n-layered foundation

$$K_n = \frac{1}{\sum_{i=1}^{n} h_i A_{33}^{(i)}} \tag{5.4}$$

where h_i is the thickness of layer i, and the elastic coefficients $A_{33}^{(i)}$ defined from (1.11).

Let us consider a more general case of non-homogeneous foundation, when the coefficient A_{33} is a function of thickness variable, namely $A_{33} = A_{33}(\gamma)$. An obvious generalization of (5.4) for the module of a non-homogeneous foundation may then be performed as

$$K = \frac{1}{\int_0^h A_{33}(\gamma)d\gamma}, \tag{5.5}$$

where h is a total thickness of the non-homogeneous foundation (compressible layer).

The same result may be obtained from the direct integration of the 3D equations of elasticity in case of a non-homogeneous foundation. Indeed, let us consider an isotropic plate $\Omega = \{(\alpha, \beta, \gamma) : \alpha, \beta \in \Sigma, -h \leq \gamma \leq h\}$, with variable elastic modulus

$$E = \frac{1}{2}(E_1 + E_2) + \frac{1}{2}(E_1 - E_2)\zeta, \quad -1 \leq \zeta \leq 1, \quad \zeta = \frac{\gamma}{h} \tag{5.6}$$

$$\nu \approx \text{const}, \quad E_1, E_2 = \text{const},$$

i.e. the Young modulus E varies linearly on thickness in the range of $[E_1, E_2]$. The solution of a spatial boundary value problem corresponding to fixed lower face $\gamma = -h$ and load of constant intensity $\sigma_{\alpha\gamma}^+, \sigma_{\beta\gamma}^+, \sigma_{\gamma\gamma}^+ = \text{const}$ acting on the face $\gamma = h$, is given by

$$\sigma_{\alpha\gamma} = \sigma_{\alpha\gamma}^+, \quad \sigma_{\beta\gamma} = \sigma_{\beta\gamma}^+, \quad \sigma_{\gamma\gamma} = -\sigma_{\gamma\gamma}^+,$$

$$\sigma_{\alpha\alpha} = -\frac{\nu}{1-\nu}\sigma_{\gamma\gamma}^+, \quad \sigma_{\beta\beta} = -\frac{\nu}{1-\nu}\sigma_{\gamma\gamma}^+, \quad \sigma_{\alpha\beta} = 0,$$

$$u_\alpha = 4h(1+\nu)K_1\sigma_{\alpha\beta}^+, \quad u_\beta = 4h(1+\nu)K_1\sigma_{\beta\gamma}^+, \tag{5.7}$$

$$u_\gamma = -\frac{2h(1+\nu)(1-2\nu)}{1-\nu}K_1\sigma_{\gamma\gamma}^+,$$

$$K_1 = \frac{1}{E_1 - E_2}\ln\frac{(E_1 - E_2)\zeta + E_1 + E_2}{2E_2}.$$

In case of only normal component of active loading $(\sigma_{\alpha\gamma}^+ = 0, \sigma_{\beta\gamma}^+ = 0)$ the solution (5.7) implies $u_\alpha = u_\beta = 0$. The relation between the normal stress at $\gamma = h$ and the appropriate displacement is written as

$$\sigma_{\gamma\gamma}(h) = Ku_\gamma(h), \quad K = K_0\frac{c-1}{\ln c}, \quad c = \frac{E_2}{E_1}, \tag{5.8}$$

where K acts a modulus of the non-homogeneous foundation, whereas

$$K_0 = \frac{(1-\nu)E_1}{(1+\nu)(1-2\nu)2h} \tag{5.9}$$

is a well-known modulus of homogeneous foundation with constant elastic modulus.

We remark that the presented formula (5.8) is valid for both $c > 1$ and $c < 1$, whereas as $c \to 1$, $K \to K_0$.

In case of variable external loading, i.e. depending on ξ, the relation (5.8) is satisfied only in approximate sense due to additional terms in the exact solution along with non-zero u_α, u_β. However, since the arising terms are of higher order compared to these within the Winkler model, these could be neglected. For example, if $\sigma_{\gamma\gamma}(h) = -(\sigma_1\xi + \sigma_2)$ is specified at $\gamma = h$, where $\xi = \frac{\alpha}{l}$, and l is a typical scale for the plate, then

$$u_\alpha = -\frac{h}{l}\sigma_1\left(\frac{4\nu(1+\nu)}{1-\nu}hK_1 + ...\right), \quad u_\beta = 0,$$

$$u_\gamma = -\frac{(1+\nu)(1-2\nu)2h}{1-\nu}K_1(\sigma_1\xi + \sigma_2),$$

$$\sigma_{\alpha\alpha} = \sigma_{\beta\beta} = -\frac{\nu}{1-\nu}(\sigma_1\xi + \sigma_2), \quad \sigma_{\gamma\gamma}(h) = Ku_\gamma(h), \tag{5.10}$$

$$\sigma_{\gamma\gamma} = -(\sigma_1\xi + \sigma_2), \quad \sigma_{\alpha\beta} = \sigma_{\beta\gamma} = 0,$$

$$\sigma_{\alpha\gamma} = -\frac{h}{l}\sigma_1\frac{\nu}{1-\nu}(1-\zeta).$$

In case of $\sigma_{\gamma\gamma}(h) = -\sigma_3 \left(\xi - \frac{1}{2}\right)^2$ the solution may be written as

$$\sigma_{\alpha\beta} = \sigma_{\beta\gamma} = 0,$$

$$\sigma_{\gamma\gamma} = -\sigma_3 \left(\xi - \frac{1}{2}\right)^2 + \sigma_3 \frac{h^2}{l^2} \frac{\nu}{1-\nu}(1-\zeta)^2,$$

$$\sigma_{\alpha\gamma} = -\frac{\nu}{1-\nu}\sigma_3 \frac{h}{l}(2\xi - 1)(1 - \zeta),$$

$$\sigma_{\alpha\alpha} = \sigma_{\beta\beta} = -\frac{\nu}{1-\nu}\sigma_3 \left(\xi - \frac{1}{2}\right)^2 - 2\sigma_3 \frac{h^2}{l^2}(1 + ...), \qquad (5.11)$$

$$u_\alpha = -\frac{h}{l}\sigma_3(2\xi - 1)\left(\frac{4h\nu(1+\nu)}{1-\nu}K_1 + ...\right), \quad u_\beta = 0,$$

$$u_\gamma = -\sigma_3 \left(\xi - \frac{1}{2}\right)^2 \frac{(1+\nu)(1-2\nu)2h}{1-\nu}K_1$$

$$-\sigma_3 \frac{h^2}{l^2}\left(\frac{4h(1-2\nu)(1+\nu)}{(1-\nu)^2}K_1(1-\zeta) + ...\right).$$

It follows from the presented solutions (5.10), (5.11) that the Winkler-Fuss hypothesis is valid to $O(\varepsilon)$, $\varepsilon = \frac{h}{l}$ for non-homogeneous foundation, but with effective value of the modulus of the foundation K, with the validity of condition (5.8) confirmed up to $O(\varepsilon^2)$.

Let us now consider exponential law for the elastic modulus

$$E = E_1 \exp(-m(\zeta - 1)), \quad \nu \approx \text{const}, \quad -1 \le \zeta \le 1 \qquad (5.12)$$

then the solution for the same boundary value problem may be obtained as

$$\sigma_{\alpha\gamma} = \sigma_{\alpha\gamma}^+, \quad \sigma_{\beta\gamma} = \sigma_{\beta\gamma}^+, \quad \sigma_{\gamma\gamma} = -\sigma_{\gamma\gamma}^+, \quad \sigma_{ij}^+ = \text{const},$$

$$\sigma_{\alpha\alpha} = -\frac{\nu}{1-\nu}\sigma_{\gamma\gamma}^+, \quad \sigma_{\beta\beta} = -\frac{\nu}{1-\nu}\sigma_{\gamma\gamma}^+, \quad \sigma_{\alpha\beta} = 0, \qquad (5.13)$$

$$u_\alpha = 2(1+\nu)hK_2\sigma_{\alpha\gamma}^+, \quad u_\beta = 2(1+\nu)hK_2\sigma_{\beta\gamma}^+,$$

$$u_\gamma = -\frac{(1+\nu)(1-2\nu)}{1-\nu}hK_2\sigma_{\gamma\gamma}^+, \quad K_2 = \frac{1}{E_1 m}\left(e^{m(\zeta-1)} - e^{-2m}\right).$$

It follows from the latter that

$$\sigma_{\gamma\gamma}(h) = Ku_\gamma(h), \quad K = K_0 \frac{2m}{1 - e^{-2m}}. \qquad (5.14)$$

Thus, the Winkler-Fuss model may still be justified for the case of non-homogeneous foundation, provided that the external load does not possess large variability. However, the value of the modulus of a foundation should be modified in accordance with (5.5), (5.8), and (5.14).

In conclusion of this chapter, we note that the obtained values of modules of foundations may be used in calculations within the framework of the Winkler-Zimmermann-Fuss model. However, the drawbacks of this model have been mentioned in earlier chapters, including the question of the stress-strain field of the foundation, which is ignored by the Winkler model. In order to take it into consideration, another model should be utilized, for example, a model of a compressible layer. Moreover, application of the model of compressible layer allows solutions, similar to these obtained in this and previous chapters. We remark that the solutions derived within the model of a compressible layer are more mathematically rigorous and clearly structured than that of the Winkler-Fuss model.

Chapter 8

Asymptotic Analysis of the Outer Problem for an Orthotropic Shell

8.1 Governing Equations and Boundary Conditions

The asymptotic method described in the previous chapters may be applied to objects of more sophisticated geometry, in particular, to anisotropic shells.

The asymptotic method allows rigorous justification of the classical shell theory, along with evaluation of the error of the 2D approximate shell theory depending on the parameters of variability and anisotropy. The method also enables formulation of the optimal boundary value problem, consistent with the classical Kirchhoff-Love assumptions, which leads to significant improvements of the results within the classical shell theory. This is achieved through modification of the governing equations and general relations. The method also provides ideas for further refined analysis of the stress-strain field.

Consider an orthotropic elastic shell of constant thickness $2h$ as a spatial body. Let the mid-surface of the shell contains the 2D curvilinear system α, β, then the position of the point belonging to the shell may be described by α, β, γ, where $|\gamma|$ is the distance from the point along the normal to the mid-surface. Here we assume that the principal directions of anisotropy coincide with the directions of the coordinate lines. In order to reduce the lengthy computations we shall use the components of a non-symmetric stress tensor, which are related to the symmetric stress tensor by

$$\sigma_\alpha = \left(1 + \frac{\gamma}{R_2}\right)\sigma'_\alpha, \quad \tau_{\alpha\beta} = \left(1 + \frac{\gamma}{R_2}\right)\tau'_{\alpha\beta}, \qquad (\alpha, \beta; 1, 2), \qquad (1.1)$$

$$\tau_{\alpha\gamma} = \left(1 + \frac{\gamma}{R_2}\right)\tau'_{\alpha\gamma}, \quad (\alpha, \beta), \quad \sigma_\gamma = \left(1 + \frac{\gamma}{R_1}\right)\left(1 + \frac{\gamma}{R_2}\right)\sigma'_\gamma,$$

where the prime denotes components of the symmetric tensor, for more details see Green and Zerna (1954); Goldenveizer (1961, 1976); Aghalovyan (1972b).

The governing equations of 3D elasticity in anisotropic media (Lekhnitskii, 1977) may be written in the chosen coordinate system in terms of non-symmetric stress tensor as follows:

the equilibrium equations

$$\frac{1}{B}\partial_\alpha\left(B\sigma_\alpha\right) - k_\beta\sigma_\beta + \frac{1}{A}\partial_\beta\left(A\tau_{\beta\alpha}\right) + k_\alpha\tau_{\alpha\beta}$$

$$+ \left(1 + \frac{\gamma}{R_1}\right)\partial_3\tau_{\alpha\gamma} + \frac{2\tau_{\alpha\gamma}}{R_1} = 0, \quad (\alpha,\beta),$$

$$\partial_3\sigma_\gamma - \left(\frac{\sigma_\alpha}{R_1} + \frac{\sigma_\beta}{R_2}\right) + \partial_\alpha\tau_{\alpha\gamma} + \partial_\beta\tau_{\beta\gamma} + k_\beta\tau_{\alpha\gamma} + k_\alpha\tau_{\beta\gamma} = 0, \quad (1.2)$$

$$\left(1 + \frac{\gamma}{R_1}\right)\tau_{\alpha\beta} = \left(1 + \frac{\gamma}{R_2}\right)\tau_{\beta\alpha} \quad \text{(symmetry condition } \tau'_{\alpha\beta} = \tau'_{\beta\alpha})$$

the constitutive relations

$$\left(1 + \frac{\gamma}{R_2}\right)\left(\partial_\alpha u + k_\alpha u_\beta + \frac{u_\gamma}{R_1}\right)$$

$$= a_{11}\left(1 + \frac{\gamma}{R_1}\right)\sigma_\alpha + a_{12}\left(1 + \frac{\gamma}{R_2}\right)\sigma_\beta + a_{13}\sigma_\gamma, \quad (\alpha,\beta\,;1,2),$$

$$\left(1 + \frac{\gamma}{R_1}\right)\left(1 + \frac{\gamma}{R_2}\right)\partial_3 u_\gamma$$

$$= a_{33}\sigma_\gamma + a_{13}\sigma_\alpha + a_{23}\sigma_\beta + \gamma\left(a_{13}\frac{\sigma_\alpha}{R_1} + a_{23}\frac{\sigma_\beta}{R_2}\right), \quad (1.3)$$

$$\left(1 + \frac{\gamma}{R_1}\right)\left(\partial_\beta u_\alpha - k_\beta u_\beta\right) + \left(1 + \frac{\gamma}{R_2}\right)\left(\partial_\alpha u_\beta - k_\alpha u_\alpha\right)$$

$$= \left(1 + \frac{\gamma}{R_1}\right)a_{66}\tau_{\alpha\beta},$$

$$\left(1 + \frac{\gamma}{R_1}\right)\left(1 + \frac{\gamma}{R_2}\right)\partial_3 u_\alpha - \left(1 + \frac{\gamma}{R_2}\right)\frac{u_\alpha}{R_1}$$

$$+ \left(1 + \frac{\gamma}{R_2}\right)\partial_\alpha u_\gamma = \left(1 + \frac{\gamma}{R_1}\right)a_{55}\tau_{\alpha\gamma}, \quad (\alpha,\beta;1,2;4,5),$$

face boundary conditions

$$\sigma'_\gamma = \pm\frac{1}{2}p_\gamma - \frac{1}{2}m, \quad \tau'_{\alpha\gamma} = \pm\frac{1}{2}p_\alpha + \frac{1}{2}m_\beta, \quad (\alpha,\beta) \tag{1.4}$$

or in terms of the non-symmetric stress tensor

$$\sigma_\gamma = \pm\frac{1}{2}\left(1\pm\frac{h}{R_1}\right)\left(1\pm\frac{h}{R_2}\right)p_\gamma - \frac{1}{2}\left(1\pm\frac{h}{R_1}\right)\left(1\pm\frac{h}{R_2}\right)m, \tag{1.5}$$

$$\tau_{\alpha\gamma} = \left(1\pm\frac{h}{R_2}\right)\left(\pm\frac{1}{2}p_\alpha + \frac{1}{2}m_\beta\right), \quad (\alpha,\beta) \quad \text{at} \quad \gamma = \pm h.$$

Here

$$\partial_\alpha = \frac{1}{A}\frac{\partial}{\partial\alpha} \quad (\alpha,\beta), \quad \partial_3 = \frac{\partial}{\partial\gamma}, \quad k_\alpha = \frac{1}{AB}\frac{\partial A}{\partial\beta}, \quad (\alpha,\beta). \tag{1.6}$$

The components of surface loading are related to the appropriate components used in Ambartsumyan (1961) through

$$p_\gamma = Z^+ + Z^-, \quad p_\alpha = X^+ + X^-, \quad (\alpha,\beta;X,Y), \tag{1.7}$$

$$m = -(Z^+ - Z^-), \quad m_\beta = (X^+ - X^-), \quad (\alpha,\beta;X,Y).$$

The general formulation of the boundary value problem involves the system (1.2), (1.3) along with the boundary conditions (1.4) and other boundary conditions imposed at the edges of the shell. Let the edge of the shell be defined by $\alpha = \alpha_0$. The following types of boundary conditions will be discussed

$$\sigma_\alpha = \tau_{\alpha\beta} = \tau_{\alpha\gamma} = 0, \tag{1.8}$$

$$\sigma_\alpha = \tau_{\alpha\beta} = u_\gamma = 0, \tag{1.9}$$

$$\sigma_\alpha = u_\beta = u_\gamma = 0 \quad \text{at} \quad \alpha = \alpha_0, \tag{1.10}$$

$$u_\alpha = u_\beta = u_\gamma = 0, \tag{1.11}$$

$$u_\alpha = \tau_{\alpha\beta} = u_\gamma = 0. \tag{1.12}$$

Similar conditions may be imposed on $\beta = \beta_0$, along with other types of boundary conditions.

In this chapter we focus our attention on the construction of the iterative process allowing satisfaction of the conditions (1.2), (1.3), and (1.4) to any degree of accuracy. In addition, the relation of this procedure with the classical shell theory is revealed. Here and below under classical shell theories we understand the Love type 2D theories, which are introduced in more details in Ambartsumyan (1961, 1974). A comprehensive asymptotic analysis of the problem (1.2)-(1.4), (1.8)-(1.12) will require construction of the boundary layer solution along with investigation of its interaction with the iterative process for the outer problem.

8.2 The Long Wave Outer Solution

As before in case of elastic plates, here we refer to the solution propagating into the interior out of the edge, as the outer solution or the outer stress-strain field. In order to perform asymptotic analysis of the outer problem, let us introduce the following dimensionless scaling

$$\alpha = R\lambda^{-p}\xi, \quad \beta = R\lambda^{-p}\eta, \quad \gamma = h\zeta = R\lambda^{-q}\zeta, \tag{2.1}$$

where R is a typical radius of curvature of the shell, $\lambda = \left(\frac{h}{R}\right)^{-1/q} = \varepsilon^{-1/q}$ is a large asymptotic parameter, and p and q are integers. Then

$$R\frac{\partial}{\partial\alpha} = \lambda^p\frac{\partial}{\partial\xi}, \quad R\frac{\partial}{\partial\beta} = \lambda^p\frac{\partial}{\partial\eta}, \quad R\frac{\partial}{\partial\gamma} = \lambda^q\frac{\partial}{\partial\zeta}. \tag{2.2}$$

In the following analysis it is assumed that differentiation with respect to the scaled variables ξ, η and ζ does not change the asymptotic order of the quantity, which is equivalent to the fact that differentiation with respect to α, β and γ the order of the quantity is multiplied by λ^p and λ^q, thus the sought-for stress-strain field has an index of variability $t = \frac{p}{q}$, which is non-zero in general case.

Transforming equations (1.2), (1.3) in view of (2.1) and (2.2), the solutions are sought for in the form

$$\sigma_\alpha = \lambda^{q+d} \sum_{s=0}^{S} \lambda^{-s} \sigma_\alpha^{(s)}, \quad (\alpha, \beta), \quad \tau_{\alpha\beta} = \lambda^{q+d} \sum_{s=0}^{S} \lambda^{-s} \tau_{\alpha\beta}^{(s)},$$

$$\tau_{\alpha\gamma} = \lambda^{p+d} \sum_{s=0}^{S} \lambda^{-s} \tau_{\alpha\gamma}^{(s)}, \quad (\alpha, \beta), \quad \sigma_\gamma = \lambda^{c+d} \sum_{s=0}^{S} \lambda^{-s} \sigma_\gamma^{(s)}, \qquad (2.3)$$

$$u_\alpha = \lambda^{q-p+d} \sum_{s=0}^{S} \lambda^{-s} u_\alpha^{(s)}, \quad (\alpha, \beta), \quad u_\gamma = \lambda^{q-c+d} \sum_{s=0}^{S} \lambda^{-s} u_\gamma^{(s)}.$$

Here $c = 0$, $t = \frac{p}{q} \leq \frac{1}{2}$, $c = 2p - q$ for $\frac{1}{2} \leq t < 1$. The value of d depends on the intensity of the external loading, which should be chosen in order to satisfy the boundary conditions (1.4), i.e. if $\sigma_\gamma \neq 0$ at $\gamma = \pm h$, then $d = -c$, if $\sigma_\gamma = 0$ and simultaneously $\tau_{\alpha\gamma} \neq 0$ at $\gamma = \pm h$, then $d = -p$.

The expansion similar to (2.3) has been applied by Goldenveizer (1961, 1963, 1976) for asymptotic analysis of the outer solution of isotropic shells at $t < 1$. It has been shown by the author in Aghalovyan (1966a) that the same approach may be used for anisotropic shells.

Substituting expansions (2.3) into equations (1.2), (1.3) transformed through (2.1), (2.2) and equating to zero the coefficients in powers of λ, we arrive at the following system for unknown coefficients of (2.3)

$$\frac{1}{B} \partial_\xi \left(B \sigma_\alpha^{(s)} \right) - R k_\beta \sigma_\beta^{(s-p)} + \frac{1}{A} \partial_\eta \left(A \tau_{\beta\alpha}^{(s)} \right) + R k_\alpha \tau_{\alpha\beta}^{(s-p)}$$

$$+ \partial_\zeta \tau_{\alpha\gamma}^{(s)} + \frac{\zeta}{r_1} \partial_\xi \tau_{\alpha\gamma}^{(s-q)} + \frac{2}{r_1} \tau_{\alpha\gamma}^{(s-q)} = 0, \quad (\alpha, \beta; 1, 2),$$

$$\partial_\zeta \sigma_\gamma^{(s)} - \left(\frac{\sigma_\alpha^{(s-c)}}{r_1} + \frac{\sigma_\beta^{(s-c)}}{r_2} \right) + \partial_\zeta \tau_{\alpha\gamma}^{(s-q+2p-c)} + \partial_\eta \tau_{\beta\gamma}^{(s-q+2p-c)} \qquad (2.4)$$

$$+ R k_\beta \tau_{\alpha\gamma}^{(s-q+p-c)} + R k_\alpha \tau_{\beta\gamma}^{(s-q+p-c)} = 0,$$

$$\tau_{\alpha\beta}^{(s)} + \frac{\zeta}{r_1} \tau_{\alpha\beta}^{(s-q)} = \tau_{\beta\alpha}^{(s)} + \frac{\zeta}{r_2} \tau_{\beta\alpha}^{(s-q)}$$

and

$$\partial_\xi u_\alpha^{(s)} + R k_\alpha u_\beta^{(s-p)} + \frac{u_\gamma^{(s-c)}}{r_1} + \frac{\zeta}{r_2} \left(\partial_\xi u_\alpha^{(s-q)} + R k_\alpha u_\beta^{(s-q-p)} + \frac{u_\gamma^{(s-q-c)}}{r_1} \right)$$

$$= R \left(a_{11} \sigma_\alpha^{(s)} + a_{12} \sigma_\beta^{(s)} + a_{13} \sigma_\gamma^{(s-q+c)} + \zeta \left(a_{11} \frac{\sigma_\alpha^{(s-q)}}{r_1} + a_{12} \frac{\sigma_\beta^{(s-q)}}{r_2} \right) \right), \quad (\alpha, \beta),$$

$$\partial_\zeta u_\gamma^{(s)} + \zeta\left(\frac{1}{r_1} + \frac{1}{r_2}\right)\partial_\zeta u_\gamma^{(s-q)} + \frac{\zeta^2}{r_1 r_2}\partial_\zeta u_\gamma^{(s-2q)}$$

$$= R\left(a_{13}\sigma_\alpha^{(s-q+c)} + a_{23}\sigma_\beta^{(s-q+c)} + a_{33}\sigma_\gamma^{(s-2q+2c)}\right.$$

$$\left.+ \zeta\left(a_{13}\frac{\sigma_\alpha^{(s-2q+c)}}{r_1} + a_{23}\frac{\sigma_\beta^{(s-2q+c)}}{r_2}\right)\right), \qquad (2.5)$$

$$\partial_\eta u_\alpha^{(s)} + \partial_\xi u_\beta^{(s)} - R\left(k_\alpha u_\alpha^{(s-p)} + k_\beta u_\beta^{(s-p)}\right)$$

$$+ \frac{\zeta}{r_1}\partial_\eta u_\alpha^{(s-q)} + \frac{\zeta}{r_2}\partial_\xi u_\beta^{(s-q)}$$

$$- R\zeta\left(\frac{k_\alpha}{r_2}u_\alpha^{(s-p-q)} + \frac{k_\beta}{r_1}u_\beta^{(s-q-p)}\right) = Ra_{66}\tau_{\alpha\beta}^{(s)} + \frac{\zeta}{r_1}Ra_{66}\tau_{\alpha\beta}^{(s-q)},$$

$$\partial_\zeta u_\alpha^{(s)} + \partial_\xi u_\gamma^{(s-q+2p-c)} - \frac{u_\alpha^{(s-q)}}{r_1} + \zeta\left(\frac{1}{r_1} + \frac{1}{r_2}\right)\partial_\zeta u_\alpha^{(s-q)}$$

$$+ \frac{\zeta^2}{r_1 r_2}\partial_\zeta u_\alpha^{(s-2q)} - \frac{\zeta}{r_1 r_2}u_\alpha^{(s-2q)} + \frac{\zeta}{r_2}\partial_\xi u_\gamma^{(s-2q+2p-c)}$$

$$= Ra_{55}\tau_{\alpha\gamma}^{(s-2q+2p)} + Ra_{55}\frac{\zeta}{r_1}\tau_{\alpha\gamma}^{(s-3q+2p)}, \quad (\alpha,\beta;1,2;4,5),$$

$$\partial_\xi = \frac{1}{A}\frac{\partial}{\partial\xi}\quad(\xi,\eta), \quad \partial_\zeta = \frac{\partial}{\partial\zeta}, \quad r_{.1} = \frac{R_1}{R}, \quad r_2 = \frac{R_2}{R}.$$

We remark that the quantities of (2.4) and (2.5) with negative superscript should be assigned zero.

Equations (2.4) and (2.5) along with the boundary conditions (1.5) transformed in line with (2.3), form a complete system for sequential evaluation of all of the involved unknown quantities. It may be shown that the process of solving is iterative. Let us now restrict our attention to approximations corresponding to the classical shell theory. In order to make our considerations more lucid we assume the field zero variability. However, the analysis presented below could be easily generalized to the case of $0 < t < 1$. It is well known that the main assumptions of the classical shell theory rely on neglecting the normal stress σ_γ in the third equation, along with the tangential stresses $\tau_{\alpha\gamma}$ and $\tau_{\beta\gamma}$ in the last two equations during the derivation of the governing equations (1.3), for more details see Goldenveizer (1961); Novozhilov (1962); Goldenveizer (1976). As follows from the asymptotic integration procedure, the above-mentioned assumptions are satisfied when $s < 2q - 2p$. Therefore, the classical results should correspond to these obtained through the asymptotic integration of the equations of 3D elasticity up to and including the orders of $s = 2q - 2p - 1$ (the affect of anisotropy is discussed in Section 8.5). In particular for shells with zero variability ($p = 0$, $q = 1$) the corresponding orders of approximation are $s = 0, 1$. Let us consider these approximations in case of zero variability. At leading order $s = 0$ it may be deduced from the relations (2.4), (2.5) that

$$\partial_\zeta u_\gamma^{(0)} = 0, \quad \partial_\zeta u_\alpha^{(0)} = 0,$$

$$\partial_\xi u_\alpha^{(0)} + Rk_\alpha u_\beta^{(0)} + \frac{u_\gamma^{(0)}}{r_1} = R\left(a_{11}\sigma_\alpha^{(0)} + a_{12}\sigma_\beta^{(0)}\right), \quad (\alpha,\beta),$$

$$\partial_\eta u_\alpha^{(0)} + \partial_\xi u_\beta^{(0)} - R\left(k_\alpha u_\alpha^{(0)} + k_\beta u_\beta^{(0)}\right) = Ra_{66}\tau_{\alpha\beta}^{(0)}, \quad (2.6)$$

$$\frac{1}{B}\partial_\xi\left(B\sigma_\alpha^{(0)}\right) - Rk_\beta\sigma_\beta^{(0)} + \frac{1}{A}\partial_\eta\left(A\tau_{\beta\alpha}^{(0)}\right) + Rk_\alpha\tau_{\alpha\beta}^{(0)} + \partial_\zeta\tau_{\alpha\gamma}^{(0)} = 0, \quad (\alpha,\beta),$$

$$\partial_\zeta\sigma_\gamma^{(0)} - \left(\frac{\sigma_\alpha^{(0)}}{r_1} + \frac{\sigma_\beta^{(0)}}{r_2}\right) = 0, \quad \tau_{\alpha\beta}^{(0)} = \tau_{\beta\alpha}^{(0)}.$$

The boundary conditions imposed on the surfaces $\zeta = \pm 1$ corresponding to (2.6), may be formulated as

$$\sigma_\gamma^{(0)} = \pm\frac{1}{2}p_\gamma - \frac{1}{2}m, \tag{2.7}$$

$$\tau_{\alpha\gamma}^{(0)} = \pm\frac{1}{2}p_\alpha + \frac{1}{2}m_\beta \quad (\alpha,\beta) \quad \text{at} \quad \zeta = \pm 1.$$

The relations (2.6), (2.7) form a complete system for evaluation of the unknown leading order stress and displacement components. Integrating with respect to ζ and making use of (2.7), one arrives at

$$u_\gamma^{(0)} = w^{(0)}(\alpha,\beta), \quad u_\alpha^{(0)} = v_\alpha^{(0)}(\alpha,\beta), \quad (\alpha,\beta),$$

$$\frac{1}{B}\partial_\alpha\left(B\sigma_\alpha^{(0)}\right) + \frac{1}{A}\partial_\beta\left(A\tau_{\beta\alpha}^{(0)}\right) - k_\beta\sigma_\beta^{(0)} + k_\alpha\tau_{\alpha\beta}^{(0)} = -\frac{p_\alpha}{2R}, \quad (\alpha,\beta), \tag{2.8}$$

$$\frac{\sigma_\alpha^{(0)}}{R_1} + \frac{\sigma_\beta^{(0)}}{R_2} = \frac{1}{2R}p_\gamma, \quad \tau_{\alpha\beta}^{(0)} = \tau_{\beta\alpha}^{(0)},$$

$$\sigma_\alpha^{(0)} = B_{11}\varepsilon_1^{(0)} + B_{12}\varepsilon_2^{(0)}, \quad (\alpha,\beta), \quad \tau_{\alpha\beta}^{(0)} = B_{66}\omega^{(0)}, \tag{2.9}$$

$$\tau_{\alpha\gamma}^{(0)} = \frac{1}{2}m_\beta + \frac{1}{2}p_\alpha\zeta, \quad (\alpha,\beta), \quad \sigma_\gamma^{(0)} = -\frac{1}{2}m + \frac{1}{2}p_\gamma\zeta, \tag{2.10}$$

where

$$B_{11} = \frac{a_{22}}{\Omega}, \quad B_{22} = \frac{a_{11}}{\Omega}, \quad B_{12} = -\frac{a_{12}}{\Omega}, \quad \Omega = a_{11}a_{22} - a_{12}^2,$$

and the quantities $\varepsilon_1^{(0)}$, $\varepsilon_2^{(0)}$, $\omega^{(0)}$ have been expressed through $w^{(0)}(\alpha,\beta)$, $v_\alpha^{(0)}(\alpha,\beta)$, $v_\beta^{(0)}(\alpha,\beta)$ and traditional relations of the classical shell theory for tangential strains, meaning essentially the relative extensional and shear deformations of the mid-surface of the shell. As will be shown later, equations (2.8) and (2.9) coincide with the equations of the membrane shell theory. However, in our case some more features are involved, since the expressions for $\tau_{\alpha\gamma}, \tau_{\beta\gamma}$ and σ_γ have been obtained, with the latter usually not determined within the framework of the classical shell theory.

Now using the results of the leading order approximation, at next order $s = 1$ we get

$$R\left(\frac{1}{B}\partial_\alpha\left(B\sigma_\alpha^{(1)}\right) - k_\beta\sigma_\beta^{(1)} + \frac{1}{A}\partial_\beta\left(A\tau_{\beta\alpha}^{(1)}\right) + k_\alpha\tau_{\alpha\beta}^{(1)}\right)$$

$$+\partial_\zeta\tau_{\alpha\gamma}^{(1)} = -\frac{1}{r_1}\left(m_\beta + \frac{3}{2}\zeta p_\alpha\right), \quad (\alpha,\beta),$$

$$\frac{1}{R}\partial_\zeta\sigma_\gamma^{(1)} - \left(\frac{\sigma_\alpha^{(1)}}{R_1} + \frac{\sigma_\beta^{(1)}}{R_2}\right) = -\frac{1}{AB}\left(\frac{\partial B\tau_{\alpha\gamma}^{(0)}}{\partial\alpha} + \frac{\partial A\tau_{\beta\gamma}^{(0)}}{\partial\beta}\right),$$

$$\tau_{\alpha\beta}^{(1)} + \frac{\zeta}{r_1}\tau_{\alpha\beta}^{(0)} = \tau_{\beta\alpha}^{(1)} + \frac{\zeta}{r_2}\tau_{\beta\alpha}^{(0)},$$

$$u_\gamma^{(1)} = w^{(1)}(\alpha,\beta) + \zeta R\left(a_{13}\sigma_\alpha^{(0)} + a_{23}\sigma_\beta^{(0)}\right), \quad (2.11)$$

$$u_\alpha^{(1)} = v_\alpha^{(1)}(\alpha,\beta) - \zeta R\gamma_1^{(0)}, \quad (\alpha,\beta),$$

$$\varepsilon_1^{(1)} + \zeta R\left(\chi_1^{(0)} + \frac{\varepsilon_1^{(0)}}{R_2} + \frac{1}{R_1}\left(a_{13}\sigma_\alpha^{(0)} + a_{23}\sigma_\beta^{(0)}\right)\right)$$

$$= a_{11}\sigma_\alpha^{(1)} + a_{12}\sigma_\beta^{(1)} + a_{13}\sigma_\gamma^{(0)} + \zeta\left(a_{11}\frac{\sigma_\alpha^{(0)}}{r_1} + a_{12}\frac{\sigma_\beta^{(0)}}{r_2}\right), \quad (\alpha,\beta;1,2),$$

$$\omega^{(1)} + 2\zeta R\tau^{(0)} = a_{66}\tau_{\alpha\beta}^{(1)} + \frac{\zeta}{r_1}a_{66}\tau_{\alpha\beta}^{(0)}$$

with the meaning of the parameters

$$\gamma_1^{(0)} = \partial_\alpha u_\gamma^{(0)} - \frac{u_\alpha^{(0)}}{R_1}, \quad (\alpha,\beta;1,2),$$

$$\chi_1^{(0)} = -\frac{1}{A}\frac{\partial\gamma_1^{(0)}}{\partial\alpha} - k_\alpha\gamma_2^{(0)}, \quad (\alpha,\beta;1,2), \quad (2.12)$$

$$\tau^{(0)} = -\partial_\alpha\gamma_2^{(0)} + k_\alpha\gamma_1^{(0)} + \frac{\omega_2^{(0)}}{R_1} = -\partial_\beta\gamma_1^{(0)} + k_\beta\gamma_2^{(0)} + \frac{\omega_1^{(0)}}{R_2},$$

$$\omega_1^{(0)} = \partial_\alpha u_\beta^{(0)} - k_\alpha u_\alpha^{(0)}, \quad (\alpha,\beta;1,2), \quad \omega^{(0)} = \omega_1^{(0)} + \omega_2^{(0)}$$

being basically the same as that usually introduced in the classical shell theory.

It follows from (2.11) that

$$\sigma_\alpha^{(1)} = \sigma_{\alpha0}^{(1)} + \zeta\sigma_{\alpha1}^{(1)}, \quad (\alpha,\beta), \quad \tau_{\alpha\beta}^{(1)} = \tau_{\alpha\beta0}^{(1)} + \zeta\tau_{\alpha\beta1}^{(1)},$$

$$\tau_{\alpha\gamma}^{(1)} = \tau_{\alpha\gamma0}^{(1)} + \zeta\tau_{\alpha\gamma1}^{(1)} + \zeta^2\tau_{\alpha\gamma2}^{(1)}, \quad (\alpha,\beta), \quad (2.13)$$

$$\sigma_\gamma^{(1)} = \sigma_{\gamma0}^{(1)} + \zeta\sigma_{\gamma1}^{(1)} + \zeta^2\sigma_{\gamma2}^{(1)}.$$

In order to determine the unknown coefficients in relations (2.13), the latter should be substituted into (2.11). The coefficients within the same powers of ζ should then be equated, and the boundary conditions at $\zeta = \pm 1$ of the next order

$s = 1$ should be satisfied. These boundary conditions may be obtained from (1.5) by taking into consideration (2.3) and (2.13), giving

$$\sigma_\gamma^{(1)} = \frac{1}{2}\left(\frac{1}{r_1} + \frac{1}{r_2}\right)(p_\gamma \mp m),\tag{2.14}$$

$$\tau_{\alpha\gamma}^{(1)} = \frac{1}{2r_2}(\pm p_\alpha + m_\beta), \quad (\alpha, \beta) \quad \text{at} \quad \zeta = \pm 1.$$

As a result, the following system for evaluation of $\sigma_{\alpha 0}^{(1)}$, $\sigma_{\beta 0}^{(1)}$, $\tau_{\alpha\beta 0}^{(1)}$, $v_\alpha^{(1)}$, $w^{(1)}$ is obtained

$$\frac{1}{B}\partial_\alpha\left(B\sigma_{\alpha 0}^{(1)}\right) - k_\beta\sigma_{\beta 0}^{(1)} + \frac{1}{A}\partial_\beta\left(A\tau_{\beta\alpha 0}^{(1)}\right) + k_\alpha\tau_{\alpha\beta 0}^{(1)}$$

$$= -\left(\frac{1}{R_1} + \frac{1}{2R_2}\right)m_\beta, \quad (\alpha, \beta),$$

$$\frac{\sigma_{\alpha 0}^{(1)}}{R_1} + \frac{\sigma_{\beta 0}^{(1)}}{R_2} = -\frac{m}{2}\left(\frac{1}{R_1} + \frac{1}{R_2}\right) + \frac{1}{2AB}\left(\frac{\partial Bm_\beta}{\partial\alpha} + \frac{\partial Am_\alpha}{\partial\beta}\right),\tag{2.15}$$

$$\tau_{\alpha\beta 0}^{(1)} = \tau_{\beta\alpha 0}^{(1)},$$

$$\varepsilon_1^{(1)} = a_{11}\sigma_{\alpha 0}^{(1)} + a_{12}\sigma_{\beta 0}^{(1)} - \frac{1}{2}a_{13}m, \quad (\alpha, \beta; 1, 2), \quad w^{(1)} = a_{66}\tau_{\alpha\beta 0}^{(1)}$$

where $\varepsilon_1^{(1)}, \varepsilon_2^{(1)}$, $w^{(1)}$ are the next order $(s = 1)$ components of tangential strains of the mid-surface of the shell.

It is possible to obtain in respect of the stress components corresponding to the pure bending part of the stress and strain field,

$$\sigma_{\alpha 1}^{(1)} = R\left(B_{11}\chi_1^{(0)} + B_{12}\chi_2^{(0)} + \left(\frac{B_{11}}{R_1} + \frac{B_{12}}{R_2}\right)(A_1\varepsilon_1^{(0)} + A_2\varepsilon_2^{(0)})\right.$$

$$\left. + \left(\frac{1}{R_2} - \frac{1}{R_1}\right)B_{11}\varepsilon_1^{(0)}\right) - \frac{1}{2}A_1p_\gamma, \quad (\alpha, \beta; 1, 2),\tag{2.16}$$

$$\tau_{\alpha\beta 1}^{(1)} = 2B_{66}R\left(\tau^{(0)} - \frac{w^{(0)}}{2R_1}\right), \quad \tau_{\alpha\beta 1}^{(1)} = \tau_{\beta\alpha 1}^{(1)} + \left(\frac{1}{r_2} - \frac{1}{r_1}\right)\tau_{\beta\alpha}^{(0)}$$

with the transverse tangential and normal stress components given by

$$\tau_{\alpha\gamma 0}^{(1)} = \frac{p_\alpha}{2r_2} - \tau_{\alpha\gamma 2}^{(1)}, \quad \tau_{\alpha\gamma 1}^{(1)} = \frac{m_\beta}{2r_2}, \quad (\alpha, \beta),$$

$$\tau_{\alpha\gamma}^{(1)} = \frac{1}{2r_2}(p_\alpha + \zeta m_\beta) + (\zeta^2 - 1)\tau_{\alpha\gamma 2}^{(1)}, \quad (\alpha, \beta),\tag{2.17}$$

$$\sigma_{\gamma 1}^{(1)} = -\frac{m}{2}\left(\frac{1}{r_1} + \frac{1}{r_2}\right), \quad \sigma_{\gamma 0}^{(1)} = \frac{1}{2}p_\gamma\left(\frac{1}{r_1} + \frac{1}{r_2}\right) - \sigma_{\gamma 2}^{(1)},$$

$$\sigma_\gamma^{(1)} = \frac{1}{2}\left(\frac{1}{r_1} + \frac{1}{r_2}\right)(p_\gamma - m\zeta) - (1 - \zeta^2)\sigma_{\gamma 2}^{(1)}.$$

Here the quantities $\tau_{\alpha\gamma2}^{(1)}$ and $\sigma_{\gamma2}^{(1)}$ are governed by

$$\frac{1}{B}\partial_\alpha\left(B\sigma_{\alpha1}^{(1)}\right) - k_\beta\sigma_{\beta1}^{(1)} + \frac{1}{A}\partial_\beta\left(A\tau_{\beta\alpha1}^{(1)}\right) + k_\alpha\tau_{\alpha\beta1}^{(1)}$$

$$+\frac{2}{R}\tau_{\alpha\gamma2}^{(1)} = -\frac{3}{2}\frac{p_\alpha}{R_1}, \quad (\alpha,\beta), \tag{2.18}$$

$$\frac{2}{R}\sigma_{\gamma2}^{(1)} - \left(\frac{\sigma_{\alpha1}^{(1)}}{R_1} + \frac{\sigma_{\beta1}^{(1)}}{R_2}\right) = -\frac{1}{2AB}\left(\frac{\partial Bp_\alpha}{\partial\alpha} + \frac{\partial Ap_\beta}{\partial\beta}\right).$$

As may be seen from (2.16), the stresses $\sigma_{\alpha1}^{(1)}, \sigma_{\beta1}^{(1)}, \tau_{\alpha\beta1}^{(1)}$ are fully determined through leading order quantities. Then, with these stresses known, the quantities $\sigma_{\gamma2}^{(1)}$ and $\tau_{\alpha\gamma2}^{(1)}$ are found from (2.18), with tangential and normal stresses in the cross sections governed by (2.17). Thus, all of the quantities of the next-order approximations may be determined.

In case of non-zero variability it may be inferred from the system (2.4), (2.5) that at $s \in [0, q - 2p + c)$ the structure of the governing system is that of (2.6), whereas in case of $s \in [q - 2p + c, 2q - 2p - 1]$ the structure is similar to (2.11). Therefore the form of solutions of (2.4) and (2.5) will be analogous to that obtained earlier for $s = 0, 1; p = 0, q = 1$. In order to satisfy the boundary conditions at $\zeta = \pm 1$ we also require

$$\sigma_\gamma^{(s)} = \pm\frac{1}{2}p_\gamma^{(s)} - \frac{1}{2}m^{(s)}, \quad \tau_{\alpha\gamma}^{(s)} = \pm\frac{1}{2}p_\alpha^{(s)} + \frac{1}{2}m_\beta^{(s)}, \quad (\alpha,\beta) \tag{2.19}$$

where if $p_\gamma \neq 0$, $m \neq 0$, then $d = -c$, and

$$p_\gamma^{(0)} = p_\gamma, \quad m^{(0)} = m, \quad p_\gamma^{(s)} = m^{(s)} = 0, \quad (s \neq 0) \tag{2.20}$$

$$p_\alpha^{(p)} = p_\alpha, \quad m_\beta^{(p)} = m_\beta, \quad p_\alpha^{(s)} = m_\beta^{(s)} = 0, \quad (s \neq p), \quad (\alpha,\beta).$$

In case of $p_\gamma = m = 0$, and $p_\alpha \neq 0$, $m_\beta \neq 0$ then $d = -p$, therefore

$$p_\alpha^{(0)} = p_\alpha, \quad m_\beta^{(0)} = m_\beta, \quad (\alpha,\beta), \tag{2.21}$$

$$p_\gamma^{(s)} = m^{(s)} = p_\alpha^{(s)} = m_\beta^{(s)} = 0 \quad (s \neq 0), \quad (\alpha,\beta).$$

In view of the above stated, if we restrict ourselves to the approximations of order $s \in [0, 2q - 2p - 1]$, the stress and displacement components may be generally represented as $(p_\gamma \neq 0, m \neq 0)$

$$\sigma_\alpha = \lambda^{q-s}(\overset{0}{\sigma}{}^{(s)}_\alpha + \zeta\overset{1}{\sigma}{}^{(s)}_\alpha), \quad \tau_{\alpha\beta} = \lambda^{q-s}(\overset{0}{\tau}{}^{(s)}_{\alpha\beta} + \zeta\overset{1}{\tau}{}^{(s)}_{\alpha\beta}),$$

$$\tau_{\alpha\gamma} = \lambda^{p-s}(\overset{0}{\tau}{}^{(s)}_{\alpha\gamma} + \zeta\overset{1}{\tau}{}^{(s)}_{\alpha\gamma} + \zeta^2\overset{2}{\tau}{}^{(s)}_{\alpha\gamma}), \quad (\alpha,\beta), \tag{2.22}$$

$$\sigma_\gamma = \lambda^{c-s}(\overset{0}{\sigma}{}^{(s)}_\gamma + \zeta\overset{1}{\sigma}{}^{(s)}_\gamma + \zeta^2\overset{2}{\sigma}{}^{(s)}_\gamma),$$

$$u_\alpha = \lambda^{q-p-s}(\overset{0}{v}{}^{(s)}_\alpha + \zeta\overset{1}{v}{}^{(s)}_\alpha) \quad (\alpha,\beta), \quad w = \lambda^{q-c-s}(\overset{0}{w}{}^{(s)} + \zeta\overset{1}{w}{}^{(s)})$$

with summation assumed along the dummy index s in the limits from zero to $2q - 2p - 1$. We note in addition that $\overset{1}{\sigma}{}^{(s)}_\alpha = \overset{1}{\tau}{}^{(s)}_{\alpha\beta} = \overset{2}{\tau}{}^{(s)}_{\alpha\gamma} = \overset{2}{\sigma}{}^{(s)}_\gamma = 0$, $\overset{1}{v}{}^{(s)}_\alpha = \overset{1}{w}{}^{(s)} = 0$ for $s < q - 2p + c$.

We remark that more accurate analysis drives us beyond the field defined by the assumptions of the classical Kirchhoff-Love theory. Within the asymptotic approach these refinements may be obtained from the following approximations of order $s \geq 2q - 2p$.

8.3   Comparison of the Asymptotic Outer Solution with Membrane and Full Shell Theories

Let us compare the results obtained in the previous section to that of the classical shell theory. In particular, the first two terms may only be kept in expansions (2.3) in case of zero variability, because of the earlier remark that keeping these terms only does not contradict the general assumptions of the classical theory. Taking into account relations (1.1), (2.3), (2.10), (2.13), and (2.17) in respect of the forces, moments and displacements of the mid-surface of the shell, one may obtain that

$$T_1 = T_1^{(0)} + \frac{h}{R} T_1^{(1)} = \int_{-h}^{+h} \sigma_\alpha d\gamma = 2R \left(\sigma_\alpha^{(0)} + \frac{h}{R} \sigma_{\alpha 0}^{(1)} \right), \quad (\alpha, \beta), \qquad (3.1)$$

$$S_{12} = S_{12}^{(0)} + \frac{h}{R} S_{12}^{(1)} = \int_{-h}^{+h} \tau_{\alpha\beta} d\gamma = 2R \left(\tau_{\alpha\beta}^{(0)} + \frac{h}{R} \tau_{\alpha\beta 0}^{(1)} \right),$$

$$M_1 = \int_{-h}^{+h} \sigma_\alpha \gamma d\gamma = \frac{2}{3} h^2 \sigma_{\alpha 1}^{(1)}, \quad (\alpha, \beta), \qquad (3.2)$$

$$H_{12} = \int_{-h}^{+h} \tau_{\alpha\beta} \gamma \, d\gamma = \frac{2}{3} h^2 \tau_{\alpha\beta 1}^{(1)}, \quad (\alpha, \beta),$$

$$N_1 = N_1^{(n)} + N_1^{(M)} = \int_{-h}^{+h} \tau_{\alpha\gamma} d\gamma = h m_\beta + \frac{1}{R_2} h^2 p_\alpha - \frac{4}{3} \frac{h^2}{R} \tau_{\alpha\gamma 2}^{(1)}, \quad (\alpha, \beta; 1, 2). \quad (3.3)$$

Here the signs of the forces are chosen in accordance with Ambartsumyan (1961). It also follows that

$$u = u_{\alpha 0} + \frac{h}{R} u_{\alpha 1} = R h^{-1} \left(v_\alpha^{(0)} + \frac{h}{R} v_\alpha^{(1)} \right), \quad (\alpha, \beta), \qquad (3.4)$$

$$w = w_{(0)} + \frac{h}{R} w_{(1)} = R h^{-1} \left(w^{(0)} + \frac{h}{R} w^{(1)} \right).$$

Therefore,

$$T_1^{(i)} = 2R \sigma_\alpha^{(i)}, \quad S_{12}^{(i)} = 2R \tau_{\alpha\beta}^{(i)}, \quad i = 0, 1; \quad (\alpha, \beta), \qquad (3.5)$$

$$\sigma_{\alpha 1}^{(1)} = \frac{3}{2} h^{-2} M_1, \quad \tau_{\alpha\beta 1}^{(1)} = \frac{3}{2} h^{-2} H_{12}, \quad (\alpha, \beta; M_1, M_2), \qquad (3.6)$$

$$N_1^{(n)} = h m_\beta, \quad \frac{4}{3} \frac{h^2}{R} \tau_{\alpha\gamma 2}^{(1)} = -N_1^{(M)} + \frac{1}{R_2} h^2 p_\alpha, \quad (\alpha, \beta), \qquad (3.7)$$

and

$$u_{\alpha i} = Rh^{-1}v_\alpha^{(i)}, \quad (\alpha,\beta), \quad w_{(i)} = Rh^{-1}w^{(i)}, \quad i = 0,1. \tag{3.8}$$

Substitution of (3.5) and (3.8) into (2.8), (2.9), and (2.15) gives

$$\frac{\partial BT_1^{(i)}}{\partial \alpha} + \frac{\partial AS_{21}^{(i)}}{\partial \beta} - \frac{\partial B}{\partial \alpha}T_2^{(i)} + \frac{\partial A}{\partial \beta}S_{12}^{(i)} = -ABX^{(i)}, \quad (\alpha,\beta;X,Y), \tag{3.9}$$

$$\frac{T_1^{(i)}}{R_1} + \frac{T_2^{(i)}}{R_2} = Z^{(i)}, \quad S_{12}^{(i)} = S_{21}^{(i)}, \quad i = 0,1,$$

$$T_1^{(i)} = 2h\left(B_{11}\varepsilon_{1(i)} + B_{12}\varepsilon_{2(i)}\right) + m_1^{(i)}, \quad (1,2), \tag{3.10}$$

$$S_{12}^{(i)} = 2hB_{66}\omega_{(i)}, \quad i = 0,1$$

where

$$X^{(0)} = p_\alpha, \quad (\alpha,\beta;X,Y), \tag{3.11}$$

$$Z^{(0)} = p_\gamma,$$

$$X^{(1)} = \left(\frac{2}{r_1} + \frac{1}{r_2}\right)m_\beta, \quad (\alpha,\beta;X,Y), \tag{3.12}$$

$$Z^{(1)} = -\left(\frac{1}{r_1} + \frac{1}{r_2}\right)m + \frac{R}{AB}\left(\frac{\partial Bm_\beta}{\partial \alpha} + \frac{\partial Am_\alpha}{\partial \beta}\right),$$

and

$$m_1^{(0)} = 0, \quad m_1^{(1)} = A_1 mR. \quad (1,2), \tag{3.13}$$

$$\varepsilon_{i(j)} = Rh^{-1}\varepsilon_i^{(j)}, \quad \omega_{(j)} = Rh^{-1}\omega^{(j)}, \quad (i = 1,2; j = 0,1).$$

It may be observed that equations (3.9) and (3.10) for approximation $i = 0$ coincide with the formulation of the membrane shell theory. Similarly, using (2.10), one may calculate the quantities $\tau_{\alpha\gamma}$, $\tau_{\beta\gamma}$, σ_γ. Moreover, at $i = 1$ equations (3.9) and (3.10) once again correspond to the membrane shell theory, however, the external loads are now governed by (3.12).

Multiplying equations (3.9) and (3.10) at $i = 1$ by h/R and adding these to the same equations at $i = 0$ and making use of (3.1), one results in

$$\frac{\partial BT_1}{\partial \alpha} + \frac{\partial AS_{21}}{\partial \beta} - \frac{\partial B}{\partial \alpha}T_2 + \frac{\partial A}{\partial \beta}S_{12}$$

$$= -AB\left(p_\alpha + h\left(\frac{2}{R_1} + \frac{1}{R_2}\right)m_\beta\right), \quad (\alpha,\beta), \tag{3.14}$$

$$\frac{T_1}{R_1} + \frac{T_2}{R_2} = p_\gamma - h\left(\frac{1}{R_1} + \frac{1}{R_2}\right)m$$

$$+ \frac{h}{AB}\left(\frac{\partial Bm_\beta}{\partial \alpha} + \frac{\partial Am_\alpha}{\partial \beta}\right), \quad S_{12} = S_{21},$$

$$T_1 = 2h\left(B_{11}\varepsilon_1 + B_{12}\varepsilon_2\right) + A_1 mh, \quad (1,2), \tag{3.15}$$

$$S_{12} = 2hB_{66}\omega,$$

where

$$A_1 = a_{13}B_{11} + a_{23}B_{12} = -\frac{E_1}{E_3}\frac{\nu_{13} + \nu_2\nu_{23}}{1 - \nu_1\nu_2}, \quad (1,2). \tag{3.16}$$

The obtained relations (3.14) enable evaluation of the tangential forces with accuracy of $O(h_*^2)$. We remark that equations (3.14) and (3.15) are more general than the corresponding equations of the membrane shell theory, since they take into account moments and transverse compression arising from transformation of the external forces to the mid-surface. It should also be noted that in case of isotropic shells the relations (3.15) coincide with those obtained by Goldenveizer (1961, 1976) (up to a sign of parameter m). One more observation is that provided that the main curvatures are neglected in the loading terms of system (3.14), the latter then coincides with the governing system of the membrane shell theory. Moreover, it may be verified that the constitutive relations also coincide with that of the membrane shell theory provided some of the higher order terms are neglected. On the other hand, as seen from (3.16) taking into account compression could be rather significant for anisotropic shells with small extensional stiffness in the transverse direction.

Substituting (3.6) and (3.7) into equations (2.16) and (2.18), we arrive at

$$\frac{\partial BM_1}{\partial \alpha} - \frac{\partial B}{\partial \alpha}M_2 + \frac{\partial AH_{21}}{\partial \beta} + \frac{\partial A}{\partial \beta}H_{12} - ABN_1^{(M)} \tag{3.17}$$

$$= -AB\left(\frac{1}{R_1} + \frac{1}{R_2}\right)h^2 p_\alpha, \quad (\alpha, \beta),$$

$$M_1 = D_{11}\chi_1 + D_{12}\chi_2 + \left(\frac{D_{11}}{R_1} + \frac{D_{12}}{R_2}\right)(A_1\varepsilon_1 + A_2\varepsilon_2)$$

$$-D_{11}\left(\frac{1}{R_1} - \frac{1}{R_2}\right)\varepsilon_1 - \frac{1}{3}A_1 h^2 p_\gamma, \quad (1,2), \tag{3.18}$$

$$H_{12} = 2D_{66}\left(\tau - \frac{\omega}{2R_1}\right), \quad D_{ik} = \frac{2}{3}h^3 B_{ik},$$

$$H_{21} - H_{12} + \frac{h^2}{3}\left(\frac{S_{21}}{R_2} - \frac{S_{12}}{R_1}\right) = 0, \tag{3.19}$$

where the quantities χ_1, χ_2, τ, ε_1, ε_2, ω are expressed through leading order solution through conventional relations of the classical shell theory. It is worth noting that the value of τ adopted here coincides with that of Goldenveizer (1961), however, it is a half of τ according to notation in Ambartsumyan (1961).

The displacement components of an arbitrary point of the shell are given by

$$u_\alpha = u - \gamma\gamma_1, \quad (\alpha, \beta; 1, 2), \quad (u, v) \tag{3.20}$$

$$u_\gamma = w + \frac{1}{2}\zeta(a_{13}T_1^{(0)} + a_{23}T_2^{(0)}).$$

At the same time, bearing in mind (2.10), (2.17), and (3.7), one may obtain expressions for the transverse stresses in the form

$$\tau_{\alpha\gamma} = \frac{1}{2}\left(m_\beta + \zeta p_\alpha + \frac{h}{R_2}\left(p_a + \zeta m_\beta\right)\right.$$
$$\left. + \frac{3}{2}(1 - \zeta^2)\left(\frac{N_1^{(M)}}{h} - \frac{hp_\alpha}{R_2}\right)\right), \quad (\alpha, \beta), \qquad (3.21)$$
$$\sigma_\gamma = \frac{1}{2}\left(-m + \zeta p_\gamma + h\left(\frac{1}{R_1} + \frac{1}{R_2}\right)(p_\gamma - \zeta m)\right)$$
$$-\frac{1}{4}(1 - \zeta^2)\left(\frac{3}{h}\left(\frac{M_1}{R_1} + \frac{M_2}{R_2}\right) - \frac{h}{AB}\left(\frac{\partial Bp_\alpha}{\partial \alpha} + \frac{\partial Ap_\beta}{\partial \beta}\right)\right).$$

We remark that equations (3.17) and (3.18) are in fact the fourth and fifth equations of equilibrium, provided that the moments arising at projecting of tangential forces onto the mid-surface of the shell, are taken into consideration. Similarly, equations (3.18) are generalized constitutive relations. Indeed, if the higher order contribution is neglected and only two first terms are kept, the discussed equations coincide with the well-known relations presented by Ambartsumyan (1961), whereas in isotropic case these coincide with the results of Goldenveizer (1961, 1968b, 1976).

It should be noted that equation (3.19) is actually an additional equation of the shell theory, effectively replacing the sixth equilibrium equation. It may be shown that (3.19) follows from the symmetry condition (coupling of tangential stresses). Indeed, it has been demonstrated in Aghalovyan (1973b) that the sixth equilibrium equation is an implication of higher order asymptotic approximations for the symmetry condition.

It may be deduced from the remarks above that violation of the sixth equilibrium equation could lead to violation of coupling of tangential stresses (the symmetry condition). It may be observed that condition (3.19) is trivially satisfied due to the last equation of (3.14) and constitutive relations on the torsional moment (3.18). Therefore, it may be excluded from consideration (similarly to the sixth equation of the classical theory). However, as demonstrated by Goldenveizer (1961, 1976), violation of the sixth equation of equilibrium could lead to serious consequences including violation (to a small extent) of the reciprocity and uniqueness theorems. As will be shown in the next section, this effect could be easily removed by keeping higher order small terms of the tangential forces, which are beyond the framework of the classical shell theory, but nevertheless guarantee that the sixth equilibrium equation is satisfied.

Because of the solutions for ε_1, ε_2, ω are delivered by (3.9), (3.10), whereas equations (3.18) enable evaluation of the bending and torsional moments, it follows that equations (3.17) serve for determination of the pure bending part of the shear forces. This is an important result, since it allows solution procedure within the membrane shell theory for shells with non-zero variability, whereas the pure bending part could be found by the above stated method.

Investigation of the results reveals that these could be obtained through straight-forward asymptotic analysis of the governing equations of anisotropic shells provided that the equations of equilibrium and the constitutive relations are written more rigorously and apply integration of these equations in certain order. In particular, the surface forces appearing in the equations of equilibrium, should be adopted in the form

$$X = X^+ \left(1 + \frac{h}{R_1}\right)\left(1 + \frac{h}{R_2}\right) + X^- \left(1 - \frac{h}{R_1}\right)\left(1 - \frac{h}{R_2}\right)$$

$$\approx p_\alpha + h \left(\frac{1}{R_1} + \frac{1}{R_2}\right) m_\beta, \quad (\alpha, \beta),$$

$$Z = Z^+ \left(1 + \frac{h}{R_1}\right)\left(1 + \frac{h}{R_2}\right) + Z^- \left(1 - \frac{h}{R_1}\right)\left(1 - \frac{h}{R_2}\right) \qquad (3.22)$$

$$\approx p_\gamma - h \left(\frac{1}{R_1} + \frac{1}{R_2}\right) m,$$

$$F = h \left(X^+ \left(1 + \frac{h}{R_1}\right)\left(1 + \frac{h}{R_2}\right) - X^- \left(1 - \frac{h}{R_1}\right)\left(1 - \frac{h}{R_2}\right)\right)$$

$$\approx h m_\beta + \left(\frac{1}{R_1} + \frac{1}{R_2}\right) h^2 p_\alpha, \quad (\alpha, \beta; E, F).$$

The constitutive relations have to be taken as (3.15) and (3.18). The integration involves two general steps. The first stage contains representation of the shear forces N_1 and N_2 as $N_1^{(n)} + N_1^{(M)}$, $N_2^{(n)} + N_2^{(M)}$. Then the pure bending part of the shear forces should be neglected in the first three equations of equilibrium, and the values of $N_1^{(n)}$ and $N_2^{(n)}$ are evaluated from the fourth and fifth equations $(N_1^{(n)} = h m_\beta, N_2^{(n)} = h m_\alpha)$. This will lead to a refined form of the membrane shell theory (3.14), which allows evaluation of the tangential forces. The procedure of the second step is the following: the quantities ε_1, ε_2, ω, χ_1, χ_2, τ are obtained through (3.15), with the bending and torsional moments found from (3.18). Then from the fourth and fifth equations of equilibrium the pure bending parts of the shear forces $N_1^{(M)}$ and $N_2^{(M)}$ are determined. Finally, relations (3.21) are used to calculate the transverse stresses, which are usually not determined within the classical framework.

The described process may look similar to integration of the governing equations of the membrane shell theory presented in Ambartsumyan (1961, 1974); Goldenveizer (1961, 1976), and may be referred to as asymptotic integration of the governing equations of 3D elasticity. However, the second step in classical theory plays secondary role, being used for rough estimates of moments and shear forces, whereas in the described approach the second step is crucial, serving for refined evaluation of moments and shear forces.

In case of the non-zero variability the forces and moments corresponding to

(2.22), are written as

$$T_1 = 2h\lambda^{q+d-s} \overset{0}{\underset{\alpha}{\sigma}} {}^{(s)}, \quad S_{12} = 2h\lambda^{q+d-s} \overset{0}{\underset{\alpha\beta}{\tau}} {}^{(s)}, \quad (\alpha,\beta;1,2),$$

$$N_1 = 2h\lambda^{p+d-s}\left(\overset{0}{\underset{\alpha\gamma}{\tau}} {}^{(s)} + \frac{1}{3}\overset{2}{\underset{\alpha\gamma}{\tau}} {}^{(s)}\right), \quad M_1 = \frac{2}{3}h^2\lambda^{q+d-s}\overset{1}{\underset{\alpha}{\sigma}} {}^{(s)}, \qquad (3.23)$$

$$H_{12} = \frac{2}{3}h^2\lambda^{q+d-s}\overset{1}{\underset{\alpha\beta}{\tau}} {}^{(s)}, \quad (\alpha,\beta;1,2)$$

where summation along the dummy index s is tacitly assumed in the limits of $0, 2q - 2p - 1$. The governing system of equations, corresponding to (2.4) and (2.5), may be presented in terms of (3.23). It includes the modified equilibrium equations

$$\frac{\partial BT_1}{\partial\alpha} - \frac{\partial B}{\partial\alpha}T_2 + \frac{\partial AS_{21}}{\partial\beta} + \frac{\partial A}{\partial\beta}S_{12} + AB\frac{N_1}{R_1} = -ABX_\alpha, \quad (\alpha,\beta),$$

$$\frac{T_1}{R_1} + \frac{T_2}{R_2} - \frac{1}{AB}\left(\frac{\partial BN_1}{\partial\alpha} + \frac{\partial AN_2}{\partial\beta}\right) = Z_\gamma, \quad (3.24)$$

$$\frac{\partial BM_1}{\partial\alpha} - \frac{\partial B}{\partial\alpha}M_2 + \frac{\partial AH_{21}}{\partial\beta} + \frac{\partial A}{\partial\beta}H_{12} - ABN_1 = -ABF_\beta, \quad (\alpha,\beta)$$

where

$$X_\alpha = p_\alpha + h\left(\frac{1}{R_1} + \frac{1}{R_2}\right)m_\beta + \frac{h^2}{R_1 R_2}p_\alpha, \quad (\alpha,\beta),$$

$$Z_\gamma = p_\gamma - h\left(\frac{1}{R_1} + \frac{1}{R_2}\right)m + \frac{h^2}{R_1 R_2}p_\gamma, \qquad (3.25)$$

$$F_\beta = hm_\beta + h^2\left(\frac{1}{R_1} + \frac{1}{R_2}\right)p_\alpha, \quad (\alpha,\beta)$$

along with constitutive relations (3.15) and (3.18). In order to have a full statement of the boundary value problem, the boundary conditions on the side surface of the shell should be added to the above-mentioned conditions. Equations (3.24) differ from the corresponding equations of the classical theory Ambartsumyan (1961), since they take into account the moments arising during transformation of external forces to the mid-surface of a shell. The conditions (3.15) incorporate the effect of compression in addition to the traditional formulation, with the constitutive relations (3.18) exhibiting principal distinctions from the ones suggested in Ambartsumyan (1961).

We remark that the relations (3.24), (3.15) and (3.18) may be obtained through asymptotic integration in a different way, as proposed by Goldenveizer (1961, 1968b, 1976) for isotropic shells.

8.4 Optimal Constitutive Relations

Here we shall discuss the so-called optimal constitutive relations, i.e. the formulation of a 2D shell theory which does not contradict the general assumption of the Kirchhoff-Love classical theory, and at the same time provides maximal accuracy in analysis of the stress-strain field. In case of isotropic shells such conditions were derived in Goldenveizer (1961, 1976). We also impose an additional requirement, namely that the constitutive relations should imply the sixth equation of equilibrium, providing that the deformation energy is homogeneous function of the deformation components along with some other consequences.

As mentioned in Section 8.3, in order to derive the constitutive relation, guaranteeing that the sixth equation of equilibrium is fulfilled, some of the small terms which are going beyond the required asymptotic tolerance of $O(h_*^{2-2t})$, should be kept during integration of (2.4) and (2.5). It follows from the asymptotic analysis of (2.4) and (2.5) that these are the small terms arising around S_{12} and S_{21}, which occur at next order of integration.

In order to reduce lengthy algebraic manipulations we shall only present the results for S_{12} and S_{21}, namely

$$S_{12} = 2hB_{66}\omega + D_{66}\left(\frac{1}{R_2} - \frac{1}{R_1}\right)\left(\tau - \frac{\omega}{R_1}\right) \tag{4.1}$$

$$+\frac{h^2}{6}\left(\frac{G_{12}}{G_{13}}\frac{A}{B}\frac{\partial}{\partial\beta}\frac{p_\alpha}{A} + \frac{G_{12}}{G_{23}}\frac{B}{A}\frac{\partial}{\partial\alpha}\frac{p_\beta}{B}\right), \quad (1,2;\alpha,\beta).$$

Using relation (3.18) for the torsional moment along with condition (4.1), it may be verified that the sixth equation of equilibrium is satisfied. Indeed,

$$S_{12} - S_{21} + \frac{H_{12}}{R_1} - \frac{H_{21}}{R_2} = 0. \tag{4.2}$$

If one neglects the refined terms in (4.1) and keeps the first term only, then the condition (4.1) coincides with the results of Ambartsumyan (1961, 1967), whereas simplification to isotropic shell re-casts the results of Goldenveizer (1961, 1976) (we note that the current notation of S_{12}, H_{12} corresponds to S_{21}, H_{21} in the notation of A.L. Goldenveiser, and vice versa). We also note that in case of keeping the two first terms for isotropic shells the result of transformed (4.1) coincides with the result of Lourier (1940). However, within the asymptotically consistent approach the resulting expression for S_{12} and S_{21} is (4.1).

It is worth mentioning that in the isotropic case the operator applied to p_α and p_β, coincides with the operator ω. The third term in (4.1) may be of importance for shells of weak cross sectional shear stiffness.

Thus, the following formulation may be suggested as optimal 2D theory for orthotropic shells (which is within the field of classical shell theory but nevertheless provides maximal accuracy of the stress-strain analysis):

the equilibrium equations

$$\frac{\partial BT_1}{\partial\alpha} - \frac{\partial B}{\partial\alpha}T_2 + \frac{\partial AS_{21}}{\partial\beta} + \frac{\partial A}{\partial\beta}S_{12} + AB\frac{N_1}{R_1}$$

$$= -AB\left(p_\alpha + h\left(\frac{1}{R_1} + \frac{1}{R_2}\right)m_\beta + \frac{h^2}{R_1 R_2}p_\alpha\right), \quad (\alpha,\beta),$$

$$\frac{T_1}{R_1} + \frac{T_2}{R_2} - \frac{1}{AB}\left(\frac{\partial BN_1}{\partial \alpha} + \frac{\partial AN_2}{\partial \beta}\right)$$

$$= p_\gamma - h\left(\frac{1}{R_1} + \frac{1}{R_2}\right)m + \frac{h^2}{R_1 R_2}p_\gamma, \qquad (4.3)$$

$$\frac{\partial BM_1}{\partial \alpha} - \frac{\partial B}{\partial \alpha}M_2 + \frac{\partial AH_{21}}{\partial \beta} + \frac{\partial A}{\partial \beta}H_{12} - ABN_1$$

$$= -AB\left(hm_\beta + h^2\left(\frac{1}{R_1} + \frac{1}{R_2}\right)p_\alpha\right), \quad (\alpha,\beta)$$

constitutive relations

$$T_1 = 2h\left(B_{11}\varepsilon_1 + B_{12}\varepsilon_2\right) + A_1 mh, \quad (1,2),$$

$$S_{12} = 2hB_{66}\omega + D_{66}\left(\frac{1}{R_2} - \frac{1}{R_1}\right)\left(\tau - \frac{\omega}{R_1}\right)$$

$$+\frac{1}{6}h^2\left(\frac{G_{12}}{G_{13}}\frac{A}{B}\frac{\partial}{\partial\beta}\frac{p_\alpha}{A} + \frac{G_{12}}{G_{23}}\frac{B}{A}\frac{\partial}{\partial\alpha}\frac{p_\beta}{B}\right), \quad (1,2;\alpha,\beta), \qquad (4.4)$$

$$M_1 = D_{11}\chi_1 + D_{12}\chi_2 + \left(\frac{D_{11}}{R_1} + \frac{D_{12}}{R_2}\right)(A_1\varepsilon_1 + A_2\varepsilon_2)$$

$$-D_{11}\left(\frac{1}{R_1} - \frac{1}{R_2}\right)\varepsilon_1 - \frac{1}{3}A_1 h^2 p_\gamma, \quad (1,2),$$

$$H_{12} = 2D_{66}\left(\tau - \frac{\omega}{2R_1}\right), \qquad (1,2),$$

where

$$A_1 = -\frac{E_1}{E_3}\frac{\nu_{13} + \nu_2\nu_{23}}{1 - \nu_1\nu_2} \quad (1,2), \quad D_{ik} = \frac{2}{3}h^3 B_{ik} \qquad (4.5)$$

expressions for the components of tangential strain-displacement

$$\varepsilon_1 = \partial_\alpha u + k_\alpha v + \frac{w}{R_1}, \quad \varepsilon_2 = \partial_\beta v + k_\beta u + \frac{w}{R_2},$$

$$\omega = \frac{A}{B}\frac{\partial}{\partial\beta}\left(\frac{u}{A}\right) + \frac{B}{A}\frac{\partial}{\partial\alpha}\left(\frac{v}{B}\right), \qquad (4.6)$$

$$\partial_\alpha = \frac{1}{A}\frac{\partial}{\partial\alpha}, \ \partial_\beta = \frac{1}{B}\frac{\partial}{\partial\beta}, \ k_\alpha = \frac{1}{AB}\frac{\partial A}{\partial\beta}, \ k_\beta = \frac{1}{AB}\frac{\partial B}{\partial\alpha},$$

the relations for components of bending deformation and angles of rotations

$$\chi_1 = -\partial_\alpha\gamma_1 - k_\alpha\gamma_2, \quad \chi_2 = -\partial_\beta\gamma_2 - k_\beta\gamma_1, \qquad (4.7)$$

$$\tau = -\partial_\alpha\gamma_2 + k_\alpha\gamma_1 + \frac{\omega_2}{R_1} = -\partial_\beta\gamma_1 + k_\beta\gamma_2 + \frac{\omega_1}{R_2},$$

formulas for angles of rotation and displacements

$$\gamma_1 = \partial_\alpha w - \frac{u}{R_1}, \quad \gamma_2 = \partial_\beta w - \frac{v}{R_2}, \qquad (4.8)$$

$$\omega_1 = \partial_\alpha v - k_\alpha u, \quad \omega_2 = \partial_\beta u - k_\beta v,$$

where u, v, w are the components of the displacement vector of the points on the mid-surface of the shell.

We remark that the constitutive relations are non-homogeneous, however, they may be made homogeneous by means of a linear transformation, with associated changes arising on the right-hand sides of the equilibrium equations (4.3).

The resulting 2D formulation includes equations of equilibrium (4.3), the constitutive relations (4.4) along with the relations between strains and displacements (4.6)-(4.8) and the corresponding effective boundary conditions on the side surface of the shell. These conditions being an effective substitution of the original boundary condition within the 3D formulation, are derived in Chapter 9. The presented 2D formulation is perhaps the optimal possible within the framework of the classical shell theory and 2D refined theories.

In order to increase the accuracy of analysis of the stress-strain field of elastic shell one needs to go beyond the classical assumptions. In terms of asymptotic method this means consideration of higher orders of approximation for the outer problem, construction of the boundary layer solution and its matching with the outer solution for given boundary conditions of specified problem of 3D elasticity. We note that the alternative approaches to the problem usually result in a system of differential equations of higher order than the conventional classical one.

8.5   The Effect of Anisotropy on the Accuracy of Kirchhoff-Love Theory

So far no restrictions on the parameters of material have been discussed, however, it is known from Ambartsumyan (1961, 1964, 1968) that the assumptions of the classical theory, though initially formulated as geometrical, have a physical aspect as well. Ignoring the transverse tangential and normal stresses in case of strongly anisotropic shells could lead to crucial errors in computations. The asymptotic analysis reveals the limits of applicability of governing equations for anisotropic shells.

It may be deduced from the obtained relations (3.15) and (4.1) that if $E_1 \gg E_3$ or $G_{12} \gg G_{13}$, $G_{12} \gg G_{23}$, then the refining terms are of paramount importance, in the opposite case these may be neglected. In other words, the possibility of severe errors occurs when the modules of elasticity and shear for transverse direction are small compared to that of tangential directions. At the same time, as follows from numerous theoretical and experimental investigations, the absolute majority of traditional materials including reinforced plastics, exhibit this type of behavior.

Let us study the limits of relations of elastic parameters for which the general assumptions of the classical theory for anisotropic shells are still valid. Here we consider the case when the elastic parameters of the material do not vary rapidly along the mid-surface of a shell, though could differ sharply from the values of parameters in transverse directions. Let

$$\frac{E_3}{E_1} = O(h_*^{t_1}), \quad \frac{G_{13}}{G_{12}} = O(h_*^{t_2}), \quad \frac{G_{23}}{G_{12}} = O(h_*^{t_2}) \tag{5.1}$$

where $h_* = \dfrac{h}{R}$, $t_1 = \dfrac{l_1}{q}$, $t_2 = \dfrac{l_2}{q}$ will be referred to as anisotropic indices of extension and shear. Therefore

$$E_3 = \overline{E}_3\lambda^{-l_1}, \quad G_{13} = \overline{G}_{13}\lambda^{-l_2}, \quad a_{33} = \bar{a}_{33}\lambda^{l_1}, \tag{5.2}$$

$$a_{44} = \bar{a}_{44}\lambda^{l_2}, \quad a_{55} = \bar{a}_{55}\lambda^{l_2}.$$

It is also assumed that the quantities $\overline{E}_3, \overline{G}_{13}, \bar{a}_{33}, \bar{a}_{44}, \bar{a}_{55}$ are of the same asymptotic order as E_1, E_2; G_{12}; a_{11}, a_{22}; a_{66}.

In view of (5.1) the asymptotic integration of (1.2) and (1.3) leads again to (2.4), with a variation occurring on the right-hand sides of the third and the last two equations of (2.5). In particular, in case of the third equation instead of $\sigma_\gamma^{(s-2q+2c)}$ one obtains $\sigma_\gamma^{(s-2q+2c+l_1)}$, whereas in the fourth equation - $\tau_{\alpha\gamma}^{(s-2q+2p+l_2)}$ instead of $\tau_{\alpha\gamma}^{(s-2q+2p)}$. Since these terms are neglected within the framework of the classical theory, the formal asymptotic error arising from neglecting σ_γ may be estimated as

$$\delta_1 = O(\lambda^{-2q+2c+l_1}) = \begin{cases} O(h_*^{2-t_1}) & \text{at } t \le \frac{1}{2} \\ O(h_*^{4-4t-t_1}) & \text{at } t \ge \frac{1}{2} \end{cases} \tag{5.3}$$

with the asymptotic error corresponding to neglecting $\tau_{\alpha\gamma}$, is

$$\delta_2 = O(\lambda^{-2q+2p+l_2}) = O(h_*^{2-2t-t_2}). \tag{5.4}$$

Hence, the formal asymptotic error of the classical theory is $\varepsilon = \max(\delta_1, \delta_2)$.

This result is a generalization of the previously known estimates of A.L. Goldenveiser for isotropic shells, and reduce to the latter in case of $t_1 = t_2 = 0$.

As may be observed from (5.3) and (5.4), the error of the classical theory for anisotropic shells depends not only on the usual variability index, but also on the index of anisotropy.

Let us also discuss a traditional opinion that ignoring of tangential stresses could lead to more significant errors than that of the normal stress σ_γ. This is generally true for isotropic shells of zero variability, however, this postulate could not be true for anisotropic shells. For example, if the variability index is $t = 0$, whereas $t_1 > t_2$, it follows from (5.3) and (5.4) that the asymptotic error of not taking of σ_γ into consideration exceeds the error arising from neglecting $\tau_{\alpha\gamma}$. We also note that for further refinement of the results of the classical theory incorporation of both tangential and normal stresses is crucial.

8.6 Special Cases of Asymptotic Behavior of Elastic Shells

It is well known that it is hardly possible to construct a 2D shell theory describing universal stress states for arbitrary shells. Indeed, every shell theory as any physical theory has its area of applicability, which could be wide but still bounded. The same statement is true for the theory presented above, for example, it does not describe all stress states including pure bending and twisting stress states along with generalized boundary effects. Following the terminology of Goldenveizer (1961, 1976), we shall refer to these states as states of special asymptotic behavior.

In case of the pure bending and twisting stress state the stresses σ_T caused by tangential forces are relatively small compared to σ_M, the stresses occurring due to moments, i.e.

$$\frac{\sigma_T}{\sigma_M} = O(\lambda^{-b}) \quad (b = q - 2p > 0). \tag{6.1}$$

Let us now describe briefly the generalized boundary effects. This terms corresponds to the stress state which is localized near the asymptotic line of distortion. The asymptotic directions on the mid-surface are quasi-stationary. Therefore if α coincides with the asymptotic lines, whereas β is not, differentiation with respect to α causes less increase of the original quantities than differentiation along β. Therefore, it is also an example of singular asymptotic behavior, due to various increase of the sought-for quantities while differentiation with respect to α and β. For some more details on the subject the reader is referred to Goldenveizer (1961, 1976). It may be shown that in both cases of singular asymptotic behavior, the following iterative procedure may be constructed following from (1.2)-(1.4) using

$$\sigma_\alpha = \lambda^{q+d-s}\sigma_\alpha^{(s)}, \quad \tau_{\alpha\beta} = \lambda^{q+d-s}\tau_{\alpha\beta}^{(s)},$$

$$\tau_{\alpha\gamma} = \lambda^{p+d-s}\tau_{\alpha\gamma}^{(s)}, \quad \sigma_\gamma = \lambda^{d-s}\sigma_\gamma^{(s)}, \tag{6.2}$$

$$u_\alpha = \lambda^{2q-3p+d-s}u_\alpha^{(s)}, \quad u_\gamma = \lambda^{2q-2p+d-s}u_\gamma^{(s)}, \quad (\alpha, \beta).$$

Due to the fact that in respect of the stresses the expression (6.2) coincides with (2.3) for $t \leq 1/2$ ($c = 0$), the system for sought-for quantities may be once again written as (2.4), whereas instead of (2.5) we have

$$\partial_\xi u_\alpha^{(s)} + Rk_\alpha u_\beta^{(s-p)} + \frac{u_\gamma^{(s)}}{r_1} = R(a_{11}\sigma_\alpha^{(s-q+2p)} + a_{12}\sigma_\beta^{(s-q+2p)}$$

$$+ a_{13}\sigma_\gamma^{(s-2q+2p)}) + \zeta R \left(a_{11}\frac{\sigma_\alpha^{(s-2q+2p)}}{r_1} + a_{12}\frac{\sigma_\beta^{(s-2q+2p)}}{r_2} \right)$$

$$- \frac{\zeta}{r_2} \left(\partial_\xi u_\alpha^{(s-q)} + \frac{u_\gamma^{(s-q)}}{r_1} + Rk_\alpha u_\beta^{(s-q-p)} \right),$$

$$\frac{\partial u_\gamma^{(s)}}{\partial \zeta} = -\zeta \left(\frac{1}{r_1} + \frac{1}{r_2} \right) \frac{\partial u_\gamma^{(s-q)}}{\partial \zeta} - \frac{\zeta^2}{r_1 r_2} \frac{\partial u_\gamma^{(s-2q)}}{\partial \zeta} + R(a_{33}\sigma_\gamma^{(s-3q+2p)}$$

$$+ a_{13}\sigma_\alpha^{(s-2q+2p)} + a_{23}\sigma_\beta^{(s-2q+2p)})$$

$$+ \zeta \left(a_{13}\frac{\sigma_\alpha^{(s-3q+2p)}}{r_1} + a_{23}\frac{\sigma_\beta^{(s-3q+2p)}}{r_2} \right),$$

$$\partial_\eta u_\alpha^{(s)} - Rk_\beta u_\beta^{(s-p)} + \partial_\xi u_\beta^{(s)} - Rk_\alpha u_\alpha^{(s-p)}$$

$$= Ra_{66}\tau_{\alpha\beta}^{(s-q+2p)} + \zeta\frac{R}{r_1}a_{66}\tau_{\alpha\beta}^{(s-2q+2p)} \tag{6.3}$$

$$-\frac{\zeta}{r_1}(\partial_\eta u_\alpha^{(s-q)} - Rk_\beta u_\beta^{(s-q-p)}) - \frac{\zeta}{r_2}(\partial_\xi u_\beta^{(s-q)} - Rk_\alpha u_\alpha^{(s-q-p)}),$$

$$\frac{\partial u_\alpha^{(s)}}{\partial\zeta} + \partial_\xi u_\gamma^{(s-q+2p)} = Ra_{55}\tau_{\alpha\gamma}^{(s-3q+4p)} + R\frac{\zeta}{r_1}a_{55}\tau_{\alpha\gamma}^{(s-4q+4p)}$$

$$-\zeta\left(\frac{1}{r_1}+\frac{1}{r_2}\right)\frac{\partial u_\alpha^{(s-q)}}{\partial\zeta} - \frac{1}{r_1r_2}\zeta^2\frac{\partial u_\alpha^{(s-2q)}}{\partial\zeta} + \frac{u_\alpha^{(s-q)}}{r_1}$$

$$+\frac{1}{r_1r_2}\zeta\frac{\partial u_\alpha^{(s-2q)}}{\partial\zeta} - \frac{\zeta}{r_2}\partial_\xi u_\gamma^{(s-2q+2p)}, \quad (\alpha,\beta;4,5).$$

It follows from (6.3) that the classical assumptions are valid, the number of asymptotic approximations is restricted to $s \in [0, 3q - 4p - 1]$ for the last two equations, and is in the limits of $s \in [0, 2q - 2p]$ for the rest of the equations. Integrating the systems (2.4) and (6.3) with tolerance of $O(\lambda^{-3q+4p})$ and $O(\lambda^{-2q+2p})$ for displacements and stresses, respectively, one arrives at

$$\sigma_\alpha^{(s)} = \overset{0}{\sigma}{}_\alpha^{(s)} + \zeta\overset{1}{\sigma}{}_\alpha^{(s)} + \zeta^2\overset{2}{\sigma}{}_\alpha^{(s)}, \quad \tau_{\alpha\beta}^{(s)} = \overset{0}{\tau}{}_{\alpha\beta}^{(s)} + \zeta\overset{1}{\tau}{}_{\alpha\beta}^{(s)} + \zeta^2\overset{2}{\tau}{}_{\alpha\beta}^{(s)}, \tag{6.4}$$

$$u_\alpha^{(s)} = \overset{0}{u}{}_\alpha^{(s)} + \zeta\overset{1}{u}{}_\alpha^{(s)} + \zeta^2\overset{2}{u}{}_\alpha^{(s)}, \quad u_\gamma^{(s)} = w_{(\xi,\eta)}^{(s)} + \zeta\overset{1}{u}{}_\gamma^{(s)} + \zeta^2\overset{2}{u}{}_\gamma^{(s)}$$

where $\overset{0}{\sigma}{}_\alpha^{(s)} = \overset{2}{\sigma}{}_\alpha^{(s)} = \overset{0}{\tau}{}_{\alpha\beta}^{(s)} = \overset{2}{\tau}{}_{\alpha\beta}^{(s)} = 0$ for $s \in [0, q - 2p)$, while a recurrent system is obtained for the remaining quantities.

We remark that within adopted tolerance, in order to determine the sought-for quantities, the equations of equilibrium (4.3) should be satisfied for forces and moments, with the constitutive relations taking form

$$T_1 = 2h(B_{11}\varepsilon_1 + B_{12}\varepsilon_2) + \frac{1}{2}\left(\frac{D_{11}}{R_1} + \frac{D_{12}}{R_2}\right)(A_1\chi_1 + A_2\chi_2)$$

$$+D_{11}\left(\frac{1}{R_2} - \frac{1}{R_1}\right)\chi_1 + A_1 mh, \quad (1,2), \tag{6.5}$$

$$S_{12} = 2hB_{66}\omega + D_{66}\left(\frac{1}{R_2} - \frac{1}{R_1}\right)\tau, \quad M_1 = D_{11}\chi_1 + D_{12}\chi_2,$$

$$H_{12} = 2D_{66}\tau, \quad D_{ik} = \frac{2}{3}h^3 B_{ik}, \quad (1,2).$$

It is emphasized that the relations (6.5) for special asymptotic behavior are different from the previously noted constitutive relations (4.4) for conventional asymptotic behavior. As may be expected, the dominant role in case of special asymptotic behaviour is played by the components of bending origin.

Both of these situations may be united by keeping the terms which appear in any of the relations (4.4) and (6.5). Therefore, the new constitutive relations would describe a wide class of stress-strain fields with sufficient accuracy, covering the real-life engineering applications. This class includes in particular the stress states of conventional asymptotic behavior along with that of special behavior arising either for pure bending and twisting stress-strain field or in case of generalized boundary effects. These universal constitutive relations are written as

$$T_1 = 2h(B_{11}\varepsilon_1 + B_{12}\varepsilon_2) + \frac{1}{2}\left(\frac{D_{11}}{R_1} + \frac{D_{12}}{R_2}\right)(A_1\chi_1 + A_2\chi_2)$$

$$+ D_{11}\left(\frac{1}{R_2} - \frac{1}{R_1}\right)\chi_1 + A_1 mh, \quad (1,2),$$

$$S_{12} = 2hB_{66}\omega + D_{66}\left(\frac{1}{R_2} - \frac{1}{R_1}\right)\left(\tau - \frac{\omega}{R_1}\right)$$

$$+ \frac{h^2}{6}\left(\frac{G_{12}}{G_{13}}\frac{A}{B}\frac{\partial}{\partial\beta}\frac{p_\alpha}{A} + \frac{G_{12}}{G_{23}}\frac{B}{A}\frac{\partial}{\partial\alpha}\frac{p_\alpha}{B}\right), \quad (1,2;\alpha,\beta), \qquad (6.6)$$

$$M_1 = D_{11}\chi_1 + D_{12}\chi_2 + \left(\frac{D_{11}}{R_1} + \frac{D_{12}}{R_2}\right)(A_1\varepsilon_1 + A_2\varepsilon_2)$$

$$- D_{11}\left(\frac{1}{R_1} - \frac{1}{R_2}\right)\varepsilon_1 - \frac{1}{3}h^2 A_1 p_\gamma, \quad (1,2),$$

$$H_{12} = 2D_{66}\left(\tau - \frac{\omega}{2R_1}\right), \quad (1,2).$$

Chapter 9

Boundary Layer in Orthotropic Shells

9.1　Boundary Layer

Similarly to the case of a plate, it is impossible to fulfil edge boundary conditions (8.1.8)-(8.1.12) of a shell by the outer solution only (which is given by (8.2.3) in case of conventional asymptotic behavior). In order to progress a novel solution should be introduced, namely, the boundary layer.

Let us make a few remarks on the notation. Let the components of the non-symmetric stress tensor of the boundary layer be denoted by $\sigma_1, \sigma_2, \sigma_3, \sigma_{12}, \sigma_{13}, \sigma_{23}$, with the components of dimensionless displacement field introduced as $u = u_1/h$, $v = u_2/h$, $w = u_3/h$, and $\alpha = \alpha_0$ is the side surface of the shell under consideration. In order to construct the boundary layer solution in the vicinity of the edge $\alpha = \alpha_0$ we introduce the following scaling

$$\alpha - \alpha_0 = R\lambda^{-q}\xi_1, \quad \beta = R\lambda^{-p}\eta, \quad \gamma = h\zeta = R\lambda^{-q}\zeta, \tag{1.1}$$

hence,

$$\frac{\partial}{\partial\alpha} = \frac{\lambda^q}{R}\frac{\partial}{\partial\xi_1}, \quad \frac{\partial}{\partial\beta} = \frac{\lambda^p}{R}\frac{\partial}{\partial\eta}, \quad \frac{\partial}{\partial\gamma} = \frac{\lambda^q}{R}\frac{\partial}{\partial\zeta}. \tag{1.2}$$

Let us now expand the quantities

$$\frac{1}{A}, \frac{1}{B}, k_\alpha = \frac{1}{AB}\frac{\partial A}{\partial\beta}, \quad k_\beta = \frac{1}{AB}\frac{\partial B}{\partial\alpha}, \frac{1}{R_1}, \frac{1}{R_2}, \tag{1.3}$$

$$2H = \frac{1}{R_1} + \frac{1}{R_2}, \quad K = \frac{1}{R_1 R_2}$$

as Taylor series around $\alpha = \alpha_0$ assuming that it is possible. Then if Q is any of these quantities,

$$Q = Q_n(\alpha - \alpha_0)^n = Q_n R^n \lambda^{-qn}\xi_1^n, \tag{1.4}$$

where Q_n is the coefficient of Taylor expansion, and summation is assumed along n in the limits $[0, +\infty)$. We also denote $R^n(1/R_1)_n\xi_1^n = R_{1n}$, $R^n(1/R_2)_n\xi_1^n =$

257

R_{2n}, $\left(\underset{n}{\overset{=}{\Sigma}}\right)$, then

$$\frac{1}{A}\frac{\partial}{\partial\alpha} = \frac{\lambda^q}{R}\lambda^{-qn}\xi_1^n d_{1n}, \qquad d_{1n} = \left(\frac{1}{A}\right)_n R^n \frac{\partial}{\partial\xi_1}, \qquad (1.5)$$

$$\frac{1}{B}\frac{\partial}{\partial\beta} = \frac{\lambda^p}{R}\lambda^{-qn}\xi_1^n d_{2n}, \qquad d_{2n} = \left(\frac{1}{B}\right)_n R^n \frac{\partial}{\partial\eta}.$$

In view of (1.1)-(1.5) the equilibrium equations (8.1.2) and the constitutive relations (8.1.3) may be rewritten as

$$\frac{\lambda^q}{R}(\lambda^{-qn}\xi_1^n d_{1n})\sigma_1 + \frac{\lambda^p}{R}(\lambda^{-qn}\xi_1^n d_{2n})\sigma_{21} + \frac{\lambda^q}{R}\left(1 + R_{1n}\lambda^{-qn-q}\zeta\right)\frac{\partial\sigma_{13}}{\partial\zeta}$$

$$+k_{\beta n}R^n\lambda^{-qn}\xi_1^n(\sigma_1 - \sigma_2) + k_{\alpha n}R^n\lambda^{-qn}\xi_1^n(\sigma_{12} + \sigma_{21}) + 2\sigma_{13}R_{1n}\lambda^{-qn} = 0,$$

$$\frac{\lambda^q}{R}\frac{\partial\sigma_3}{\partial\zeta} + \frac{\lambda^q}{R}(\lambda^{-qn}\xi_1^n d_{1n})\sigma_{13} + \frac{\lambda^p}{R}(\lambda^{-qn}\xi_1^n d_{2n})\sigma_{23}$$

$$-(R_{1n}\sigma_1 + R_{2n}\sigma_2)\lambda^{-qn} + k_{\beta n}R^n\lambda^{-qn}\xi_1^n\sigma_{13} + k_{\alpha n}R^n\lambda^{-qn}\xi_1^n\sigma_{23} = 0,$$

$$(1 + RR_{2n}\lambda^{-qn-q}\zeta)((\lambda^{-qn}\xi_1^n d_{1n})u$$

$$+Rk_{\alpha n}R^n\lambda^{-qn-q}\xi_1^n v + RR_{1n}\lambda^{-qn-q}w)$$

$$= a_{11}(1 + RR_{1n}\lambda^{-qn-q}\zeta)\sigma_1 + a_{12}(1 + RR_{2n}\lambda^{-qn-q}\zeta)\sigma_2 + a_{13}\sigma_3,$$

$$\left(1 + 2RH_n R^n\lambda^{-qn-q}\xi_1^n\zeta + K_n R^{n+2}\lambda^{-qn-2q}\xi_1^n\zeta^2\right)\frac{\partial w}{\partial\zeta}$$

$$= a_{33}\sigma_3 + a_{13}\sigma_1 + a_{23}\sigma_2 + R\zeta(a_{13}R_{1n}\sigma_1 + a_{23}R_{2n}\sigma_2)\lambda^{-qn-q},$$

$$\left(1 + 2H_n R^{n+1}\lambda^{-qn-q}\xi_1^n\zeta + K_n R^{n+2}\lambda^{-qn-2q}\xi_1^n\zeta^2\right)\frac{\partial u}{\partial\zeta}$$

$$-(1 + RR_{2n}\lambda^{-qn-q}\zeta)RR_{1n}\lambda^{-qn-q}u$$

$$+(1 + RR_{2n}\lambda^{-qn-q}\zeta)(\lambda^{-qn}\xi_1^n d_{1n})w = a_{55}(1 + RR_{1n}\lambda^{-qn-q}\zeta)\sigma_{13}, \quad (1.6)$$

$$(1 + RR_{1n}\lambda^{-qn-q}\zeta)\left((\lambda^{-qn-q+p}\xi_1^n d_{2n})v\right.$$

$$+k_{\beta n}R^{n+1}\lambda^{-qn-q}\xi_1^n u + RR_{2n}\lambda^{-qn-q}w)$$

$$= a_{12}(1 + RR_{1n}\lambda^{-qn-q}\zeta)\sigma_1 + a_{22}(1 + RR_{2n}\lambda^{-qn-q}\zeta)\sigma_2 + a_{23}\sigma_3,$$

$$\frac{\lambda^q}{R}(\lambda^{-qn}\xi_1^n d_{1n})\sigma_{12} + \frac{\lambda^p}{R}(\lambda^{-qn}\xi_1^n d_{2n})\sigma_2 + \frac{\lambda^q}{R}\left(1 + RR_{2n}\lambda^{-qn-q}\zeta\right)\frac{\partial\sigma_{23}}{\partial\zeta}$$

$$+k_{\beta n}R^n\lambda^{-qn}\xi_1^n(\sigma_{12} + \sigma_{21}) + k_{\alpha n}R^n\lambda^{-qn}\xi_1^n(\sigma_2 - \sigma_1) + 2\sigma_{23}R_{2n}\lambda^{-qn} = 0,$$

$$(1 + RR_{1n}\lambda^{-qn-q}\zeta)\sigma_{12} = (1 + RR_{2n}\lambda^{-qn-q}\zeta)\sigma_{21},$$

$$\left(1 + 2H_n R^{n+1}\lambda^{-qn-q}\xi_1^n\zeta + K_n R^{n+2}\lambda^{-qn-2q}\xi_1^n\zeta^2\right)\frac{\partial v}{\partial\zeta}$$

$$-(1 + RR_{1n}\lambda^{-qn-q}\zeta)RR_{2n}\lambda^{-qn-q}v$$

$$+(1 + RR_{1n}\lambda^{-qn-q}\zeta)(\lambda^{-qn-q+p}\xi_1^n d_{2n})w = a_{44}(1 + RR_{2n}\lambda^{-qn-q}\zeta)\sigma_{23},$$

$$(1 + RR_{1n}\lambda^{-qn-q}\zeta)\left((\lambda^{-qn-q+p}\xi_1^n d_{2n})u - k_{\beta n}R^{n+1}\lambda^{-qn-q}\xi_1^n v\right)$$

$$+(1 + RR_{2n}\lambda^{-qn-q}\zeta)\left(\lambda^{-qn}\xi_1^n d_{1n}v - k_{\alpha n}R^{n+1}\lambda^{-qn-q}\xi_1^n u\right)$$

$$= a_{66}(1 + RR_{1n}\lambda^{-qn-q}\zeta)\sigma_{12}.$$

The sought-for solution of (1.6) should decay away from the edge $\alpha = \alpha_0$ and also satisfy the face boundary conditions

$$\sigma_{13} = \sigma_{23} = \sigma_3 = 0 \quad \text{at} \quad \zeta = \pm 1. \tag{1.7}$$

Let Ω_i be any of the stresses or displacements of the boundary layer solution. The solution of (1.6) is sought in the form

$$\Omega_i = \lambda^{\chi_i} \sum_{s=0}^{S} \lambda^{-s} \Omega_i^{(s)}. \tag{1.8}$$

It may be shown that on substitution of (1.8) into (1.6) one arrives at a consistent system in respect of $\Omega_i^{(s)}$ provided that $\chi_{\sigma_i} = \chi_{u_i} = -q + p$. After some transformation the governing system may be represented as

$$\frac{1}{A_0} \frac{\partial \sigma_1^{(s)}}{\partial \xi_1} + \frac{\partial \sigma_{13}^{(s)}}{\partial \zeta} = R_\alpha^{(s+p-q)},$$

$$\frac{1}{A_0} \frac{\partial \sigma_{13}^{(s)}}{\partial \xi_1} + \frac{\partial \sigma_3^{(s)}}{\partial \zeta} = R_\gamma^{(s+p-q)},$$

$$\frac{1}{A_0} \frac{\partial u^{(s)}}{\partial \xi_1} = b_{11}\sigma_1^{(s)} + b_{13}\sigma_3^{(s)} + R_u^{(s-q)} + \frac{a_{12}}{a_{22}} R_v^{(s+p-q)}, \tag{1.9}$$

$$\frac{\partial w^{(s)}}{\partial \zeta} = b_{13}\sigma_1^{(s)} + b_{33}\sigma_3^{(s)} + R_w^{(s-q)} + \frac{a_{23}}{a_{22}} R_v^{(s+p-q)},$$

$$\frac{1}{A_0} \frac{\partial w^{(s)}}{\partial \xi_1} + \frac{\partial u^{(s)}}{\partial \zeta} = a_{55}\sigma_{13}^{(s)} + R_{uw}^{(s-q)},$$

$$\sigma_2^{(s)} = -\frac{a_{12}}{a_{22}}\sigma_1^{(s)} - \frac{a_{23}}{a_{22}}\sigma_3^{(s)} + \frac{1}{a_{22}} R_v^{(s+p-q)},$$

$$b_{ij} = a_{ij} - \frac{a_{i2}a_{j2}}{a_{22}}, \quad (i,j = 1,3)$$

and

$$\frac{1}{A_0} \frac{\partial \sigma_{12}^{(s)}}{\partial \xi_1} + \frac{\partial \sigma_{23}^{(s)}}{\partial \zeta} = R_\tau^{(s+p-q)}, \tag{1.10}$$

$$\sigma_{12}^{(s)} + \zeta RR_{1m}\sigma_{12}^{(s-qm-q)} = \sigma_{21}^{(s)} + \zeta RR_{2m}\sigma_{21}^{(s-qm-q)},$$

$$\frac{\partial v^{(s)}}{\partial \zeta} = a_{44}\sigma_{23}^{(s)} + R_\beta^{(s+p-q)}, \quad \frac{1}{A_0}\frac{\partial v^{(s)}}{\partial \xi_1} = a_{66}\sigma_{12}^{(s)} + R_{\alpha\beta}^{(s+p-q)},$$

where

$$R_\alpha^{(s-q+p)} = -\xi_1^n d_{1n}\sigma_1^{(s-qn)} - \xi_1^m d_{2m}\sigma_{21}^{(s+p-q-qm)} - RR_{1m}\zeta\frac{\partial \sigma_{13}^{(s-qm-q)}}{\partial \zeta}$$

$$-k_{\beta m}R^{m+1}\xi_1^m(\sigma_1 - \sigma_2)^{(s-qm-q)}$$

$$-k_{\alpha m}R^{m+1}\xi_1^m(\sigma_{21} + \sigma_{12})^{(s-qm-q)} - 2RR_{1m}\sigma_{13}^{(s-qm-q)},$$

$$R_\gamma^{(s-q+p)} = -\xi_1^n d_{1n}\sigma_{13}^{(s-qn)} - \xi_1^m d_{2m}\sigma_{23}^{(s+p-q-qm)}$$

$$+R(R_{1m}\sigma_1 + R_{2m}\sigma_2)^{(s-qm-q)} - R^{m+1}\xi_1^m(k_{\beta m}\sigma_{13} + k_{\alpha m}\sigma_{23})^{(s-qm-q)},$$

$$R_u^{(s-q)} = \zeta R(a_{11}R_{1m}\sigma_1 + a_{22}R_{2m}\sigma_2)^{(s-qm-q)} - \xi_1^n d_{1n}u^{(s-qn)}$$

$$-RR_{2m}\xi_1^i\zeta d_{1i}u^{(s-qm-q-qi)} - Rk_{\alpha m}R^m\xi_1^m v^{(s-qm-q)} - RR_{1m}w^{(s-qm-q)},$$

$$R_v^{(s-q+p)} = \xi_1^m d_{2m} v^{(s+p-qm-q)} + RR_{1m}\xi_1^i \zeta d_{2i} v^{(s+p-2q-qm-qi)}$$
$$+ k_{\beta m} R^{m+1}\xi_1^m u^{(s-qm-q)}$$
$$+ RR_{2m} w^{(s-qm-q)} - \zeta R(a_{12}R_{1m}\sigma_1 + a_{22}R_{2m}\sigma_2)^{(s-qm-q)},$$
$$R_w^{(s-q)} = \zeta R(a_{13}R_{1m}\sigma_1 + a_{23}R_{2m}\sigma_2)^{(s-qm-q)} \quad (1.11)$$
$$-2H_m R^{m+1}\zeta\xi_1^m \frac{\partial w^{(s-qm-q)}}{\partial\zeta} - K_m R^{m+2}\zeta^2\xi_1^m \frac{\partial w^{(s-qm-2q)}}{\partial\zeta},$$
$$R_{uw}^{(s-q)} = a_{55}\zeta RR_{1m}\sigma_{13}^{(s-qm-q)} - 2H_m R^{m+1}\zeta\xi_1^m \frac{\partial u^{(s-qm-q)}}{\partial\zeta}$$
$$-K_m R^{m+2}\zeta^2\xi_1^m \frac{\partial u^{(s-qm-2q)}}{\partial\zeta} + RR_{1m}u^{(s-qm-q)}$$
$$+\zeta R^2 R_{2m}R_{1i}u^{(s-qm-2q-qi)}$$
$$-\xi_1^n d_{1n}w^{(s-qn)} - \zeta RR_{2m}\xi_1^i d_{1i}w^{(s-qm-q-qi)},$$
$$R_\tau^{(s-q+p)} = -\xi_1^m d_{2m}\sigma_2^{(s+p-q-qm)} - \xi_1^n d_{1n}\sigma_{12}^{(s-qn)}$$
$$-RR_{2m}\zeta\frac{\partial\sigma_{23}^{(s-qm-q)}}{\partial\zeta} - R^{m+1}\xi_1^m(k_{\beta m}(\sigma_{12}+\sigma_{21})$$
$$+k_{\alpha m}(\sigma_2 - \sigma_1))^{(s-qm-q)} - 2RR_{2m}\sigma_{23}^{(s-qm-q)},$$
$$R_\beta^{(s-q+p)} = -\xi_1^m d_{2m}w^{(s+p-qm-q)} - \zeta RR_{1m}\xi_1^i d_{2i}w^{(s+p-2q-qm-qi)}$$
$$-2H_m R^{m+1}\zeta\xi_1^m \frac{\partial v^{(s-qm-q)}}{\partial\zeta}$$
$$-K_m R^{m+2}\zeta^2\xi_1^m \frac{\partial v^{(s-qm-2q)}}{\partial\zeta} + RR_{2m}v^{(s-qm-q)}$$
$$+\zeta R^2 R_{1m}R_{2i}v^{(s-qm-2q-qi)} + a_{44}\zeta RR_{2m}\sigma_{23}^{(s-q-qm)},$$
$$R_{\alpha\beta}^{(s-q+p)} = -\xi_1^m d_{2m}u^{(s+p-qm-q)}$$
$$-\zeta RR_{1m}\xi_1^i d_{2i}u^{(s+p-2q-qm-qi)} - \xi_1^n d_{1n}v^{(s-qn)}$$
$$-\zeta RR_{2m}\xi_1^i d_{1i}v^{(s-qm-q-qi)} + R^{m+1}\xi_1^m(k_{\alpha m}u + k_{\beta m}v)^{(s-qm-q)}$$
$$+R^2\zeta\xi_1^i(R_{2m}k_{\alpha i}u + R_{1m}k_{\beta i}v)^{(s-2q-qm-qi)} + a_{66}\zeta RR_{1m}\sigma_{12}^{(s-q-qm)},$$

where $\Omega^{(s)} \equiv 0$ for $s < 0$ is assumed along with summation over repeated indices n, m, i in the limits $n \in [1,s]$, m, $i \in [0,s]$. The upper index of $R_i^{(s)}$ denotes that the first non-zero term (leading order term) appears in $R_i^{(s)}$ starting from approximation of order s, with the notation $(...)^{(s)}$ meaning that the quantities inside the brackets correspond to approximation of order s.

The expressions (1.9) and (1.10) describe the plane and anti-plane boundary layers, respectively. We remark that at $s < q - p$ these equations are homogeneous and may be shown to coincide with equations (3.5.7) and (3.5.9) ($s = 0$) for orthotropic plates. Therefore a decaying solution may be found through the same approach described in Section 3.5. In case of $s > q - p$ equations (1.9), (1.10) are non-homogeneous and differ from (3.5.7) and (3.5.9) by the form of their right-hand sides R_α, R_β, R_u, ..., $R_{\alpha\beta}$. Similarly to the case of plates, two types of the stress-strain field may be pointed out, namely $\overset{f}{\Omega}$ and $\overset{a}{\Omega}$. In case of the first time the main contribution is given by the quantities $\overset{f}{\sigma}_1$, $\overset{f}{\sigma}_{13}$, $\overset{f}{\sigma}_3$, $\overset{f}{u}$, $\overset{f}{w}$ (with any of these quantities denoted by $\overset{f}{P}$, corresponding to the plane boundary layer). These quantities are accompanied by a certain anti-plane boundary layer which is denoted by $\overset{f}{\sigma}_{12}$, $\overset{f}{\sigma}_{21}$, $\overset{f}{\sigma}_{23}$, $\overset{f}{v}$ (with any of the latter denoted by $\overset{f}{Q}$). Thus, for $\overset{f}{\Omega} = \left\{ \overset{f}{P}, \overset{f}{Q} \right\}$ we have

$$\overset{f}{P}^{(s)} \neq 0, \qquad \overset{f}{Q}^{(s)} \equiv 0 \qquad \text{at} \qquad s < q - p. \tag{1.12}$$

In case of the second type of the stress-strain field $\overset{a}{\Omega}$ the major role is played by the quantities $\overset{a}{\sigma}_{12}$, $\overset{a}{\sigma}_{21}$, $\overset{a}{\sigma}_{23}$, $\overset{a}{v}$, i.e. $(\overset{a}{Q})$, with the corresponding boundary layer $\overset{a}{\sigma}_1$, $\overset{a}{\sigma}_{13}$, ..., $\overset{a}{w}$, namely $\overset{a}{P}$. Thus, for $\overset{a}{\Omega} = \left\{ \overset{a}{P}, \overset{a}{Q} \right\}$ we obtain

$$\overset{a}{Q}^{(s)} \neq 0, \qquad \overset{a}{P}^{(s)} \equiv 0 \qquad \text{at} \qquad s < q - p. \tag{1.13}$$

We note that the choice of $\overset{f}{\Omega}$ or $\overset{a}{\Omega}$ depends on the related choice of the system (1.9) or (1.10) as a basic system at $s = 0$. In order to obtain the full solution, the systems (1.9), (1.10) with the appropriate superscripts f or a assigned to all of the appropriate quantities.

The conditions of existence of decaying solutions of the systems (1.9), (1.10) may be derived analogously to the previously case of elastic plates. These conditions may be formulated as

$$\int_{-1}^{+1} \overset{f}{\sigma}_1{}^{(s)}(t = 0)d\zeta = -\int_{-1}^{+1} d\zeta \int_0^\infty \overset{f}{R}_\alpha{}^{(s-q+p)}dt, \qquad (A_0\xi_1 = t),$$

$$\int_{-1}^{+1} \zeta\, \overset{f}{\sigma}_1{}^{(s)}(t = 0)d\zeta$$
$$= -\int_{-1}^{+1} \zeta d\zeta \int_0^\infty \overset{f}{R}_\alpha{}^{(s-q+p)}dt + \int_{-1}^{+1} d\zeta \int_0^\infty \overset{f}{R}_\gamma{}^{(s-q+p)}t\,dt,$$

$$\int_{-1}^{+1} \overset{f}{\sigma}_{13}{}^{(s)}(t = 0)d\zeta = -\int_{-1}^{+1} d\zeta \int_0^\infty \overset{f}{R}_\gamma{}^{(s-q+p)}dt, \tag{1.14}$$

$$\int_{-1}^{+1} \overset{f}{\sigma}_{12}{}^{(s)}(t = 0)d\zeta = -\int_{-1}^{+1} d\zeta \int_0^\infty \overset{f}{R}_\tau{}^{(s-q+p)}dt,$$

$$\int_{-1}^{+1} \overset{f}{v}{}^{(s)}(t = 0)d\zeta$$
$$= a_{66} \int_0^\infty t\,dt \int_{-1}^{+1} \overset{f}{R}_\tau{}^{(s-q+p)}d\zeta - \int_{-1}^{+1} d\zeta \int_0^\infty \overset{f}{R}_{\alpha\beta}{}^{(s-q+p)}dt,$$

$$(f, a).$$

As shown in Chapter 3, the main distinction between $\overset{f}{\Omega}$ and $\overset{a}{\Omega}$ is the different velocity of attenuation. Since the boundary value problem in respect of the boundary layer is homogeneous, its solution is defined up to an arbitrary constant factor, therefore, if $\overset{f}{\Omega}$ and $\overset{a}{\Omega}$ are solutions, then $\lambda^\mu \overset{f}{\Omega}$ and $\lambda^\chi \overset{a}{\Omega}$ are also solutions along with the appropriate linear combinations. Now using (1.8), we have

$$\Omega = \lambda^{-q+p+\chi-s} \overset{a}{\Omega}{}^{(s)} + \lambda^{-q+p+\mu-s} \overset{f}{\Omega}{}^{(s)} \tag{1.15}$$

with summation assumed along the dummy index s in the limits between zero and S, with constants χ and μ characterizing the intensities of the anti-plane and plane boundary layer solutions, respectively. These constants are to be determined from the matching procedure of the outer solution and the boundary layer, depending on the type of edge boundary conditions. The integral of the problem may be presented in the form

$$I = Q^{out} + \lambda^{-q+p+\chi-s} \overset{a}{\Omega}{}^{(s)} + \lambda^{-q+p+\mu-s} \overset{f}{\Omega}{}^{(s)} \tag{1.16}$$

containing sufficient number of arbitrary constants (functions of η) in order to satisfy the spatial boundary conditions on the edge of a shell.

9.2   Matching of the Outer Solution with Boundary Layer

Let us discuss the matters of the boundary conditions imposed on the side of a shell. We shall consider the conditions (8.1.8)-(8.1.12).

In case of the first boundary value problem the edge boundary conditions at $\alpha = \alpha_0$ are given by

$$\sigma_a = \sigma_{\alpha\beta} = \sigma_{\alpha\gamma} = 0 \quad \text{at} \quad \alpha = \alpha_0. \tag{2.1}$$

Using (8.2.3) and (1.16), we obtain

$$\begin{aligned}
\sigma_a &= \lambda^{q+d-s}\sigma_\alpha^{(s)} + \lambda^{-q+p+\chi-s}\overset{a}{\sigma}{}_1^{(s)} + \lambda^{-q+p+\mu-s}\overset{f}{\sigma}{}_1^{(s)}, \\
\sigma_{a\beta} &= \lambda^{q+d-s}\tau_{\alpha\beta}^{(s)} + \lambda^{-q+p+\chi-s}\overset{a}{\sigma}{}_{12}^{(s)} + \lambda^{-q+p+\mu-s}\overset{f}{\sigma}{}_{12}^{(s)}, \\
\sigma_{a\gamma} &= \lambda^{p+d-s}\tau_{\alpha\gamma}^{(s)} + \lambda^{-q+p+\chi-s}\overset{a}{\sigma}{}_{13}^{(s)} + \lambda^{-q+p+\mu-s}\overset{f}{\sigma}{}_{13}^{(s)}.
\end{aligned} \tag{2.2}$$

In view of (8.2.22), the following relations also take place for $(0 \le s \le 2q - 2p - 1)$

$$\sigma_\alpha^{(s)} = \overset{0}{\sigma}{}_\alpha^{(s)} + \zeta \overset{1}{\sigma}{}_\alpha^{(s)}, \quad \tau_{\alpha\beta}^{(s)} = \overset{0}{\tau}{}_{\alpha\beta}^{(s)} + \zeta \overset{1}{\tau}{}_{\alpha\beta}^{(s)},$$

$$\tau_{\alpha\gamma}^{(s)} = \overset{0}{\tau}{}_{\alpha\gamma}^{(s)} + \zeta \overset{1}{\tau}{}_{\alpha\gamma}^{(s)} + \zeta^2 \overset{2}{\tau}{}_{\alpha\gamma}^{(s)}, \tag{2.3}$$

$$\overset{1}{\sigma}{}_\alpha^{(s)} = \overset{1}{\tau}{}_{\alpha\beta}^{(s)} = \overset{2}{\tau}{}_{\alpha\gamma}^{(s)} \equiv 0 \quad \text{at} \quad s < q - 2p + c.$$

Substituting (2.2) into (2.1), we equate the coefficients within the same powers of λ starting from the maximal power, thus obtaining the values of χ and μ for which a consistent asymptotic procedure may be constructed. It follows from (2.1), (2.2) along with (1.12) and (1.13) that

$$\chi = q + p - c + d, \qquad \mu = q + d. \tag{2.4}$$

The relations corresponding to (2.4) are written as

$$\sigma_\alpha^{(s)} + \overset{a}{\sigma}\,\overset{(s-q+2p-c)}{{}_1} + \overset{f}{\sigma}\,\overset{(s-q+p)}{{}_1} = 0,$$

$$\tau_{\alpha\beta}^{(s)} + \overset{a}{\sigma}\,\overset{(s-q+2p-c)}{{}_{12}} + \overset{f}{\sigma}\,\overset{(s-q+p)}{{}_{12}} = 0 \quad \text{at} \quad \alpha = \alpha_0 \ (\xi_1 = 0), \tag{2.5}$$

$$\tau_{\alpha\gamma}^{(s)} + \overset{a}{\sigma}\,\overset{(s+p-c)}{{}_{13}} + \overset{f}{\sigma}\,\overset{(s)}{{}_{13}} = 0$$

from where

$$\sigma_\alpha^{(s)}(\alpha_0) = 0, \quad 0 \le s < q - p \tag{2.6}$$

$$\tau_{\alpha\beta}^{(s)}(\alpha_0) = 0, \quad 0 \le s < q - 2p + c.$$

All of the quantities in (2.5) may then be expressed through $\overset{f}{\sigma}\,\overset{(s)}{{}_1}$, $\overset{f}{\sigma}\,\overset{(s)}{{}_{13}}$, $\overset{a}{\sigma}\,\overset{(s)}{{}_{12}}$. Bearing in mind the decay conditions (1.14), these may now be presented as

$$\int_{-1}^{+1} \sigma_\alpha^{(s)}(\alpha_0)d\zeta = \int_{-1}^{+1} d\zeta \int_0^\infty \left(\overset{f}{R}\,\overset{(s-2q+2p)}{{}_\alpha} + \overset{a}{R}\,\overset{(s+3p-2q-c)}{{}_\alpha} \right) dt,$$

$$\int_{-1}^{+1} \zeta\sigma_\alpha^{(s)}(\alpha_0)d\zeta = \int_{-1}^{+1} \zeta d\zeta \int_0^\infty \left(\overset{f}{R}\,\overset{(s-2q+2p)}{{}_\alpha} + \overset{a}{R}\,\overset{(s+3p-2q-c)}{{}_\alpha} \right) dt$$

$$- \int_{-1}^{+1} d\zeta \int_0^\infty \left(\overset{f}{R}\,\overset{(s-2q+2p)}{{}_\gamma} + \overset{a}{R}\,\overset{(s+3p-2q-c)}{{}_\gamma} \right) tdt, \tag{2.7}$$

$$\int_{-1}^{+1} \tau_{\alpha\gamma}^{(s)}(\alpha_0)d\zeta = \int_{-1}^{+1} d\zeta \int_0^\infty \left(\overset{f}{R}\,\overset{(s+p-q)}{{}_\gamma} + \overset{a}{R}\,\overset{(s+2p-q-c)}{{}_\gamma} \right) dt,$$

$$\int_{-1}^{+1} \tau_{\alpha\beta}^{(s)}(\alpha_0)d\zeta = \int_{-1}^{+1} d\zeta \int_0^\infty \left(\overset{a}{R}\,\overset{(s+3p-2q-c)}{{}_\tau} + \overset{f}{R}\,\overset{(s+4p-2q-2c)}{{}_\tau} \right) dt.$$

The conditions for the outer problem follow from (2.7) in view of (2.3), (2.6) and (1.11), given by

$$\overset{0}{\sigma}\,\overset{(s)}{{}_\alpha}(\alpha_0) = 0, \quad \overset{0}{\tau}\,\overset{(s)}{{}_{\alpha\beta}}(\alpha_0) = 0,$$

$$\frac{2}{3}\overset{1}{\sigma}\,\overset{(s)}{{}_\alpha}(\alpha_0) = -\int_{-1}^{+1} \zeta d\zeta \int_0^\infty d_{20}\,\overset{a}{\sigma}\,\overset{(s+3p-2q-c)}{{}_{21}} dt$$

$$+ \int_{-1}^{+1} d\zeta \int_0^\infty d_{20}\,\overset{a}{\sigma}\,\overset{(s+3p-2q-c)}{{}_{23}} tdt, \tag{2.8}$$

$$2\left(\overset{0}{\tau}\,\overset{(s)}{{}_{\alpha\gamma}} + \frac{1}{3}\overset{2}{\tau}\,\overset{(s)}{{}_{\alpha\gamma}} \right)$$

$$= -\int_{-1}^{+1} d\zeta \int_0^\infty \left(d_{20}\,\overset{a}{\sigma}\,\overset{(s+2p-q-c)}{{}_{23}} + Rk_{\alpha0}\,\overset{a}{\sigma}\,\overset{(s+p-q-c)}{{}_{23}} \right) dt$$

$$0 \le s \le 2q - 2p - 1.$$

In order to derive the boundary conditions for the plane boundary layer, let us replace s by "$s + q - p$" in the first and third conditions of (2.5) and restrict ourselves to approximations up to $s = q - 2p + c$, which generally corresponds to approximation orders $0 \le s < 2q - 3p + c$. Then the plane boundary layer is constructed up to $\varepsilon_f = O\left(\lambda^{-q+2p-c}\right)$. Then the expressions on the right-hand sides of equation (2.8) vanish, hence, using (2.3) and (2.8), we deduce the following conditions

$$\overset{f}{\sigma}{}^{(s)}_{1} = 0, \quad \overset{f}{\sigma}{}^{(s)}_{13} = -\zeta\tau^{(s)}_{\alpha\gamma}(\alpha_0) \quad \text{at} \quad \xi_1 = 0. \tag{2.9}$$

It may be shown that the obtained conditions (2.9) coincide with these obtained by Goldenveizer (1976) for isotropic shells. It will also be demonstrated below that the case of anti-plane boundary layer may be treated even with higher accuracy.

Now if we replace s by "$s + q - 2p + c$" in the second equation of (2.5), then the orders $s < 2q - 2p$ of the outer solution correspond to the following orders $s < q - c$ of the anti-plane boundary layer solution. Then, assuming the following asymptotic accuracy

$$\varepsilon_a = O\left(\lambda^{-q+c}\right) = \begin{cases} O(h^1_*) & \text{at} \quad t \le 1/2 \\ O(\lambda^{-2q+2p}) = O(h^{2-2t}_*) & \text{at} \quad t \ge 1/2 \end{cases} \tag{2.10}$$

and making use of (2.5), the second relation of (2.8), (1.12), and (2.3), the following boundary conditions may be formulated

$$\overset{a}{\sigma}{}^{(s)}_{12}(t = 0) = -\zeta\overset{1}{\tau}{}^{(s+q-2p+c)}_{\alpha\beta}(\alpha_0), \quad (0 \le s \le q - c). \tag{2.11}$$

The anti-plane boundary layer solution, corresponding to (2.11), is then given by (4.2.13), where

$$\Phi^{(s)} = -\overset{1}{\tau}{}^{(s+q-2p+c)}_{\alpha\beta}(\alpha_0)\Phi_*. \tag{2.12}$$

It may be observed from (2.2) and (2.4) that the intensities of the anti-plane and plane boundary layer solutions are $O(\lambda^{(d+2p-c)})$ and $O(\lambda^{(p+d)})$, respectively. Therefore the anti-plane boundary layer is more intensive than the plane one in near free edge of a shell, i.e.

$$\overset{f}{P} = O\left(\lambda^{-p+c}\right)\overset{a}{Q}. \tag{2.13}$$

Let us evaluate the integrals involved in (2.8). Applying the following operator $\int_{-1}^{+1}\zeta d\zeta \int_{0}^{\infty} tdt$ to both sides of equation (1.10), we deduce

$$\int_{-1}^{+1} d\zeta \int_{0}^{\infty} \overset{a}{\sigma}{}^{(s+3p-2q-c)}_{23} tdt = -\int_{-1}^{+1}\zeta d\zeta \int_{0}^{\infty}\overset{a}{\sigma}{}^{(s+3p-2q-c)}_{21}dt = -I^*, \tag{2.14}$$

where from (2.11), (4.2.13), and (2.12)

$$I^* = -\frac{\bar{A}}{2}\sqrt{\frac{a_{44}}{a_{66}}}\frac{2}{3}\tau^{(s+3p-2q-c)}_{\alpha\beta}(\alpha_0). \tag{2.15}$$

Application of the operator $\int_{-1}^{+1} \zeta \, d\zeta \int_0^\infty dt$ to (1.10) and use of (2.11) leads to

$$
I_1 = \int_{-1}^{+1} d\zeta \int_0^\infty \left(d_{20} \overset{a}{\underset{23}{\sigma}}{}^{(s+2p-q-c)} + Rk_{\alpha 0} \overset{a}{\underset{23}{\sigma}}{}^{(s+p-q-c)} \right) dt
$$

$$
= \frac{1}{B_0} \frac{\partial}{\partial \eta} \left(\int_{-1}^{+1} d\zeta \int_0^\infty \overset{a}{\underset{23}{\sigma}}{}^{(s+2p-q-c)} dt \right) \quad (2.16)
$$

$$
= \frac{1}{B_0} \frac{\partial}{\partial \eta} \left(-\int_{-1}^{+1} \zeta \overset{a}{\underset{12}{\sigma}}{}^{(s+2p-q-c)}(t=0) d\zeta \right) = \frac{1}{B_0} \frac{\partial}{\partial \eta} \left(\frac{2}{3} \overset{1}{\underset{12}{\tau}}{}^{(s)}(\alpha_0) \right).
$$

The conditions (2.8) may then be expressed as

$$
\overset{0}{\underset{\alpha}{\sigma}}{}^{(s)}(\alpha_0) = 0, \quad \overset{0}{\underset{\alpha\beta}{\tau}}{}^{(s)}(\alpha_0) = 0,
$$

$$
\frac{2}{3} \overset{1}{\underset{\alpha}{\sigma}}{}^{(s)}(\alpha_0) = \bar{A} \sqrt{\frac{G_{12}}{G_{23}}} \frac{1}{B_0} \frac{\partial}{\partial \eta} \left(\frac{2}{3} \overset{1}{\underset{\alpha\beta}{\tau}}{}^{(s+3p-2q-c)}(\alpha_0) \right), \quad (2.17)
$$

$$
2 \left(\overset{0}{\underset{\alpha\gamma}{\tau}}{}^{(s)} + \frac{1}{3} \overset{2}{\underset{\alpha\gamma}{\tau}}{}^{(s)} \right) = -\frac{1}{B_0} \frac{\partial}{\partial \eta} \left(\frac{2}{3} \overset{1}{\underset{12}{\tau}}{}^{(s)}(\alpha_0) \right).
$$

$$
0 \le s \le 2q - 2p - 1.
$$

We note that the formulated conditions (2.17) correspond to the outer stress field determined up to $\varepsilon_{out} = O\left(h_*^{2-2t}\right)$.

The plane boundary layer solution may also be refined, if one assumes that the outer solution used in (2.5), is governed from (2.17). It follows then from the first and third conditions of (2.5) that

$$
\overset{f}{\underset{1}{\sigma}}{}^{(s)} = -\zeta \bar{A} \sqrt{\frac{G_{12}}{G_{23}}} \frac{1}{B_0} \frac{\partial}{\partial \eta} \left(\overset{1}{\underset{\alpha\beta}{\tau}}{}^{(s+2p-q-c)}(\alpha_0) \right)
$$

$$
- \overset{a}{\underset{1}{\sigma}}{}^{(s+p-c)}(t=0) \quad \text{at} \quad t = 0(\alpha = \alpha_0),
$$

$$
\overset{f}{\underset{13}{\sigma}}{}^{(s)} = (3\zeta^2 - 1) \overset{0}{\underset{\alpha\gamma}{\tau}}{}^{(s)}(\alpha_0) - \zeta \overset{1}{\underset{\alpha\gamma}{\tau}}{}^{(s)}(\alpha_0) \quad (2.18)
$$

$$
+ \zeta^2 \frac{1}{B_0} \frac{\partial}{\partial \eta} \overset{1}{\underset{12}{\tau}}{}^{(s)}(\alpha_0) - \overset{a}{\underset{13}{\sigma}}{}^{(s+p-c)}(t=0),
$$

$$
(0 \le s < q - p).
$$

The refined conditions (2.18) allow evaluation of the plane boundary layer up to $\varepsilon_f = O(\lambda^{-q+p}) = O\left(h_*^{1-t}\right)$.

In case of mixed boundary conditions (pinned edge) we have

$$
\sigma_a = u_\beta = u_\gamma = 0 \quad \text{at} \quad \alpha = \alpha_0. \quad (2.19)
$$

In view of (8.2.3), (1.8), and (1.16)

$$
\sigma_a = \lambda^{q+d-s} \sigma_\alpha^{(s)} + \lambda^{-q+p+\chi-s} \overset{a}{\underset{1}{\sigma}}{}^{(s)} + \lambda^{-q+p+\mu-s} \overset{f}{\underset{1}{\sigma}}{}^{(s)} = 0,
$$

$$
u_\gamma = \lambda^{q-c+d-s} u_\gamma^{(s)} + R\lambda^{-2q+p+\chi-s} \overset{a}{w}{}^{(s)} + R\lambda^{-2q+p+\mu-s} \overset{f}{w}{}^{(s)} = 0, \quad (2.20)
$$

$$
u_\beta = \lambda^{q-p+d-s} u_\beta^{(s)} + R\lambda^{-2q+p+\chi-s} \overset{a}{v}{}^{(s)} + R\lambda^{-2q+p+\mu-s} \overset{f}{v}{}^{(s)} = 0.
$$

The values of χ and μ may be determined from the requirement of consistency of the asymptotic procedure, giving

$$\chi = q + d, \qquad \mu = 2q - p + d. \tag{2.21}$$

Therefore, substituting (2.20) into (2.19), we result in

$$\sigma_\alpha^{(s)} + \overset{a}{\sigma}{}_1^{(s-q+p)} + \overset{f}{\sigma}{}_1^{(s)} = 0,$$

$$u_\gamma^{(s)} + R\,\overset{a}{w}{}^{(s-2q+p+c)} + R\,\overset{f}{w}{}^{(s-q+c)} = 0 \qquad \text{at} \qquad \alpha = \alpha_0 \ (\xi_1 = 0), \tag{2.22}$$

$$u_\beta^{(s)} + R\,\overset{a}{v}{}^{(s-2q+2p)} + R\,\overset{f}{v}{}^{(s-q+p)} = 0.$$

Here and below we consider approximations of order $s \in [0, 2q - 2p - 1]$, corresponding to analysis of the outer solution up to $O(\lambda^{-2q+2p}) = O\left(h_*^{2-2t}\right)$. It should be noted that this accuracy is within the validity of the main assumption of the classical shell theory, that the normal direction is preserved throughout the deformation process.

Bearing in mind (8.2.22), (1.12), and (1.13), the system (2.22) implies

$$\overset{0}{w}{}^{(s)}(\alpha_0) = 0 \ \ s \in [0, q - c), \quad u_\beta^{(s)}(\alpha_0) = 0 \ \ s \in [0, 2q - 2p),$$

$$\overset{f}{\sigma}{}_1^{(s)} + \sigma_\alpha^{(s)} = 0, \tag{2.23}$$

$$R\,\overset{f}{w}{}^{(s)} = -\,\overset{0}{w}{}^{(s+q-c)} - R\zeta \left(a_{13}\,\overset{0}{\sigma}{}_\alpha^{(s)} + a_{23}\,\overset{0}{\sigma}{}_\beta^{(s)}\right) \quad \text{at } t = 0 \ (\alpha = \alpha_0)$$

$$s \in [0, q - 2p + c).$$

Now $\overset{f}{\sigma}{}_1^{(s)}\ (t = 0)$ may be expressed through $\sigma_\alpha^{(s)}$, then it may be deduced from (1.14) that

$$\int_{-1}^{+1} \sigma_\alpha^{(s)}(\alpha_0)d\zeta = \int_{-1}^{+1} d\zeta \int_0^\infty \overset{f}{R}{}_\alpha^{(s+p-q)} dt, \tag{2.24}$$

$$\int_{-1}^{+1} \zeta\sigma_\alpha^{(s)}(\alpha_0)d\zeta$$

$$= \int_{-1}^{+1} \zeta d\zeta \int_0^\infty \overset{f}{R}{}_\alpha^{(s+p-q)} dt - \int_{-1}^{+1} d\zeta \int_0^\infty \overset{f}{R}{}_\gamma^{(s+p-q)} t\,dt,$$

leading to a conclusion that $\overset{0}{\sigma}{}_\alpha^{(s)}(\alpha_0) = 0 \ (s < q - p)$. Hence, since $q - 2p + c < q - p$, the conditions for the plane boundary layer follow from (2.23) in the form

$$\overset{f}{\sigma}{}_1^{(s)} = 0, \quad R\,\overset{f}{w}{}^{(s)} = -\,\overset{0}{w}{}^{(s+q-c)}(\alpha_0) - R\zeta a_{23}\,\overset{0}{\sigma}{}_\beta^{(s)}(\alpha_0) \quad \text{at} \quad t = 0$$

$$(0 \le s < q - 2p + c). \tag{2.25}$$

The plane boundary layer solution may be presented as a sum of bending and extensional components. In order to have decaying solution, the right-hand sides of (2.25) should satisfy

$$(b_{13} + a_{55}) \int_0^1 \zeta^3 f_1 d\zeta + 3 \int_0^1 (1 - \zeta^2) f_2 d\zeta = 0. \tag{2.26}$$

Since in our case $f_1 = 0$, $f_2 = -\overset{0}{w}{}^{(s+q-c)}(\alpha_0)$, along with $\overset{0}{w}{}^{(s+q-c)}(\alpha_0) = 0$ following from (2.26). Therefore, taking into consideration (2.23), we obtain

$$\overset{0}{w}{}^{(s)}(\alpha_0) = 0, \quad \overset{0}{v}{}^{(s)}(\alpha_0) = 0, \quad s \in [0, 2q - 2p). \tag{2.27}$$

Returning back to (2.23), we deduce zero boundary condition for the bending component of the plane boundary layer, hence $\overset{f}{P}{}_b^{(s)} = 0$, $s \in [0, q - 2p + c)$. The symmetric plane boundary layer is governed by the following conditions

$$\overset{f}{\sigma}{}_1^{(s)} = 0, \quad R\overset{f}{w}{}^{(s)} = -\zeta R a_{23} \overset{0}{\sigma}{}_\beta^{(s)}(\alpha_0) \quad \text{at} \quad t = 0. \tag{2.28}$$

The associated solution may be presented in the form

$$\overset{f}{P}{}^{(s)} = -a_{23} \overset{0}{\sigma}{}_\beta^{(s)}(\alpha_0) \overset{f}{P}{}^{[1]}, \quad (0 \le s < q - 2p + c) \tag{2.29}$$

where $\overset{f}{P}{}^{[1]}$ is the solution of the same problem subject to conditions $\overset{f}{\sigma}{}_1^{[1]} = 0$, $\overset{f}{w}{}^{[1]} = \zeta$ at $t = 0$. The associated solution may be found through a method described in Chapters 1 and 4, namely

$$\overset{f}{\sigma}{}_1^{[1]} = A_k^{[1]} F_k'' \exp(-\lambda_k t), \quad \overset{f}{\sigma}{}_{13}^{[1]} = A_k^{[1]} \lambda_k F_k' \exp(-\lambda_k t),$$

$$\overset{f}{\sigma}{}_3^{[1]} = A_k^{[1]} \lambda_k^2 F_k \exp(-\lambda_k t),$$

$$\overset{f}{u}{}^{[1]} = -A_k^{[1]} \left(\frac{b_{11}}{\lambda_k} F_k'' + b_{13} \lambda_k F_k \right) \exp(-\lambda_k t), \tag{2.30}$$

$$\overset{f}{w}{}^{[1]} = -A_k^{[1]} \left(\frac{b_{11}}{\lambda_k^2} F_k''' + (b_{13} + a_{55}) F_k' \right) \exp(-\lambda_k t),$$

$$A_k^{[1]} = -\frac{1}{b_{11}} \frac{\lambda_k^2}{\Delta_k} \int_0^1 F_k \, d\zeta \ (\bar{\Sigma}_{(k)}), \quad \Delta_k = \int_0^1 \left((F_k'')^2 - \frac{b_{33}}{b_{11}} \lambda_k^4 F_k^2 \right) d\zeta.$$

Since the integrals, corresponding to appropriate terms $\overset{f}{\sigma}{}_1^{(s-q)}$, $\overset{f}{\sigma}{}_{21}^{(s+p-q)}$, $(\overset{f}{\sigma}{}_{12} + \overset{f}{\sigma}{}_{21})^{(s-q)}$, $\zeta \dfrac{\partial \overset{f}{\sigma}{}_{13}^{(s-q)}}{\partial \zeta}$, $\overset{f}{\sigma}{}_{13}^{(s-q)}$ in (2.24) are zeros (see also (1.11)), we may use the second equation of equilibrium (1.9) along with (2.29) to obtain

$$\int_{-1}^{+1} d\zeta \int_0^\infty \overset{f}{R}{}_\alpha^{(s+p-q)} dt = \int_{-1}^{+1} d\zeta \int_0^\infty k_{\beta 0} R \overset{f}{\sigma}{}_2^{(s-q)} dt$$

$$= -k_{\beta 0} R \frac{a_{23}}{a_{22}} \int_{-1}^{+1} d\zeta \int_0^\infty \overset{f}{\sigma}{}_3^{(s-q)} dt$$

$$= k_{\beta 0} R \frac{a_{23}}{a_{22}} \int_{-1}^{+1} \zeta \overset{f}{\sigma}{}_{13}^{(s-q)}(t = 0) d\zeta = -k_{\beta 0} R B_1 D_2 (2 \overset{0}{\sigma}{}_\beta^{(s-q)}(\alpha_0)), \tag{2.31}$$

$$B_1 = \frac{a_{23}^2}{a_{11} a_{22} - a_{12}^2} = \frac{E_1}{E_3} \frac{\nu_{23} \nu_{32}}{1 - \nu_{12} \nu_{21}}, \quad D_2 = \sum_{k=1}^\infty \frac{\lambda_k^3}{\Delta_k} \left(\int_0^{+1} F_k \, d\zeta \right)^2.$$

268   *Asymptotic Theory of Anisotropic Plates and Shells*

In case of the second condition of (2.24) we may deduce

$$\int_{-1}^{+1} \zeta \, d\zeta \int_0^\infty \overset{f}{R}_\alpha{}^{(s+p-q)} dt = 0, \quad s \in [0, 2q - 2p)$$

since the integrals corresponding to $\overset{f}{\sigma}_1{}^{(s-q)}$, $(\overset{f}{\sigma}_1 - \overset{f}{\sigma}_2)^{(s-q)}$ are zeros due to the fact that the integrands are odd functions, accordingly, the terms $\zeta^2 \dfrac{\partial \overset{f}{\sigma}_{13}{}^{(s-q)}}{\partial \zeta}$, $\zeta \overset{f}{\sigma}_{13}{}^{(s-q)}$ add up to give zero, and finally, the terms corresponding to $\overset{f}{\sigma}_{21}{}^{(s+p-q)}$, $\overset{f}{\sigma}_{21}{}^{(s-q)}$ are zeros due to $s - q < s - q + p < q - p$. Similarly, it may be shown that

$$\int_{-1}^{+1} d\zeta \int_0^\infty \overset{f}{R}_\gamma{}^{(s+p-q)} t \, dt = 0,$$

due to odd nature of $\overset{f}{\sigma}_{13}{}^{(s-q)}$ and $s+p-q < q-p$ for $\overset{f}{\sigma}_{23}{}^{(s+p-q)}$. Applying the operator $\int_0^\infty t^2 dt \int_{-1}^{+1} d\zeta$ to the first equation of (1.9), it may be proven that the integral, corresponding to $\overset{f}{\sigma}_1{}^{(s-q)}$, is also zero. Now applying the operator $\int_{-1}^{+1} \zeta \, d\zeta \int_0^\infty t \, dt$ to the second equation of (1.9) and using (2.28), we obtain

$$\int_{-1}^{+1} d\zeta \int_0^\infty \overset{f}{\sigma}_2{}^{(s-q)} t \, dt = -\frac{a_{23}}{a_{22}} \int_{-1}^{+1} d\zeta \int_0^\infty \overset{f}{\sigma}_3{}^{(s-q)} t \, dt$$

$$= \frac{a_{23}}{a_{22}} \int_0^\infty dt \int_{-1}^{+1} \zeta \overset{f}{\sigma}_{13}{}^{(s-q)} d\zeta$$

$$= -\frac{a_{23}}{2a_{22}} \int_0^\infty dt \int_{-1}^{+1} \zeta^2 \frac{\partial \overset{f}{\sigma}_{13}{}^{(s-q)}}{\partial \zeta} d\zeta \qquad (2.32)$$

$$= \frac{a_{23}}{2a_{22}} \int_0^\infty dt \int_{-1}^{+1} \zeta^2 \frac{\partial \overset{f}{\sigma}_1{}^{(s-q)}}{\partial t} d\zeta$$

$$= -\frac{a_{23}}{2a_{22}} \int_{-1}^{+1} \zeta^2 \overset{f}{\sigma}_1{}^{(s-q)}(t = 0) d\zeta = 0.$$

In other words, we have shown that the right-hand side of the second condition of (2.24) at $s \in [0, 2q - 2p)$ is zero.

Thus, if the outer stress field is analyzed up to $O(\lambda^{-2q+2p}) = O\left(h_*^{2-2t}\right)$, it is governed by the following conditions

$$2 \overset{0}{\sigma}_\alpha{}^{(s)} + Rk_\beta B_1 D_2(2 \overset{0}{\sigma}_\beta{}^{(s-q)}) = 0, \quad \frac{2}{3} \overset{1}{\sigma}_\alpha{}^{(s)} = 0, \qquad (2.33)$$

$$\overset{0}{v}{}^{(s)} = 0, \quad \overset{0}{w}{}^{(s)} = 0 \quad s \in [0, 2q - 2p) \quad \text{at} \quad \alpha = \alpha_0.$$

It should be noted that the constant D_2 arising in (2.33) depends on the elastic properties of material only and may be calculated through (2.31) or by formula (4.4.29) or (4.4.30) for isotropic or orthotropic materials respectively. The values of D_2 for some standard materials has been listed in Table 10.

We remark that the components of the anti-plane boundary layer are of higher order than the accepted tolerance, and is therefore assumed zero here. Thus, the boundary layer solution is governed by the plane component only.

Let us now study the case of mixed boundary conditions of second type (pinned edge) specified in the form

$$\sigma_a = \sigma_{\alpha\beta} = u_\gamma = 0 \quad \text{at} \quad \alpha = \alpha_0. \tag{2.34}$$

The associated representation (1.16) is written explicitly as

$$\sigma_a = \lambda^{q+d-s}\sigma_\alpha^{(s)} + \lambda^{-q+p+\chi-s}\overset{a}{\sigma}\,_1^{(s)} + \lambda^{-q+p+\mu-s}\overset{f}{\sigma}\,_1^{(s)},$$

$$\sigma_{\alpha\beta} = \lambda^{q+d-s}\tau_{\alpha\beta}^{(s)} + \lambda^{-q+p+\chi-s}\overset{a}{\sigma}\,_{12}^{(s)} + \lambda^{-q+p+\mu-s}\overset{f}{\sigma}\,_{12}^{(s)}, \tag{2.35}$$

$$u_\gamma = \lambda^{q-c+d-s}u_\gamma^{(s)} + R\lambda^{-2q+p+\chi-s}\overset{a}{w}\,^{(s)} + R\lambda^{-2q+p+\mu-s}\overset{f}{w}\,^{(s)}.$$

Substituting (2.35) into (2.34) leads to the following values of consistent χ and μ

$$\chi = q + p - c + d, \quad \mu = 2q - p + d, \tag{2.36}$$

for which

$$\sigma_\alpha^{(s)} + \overset{a}{\sigma}\,_1^{(s-q+2p-c)} + \overset{f}{\sigma}\,_1^{(s)} = 0,$$

$$\tau_{\alpha\beta}^{(s)} + \overset{a}{\sigma}\,_{12}^{(s-q+2p-c)} + \overset{f}{\sigma}\,_{12}^{(s)} = 0 \quad \text{at} \quad \alpha = \alpha_0 \ (t = 0) \tag{2.37}$$

$$u_\gamma^{(s)} + R\overset{a}{w}\,^{(s-2q+2p)} + R\overset{f}{w}\,^{(s-q+c)} = 0.$$

Expressing the quantities $\overset{f}{\sigma}\,_1^{(s)}$, $\overset{f}{\sigma}\,_{12}^{(s)}$ through the remaining ones and substituting the results into (1.14), we obtain

$$\int_{-1}^{+1} \sigma_\alpha^{(s)}(\alpha_0)d\zeta = \int_{-1}^{+1} d\zeta \int_0^\infty \left(\overset{f}{R}\,_\alpha^{(s+p-q)} + \overset{a}{R}\,_\alpha^{(s-2q+3p-c)} \right) dt,$$

$$\int_{-1}^{+1} \zeta\sigma_\alpha^{(s)}(\alpha_0)d\zeta = \int_{-1}^{+1} \zeta d\zeta \int_0^\infty \left(\overset{f}{R}\,_\alpha^{(s+p-q)} + \overset{a}{R}\,_\alpha^{(s-2q+3p-c)} \right) dt$$

$$- \int_{-1}^{+1} d\zeta \int_0^\infty \left(\overset{f}{R}\,_\gamma^{(s+p-q)} + \overset{a}{R}\,_\gamma^{(s+3p-2q-c)} \right) t\,dt, \tag{2.38}$$

$$\int_{-1}^{+1} \tau_{\alpha\beta}^{(s)}(\alpha_0)d\zeta = \int_{-1}^{+1} d\zeta \int_0^\infty \left(\overset{a}{R}\,_\tau^{(s-2q+3p-c)} + \overset{f}{R}\,_\tau^{(s+p-q)} \right) dt.$$

It then follows from (2.37), (2.38) and (2.26) that

$$\overset{0}{\sigma}\,_\alpha^{(s)}(\alpha_0) = 0, \quad \overset{0}{\tau}\,_{\alpha\beta}^{(s)}(\alpha_0) = 0 \quad s \in [0, q - 2p + c) \tag{2.39}$$

$$\overset{0}{w}\,^{(s)}(\alpha_0) = 0, \quad s \in [0, 2q - 2p), \quad \overset{f}{P}\,_{bend}^{(s)} \equiv 0 \ s \in [0, q - 2p + c),$$

with the boundary conditions (2.28) for the plane symmetric boundary layer.

It may be shown for the third condition of (2.38) using (1.11) and (2.29) that the integral corresponding to $\overset{a}{R}\,_\tau^{(s-2q+3p-c)}$ is zero, and that evaluation of the integral associated with $\overset{f}{R}\,_\tau^{(s+p-q)}$ leads to the following condition

$$2\overset{0}{\tau}\,_{\alpha\beta}^{(s)} - B_1 D_2 d_{20}(2\overset{0}{\sigma}\,_\beta^{(s-q)}) = 0 \quad \text{at} \quad \alpha = \alpha_0 \, s \in [0, 2q - 2p).$$

Returning back to the second equation of (2.37), we are now in position to formulate the boundary conditions for the anti-plane boundary layer as

$$\overset{f}{\sigma}\,\overset{(s)}{_{12}}(t=0) = -\zeta\,\overset{1}{\tau}\,\overset{(s+q-2p+c)}{_{\alpha\beta}}, \quad s \in [0, q-c). \tag{2.40}$$

The solution associated with (2.40) may then be found through (4.2.13) and (2.12).

In case of the first equation of (2.38) the integral corresponding to $\overset{a}{R}\,\overset{(s+3p-2q-c)}{_{\alpha}}$ is zero, hence treating the problem similarly to the previously considered case of the first type of mixed boundary conditions, we arrive at

$$2\,\overset{0}{\sigma}\,\overset{(s)}{_{\alpha}} + Rk_\beta B_1 D_2(2\,\overset{0}{\sigma}\,\overset{(s-q)}{_{\beta}}) = 0 \quad \text{at} \quad \alpha = \alpha_0.$$

As for the second relation of (2.38), it may be shown that the integrals associated with $\overset{f}{R}\,\overset{(s+p-q)}{_{\alpha}}$, $\overset{f}{R}\,\overset{(s+p-q)}{_{\tau}}$, are indeed zeros, whereas

$$J_2 = \int_{-1}^{+1}\zeta d\zeta \int_0^\infty \overset{a}{R}\,\overset{(s+3p-2q-c)}{_{\alpha}}dt = -\int_{-1}^{+1}\zeta d\zeta \int_0^\infty (d_{20}\,\overset{a}{\sigma}\,\overset{(s+3p-2q-c)}{_{21}}$$
$$+k_{\alpha 0}R(\overset{a}{\sigma}\,_{12} + \overset{a}{\sigma}\,_{21})^{(s+2p-2q-c)})dt = -d_{20}\left(\int_{-1}^{+1}\zeta d\zeta \int_0^\infty \overset{a}{\sigma}\,\overset{(s+3p-2q-c)}{_{21}}dt\right)$$
$$-Rk_{\alpha 0}\int_{-1}^{+1}\zeta d\zeta \int_0^\infty \overset{a}{\sigma}\,\overset{(s+2p-2q-c)}{_{21}}dt,$$
$$J_3 = -\int_{-1}^{+1}d\zeta \int_0^\infty \overset{a}{R}\,\overset{(s-2q+3p-c)}{_{\gamma}}tdt$$
$$= \int_{-1}^{+1}d\zeta \int_0^\infty (d_{20}\,\overset{a}{\sigma}\,\overset{(s+3p-2q-c)}{_{23}} + Rk_{\alpha 0}\,\overset{a}{\sigma}\,\overset{(s+2p-2q-c)}{_{23}})tdt$$
$$= d_{20}\int_{-1}^{+1}d\zeta \int_0^\infty \overset{a}{\sigma}\,\overset{(s+3p-2q-c)}{_{23}}tdt - Rk_{\alpha 0}\int_{-1}^{+1}d\zeta \int_0^\infty \overset{a}{\sigma}\,\overset{(s+2p-2q-c)}{_{23}}tdt.$$

In view of (2.14), (2.15), and (2.40)

$$J_2 + J_3 = \bar{A}\sqrt{\frac{a_{44}}{a_{66}}}\frac{1}{B_0}\frac{\partial}{\partial\eta}\left(\frac{2}{3}\,\overset{1}{\tau}\,\overset{(s+p-q)}{_{\alpha\beta}}(\alpha_0)\right), \tag{2.41}$$

$$\frac{2}{3}\,\overset{0}{\sigma}\,\overset{(s)}{_{\alpha}}(\alpha_0) = J_2 + J_3 \quad s \in [0, 2q - 2p).$$

Thus, in case of (2.34) reduction of the original 3D boundary value problem to a 2D one leads to the following formulation of the outer stress-strain field

$$2\,\overset{0}{\sigma}\,\overset{(s)}{_{\alpha}} + Rk_\beta B_1 D_2(2\,\overset{0}{\sigma}\,\overset{(s-q)}{_{\beta}}) = 0,$$

$$\frac{2}{3}\,\overset{1}{\sigma}\,\overset{(s)}{_{\alpha}} - \bar{A}\sqrt{\frac{a_{44}}{a_{66}}}d_{20}\left(\frac{2}{3}\,\overset{1}{\tau}\,\overset{(s+p-q)}{_{\alpha\beta}}(\alpha_0)\right) = 0 \quad \text{at} \quad \alpha = \alpha_0, \tag{2.42}$$

$$2\,\overset{0}{\tau}\,\overset{(s)}{_{\alpha\beta}} - B_1 D_2 d_{20}\left(2\,\overset{0}{\sigma}\,\overset{(s-q)}{_{\beta}}(\alpha_0)\right) = 0,$$

$$\overset{0}{w}\,\overset{(s)}{} = 0, \quad s \in [0, 2q - 2p).$$

We note that in case of the second type of mixed boundary conditions both plane symmetric and anti-plane boundary layer solutions are zeros, whereas for the previously considered first type of mixed boundary conditions only plane symmetric boundary layer vanished. Due to (2.35), (2.36) the intensities of stresses of plane and anti-plane boundary layers are $O(\lambda^{q+d})$, $O(\lambda^{2p-c+d})$, respectively. Therefore, the conditions (2.19) may be thought of as preferable to (2.34) in a sense of low stressed edge.

Let us now present results of analysis for clamped edge, when

$$u_a = u_\beta = u_\gamma = 0 \quad \text{at} \quad \alpha = \alpha_0. \tag{2.43}$$

The representation for displacements u_a, u_β, u_γ is then given by

$$u_a = \lambda^{q-p+d-s} u_\alpha^{(s)} + R\lambda^{-2q+p+\chi-s}\, \overset{a}{u}\, {}^{(s)} + R\lambda^{-2q+p+\mu-s}\, \overset{f}{u}\, {}^{(s)},$$

$$u_\beta = \lambda^{q-p+d-s} u_\beta^{(s)} + R\lambda^{-2q+p+\chi-s}\, \overset{a}{v}\, {}^{(s)} + R\lambda^{-2q+p+\mu-s}\, \overset{f}{v}\, {}^{(s)}, \tag{2.44}$$

$$u_\gamma = \lambda^{q-c+d-s} u_\gamma^{(s)} + R\lambda^{-2q+p+\chi-s}\, \overset{a}{w}\, {}^{(s)} + R\lambda^{-2q+p+\mu-s}\, \overset{f}{w}\, {}^{(s)}.$$

It follows from (8.2.4), (8.2.5), and (8.2.22) that

$$u_\alpha^{(s)} = \overset{0}{u}{}_\alpha^{(s)}(\xi,\eta) + \zeta\,\overset{1}{u}{}_\alpha^{(s)}(\xi,\eta) = \overset{0}{u}{}_\alpha^{(s)} - \zeta\gamma_1^{(s)},$$

$$u_\beta^{(s)} = \overset{0}{v}{}_\beta^{(s)}(\xi,\eta) + \zeta\,\overset{1}{v}{}_\beta^{(s)}(\xi,\eta) = \overset{0}{v}{}_\beta^{(s)} - \zeta\gamma_2^{(s)},$$

$$w^{(s)} = \overset{0}{w}{}^{(s)}(\xi,\eta) + \zeta\,\overset{1}{w}{}^{(s)} = \overset{0}{w}{}^{(s)}(\xi,\eta) \tag{2.45}$$
$$+\zeta R\left(a_{13}\,\overset{0}{\sigma}{}_\alpha^{(s-q+c)} + a_{23}\,\overset{0}{\sigma}{}_\beta^{(s-q+c)}\right),$$

$$\gamma_1^{(s)} = \partial_\xi\,\overset{0}{w}{}^{(s-q+2p-c)} - \frac{\overset{0}{u}{}_\alpha^{(s-q)}}{r_1} \quad (1,2;\xi,\eta;u,v) \quad s \in [0, 2q-2p).$$

From the consistency of the asymptotic procedure we deduce

$$\chi = q + d, \qquad \mu = 2q - p + d \tag{2.46}$$

for which the conditions (2.43), (2.44) imply

$$u_\alpha^{(s)} + R\,\overset{a}{u}\,{}^{(s-2q+2p)} + R\,\overset{f}{u}\,{}^{(s-q+p)} = 0,$$

$$u_\gamma^{(s)} + R\,\overset{a}{w}\,{}^{(s-2q+p+c)} + R\,\overset{f}{w}\,{}^{(s-q+c)} = 0 \quad \text{at} \quad \alpha = \alpha_0\ (\xi_1 = 0), \tag{2.47}$$

$$u_\beta^{(s)} + R\,\overset{a}{v}\,{}^{(s-2q+2p)} + R\,\overset{f}{v}\,{}^{(s-q+p)} = 0.$$

It may also be inferred from (2.47) that

$$\overset{0}{u}{}_\alpha^{(s)} = 0, \quad \gamma_1^{(s)} = 0 \quad s \in [0, q-p) \quad \text{at} \quad \alpha = \alpha_0, \tag{2.48}$$

$$\overset{0}{w}{}^{(s)} = 0 \quad s \in [0, q-c), \quad \overset{0}{v}{}_\beta^{(s)} = 0 \quad s \in [0, 2q-2p).$$

We note that the last condition of (2.47) leads to an identity $\gamma_2^{(s)}(\alpha_0) = 0,\ s \in [0, 2q-2p)$.

If we restrict the analysis of the outer problem to approximations of order $0 \le s < 2q - 2p$, then the conditions (2.47) allow evaluation of the boundary layer up to $O(\lambda^{-q+2p-c})$. In order to determine the boundary layer solution, the first conditions of (2.47) are rewritten in view of (1.12), (1.13), and (2.45) as

$$R\,\overset{f}{u}\,{}^{(s)} = -\,\overset{0}{u}{}_\alpha^{(s+q-p)}(\alpha_0) + \zeta\gamma_1^{(s+q-p)}(\alpha_0)$$

$$R\,\overset{f}{w}\,{}^{(s)} = -\,\overset{0}{w}{}^{(s+q-c)}(\alpha_0) \tag{2.49}$$

$$-\zeta R\left(a_{13}\,\overset{0}{\sigma}{}_\alpha^{(s)}(\alpha_0) + a_{23}\,\overset{0}{\sigma}{}_\beta^{(s)}(\alpha_0)\right), \quad s \in [0, q-2p+c) \quad \text{at} \quad t = 0.$$

The functions appearing on the right-hand sides of (2.49) should satisfy the decay conditions which could be derived similarly to elastic plates. Let us introduce the following auxiliary problems, namely, to find solutions of homogeneous equations of elasticity theory (1.9) in the domain $\Omega = \{(\zeta,t),\ 0 \le t \le \ell,\ -1 \le \zeta \le +1\}$, satisfying $\overset{f}{\sigma}_{13} = \overset{f}{\sigma}_{3} = 0$ at $\zeta = \pm 1$ and $\overset{f}{u} = \overset{f}{w} = 0$ at $t = \ell$ along with

$$
\begin{aligned}
R\,\overset{f}{u}\,^{[1]}(t=0) &= \zeta, & R\,\overset{f}{w}\,^{[1]}(t=0) &= 0 & &\text{(problem 1)} \\[4pt]
R\,\overset{f}{u}\,^{[2]}(t=0) &= 0, & R\,\overset{f}{w}\,^{[2]}(t=0) &= 1 & &\text{(problem 2)} \\[4pt]
R\,\overset{f}{u}\,^{[3]}(t=0) &= 1, & R\,\overset{f}{w}\,^{[3]}(t=0) &= 0 & &\text{(problem 3)} \\[4pt]
R\,\overset{f}{u}\,^{[4]}(t=0) &= 0, & R\,\overset{f}{w}\,^{[4]}(t=0) &= -\zeta & &\text{(problem 4)}.
\end{aligned}
\qquad (2.50)
$$

The solution procedure for these auxiliary problems has been presented in Chapter 1 within more general framework.

We remark that problems 1 and 2 correspond to bending, hence at the bending moment $\overset{f}{M}\,^{[i]}$ and the shear force $\overset{f}{N}\,^{[i]}$ $(i = 1,2)$ are expected to emerge at $t = \ell$. Here ℓ is the typical scale of decaying region or a number corresponding to the opposite edge, which could also be adopted as $\ell = +\infty$, see Goldenveizer (1976); Green (1962a,b). Therefore, if we impose a requirement of zero total bending moment and shear force, then the load is self-equilibrated at $t = 0$, hence the stress field decays away from $t = 0$ into the interior of Ω according to the Saint-Venant principle. In case of bending the solution corresponding to (2.49) may be written as

$$
Q^{(s)}_{bend} = \gamma_1^{(s+q-p)}(\alpha_0)Q^{[1]} - \overset{0}{w}\,^{(s+q-c)}(\alpha_0)Q^{[2]}, \quad s \in [0, q - 2p + c).
\qquad (2.51)
$$

The condition of zero bending moment and shear force

$$
\gamma_1^{(s+q-p)}(\alpha_0) = 0, \quad \overset{0}{w}\,^{(s+q-c)}(\alpha_0) = 0
$$

along with (2.48) provide the conditions for outer problem, given by

$$
\begin{aligned}
&\overset{0}{w}\,^{(s)} = 0, \quad \overset{0}{v}_\beta^{(s)} = 0, \quad \gamma_1^{(s)} = 0 \qquad \text{at} \qquad \alpha = \alpha_0 \\
&s \in [0, 2q - 2p).
\end{aligned}
$$

Therefore, from (2.49) we infer that the bending component of the plane boundary layer is zero, $\overset{f}{P}_{bend} = 0$.

The problems 3 and 4 correspond to symmetric stress-strain fields (with $\overset{f}{\sigma}_1,\ \overset{f}{u},\ \overset{f}{\sigma}_3$ even and $\overset{f}{\sigma}_{13},\ \overset{f}{w}$ being odd functions of ζ), therefore a horizontal force $\overset{f}{T}$ only occurs at $t = \ell$. Hence, in order to ensure decay, this horizontal force should vanish

$$
\overset{f}{T}\,^{(s)} = -\overset{0}{u}_\alpha^{(s+q-p)}(\alpha_0)T^{[3]} + R\left(a_{13}\,\overset{0}{\sigma}_\alpha^{(s)}(\alpha_0) + a_{23}\,\overset{0}{\sigma}_\beta^{(s)}(\alpha_0)\right)T^{[4]}.
\qquad (2.52)
$$

It follows from $\overset{f}{T}{}^{(s)} = 0$ that

$$\overset{0}{\underset{\alpha}{u}}{}^{(s+q-p)}(\alpha_0) = m_1 R\left(a_{13}\,\overset{0}{\underset{\alpha}{\sigma}}{}^{(s)}(\alpha_0) + a_{23}\,\overset{0}{\underset{\beta}{\sigma}}{}^{(s)}(\alpha_0)\right), \quad m_1 = T^{[4]}/T^{[3]}$$

which serves as the fourth condition for the outer problem. The plane boundary layer may then be evaluated from (2.49). Thus, if the asymptotic analysis involves the orders $s \in [0, 2q - 2p)$, the outer solution is described by the following conditions

$$\overset{0}{\underset{\alpha}{u}}{}^{(s)} - m_1 R\left(a_{13}\,\overset{0}{\underset{\alpha}{\sigma}}{}^{(s-q+p)}(\alpha_0) + a_{23}\,\overset{0}{\underset{\beta}{\sigma}}{}^{(s-q+p)}(\alpha_0)\right) = 0 \quad \text{at} \quad \alpha = \alpha_0 \quad (2.53)$$

$$\overset{0}{\underset{\beta}{u}}{}^{(s)}, \quad \overset{0}{w}{}^{(s)} = 0, \quad \gamma_1^{(s)} = 0.$$

The contribution of the anti-plane and the bending component of plane boundary layer to total stress-strain field is of higher orders than the required tolerance, and so is assumed zero. The non-zero symmetric plane boundary layer is determined from

$$R\overset{f}{u}{}^{(s)}(t = 0) = -m_1 R\left(a_{13}\,\overset{0}{\underset{\alpha}{\sigma}}{}^{(s)}(\alpha_0) + a_{23}\,\overset{0}{\underset{\beta}{\sigma}}{}^{(s)}(\alpha_0)\right), \tag{2.54}$$

$$R\overset{f}{w}{}^{(s)}(t = 0) = -\zeta R\left(a_{13}\,\overset{0}{\underset{\alpha}{\sigma}}{}^{(s)}(\alpha_0) + a_{23}\,\overset{0}{\underset{\beta}{\sigma}}{}^{(s)}(\alpha_0)\right),$$

hence,

$$\overset{f}{P}{}^{(s)}_{sym} = R\left(a_{13}\,\overset{0}{\underset{\alpha}{\sigma}}{}^{(s)}(\alpha_0) + a_{23}\,\overset{0}{\underset{\beta}{\sigma}}{}^{(s)}(\alpha_0)\right) P^{[5]} \tag{2.55}$$

where $P^{[5]}$ is a solution of the same problem subject to the following conditions

$$R\overset{f}{u} = -m_1, \quad R\overset{f}{w} = -\zeta \quad \text{at} \quad t = 0.$$

In case of the third type of mixed boundary conditions

$$u_a = \sigma_{\alpha\beta} = u_\gamma = 0 \quad \text{at} \quad \alpha = \alpha_0. \tag{2.56}$$

Using expressions (2.2) for $\sigma_{\alpha\beta}$ and (2.44) for u_a, u_γ, we obtained the consistent values of χ and μ in the form

$$\chi = q + p - c + d, \quad \mu = 2q - p + d. \tag{2.57}$$

Then, the boundary conditions are represented as

$$u_\alpha^{(s)} + R\overset{a}{u}{}^{(s-2q+3p-c)} + R\overset{f}{u}{}^{(s-q+p)} = 0,$$

$$u_\gamma^{(s)} + R\overset{a}{w}{}^{(s-2q+2p)} + R\overset{f}{w}{}^{(s-q+c)} = 0 \quad \text{at} \quad \alpha = \alpha_0 \ (t = 0) \tag{2.58}$$

$$\tau_{\alpha\beta}^{(s)} + \overset{a}{\underset{12}{\sigma}}{}^{(s-q+2p-c)} + \overset{f}{\underset{12}{\sigma}}{}^{(s)} = 0.$$

Following the same procedure as for previously considered case of a clamped edge (2.43), we use the first two conditions of (2.58), leading to the conditions for the outer problem

$$\overset{0}{\underset{\alpha}{u}}{}^{(s)} - m_1 R\left(a_{13}\,\overset{0}{\underset{\alpha}{\sigma}}{}^{(s-q+p)} + a_{23}\,\overset{0}{\underset{\beta}{\sigma}}{}^{(s-q+p)}\right) = 0 \quad \text{at} \quad \alpha = \alpha_0 \tag{2.59}$$

$$\overset{0}{w}{}^{(s)} = 0, \quad \gamma_1^{(s)} = 0, \quad 0 \le s < 2q - 2p.$$

The fourth condition for the outer problem follows from the last condition of (2.58), if $\overset{a}{\sigma}\,\overset{(s)}{12}$ is expressed through remaining quantities and substituted into the decay conditions (1.14). As a result, we have

$$\int_{-1}^{+1} \tau_{\alpha\beta}^{(s)}(\alpha_0)d\zeta = \int_{-1}^{+1} d\zeta \int_0^\infty \left(\overset{a}{R}\,\overset{(s-2q+3p-c)}{\tau} + \overset{f}{R}\,\overset{(s-q+p)}{\tau} \right) dt. \qquad (2.60)$$

Due to (1.11)-(1.13) and $s \in [0, 2q-2p)$, the integral corresponding to $\overset{a}{R}\,\overset{(s-2q+3p-c)}{\tau}$ is zero, whereas the only non-zero terms corresponding to $\overset{f}{R}\,\overset{(s-q+p)}{\tau}$ are these associated with $\overset{f}{\sigma}\,_2$. Using (2.55), we obtain

$$\int_{-1}^{+1} d\zeta \int_0^\infty \overset{a}{R}\,\overset{(s-q+p)}{\tau} dt$$

$$= -\int_{-1}^{+1} d\zeta \int_0^\infty (d_{20} \overset{f}{\sigma}\,\overset{(s-q+p)}{2} + k_{\alpha 0} R \overset{f}{\sigma}\,\overset{(s-q)}{2}) dt$$

$$= \frac{a_{23}}{a_{22}} \int_{-1}^{+1} d\zeta \int_0^\infty (d_{20} \overset{f}{\sigma}\,\overset{(s-q+p)}{3} + k_{\alpha 0} R \overset{f}{\sigma}\,\overset{(s-q)}{3}) dt$$

$$= \frac{a_{23}}{a_{22}} d_{20} \left(\int_{-1}^{+1} d\zeta \int_0^\infty \overset{f}{\sigma}\,\overset{(s-q+p)}{3} dt \right) \qquad (2.61)$$

$$= \frac{a_{23}}{a_{22}} d_{20} \left(a_{13} \overset{0}{\sigma}\,\overset{(s-q+p)}{\alpha}(\alpha_0) + a_{23} \overset{0}{\sigma}\,\overset{(s-q+p)}{\beta}(\alpha_0) \right) m_2,$$

$$m_2 = \int_0^1 d\zeta \int_0^\infty \overset{f}{\sigma}\,\overset{[5]}{3} dt.$$

The conditions for the outer problem are given by

$$\overset{0}{u}\,\overset{(s)}{\alpha} - m_1 R \left(a_{13} \overset{0}{\sigma}\,\overset{(s-q+p)}{\alpha} + a_{23} \overset{0}{\sigma}\,\overset{(s-q+p)}{\beta} \right) = 0,$$

$$2 \overset{0}{\tau}\,\overset{(s)}{\alpha\beta} + m_2 \nu_{32} d_{20} \left(a_{13} \left(2 \overset{0}{\sigma}\,\overset{(s-q+p)}{\alpha} \right) \right.$$

$$\left. + a_{23} \left(2 \overset{0}{\sigma}\,\overset{(s-q+p)}{\beta} \right) \right) = 0 \quad \text{at} \quad \alpha = \alpha_0, \qquad (2.62)$$

$$\overset{0}{w}\,^{(s)} = 0, \quad \gamma_1^{(s)} = 0, \quad (0 \le s < 2q - 2p).$$

The bending plane boundary layer is zero $\left(\overset{f}{P}\,_{bend} = 0 \right)$, with the non-zero symmetric plane boundary layer determined by (2.55). As follows from (2.58) and (2.62) the anti-plane boundary layer associated with symmetric problem, is zero, hence it is related to anti-symmetric stress-strain field and is governed by the conditions

$$\overset{a}{\sigma}\,\overset{(s)}{12}(t = 0) = -\zeta \overset{1}{\tau}\,\overset{(s+q-2p+c)}{\alpha\beta}(\alpha_0) \quad 0 \le s < q - c \qquad (2.63)$$

along with formulae (4.2.13) and (2.12). The stresses of plane and anti-plane boundary layer solutions are of order $O(\lambda^{q+d})$ and $O(\lambda^{2p-c+d})$, respectively.

In this section we restricted our consideration to analysis of orders $s \in [0, 2q-2p)$ which are typically important for practical problems. We remark that the results may be extended to higher orders of approximation.

9.3 Refinement of 2D Boundary Conditions

It may be shown that the conditions for the outer problem derived in the previous Section 9.2 are in fact closely related to the boundary conditions of the classical shell theory, often providing refinement of the latter. In order to reveal the underlying links let us rewrite the results in terms of moments, forces, and displacements of the mid-surface of a shell u, v, w through

$$T_1 = 2h\lambda^{q+d-s} \overset{0}{\sigma}{}^{(s)}_{\alpha}, \quad S_{12} = 2h\lambda^{q+d-s} \overset{0}{\tau}{}^{(s)}_{\alpha\beta},$$

$$N_1 = 2h\lambda^{p+d-s}\left(\overset{0}{\tau}{}^{(s)}_{\alpha\gamma} + \frac{1}{3}\overset{2}{\tau}{}^{(s)}_{\alpha\gamma}\right), \quad M_1 = \frac{2}{3}h^2\lambda^{q+d-s}\overset{1}{\sigma}{}^{(s)}_{\alpha},$$

$$H_{12} = \frac{2}{3}h^2\lambda^{q+d-s}\overset{1}{\tau}{}^{(s)}_{\alpha\beta}, \quad (\alpha,\beta;1,2), \qquad (3.1)$$

$$u = \lambda^{q-p+d-s}\overset{0}{u}{}^{(s)}_{\alpha}, \quad v = \lambda^{q-p+d-s}\overset{0}{v}{}^{(s)}_{\beta}$$

$$w = \lambda^{q-c+d-s}\overset{0}{w}{}^{(s)}, \quad s \in [0, 2q - 2p - 1].$$

As before, summation convention is assumed in (3.1) with index s in the limits $s \in [0, 2q - 2p - 1]$.

The conditions of the first boundary value problem ($\sigma_\alpha = \sigma_{\alpha\beta} = \sigma_{\alpha\gamma} = 0$ at $\alpha = \alpha_0$) may be rewritten in terms of notation (3.1) as

$$T_1 = 0, \quad S_{12} = 0, \qquad (3.2)$$

$$M_1 - \bar{A}h\sqrt{\frac{G_{12}}{G_{23}}}\frac{1}{B}\frac{\partial H_{12}}{\partial \beta} = 0, \quad N_1 + \frac{1}{B}\frac{\partial H_{12}}{\partial \beta} = 0 \quad \text{at} \quad \alpha = \alpha_0$$

where $\bar{A} \approx 1.26$ is a constant of Goldenveizer-Kolos. We note that the underlined term does not appear in the classical conditions. As may be expected, for shells with large shear stiffness in the transverse directions ($G_{23} = +\infty$) the conditions (3.2) coincide with the classical conditions. However, in case of small transverse shear stiffness, which is typical for some modern anisotropic materials, the underlined term may be of significant importance. Thus, the conditions (3.2) are oriented firstly towards shells with small shear stiffness in the cross-sections.

In case of the mixed boundary value problem of the first type $\sigma_\alpha = u_\beta = u_\gamma = 0$ at $\alpha = \alpha_0$ (pinned edge) the conditions (2.33) may be reformulated as

$$T_1 + h\frac{E_1}{E_3}\frac{\nu_{23}\nu_{32}}{1 - \nu_{12}\nu_{21}}D_2 k_\beta T_2 = 0 \quad \left(k_\beta = \frac{1}{AB}\frac{\partial B}{\partial \alpha}\right) \qquad (3.3)$$

$$v = 0, \quad M_1 = 0, \quad w = 0 \quad \text{at} \quad \alpha = \alpha_0.$$

The conditions for second type of mixed boundary value problem $\sigma_\alpha = \sigma_{\alpha\beta} =$

$u_\gamma = 0$ at $\alpha = \alpha_0$ (pinned edge), corresponding to (2.42) take the form

$$T_1 + h\frac{E_1}{E_3}\frac{\nu_{23}\nu_{32}}{1-\nu_{12}\nu_{21}}D_2 k_\beta T_2 = 0,$$

$$S_{12} - h\frac{E_1}{E_3}\frac{\nu_{23}\nu_{32}}{1-\nu_{12}\nu_{21}}D_2\frac{1}{B}\frac{\partial T_2}{\partial \beta} = 0 \quad \text{at} \quad \alpha = \alpha_0, \qquad (3.4)$$

$$M_1 - \bar{A}h\sqrt{\frac{G_{12}}{G_{23}}}\frac{1}{B}\frac{\partial H_{12}}{\partial \beta} = 0,$$

$$w = 0.$$

In case of a clamped edge ($u_\alpha = u_\beta = u_\gamma = 0$ at $\alpha = \alpha_0$) the conditions (2.53) are represented as

$$2hu - m_1 h(a_{13}T_1 + a_{23}T_2) = 0 \quad \text{at} \quad \alpha = \alpha_0 \qquad (3.5)$$

$$v = 0, \quad w = 0, \quad \gamma_1 = 0.$$

The effective conditions (2.62) corresponding to $u_\alpha = \sigma_{\alpha\beta} = u_\gamma = 0$ at $\alpha = \alpha_0$, are written as

$$2hu - m_1 h(a_{13}T_1 + a_{23}T_2) = 0,$$

$$S_{12} + m_2 h\nu_{32}\frac{1}{B}\frac{\partial}{\partial \beta}(a_{13}T_1 + a_{23}T_2) = 0 \quad \text{at} \quad \alpha = \alpha_0, \qquad (3.6)$$

$$w = 0, \quad \gamma_1 = 0.$$

The conditions (3.2)-(3.6) are written within the same accuracy as the governing equations (8.4.3), (8.4.4) and provide evaluation of the outer stress-strain field of a shell up to $O(h_*^{2-2t})$.

It follows from the structure of conditions (3.2)-(3.6) that in passing from the three-dimensional boundary value problems of the theory of elasticity to the two dimensional problems of the theory of shells, great care must be given to formulation of static boundary conditions whereas for the kinematic boundary conditions of the classical theory are more or less subject to change. If the material of the shell has a sufficiently large stiffness to shear or tension-compression in the cross-sections of the shell, then, in general the conditions of the classical theory gives acceptable results, otherwise it is necessary to use refined conditions (3.2)-(3.6).

9.4   Accuracy of Technical Shell Theories

Based on the previous results of Chapters 8 and 9, here we discuss the possibility of some applied theories of shells and ways to further extend their applicability.

First studies devoted to estimation of the accuracy of isotropic elastic shells were made by Novozhilov and Finkelstein (1943), along with Mushtari (1947). The first of cited works has another important value, namely, that in this work it has been demonstrated directly that from the three-dimensional theory of elasticity it is possible to obtain more accurate equations for shells by expanding the unknown quantities in power series in terms of transverse coordinates. According to the proposed approach in Novozhilov and Finkelstein (1943) and in contrast to the Reissner theory, the laws of change of the tangential stresses $\sigma_{\alpha\gamma}$, $\sigma_{\beta\gamma}$ and the normal stress σ_γ with respect to the transverse coordinates are given. Then, relying upon the equations of spatial problem, approximate two-dimensional equations are derived for the remaining unknown quantities. This trend has further been developed substantially in Ambartsumyan (1974). The possibility of such a generalization for Reissner and other theories have been proved by Goldenveizer (1958).

It was shown by Novozhilov and Finkelstein (1943) that the error of the Kirchhoff-Love theory is order of $h^* = h/R$ compared to unity. This estimate does not take into account variability in the stress field and is concerned only with evaluation of the outer stress-strain field. A comprehensive investigation of this question was carried out by Goldenveizer (1961, 1968a, 1976). It was shown, in particular, that if the construction lies in the state of stress-strain in the conventional asymptotic theory, then employing a Kirchhoff-Love type hypothesis leads to errors of order $\varepsilon = O(h_*^{2-2t})$. This estimation was also obtained by Koiter using the energy methods Koiter (1966, 1970).

The classical theory of shells based on the Kirchhoff-Love assumptions may be considered as a first approximation of the iterative process for the outer stress-strain field, both for the equations and the boundary conditions. We note that adoption of these assumptions does not take into account any effects of the boundary layers. In contrast to isotropic shells, the boundary layer in the orthotropic case may propagate due to weakly localization effects depending on the degree of anisotropy of the shell material. Then the actual stress-strain field differs significantly from that obtained through the classical theory. It has been shown above that, in the orthotropic shells in the vicinity of the lateral surface two types of boundary layers appear, namely the plane and the anti-plane. The anti-plane boundary layer solution decays as

$$\exp\left(-\sqrt{\frac{G_{23}}{G_{12}}}\frac{\pi}{2}t\right), \quad t = A(\alpha_0)\xi_1 = A(\alpha_0)\frac{\alpha - \alpha_0}{h}. \tag{4.1}$$

It should be mentioned that if the length of α-line is of order R, and the following relation $\sqrt{\dfrac{G_{23}}{G_{12}}}\dfrac{\pi}{2} \sim O(h_*^1)$ holds, which is only possible for small shear stiffness in cross sections that are perpendicular to the mid-surface of the shell, then the influence of the anti-plane boundary layer has an affect at a sufficient distance from the edge $\alpha = \alpha_0$, i.e. the stress state of the anti-plane boundary layer is propagating. For most anisotropic materials (fiberglass plastic, SVAM, glass fibre laminate, etc.) $G_{23}/G_{12} < 1$, therefore anti-plane boundary layer solutions in anisotropic shells decay slower than in isotropic ones.

The plane boundary layer decays as

$$\exp\left(-Re\lambda_1 t\right) \tag{4.2}$$

where λ_1 is the smallest root of the corresponding transcendental equation with positive real part $Re\lambda > 0$. Such equations for the isotropic shells are given by

$$\sin 2\lambda \pm 2\lambda = 0. \tag{4.3}$$

In the case of anisotropic materials with elastic parameters β_1, β_2 specified by (3.5.22) we have

$$\omega \sin z \pm \sin \omega z = 0, \quad z = (\beta_1 + \beta_2)\lambda, \quad \omega = \frac{\beta_2 - \beta_1}{\beta_1 + \beta_2} \tag{4.4}$$

whereas for the shells made of materials with elastic parameters α, β given by formula (3.5.23), the mentioned transcendental equation takes the form

$$\omega \sin z \pm \sinh \omega z = 0, \quad z = 2\beta\lambda, \quad \omega = \alpha/\beta. \tag{4.5}$$

We note that in shells of actual anisotropic materials the plane boundary layers are usually localized.

It is possible to extend the classical theory of anisotropic shells in two directions without refining the ordinary concepts of the shells of Love-type, and without taking into account the transverse shear. The first of these directions leads to a more precise definition of the outer stress-strain field by starting from more consistent equilibrium equations, using more accurate constitutive relations and refined boundary conditions. To this end, one must use the equilibrium equations (8.3.24), which take into account the effect of compression and moments when reducing the spatial system of forces onto the mid-surface of the shell; more accurate constitutive relations given by (8.4.4) and the given boundary conditions (3.2)-(3.6). This system of equations allows evaluation of the outer stress-strain field of the shell within the order of $\varepsilon = \max(\delta_1, \delta_2)$, where $\delta_1 = \{O(h_*^{2-t_1})$ at $t \leq 1/2, O(h_*^{4-4t-t_1})$, at $t \geq 1/2\}$, $\delta_2 = O(h_*^{2-2t-t_2})$, with t denoting the variability index and t_1, t_2 standing for the parameters of anisotropy. This accuracy is maximal possible within the framework of the basic assumptions of the classical shell theory. In order to obtain more accurate results in the following, it is necessary to dispense with the hypothesis of the classical theory and either adopt new weakened assumptions or perform more accurate integration of the corresponding three-dimensional equations. In the former case, this would lead to considerable difficulty in the theory since the order of the solvable differential equations increases. Furthermore, this approach does not take into account boundary layer solutions (this is true for the existing technical theories of shells). We remark that within the second approach, provided that the asymptotic method is used, the repeated integration of the known quantities is required, along with repeated matching of the solutions of these equations and the boundary layer.

The second direction of verifying the results of the classical theory also involves finding the boundary stress fields. From the results of Section 9.2 it may be seen that any solution of the boundary layer must be imposed on the solution of the outer problem for each of the spatial boundary conditions. This is carried out in the same manner as in the case of plates (see Chapters 4 and 6) and we do not give the remaining details here.

To illustrate the aforementioned, we consider the problem of evaluation of the stress field of a cantilever cylindrical shell loaded at one end by normal and tangential forces, see Fig. 9.1.

Let us introduce the asymmetric stress tensor components given by formulae (8.1.1), and for the-mid surface of the shell introduce the coordinates $\alpha = R\xi$, $\beta = R\theta$, where R is the radius of the shell, ξ is the relative distance along the symmetry axis of the cylinder, and θ is the central angle. Let the principal directions of anisotropy coincide with the coordinate axis. Let us study the stress-strain field of the shell $\Omega = \{(\xi, \theta, \zeta) : 0 \leq \xi \leq \ell,\ 0 \leq \theta \leq 2\pi,\ |\zeta| \leq 1\}$ for the following boundary conditions

$$X^{\pm} = Y^{\pm} = Z^{\pm} = 0 \quad \text{at} \quad \gamma = \pm h\ (\zeta = \pm 1),$$

$$u_{\alpha} = u_{\beta} = u_{\gamma} = 0 \quad \text{at} \quad \xi = \ell, \tag{4.6}$$

$$\sigma_{\alpha} = \bar{\sigma}_a,\ \ \tau_{\alpha\beta} = \bar{\tau}_{\alpha\beta},\ \ \tau_{\alpha\gamma} = \bar{\tau}_{\alpha\gamma} \quad \text{at} \quad \xi = 0$$

where, for the sake of definiteness

$$\bar{\sigma}_a = \left(\frac{\varphi_1}{h\bar{\varphi}_1} T_1^0 + \frac{\varphi_2}{h^2\bar{\varphi}_2} M_1^0 \right) \cos m\theta,$$

$$\bar{\tau}_{\alpha\beta} = \left(\frac{\psi_1}{h\bar{\psi}_1} S_{12}^0 + \frac{\psi_2}{h^2\bar{\psi}_2} H_{12}^0 \right) \sin m\theta, \quad \tau_{\alpha\gamma} = 0, \tag{4.7}$$

$$\bar{\varphi}_1 = \int_{-1}^{+1} \varphi_1 d\zeta\ \ (\varphi_1,\ \psi_1), \quad \bar{\varphi}_2 = \int_{-1}^{+1} \zeta\varphi_2 d\zeta \qquad (\varphi_2,\ \psi_2).$$

Here m is integer, and φ_1, ψ_1 are even functions of ζ, and φ_2, ψ_2 are odd functions of ζ.

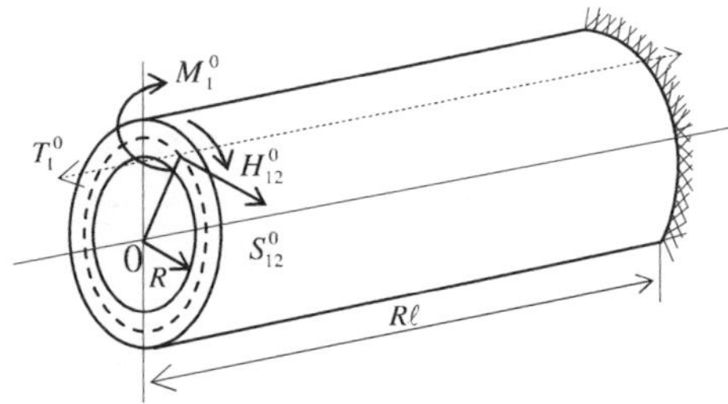

Fig. 9.1

The solution of the stated boundary value problem is composed of the outer solution and the boundary layers. The outer solution has to be determined from conditions (8.4.3) and (8.4.4) subject to effective boundary conditions corresponding to (4.6). In view of (8.3.1)-(8.3.3) along with (9.3.2), and (9.3.5) these effective boundary conditions take the form

$$T_1 = T_1^0 \cos m\theta, \quad S_{12} = S_{12}^0 \sin m\theta \qquad \text{at} \quad \xi = 0, \tag{4.8}$$

$$M_1 - \bar{A}h\sqrt{\frac{G_{12}}{G_{23}}}\frac{1}{R}\frac{\partial H_{12}}{\partial \theta} = M_1^0 \cos m\theta \qquad \text{at} \quad \xi = 0, \tag{4.9}$$

$$N_1 + \frac{1}{R}\frac{\partial H_{12}}{\partial \theta} = \bar{N}_1^0 \quad \text{where} \quad \bar{N}_1^0 = H_{12}^0\frac{m}{R}\cos m\theta,$$
$$2hu - m_1 h(a_{13}T_\alpha + a_{23}T_\beta) = 0, \quad v = 0 \quad \text{at} \quad \xi = \ell, \tag{4.10}$$

$$w = 0, \quad \gamma_1 = 0 \quad \text{at} \quad \xi = \ell. \tag{4.11}$$

The outer solution may also be decomposed into the sum of the field governed by the membrane shell theory and this described by simple edge effects (Goldenveizer, 1961, 1976). The component from the membrane shell theory is determined from equations (8.3.14) and (8.3.15) in the specified coordinate system subject to tangential conditions (4.8) and (4.10), resulting in

$$T_1 = (T_1^0 - mS_{12}^0\xi)\cos m\theta, \tag{4.12}$$
$$S_{12} = S_{12}^0 \sin m\theta, \quad T_2 = 0,$$

$$u = \frac{a_{11}R}{2h}\left(T_1^0(\xi - \ell) - \frac{1}{2}mS_{12}^0(\xi^2 - \ell^2)\right)\cos m\theta$$
$$+ \frac{1}{2}m_1a_{13}(T_1^0 - mS_{12}^0\ell)\cos m\theta,$$
$$v = a_{11}\frac{R}{4h}\left(T_1^0(\xi - \ell)^2 - mS_{12}^0(\frac{1}{3}(\xi^3 - \ell^3) - \ell^2(\xi - \ell))\right)m\sin m\theta$$
$$+ \frac{1}{2}m_1a_{13}(T_1^0 - mS_{12}^0\ell)(\xi - \ell)m\sin m\theta + a_{66}\frac{R}{2h}S_{12}^0(\xi - \ell)\sin m\theta, \tag{4.13}$$
$$w = \left(a_{11}\frac{R}{4h}\left(T_1^0(\xi - \ell)^2 - mS_{12}^0(\frac{1}{3}(\xi^3 - \ell^3) - \ell^2(\xi - \ell))\right)m^2\right.$$
$$+ a_{66}\frac{R}{2h}S_{12}^0(\xi - \ell)m + \frac{1}{2}m_1a_{13}(T_1^0 - mS_{12}^0\ell)(\xi - \ell)m^2$$
$$\left. - a_{12}\frac{R}{2h}(T_1^0 - m\xi S_{12}^0)\right)\cos m\theta.$$

Using the results (4.12), the stress field may be calculated through conventional relations.

It may be shown that stress distribution at $\xi = 0$ is different from (4.7). Therefore, the plane and anti-plane boundary layer solutions are expected in the vicinity of $\xi = 0$. In case of the anti-plane boundary layer the boundary condition is given by

$$\overset{a}{\sigma}_{\alpha\beta} = \left(\frac{\psi_1}{h\bar{\psi}_1} - \frac{1}{2h}\right)S_{12}^0 \sin m\theta \qquad \text{at} \quad \xi = 0. \tag{4.14}$$

Now using (3.5.41), (3.5.42), and (4.14), the anti-plane boundary layer solution may be obtained

$$\overset{a}{u}_\beta = \Phi(\xi,\theta,\zeta), \quad \overset{a}{\sigma}_{\alpha\beta} = G_{12}\frac{\partial\Phi}{\partial t}, \quad \overset{a}{\sigma}_{\beta\gamma} = G_{23}\frac{\partial\Phi}{\partial\zeta}, \quad t = \frac{\alpha}{h} = R\frac{\xi}{h},$$

$$\Phi = \sum_{n=1}^{\infty} A_n \cos\pi n\zeta \exp\left(-\sqrt{\frac{G_{23}}{G_{12}}}\pi n t\right)\sin m\theta, \quad (4.15)$$

$$A_n = -\frac{S_{12}^0}{\pi\sqrt{G_{12}G_{23}}}\frac{1}{n}\int_{-1}^{+1}\left(\frac{\psi_1}{h\bar{\psi}_1} - \frac{1}{2h}\right)\cos\pi n\zeta\, d\zeta.$$

As follows from (4.15), the only case when the anti-plane boundary layer in the near-edge vicinity may vanish corresponds to uniform distribution of tangential stresses along thickness ($\psi = const$). Clearly, the classical shell theory is not capable of addressing this boundary layer solution.

The plane boundary layer is determined by the following conditions

$$\overset{f}{\sigma}_\alpha = \left(\frac{\varphi_1}{h\bar{\varphi}_1}T_1^0 - \frac{T_1^0}{2h}\right)\cos m\theta, \quad \overset{f}{\sigma}_{\alpha\gamma} = 0 \quad \text{at} \quad \xi = 0. \quad (4.16)$$

The solution is given by

$$\left(\overset{f}{\sigma}_\alpha,\overset{f}{\sigma}_{\alpha\gamma},\overset{f}{\sigma}_{\gamma\gamma}\right) = \sum_{(\lambda_k)}\left(\overset{f}{\tau}_\alpha,\overset{f}{\tau}_{\alpha\gamma},\overset{f}{\tau}_\gamma\right)_k \exp(-\lambda_k t)\cos m\theta, \quad (4.17)$$

$$\left(\overset{f}{u},\overset{f}{w}\right) = R\sum_{(\lambda_k)}\left(\overset{f}{u}_k,\overset{f}{w}_k\right)_k \exp(-\lambda_k t)\cos m\theta,$$

where

$$\overset{f}{\tau}_\alpha = \frac{F_k''}{\lambda_k^2}A_k, \quad \overset{f}{\tau}_{\alpha\gamma} = \frac{F_k'}{\lambda_k}A_k, \quad \overset{f}{\tau}_\gamma = F_k A_k,$$

$$\overset{f}{u}_k = -\frac{1}{\lambda_k^3}(b_{11}F_k'' + b_{13}\lambda_k^2 F_k)A_k \quad \left(\underset{k}{\Sigma}\right), \quad (4.18)$$

$$\overset{f}{w}_k = -\frac{1}{\lambda_k^4}(b_{11}F_k''' + (b_{13}+a_{55})\lambda_k^2 F_k')A_k.$$

The functions F_n for a symmetric boundary value problem are defined by one of the formulae (3.5.24), (3.5.26), or (3.5.29), with λ_k denoting the roots of the associated transcendental equations (3.5.24), (3.5.27), (3.5.30). In case of a specified function φ_1 the values of the constants A_k may be found through the boundary collocation method, in particular, for $\varphi_1 = const$ we have trivial solution $A_k \equiv 0$.

Thus, incorporating the boundary layers into the analysis allows comprehensive solution for stress-strain field of the shell both near the edge and away from it.

In order to satisfy the conditions (4.9) and (4.11), we employ the solution for simple edge effect given by

$$M_1 = -D_{11}\frac{1}{A^2}\frac{\partial^2 w}{\partial\alpha^2}, \quad M_2 = -D_{12}\frac{1}{A^2}\frac{\partial^2 w}{\partial\alpha^2},$$

$$H_{12} = -2D_{66}\frac{1}{B}\frac{\partial}{\partial\beta}\frac{1}{A}\frac{\partial w}{\partial\alpha}, \quad N_1 = -D_{11}\frac{1}{A^3}\frac{\partial^3 w}{\partial\alpha^3}, \quad (4.19)$$

$$T_1 = 0, \quad T_2 = \frac{2E_2 h}{R}w, \quad S_{12} = S_{21} = 0,$$

where W is a solution of the following equation

$$\frac{1}{A^4}\frac{\partial^4 w}{\partial \alpha^4} + \frac{3(1-\nu_1\nu_2)}{h^2 R_2^2}\frac{E_2}{E_1}w = 0. \tag{4.20}$$

In our case, in terms of the dimensional coordinates α, β it may be shown that $A = B = 1$, $R_2 = R$. After W has been determined, it may be substituted into (4.19). Then, satisfying conditions (4.9), one obtains

$$w = e^{-k\alpha}(C_1 \cos k\alpha + C_2 \sin k\alpha)\cos m\theta,$$

$$M_1 = -2D_{11}k^2 e^{-k\alpha}(-C_2 \cos k\alpha + C_1 \sin k\alpha)\cos m\theta,$$

$$M_2 = -2D_{12}k^2 e^{-k\alpha}(-C_2 \cos k\alpha + C_1 \sin k\alpha)\cos m\theta, \tag{4.21}$$

$$H_{12} = 2D_{66}\frac{mk}{R}e^{-k\alpha}((C_2 - C_1)\cos k\alpha - (C_1 + C_2)\sin k\alpha)\sin m\theta,$$

$$N_1 = -2D_{11}k^3 e^{-k\alpha}((C_2 + C_1)\cos k\alpha + (C_2 - C_1)\sin k\alpha)\cos m\theta,$$

$$T_1 = 0, \quad S_{12} = S_{21} = 0, \quad T_2 = \frac{2E_2 h}{R}w,$$

where

$$k = \sqrt[4]{\frac{E_2}{E_1}\frac{3(1-\nu_1\nu_2)}{(2h)^2 R^2}}, \quad C_1 = \frac{1}{\Delta}(\bar{M}_1^0(k-\omega_2) - \bar{H}_{12}^0(1-\omega_1\omega_2)),$$

$$C_2 = \frac{1}{\Delta}(\bar{H}_{12}^0\omega_1\omega_2 - \bar{M}_1^0(k+\omega_2)), \quad \Delta = 2k\omega_1\omega_2 - k - \omega_2, \tag{4.22}$$

$$\omega_1 = \bar{A}h\sqrt{\frac{G_{12}}{G_{23}}}, \quad \omega_2 = \frac{D_{66}}{D_{11}}\frac{m^2}{kR^2},$$

$$\bar{M}_1^0 = \frac{M_1^0}{2k^2 D_{11}}, \quad \bar{H}_{12}^0 = -\frac{m}{R}\frac{H_{12}^0}{2k^2 D_{11}}.$$

If we adopt $\omega_1 = 0$ in the presented solution (4.21), (4.22) it coincides with the solution provided by the classical shell theory. The effect of the refined terms grows with the increase of ω_1 (corresponding to a decrease of shear stiffness in the cross-sections) and increase in frequency of the external loading m. As m increases, it causes an increase of the quantity $\omega_1\omega_2$, which leads to a necessity of taking shear factors into account.

Now if one calculates the distribution of stresses of bending type along thickness in the near-edge vicinity, it is also likely that the solution of the outer problem differs from actual distribution given by (4.7), therefore, both plane and anti-plane boundary layer solutions arise for anti-symmetric problem as well. The leading order anti-plane boundary layer is governed by the conditions

$$\overset{a}{\sigma}_{xy}(\xi = 0) = \frac{1}{h^2}\left(H_{12}^0\frac{\psi_2}{\bar{\psi}_2} - \frac{3}{2}\zeta H_{12}(\xi = 0)\right)\sin m\theta. \tag{4.23}$$

The solution is given by

$$\overset{a}{u}_\beta = \Phi_1(\xi, \theta, \zeta), \quad \overset{a}{\sigma}_{\alpha\beta} = G_{12}\frac{\partial \Phi_1}{\partial t}, \quad \overset{a}{\sigma}_{\beta\gamma} = G_{23}\frac{\partial \Phi_1}{\partial \zeta},$$

$$\Phi_1 = \sum_{n=1}^{\infty} A_n \sin(2n-1)\frac{\pi}{2}\zeta \sin m\theta \exp\left(-\sqrt{\frac{G_{23}}{G_{12}}}(2n-1)\frac{\pi}{2}\frac{\alpha}{h}\right), \qquad (4.24)$$

$$A_n = -\frac{2}{\pi\sqrt{G_{12}G_{23}}}\frac{1}{2n-1}$$

$$\times \int_{-1}^{+1}\frac{1}{h^2}\left(H_{12}^0\frac{\psi_2}{\overline{\psi}_2} - \frac{3}{2}\zeta H_{12}(\alpha_0)\right)\sin(2n-1)\frac{\pi}{2}\zeta d\zeta.$$

The plane boundary layer solution is

$$\overset{f}{\sigma}_\alpha(\xi = 0) = \frac{M_1^0}{h^2}\left(\frac{\varphi_2}{\overline{\varphi}_2} - \frac{3}{2}\zeta\right)\cos m\theta, \quad \overset{f}{\sigma}_{\alpha\gamma}(\xi = 0) = 0. \qquad (4.25)$$

The solution corresponding to (4.25) is determined by the formulae (4.17), (4.18) with the only difference that the functions $F_n(\zeta)$ correspond to the anti-symmetric problem and are given by the formulae (3.5.25), (3.5.28), (3.5.31).

Thus, the asymptotic method allows to depict a complete picture of the distribution of the stress-strain field inside the shell, both very close to the boundary and away from the boundary. Note also that, if a simple boundary effect gives a representation of the decay of boundary effects in the mid-surface of the shell, then the boundary layer allows the restoration of the spatial pattern of the decay of stress fields near the boundary.

Let us discuss some of the refined theories, using the weakened conditions compared to the classical theory. In particular we concentrate on the Reissner-type shell theories.

Applied theory of Reissner in the case of isotropic plates has been comprehensively discussed in the works (Goldenveizer, 1958; Kolos, 1966). The analysis of the corresponding equations of this theory and comparison of the obtained data with the asymptotic method led to the conclusion that in the theories of Reissner type the refinements of the classical theory take into account the transverse shear, which however occurs away from the boundary of the plate. On the contrary, the asymptotic approach takes into consideration not only the phenomenon of transverse shear, but also the influence of self-equilibrated forces (along thickness) acting on the lateral surface of the plate or shell. An approximate theory of Reissner type at best incorporates the effects of boundary torsion on the outer stress-strain field, and, therefore, does not take into account the effect of plane boundary layer. These conclusions remain true for anisotropic plates and shells even to a larger extent. It was shown above that for small shear stiffness of sections perpendicular to the mid-surface, the anti-plane boundary layer may penetrate deep enough; furthermore the contribution of the shear terms to the outer solution, in view of (8.5.4), will be more important than the terms associated with (8.5.3). Therefore, only for such plates and shells, by Reissner theory one can obtain an admissible correction to the outer stress-strain field. Reissner type theory may also reasonably accurately describe the stress state near the boundary, in the case of zero plane boundary layer or boundary layer having less intensity than the outer solution. Let us discuss the spatial problem for the five different types of boundary conditions given above. In accordance with these, we consider the structure of the boundary stress-strain fields. According to Chapter 2, for the first boundary value problem the anti-plane boundary layer has a greater intensity than that of plane one $\overset{f}{P} = O(\lambda^{-p+c})\overset{a}{Q}$. In case of the pinned edge of first type ($\sigma_\alpha = u_\beta = u_\gamma = 0$) the plane symmetric boundary layer is non-zero, with the corresponding quantities being of order $\overset{f}{P} = O(\lambda^{q+d})$. The calculation of anti-plane boundary layer exceeds the limit of accuracy of the classical theory, and the corresponding quantities have orders of $\overset{a}{Q} = O(\lambda^{-2q+3p+d})$. Thus, in this case, using the Reissner theory, we obtain incorrect results.

In case of mixed boundary conditions of the second type ($\sigma_\alpha = \sigma_{\alpha\beta} = u_\gamma = 0$) both for the plane and anti-plane boundary layers are different from zero and have intensity of order $\overset{f}{P} = O(\lambda^{q+d})$, $\overset{a}{Q} = O(\lambda^{2p-c+d})$. The Reissner theory does not apply here since it most likely will bring errors rather than elucidating the classical theory. This is because, in view of (2.42), the theory does not take into account the influence of the plane boundary layer for the outer stress-strain field. The Reissner theory, in the present case, gives admissible results only if $\dfrac{E_1}{E_3} < 1$, and at the same time $\sqrt{\dfrac{G_{12}}{G_{23}}} \gg 1$, i.e. when the cross sections perpendicular to the mid-surface have higher tensile stiffness, but lower shear stiffness.

In the case of rigidly clamped boundary condition, the calculation of boundary layer exceeds the highest accuracy with the intensity of order $O(\lambda^{-2q+3p+d})$, with the non-zero plane boundary layer solution of intensity $\overset{f}{P} = O(\lambda^{q+d})$. Thus, near a rigidly clamped boundary the Reissner type theory gives completely inaccurate results. Moreover, according to this theory we do not get any refinement of the outer solution.

As for the mixed boundary conditions of the third type ($u_\alpha = \sigma_{\alpha\beta} = u_\gamma = 0$), both anti-plane and plane boundary layers are non-zero, being of order $\overset{a}{Q} = O(\lambda^{2p-c+d})$, $\overset{f}{P} = O(\lambda^{q+d})$, respectively. Since the Reissner type theories may only incorporate the anti-plane boundary layer effects, it is not expected to provide correct results in the near-edge vicinity, since both plane and anti-plane boundary layer solutions have intensities of the same order.

The theory of Naghdi (1956), which is a Timoshenko-type theory, has been analyzed by Rogacheva (1974). It has been demonstrated that in case of shells resisting weakly to shear the Naghdi theory may be justified from asymptotic approach. This conclusion remains true for static analysis of anisotropic shells. More general results on this problem has been obtained by Goldenveizer et al. (1993).

Thus, all of the considered refined theories based on weakened assumptions, do not allow uniform accuracy (from the interior to the edge) of the transition from original 3D problem for elastic shell to approximate 2D formulation. In case of static problems of shells these theories allow refinement of the outer stress-strain field only, being valid only for shells which demonstrate weak resistance to shear in the cross sections which are perpendicular to the mid-surface of the shell. Some of these refined theories are capable of incorporating the anti-plane (torsional) boundary layer effects.

Chapter 10

Non-Classical Boundary Value Problems for Anisotropic Shells

10.1 Governing Equations and the Outer Solution of the Boundary Value Problems for an Elastic Shell

The asymptotic method may also be applied to solution of a novel class of boundary value problems for generally anisotropic shells, see Aghalovyan and Gevorkyan (1989).

Consider an anisotropic shell of thickness $2h : \Omega = \{\alpha, \beta, \gamma; \alpha, \beta \in \Omega_0, -h \leq \gamma \leq h\}$. Let the mid-surface Ω_0 be combined with the curvature lines α, β, with γ axis being perpendicular to the mid-surface, see Fig. 10.1. We also assume that the elasto-equivalent directions coincide with the coordinate lines at any point. The anisotropy of the shell is characterized by 21 elastic constants. The volume forces $F_\alpha(\alpha, \beta, \gamma)$ (α, β, γ) are also taken into consideration, with temperature variation modelled through the Duhamel-Neumann law. The variation of temperature field is described by the function $\theta(\alpha, \beta, \gamma) = T(\alpha, \beta, \gamma) - T_0(\alpha, \beta, \gamma)$ which is supposed to be known. In order to reduce excessive calculations we shall operate in terms of the non-symmetric stress tensor τ_{ij}. The relations between the symmetric tensor σ_{ij} and the non-symmetric stresses τ_{ij} are written as

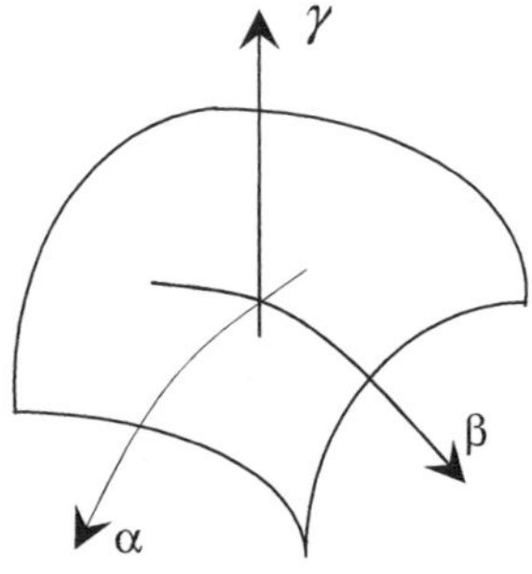

Fig. 10.1

$$\tau_{\alpha\alpha} = (1 + \gamma/R_2)\sigma_{\alpha\alpha}, \qquad \tau_{\beta\beta} = (1 + \gamma/R_1)\sigma_{\beta\beta},$$
$$\tau_{\gamma\gamma} = (1 + \gamma/R_1)(1 + \gamma/R_2)\sigma_{\gamma\gamma},$$
$$\tau_{\alpha\gamma} = (1 + \gamma/R_2)\sigma_{\alpha\gamma}, \qquad \tau_{\beta\gamma} = (1 + \gamma/R_1)\sigma_{\beta\gamma}, \tag{1.1}$$
$$\tau_{\alpha\beta} = (1 + \gamma/R_2)\sigma_{\alpha\beta}, \qquad \tau_{\beta\alpha} = (1 + \gamma/R_1)\sigma_{\alpha\beta}.$$

Since $\sigma_{\alpha\beta} = \sigma_{\beta\alpha}$, it may be deduced from (1.1) that

$$(1 + \gamma/R_1)\tau_{\alpha\beta} = (1 + \gamma/R_2)\tau_{\beta\alpha}. \tag{1.2}$$

The governing system of equations of elasticity with incorporated thermal deformations may be expressed in the chosen tri-orthogonal coordinate system as follows. The equations of equilibrium

$$\frac{1}{AB}\frac{\partial}{\partial\alpha}(B\tau_{\alpha\alpha}) - k_\beta\tau_{\beta\beta} + \frac{1}{AB}\frac{\partial}{\partial\beta}(A\tau_{\beta\alpha})$$
$$+ k_\alpha\tau_{\alpha\beta} + \left(1 + \frac{\gamma}{R_1}\right)\frac{\partial\tau_{\alpha\gamma}}{\partial\gamma} + \frac{2\tau_{\alpha\gamma}}{R_1} + F_\alpha^* = 0,$$
$$\frac{1}{AB}\frac{\partial}{\partial\beta}(A\tau_{\beta\beta}) - k_\alpha\tau_{\alpha\alpha} + \frac{1}{AB}\frac{\partial}{\partial\alpha}(B\tau_{\alpha\beta}) \tag{1.3}$$
$$+ k_\beta\tau_{\beta\alpha} + \left(1 + \frac{\gamma}{R_2}\right)\frac{\partial\tau_{\beta\gamma}}{\partial\gamma} + \frac{2\tau_{\beta\gamma}}{R_2} + F_\beta^* = 0,$$
$$\frac{\partial\tau_{\gamma\gamma}}{\partial\gamma} - \left(\frac{\tau_{\alpha\alpha}}{R_1} + \frac{\tau_{\beta\beta}}{R_2}\right) + \frac{1}{A}\frac{\partial\tau_{\alpha\gamma}}{\partial\alpha},$$
$$+ \frac{1}{B}\frac{\partial\tau_{\beta\gamma}}{\partial\beta} + k_\beta\tau_{\alpha\gamma} + k_\alpha\tau_{\beta\gamma} + F_\gamma^* = 0.$$

The constitutive relations

$$\left(1 + \frac{\gamma}{R_2}\right)\left(\frac{1}{A}\frac{\partial u_\alpha}{\partial\alpha} + k_\alpha u_\beta + \frac{u_\gamma}{R_1}\right)$$
$$= \left(1 + \frac{\gamma}{R_1}\right)(a_{11}\tau_{\alpha\alpha} + a_{15}\tau_{\alpha\gamma} + a_{16}\tau_{\alpha\beta})$$
$$+ \left(1 + \frac{\gamma}{R_2}\right)(a_{12}\tau_{\beta\beta} + a_{14}\tau_{\beta\gamma})$$
$$+ a_{13}\tau_{\gamma\gamma} + \alpha_{11}\left(1 + \gamma\left(\frac{1}{R_1} + \frac{1}{R_2}\right) + \gamma^2\frac{1}{R_1 R_2}\right)\theta,$$
$$\left(1 + \frac{\gamma}{R_1}\right)\left(\frac{1}{B}\frac{\partial u_\beta}{\partial\beta} + k_\beta u_\alpha + \frac{u_\gamma}{R_2}\right)$$
$$= \left(1 + \frac{\gamma}{R_1}\right)(a_{12}\tau_{\alpha\alpha} + a_{25}\tau_{\alpha\gamma} + a_{26}\tau_{\alpha\beta})$$
$$+ \left(1 + \frac{\gamma}{R_2}\right)(a_{22}\tau_{\beta\beta} + a_{24}\tau_{\beta\gamma})$$
$$+ a_{23}\tau_{\gamma\gamma} + \alpha_{22}\left(1 + \gamma\left(\frac{1}{R_1} + \frac{1}{R_2}\right) + \gamma^2\frac{1}{R_1 R_2}\right)\theta,$$

$$\left[1 + \gamma\left(\frac{1}{R_1} + \frac{1}{R_2}\right) + \frac{\gamma^2}{R_1 R_2}\right]\frac{\partial u_\gamma}{\partial \gamma}$$

$$= \left(1 + \frac{\gamma}{R_1}\right)(a_{13}\tau_{\alpha\alpha} + a_{35}\tau_{\alpha\gamma} + a_{36}\tau_{\alpha\beta})$$

$$+ \left(1 + \frac{\gamma}{R_2}\right)(a_{23}\tau_{\beta\beta} + a_{34}\tau_{\beta\gamma})$$

$$+ a_{33}\tau_{\gamma\gamma} + \alpha_{33}\left(1 + \gamma\left(\frac{1}{R_1} + \frac{1}{R_2}\right) + \gamma^2\frac{1}{R_1 R_2}\right)\theta,$$

$$\left(1 + \frac{\gamma}{R_1}\right)\left(\frac{1}{B}\frac{\partial u_\alpha}{\partial \beta} - k_\beta u_\beta\right)$$

$$+ \left(1 + \frac{\gamma}{R_2}\right)\left(\frac{1}{A}\frac{\partial u_\beta}{\partial \alpha} - k_\alpha u_\alpha\right)$$

$$= \left(1 + \frac{\gamma}{R_1}\right)(a_{16}\tau_{\alpha\alpha} + a_{56}\tau_{\alpha\gamma} + a_{66}\tau_{\alpha\beta})$$

$$+ a_{36}\tau_{\gamma\gamma} + \left(1 + \frac{\gamma}{R_2}\right)(a_{26}\tau_{\beta\beta} + a_{46}\tau_{\beta\gamma}) \tag{1.4}$$

$$+ \alpha_{12}\left(1 + \gamma\left(\frac{1}{R_1} + \frac{1}{R_2}\right) + \gamma^2\frac{1}{R_1 R_2}\right)\theta,$$

$$\left[1 + \gamma\left(\frac{1}{R_1} + \frac{1}{R_2}\right) + \frac{\gamma^2}{R_1 R_2}\right]\frac{\partial u_\alpha}{\partial \gamma}$$

$$- \left(1 + \frac{\gamma}{R_2}\right)\frac{u_\alpha}{R_1} + \frac{1}{A}\left(1 + \frac{\gamma}{R_2}\right)\frac{\partial u_\gamma}{\partial \alpha}$$

$$= \left(1 + \frac{\gamma}{R_1}\right)(a_{15}\tau_{\alpha\alpha} + a_{55}\tau_{\alpha\gamma} + a_{56}\tau_{\alpha\beta})$$

$$+ a_{35}\tau_{\gamma\gamma} + \left(1 + \frac{\gamma}{R_2}\right)(a_{25}\tau_{\beta\beta} + a_{45}\tau_{\beta\gamma})$$

$$+ \alpha_{13}\left(1 + \gamma\left(\frac{1}{R_1} + \frac{1}{R_2}\right) + \gamma^2\frac{1}{R_1 R_2}\right)\theta,$$

$$\left[1 + \gamma\left(\frac{1}{R_1} + \frac{1}{R_2}\right) + \frac{\gamma^2}{R_1 R_2}\right]\frac{\partial u_\beta}{\partial \gamma}$$

$$- \left(1 + \frac{\gamma}{R_1}\right)\frac{u_\beta}{R_2} + \frac{1}{B}\left(1 + \frac{\gamma}{R_1}\right)\frac{\partial u_\gamma}{\partial \beta}$$

$$= \left(1 + \frac{\gamma}{R_1}\right)(a_{14}\tau_{\alpha\alpha} + a_{45}\tau_{\alpha\gamma} + a_{46}\tau_{\alpha\beta})$$

$$+ a_{34}\tau_{\gamma\gamma} + \left(1 + \frac{\gamma}{R_2}\right)(a_{24}\tau_{\beta\beta} + a_{44}\tau_{\beta\gamma})$$

$$+ \alpha_{23}\left(1 + \gamma\left(\frac{1}{R_1} + \frac{1}{R_2}\right) + \gamma^2\frac{1}{R_1 R_2}\right)\theta,$$

where $k_\alpha = \dfrac{1}{AB}\dfrac{\partial A}{\partial \beta}$, $k_\beta = \dfrac{1}{AB}\dfrac{\partial B}{\partial \alpha}$ are the geodesic curvatures; A, B are the coefficients of the first quadratic form of the coordinate surface; R_1, R_2 are the main radii of curvature of the mid-surface; α_{ik} are coefficients of thermal expansion; θ is temperature variation, and $F_\alpha^* = \left(1 + \dfrac{\gamma}{R_1}\right)\left(1 + \dfrac{\gamma}{R_2}\right) F_\alpha(\alpha, \beta, \gamma)$, (α, β, γ).

The solution of the system (1.3), (1.4) is sought subject to the conditions at the face $\gamma = -h$ formulated in terms of displacements

$$u_\alpha(-h) = u^-, \quad u_\beta(-h) = v^-, \quad u_\gamma(-h) = w^- \tag{1.5}$$

with the boundary conditions at the opposite face $\gamma = h$ expressed in terms of either stresses

$$\sigma_{\alpha\gamma}(h) = \varepsilon^{-1}\sigma_{\alpha\gamma}^+, \quad \sigma_{\beta\gamma}(h) = \varepsilon^{-1}\sigma_{\beta\gamma}^+, \quad \sigma_{\gamma\gamma}(h) = \varepsilon^{-1}\sigma_{\gamma\gamma}^+ \tag{1.6}$$

or displacements

$$u_\alpha(h) = u^+, \quad u_\beta(h) = v^+, \quad u_\gamma(h) = w^+ \tag{1.7}$$

or in mixed form, namely

$$u_\alpha(h) = u^+, \quad u_\beta(h) = v^+, \quad \sigma_{\gamma\gamma}(h) = \varepsilon^{-1}\sigma_{\gamma\gamma}^+ \tag{1.8}$$

or

$$\sigma_{\alpha\gamma}(h) = \varepsilon^{-1}\sigma_{\alpha\gamma}^+, \quad \sigma_{\beta\gamma}(h) = \varepsilon^{-1}\sigma_{\beta\gamma}^+, \quad u_\gamma(h) = w^+. \tag{1.9}$$

The edge boundary conditions at the side surface $\partial\Omega$ of the shell are not specified at the moment, since they do not affect the outer solution, similarly to plates. We emphasize once again that the edge boundary conditions influence the boundary layer solutions only. At the same time it is worth reminding the reader that within the classical shell theory the edge boundary conditions have direct influence on the outer solution.

The formulated boundary value problems differ from these of the classical shell theory, since different boundary conditions are imposed on the faces $\gamma = \pm h$. The formulated boundary value problems will be referred to as non-classical due to qualitative difference of the solutions. One of the differences was already mentioned above, some more features will be discussed below.

Let us now introduce the dimensionless scaling

$$\alpha = R\xi, \quad \beta = R\eta, \quad \gamma = \varepsilon R\zeta = h\zeta, \tag{1.10}$$
$$u_\alpha = Ru, \quad u_\beta = Rv, \quad u_\gamma = Rw,$$

where R is a typical scale of the shell (could be taken as a minimum of the curvature radius and linear scales of the coordinate surfaces), $\varepsilon = h/R$ is a small parameter. The transformed system is singularly perturbed by a small parameter, therefore, the outer solution is now traditionally sought-for in the form

$$Q = \varepsilon^{\chi_Q + s} Q^{(s)} \qquad s = \overline{0, S} \tag{1.11}$$

where Q is any of the sought for quantities. Following the usual procedure, a consistent system for $Q^{(s)}$ may be obtained for $\chi_u = 0$ in respect of displacement components, and $\chi_\tau = -1$ for stress components. Let us assume that the contribution of the volume forces and thermal field to the total stress-strain field is comparable to that of surface forces. In other words, this corresponds to a situation when the terms arising due to volume forces and thermal variation, appear at leading order approximation. This occurs if and only if the appropriate quantities allow the following expansions

$$F_\alpha = \varepsilon^{-2+s} F_\alpha^{(s)}, \qquad (\alpha, \beta, \gamma), \tag{1.12}$$

$$\theta = \varepsilon^{-1+s} \theta^{(s)} \qquad s = \overline{0, S}.$$

Substituting expansions (1.11), (1.12) into the transformed system (1.3), (1.4), following a standard asymptotic procedure, we arrive at a consistent system in respect of $Q^{(s)}$. The solution of this system may be presented as

$$\tau_{\alpha\gamma}^{(s)}(\xi, \eta, \zeta) = \tau_{\alpha\gamma 0}^{(s)}(\xi, \eta) + \tau_{\alpha\gamma *}^{(s)}(\xi, \eta, \zeta),$$

$$\tau_{\beta\gamma}^{(s)}(\xi, \eta, \zeta) = \tau_{\beta\gamma 0}^{(s)}(\xi, \eta) + \tau_{\beta\gamma *}^{(s)}(\xi, \eta, \zeta),$$

$$\tau_{\gamma\gamma}^{(s)}(\xi, \eta, \zeta) = \tau_{\gamma\gamma 0}^{(s)}(\xi, \eta) + \tau_{\gamma\gamma *}^{(s)}(\xi, \eta, \zeta),$$

$$\tau_{\alpha\alpha}^{(s)}(\xi, \eta, \zeta) = A_{13}\tau_{\gamma\gamma 0}^{(s)} + A_{14}\tau_{\beta\gamma 0}^{(s)} + A_{15}\tau_{\alpha\gamma 0}^{(s)} + \tau_{\alpha\alpha *}^{(s)}(\xi, \eta, \zeta),$$

$$\tau_{\beta\beta}^{(s)}(\xi, \eta, \zeta) = A_{23}\tau_{\gamma\gamma 0}^{(s)} + A_{24}\tau_{\beta\gamma 0}^{(s)} + A_{25}\tau_{\alpha\gamma 0}^{(s)} + \tau_{\beta\beta *}^{(s)}(\xi, \eta, \zeta), \tag{1.13}$$

$$\tau_{\alpha\beta}^{(s)}(\xi, \eta, \zeta) = A_{63}\tau_{\gamma\gamma 0}^{(s)} + A_{64}\tau_{\beta\gamma 0}^{(s)} + A_{65}\tau_{\alpha\gamma 0}^{(s)} + \tau_{\alpha\beta *}^{(s)}(\xi, \eta, \zeta),$$

$$u^{(s)}(\xi, \eta, \zeta) = \zeta(A_{53}\tau_{\gamma\gamma 0}^{(s)} + A_{54}\tau_{\beta\gamma 0}^{(s)} + A_{55}\tau_{\alpha\gamma 0}^{(s)})$$
$$+ u_0^{(s)}(\xi, \eta) + u_*^{(s)}(\xi, \eta, \zeta),$$

$$v^{(s)}(\xi, \eta, \zeta) = \zeta(A_{43}\tau_{\gamma\gamma 0}^{(s)} + A_{44}\tau_{\beta\gamma 0}^{(s)} + A_{45}\tau_{\alpha\gamma 0}^{(s)})$$
$$+ v_0^{(s)}(\xi, \eta) + v_*^{(s)}(\xi, \eta, \zeta),$$

$$w^{(s)}(\xi, \eta, \zeta) = \zeta(A_{33}\tau_{\gamma\gamma 0}^{(s)} + A_{34}\tau_{\beta\gamma 0}^{(s)} + A_{35}\tau_{\alpha\gamma 0}^{(s)})$$
$$+ w_0^{(s)}(\xi, \eta) + w_*^{(s)}(\xi, \eta, \zeta).$$

The quantities $\tau_{\alpha\gamma 0}^{(s)}, \tau_{\beta\gamma 0}^{(s)}, \tau_{\gamma\gamma 0}^{(s)}, u_0^{(s)} v_0^{(s)}, w_0^{(s)}$ are still to be determined, whereas in respect of the quantities $Q_*^{(s)}$ we obtain the following recurrent relations

$$\tau_{\alpha\alpha *}^{(s)}(\xi, \eta, \zeta) = B_{11}P_1^{(s)} + B_{12}P_2^{(s)} + B_{16}P_3^{(s)},$$

$$\tau_{\beta\beta *}^{(s)}(\xi, \eta, \zeta) = B_{12}P_1^{(s)} + B_{22}P_2^{(s)} + B_{26}P_3^{(s)},$$

$$\tau_{\alpha\beta *}^{(s)}(\xi, \eta, \zeta) = B_{16}P_1^{(s)} + B_{26}P_2^{(s)} + B_{66}P_3^{(s)},$$

$$\tau_{\alpha\gamma *}^{(s)}(\xi, \eta, \zeta) = -\int_0^\zeta \left(\frac{1}{AB}\frac{\partial}{\partial \xi}(B\tau_{\alpha\alpha}^{(s-1)}) + \frac{1}{AB}\frac{\partial}{\partial \eta}(A\tau_{\beta\alpha}^{(s-1)}) \right.$$

$$+ R\left(k_\alpha \tau_{\alpha\beta}^{(s-1)} - k_\beta \tau_{\beta\beta}^{(s-1)} \right) + RF_\alpha^{(s)} + r_1 \left(\zeta \frac{\partial \tau_{\alpha\gamma}^{(s-1)}}{\partial \zeta} + 2\tau_{\alpha\gamma}^{(s-1)} \right)$$

$$\left. + \zeta R(r_1 + r_2)F_\alpha^{(s-1)} + \zeta^2 R r_1 r_2 F_a^{(s-2)} \right) d\zeta, \qquad (\alpha, \beta; \xi, \eta; r_1, r_2; A, B),$$

$$\tau_{\gamma\gamma*}^{(s)} = \int_0^\zeta \left(\frac{1}{A} \frac{\partial \tau_{\alpha\gamma}^{(s-1)}}{\partial \xi} + \frac{1}{B} \frac{\partial \tau_{\beta\gamma}^{(s-1)}}{\partial \eta} \right.$$

$$+R\left(k_\alpha \tau_{\beta\gamma}^{(s-1)} + k_\beta \tau_{\alpha\gamma}^{(s-1)} \right) - r_1 \tau_{\alpha\alpha}^{(s-1)} - r_2 \tau_{\beta\beta}^{(s-1)}$$

$$\left. +RF_\gamma^{(s)} + \zeta R(r_1 + r_2)F_\gamma^{(s-1)} + \zeta^2 R r_1 r_2 F_\gamma^{(s-2)} \right) d\zeta,$$

$$P_1^{(s)} = \frac{1}{A} \frac{\partial u^{(s-1)}}{\partial \xi} + Rk_\alpha v^{(s-1)} + r_1 w^{(s-1)}$$

$$-a_{13}\tau_{\gamma\gamma*}^{(s)} - a_{14}\tau_{\beta\gamma*}^{(s)} - a_{15}\tau_{\alpha\gamma*}^{(s)} - \alpha_{11}\theta^{(s)}$$

$$+\zeta r_2 \left(\frac{1}{A} \frac{\partial u^{(s-2)}}{\partial \xi} + Rk_\alpha v^{(s-2)} + r_1 w^{(s-2)} \right)$$

$$-\zeta(a_{11}r_1\tau_{\alpha\alpha}^{(s-1)} + a_{12}r_2\tau_{\beta\beta}^{(s-1)}$$

$$+a_{14}r_2\tau_{\beta\gamma}^{(s-1)} + a_{15}r_1\tau_{\alpha\gamma}^{(s-1)} + a_{16}r_1\tau_{\alpha\beta}^{(s-1)})$$

$$-\alpha_{11}\zeta(r_1 + r_2)\theta^{(s-1)} - \alpha_{11}\zeta^2 r_1 r_2 \theta^{(s-2)}, \qquad (1.14)$$

$$P_2^{(s)} = \frac{1}{B} \frac{\partial v^{(s-1)}}{\partial \eta} + Rk_\beta u^{(s-1)} + r_2 w^{(s-1)}$$

$$-a_{23}\tau_{\gamma\gamma*}^{(s)} - a_{24}\tau_{\beta\gamma*}^{(s)} - a_{25}\tau_{\alpha\gamma*}^{(s)} - \alpha_{22}\theta^{(s)}$$

$$+\zeta r_1 \left(\frac{1}{B} \frac{\partial v^{(s-2)}}{\partial \eta} + Rk_\beta u^{(s-2)} + r_2 w^{(s-2)} \right)$$

$$-\zeta(a_{11}r_1\tau_{\alpha\alpha}^{(s-1)} + a_{22}r_2\tau_{\beta\beta}^{(s-1)}$$

$$+a_{24}r_2\tau_{\beta\gamma}^{(s-1)} + a_{25}r_1\tau_{\alpha\gamma}^{(s-1)} + a_{26}r_1\tau_{\alpha\beta}^{(s-1)})$$

$$-\alpha_{22}\zeta(r_1 + r_2)\theta^{(s-1)} - \alpha_{22}\zeta^2 r_1 r_2 \theta^{(s-2)},$$

$$P_3^{(s)} = \frac{1}{B} \frac{\partial u^{(s-1)}}{\partial \eta} - Rk_\beta v^{(s-1)} + \frac{1}{A} \frac{\partial v^{(s-1)}}{\partial \xi}$$

$$-Rk_\alpha u^{(s-1)} - a_{36}\tau_{\gamma\gamma*}^{(s)} - a_{46}\tau_{\beta\gamma*}^{(s)} - a_{56}\tau_{\alpha\gamma*}^{(s)}$$

$$-\alpha_{11}\theta^{(s)} + \zeta r_1 \left(\frac{1}{B} \frac{\partial u^{(s-2)}}{\partial \eta} - Rk_\beta v^{(s-2)} \right)$$

$$+\zeta r_2 \left(\frac{1}{A} \frac{\partial v^{(s-2)}}{\partial \xi} - Rk_\alpha u^{(s-2)} \right)$$

$$-\zeta(a_{16}r_1\tau_{\alpha\alpha}^{(s-1)} + a_{26}r_2\tau_{\beta\beta}^{(s-1)}$$

$$+a_{46}r_2\tau_{\beta\gamma}^{(s-1)} + a_{56}r_1\tau_{\alpha\gamma}^{(s-1)} + a_{66}r_1\tau_{\alpha\beta}^{(s-1)})$$

$$-\alpha_{12}\zeta(r_1 + r_2)\theta^{(s-1)} - \alpha_{12}\zeta^2 r_1 r_2 \theta^{(s-2)},$$

$$u_*^{(s)} = \int_0^\zeta \left(a_{15}\tau_{\alpha\alpha*}^{(s)} + a_{25}\tau_{\beta\beta*}^{(s)} + a_{35}\tau_{\gamma\gamma*}^{(s)} \right.$$

$$\left. +a_{45}\tau_{\beta\gamma*}^{(s)} + a_{55}\tau_{\alpha\gamma*}^{(s)} + a_{56}\tau_{\alpha\beta*}^{(s)} + \alpha_{13}\theta^{(s)} \right.$$

$$-\frac{1}{A} \frac{\partial w^{(s-1)}}{\partial \xi} + r_1 u^{(s-1)} + \zeta(a_{15}r_1\tau_{\alpha\alpha}^{(s-1)} + a_{25}r_2\tau_{\beta\beta}^{(s-1)}$$

$$+a_{45}r_2\tau_{\beta\gamma}^{(s-1)} + a_{55}r_1\tau_{\alpha\gamma}^{(s-1)} + a_{56}r_1\tau_{\alpha\beta}^{(s-1)}$$

$$+(r_1+r_2)\left(\alpha_{13}\theta^{(s-1)} - \frac{\partial u^{(s-1)}}{\partial\zeta}\right)$$

$$-r_2\frac{1}{A}\frac{\partial w^{(s-2)}}{\partial\xi} + r_1r_2 u^{(s-2)})$$

$$+\zeta^2 r_1 r_2\left(\alpha_{13}\theta^{(s-2)} - \frac{\partial u^{(s-2)}}{\partial\zeta}\right)\Big)\,d\zeta,$$

$$v_*^{(s)} = \int_0^\zeta \Big(a_{14}\tau_{\alpha\alpha*}^{(s)} + a_{24}\tau_{\beta\beta*}^{(s)} + a_{34}\tau_{\gamma\gamma*}^{(s)}$$

$$+a_{44}\tau_{\beta\gamma*}^{(s)} + a_{45}\tau_{\alpha\gamma*}^{(s)} + a_{46}\tau_{\alpha\beta*}^{(s)} + \alpha_{23}\theta^{(s)}$$

$$-\frac{1}{B}\frac{\partial w^{(s-1)}}{\partial\eta} + r_2 v^{(s-1)} + \zeta(a_{14}r_1\tau_{\alpha\alpha}^{(s-1)} + a_{24}r_2\tau_{\beta\beta}^{(s-1)}$$

$$+a_{44}r_2\tau_{\beta\gamma}^{(s-1)} + a_{45}r_1\tau_{\alpha\gamma}^{(s-1)} + a_{46}r_1\tau_{\alpha\beta}^{(s-1)}$$

$$+(r_1+r_2)\left(\alpha_{23}\theta^{(s-1)} - \frac{\partial v^{(s-1)}}{\partial\zeta}\right)$$

$$-r_1\frac{1}{B}\frac{\partial w^{(s-2)}}{\partial\eta} + r_1r_2 v^{(s-2)})$$

$$+\zeta^2 r_1 r_2\left(\alpha_{23}\theta^{(s-2)} - \frac{\partial v^{(s-2)}}{\partial\zeta}\right)\Big)\,d\zeta,$$

$$w_*^{(s)} = \int_0^\zeta \Big(a_{13}\tau_{\alpha\alpha*}^{(s)} + a_{23}\tau_{\beta\beta*}^{(s)} + a_{33}\tau_{\gamma\gamma*}^{(s)}$$

$$+a_{34}\tau_{\beta\gamma*}^{(s)} + a_{35}\tau_{\alpha\gamma*}^{(s)} + a_{36}\tau_{\alpha\beta*}^{(s)} + \alpha_{33}\theta^{(s)}$$

$$+\zeta(a_{13}r_1\tau_{\alpha\alpha}^{(s-1)} + a_{23}r_2\tau_{\beta\beta}^{(s-1)}$$

$$+a_{34}r_2\tau_{\beta\gamma}^{(s-1)} + a_{35}r_1\tau_{\alpha\gamma}^{(s-1)} + a_{36}r_1\tau_{\alpha\beta}^{(s-1)}$$

$$+\zeta(r_1+r_2)\left(\alpha_{33}\theta^{(s-1)} - \frac{\partial w^{(s-1)}}{\partial\zeta}\right)$$

$$+\zeta^2 r_1 r_2\left(\alpha_{33}\theta^{(s-2)} - \frac{\partial w^{(s-2)}}{\partial\zeta}\right)\Big)\,d\zeta.$$

$$r_i = \frac{R}{R_i}, \qquad i = 1,2.$$

Here

$$B_{jk} = (a_{jl}a_{kl} - a_{jk}a_{ll})/\Delta, \qquad B_{jk} = B_{kj},$$

$$B_{ll} = (a_{jj}a_{kk} - a_{jk}^2)/\Delta, \quad (j \neq k \neq l \neq j; j, k, l = 1, 2, 6),$$

$$A_{lm} = -a_{1m}B_{l1} - a_{2m}B_{l2} - a_{6m}B_{l6}, \qquad A_{mn} \neq A_{nm}, \qquad (1.15)$$

$$A_{nm} = a_{n1}A_{1m} + a_{n2}A_{2m} + a_{n6}A_{6m} + a_{nm}, \qquad (n, m = 3, 4, 5),$$

$$\Delta = a_{11}a_{22}a_{66} + 2a_{12}a_{26}a_{16} - a_{11}a_{26}^2 - a_{22}a_{16}^2 - a_{66}a_{12}^2.$$

Using the iterative relations (1.14), it is possible to obtain solutions for all of the quantities $Q_*^{(s)}$ for all s, relying on the known quantities of the previous approximations. Thus, the general solution of the outer non-classical boundary value problem for shells is given by formulae (1.11), (1.13), and (1.14), expressed in terms of six unknown functions, which may be determined from the boundary conditions (1.5)-(1.9).

10.2 Mixed Boundary Value Problems

Since the boundary conditions (1.5) are common for all the formulated boundary value problems, let us consider these first, using (1.13), leading to

$$\tau_{\alpha\gamma}^{(s)}(\xi, \eta, \zeta) = \tau_{\alpha\gamma 0}^{(s)}(\xi, \eta) + \tau_{\alpha\gamma *}^{(s)}(\xi, \eta, \zeta),$$

$$\tau_{\beta\gamma}^{(s)}(\xi, \eta, \zeta) = \tau_{\beta\gamma 0}^{(s)}(\xi, \eta) + \tau_{\beta\gamma *}^{(s)}(\xi, \eta, \zeta),$$

$$\tau_{\gamma\gamma}^{(s)}(\xi, \eta, \zeta) = \tau_{\gamma\gamma 0}^{(s)}(\xi, \eta) + \tau_{\gamma\gamma *}^{(s)}(\xi, \eta, \zeta),$$

$$\tau_{\alpha\alpha}^{(s)} = A_{13}\tau_{\gamma\gamma 0}^{(s)} + A_{14}\tau_{\beta\gamma 0}^{(s)} + A_{15}\tau_{\alpha\gamma 0}^{(s)} + \tau_{\alpha\alpha *}^{(s)}(\xi, \eta, \zeta),$$

$$\tau_{\beta\beta}^{(s)} = A_{23}\tau_{\gamma\gamma 0}^{(s)} + A_{24}\tau_{\beta\gamma 0}^{(s)} + A_{25}\tau_{\alpha\gamma 0}^{(s)} + \tau_{\beta\beta *}^{(s)}(\xi, \eta, \zeta), \qquad (2.1)$$

$$\tau_{\alpha\beta}^{(s)} = A_{63}\tau_{\gamma\gamma 0}^{(s)} + A_{64}\tau_{\beta\gamma 0}^{(s)} + A_{65}\tau_{\alpha\gamma 0}^{(s)} + \tau_{\alpha\beta *}^{(s)}(\xi, \eta, \zeta),$$

$$u^{(s)} = (1 + \zeta)\left(A_{53}\tau_{\gamma\gamma 0}^{(s)} + A_{54}\tau_{\beta\gamma 0}^{(s)} + A_{55}\tau_{\alpha\gamma 0}^{(s)}\right)$$

$$+ u^{-(s)}(\xi, \eta) - u_*^{(s)}(\zeta = -1) + u_*^{(s)}(\xi, \eta, \zeta),$$

$$u^{-(0)} = u^-/R, \quad u^{-(s)} = 0, \quad s \neq 0,$$

$$(u, v, w; A_{5k}, A_{4k}, A_{3k}; k = 3, 4, 5).$$

Therefore, the only unknown quantities are now $\tau_{\alpha\gamma 0}^{(s)}, \tau_{\beta\gamma 0}^{(s)}, \tau_{\gamma\gamma 0}^{(s)}$, which have to be determined for each boundary value problem using the conditions (1.6)-(1.9). Let us consider these step by step.

  In case of the first boundary value problem at the face $\gamma = h$ we obtain

$$\tau_{\alpha\gamma 0}^{(s)}(\xi, \eta) = \tau_{\alpha\gamma}^{+(s)} - \tau_{\alpha\gamma *}^{(s)}(\zeta = 1),$$

$$\tau_{\beta\gamma 0}^{(s)}(\xi, \eta) = \tau_{\beta\gamma}^{+(s)} - \tau_{\beta\gamma *}^{(s)}(\zeta = 1), \qquad (2.2)$$

$$\tau_{\gamma\gamma 0}^{(s)}(\xi, \eta) = \tau_{\gamma\gamma}^{+(s)} - \tau_{\gamma\gamma *}^{(s)}(\zeta = 1),$$

where

$$\tau_{\gamma\gamma}^{+(0)} = \sigma_{\gamma\gamma}^{+}, \quad \tau_{\gamma\gamma}^{+(1)} = (r_1 + r_2)\sigma_{\gamma\gamma}^{+},$$

$$\tau_{\gamma\gamma}^{+(2)} = r_1 r_2 \sigma_{\gamma\gamma}^{+}, \quad \tau_{\gamma\gamma}^{+(s)} = 0, \quad s > 2, \tag{2.3}$$

$$\tau_{\alpha\gamma}^{+(0)} = \sigma_{\alpha\gamma}^{+}, \quad \tau_{\alpha\gamma}^{+(1)} = r_2 \sigma_{\alpha\gamma}^{+}, \quad \tau_{\alpha\gamma}^{+(s)} = 0, \quad s > 1,$$

$$(\alpha, \beta; r_1, r_2).$$

Substitution of the conditions (2.1) and (2.2) into equations (1.11) and (1.13) leads to the final solution of the outer boundary value problem (1.5), (1.6).

In case of the boundary conditions (1.7) we get

$$\tau_{\gamma\gamma 0}^{(s)} = \left(B_{35} V_\alpha^{(s)} + B_{34} V_\beta^{(s)} + B_{33} V_\gamma^{(s)} \right) / \Delta,$$

$$\tau_{\beta\gamma 0}^{(s)} = \left(B_{45} V_\alpha^{(s)} + B_{44} V_\beta^{(s)} + B_{43} V_\gamma^{(s)} \right) / \Delta, \tag{2.4}$$

$$\tau_{\alpha\gamma 0}^{(s)} = \left(B_{55} V_\alpha^{(s)} + B_{54} V_\beta^{(s)} + B_{53} V_\gamma^{(s)} \right) / \Delta,$$

where

$$V_\alpha^{(s)} = \frac{1}{2} \left(u^{+(s)} - u^{-(s)} + u_*^{(s)}(\zeta = -1) - u_*^{(s)}(\zeta = +1) \right),$$

$$u^{\pm(0)} = u^{\pm}/R, \quad u^{\pm(s)} = 0, \quad s \neq 0, \quad (\alpha, \beta, \gamma; u, v, w),$$

$$B_{jk} = A_{jk} A_{ll} - A_{jl} A_{kl}, \quad B_{ll} = A_{jk} A_{kj} - A_{jj} A_{kk}, \tag{2.5}$$

$$B_{jk} \neq B_{kj} \quad j \neq k \neq l \neq j, \quad j, k, l = 3, 4, 5,$$

$$\Delta = A_{33} B_{33} + A_{34} B_{43} + A_{35} B_{53}.$$

The solution, corresponding to the conditions (1.5), (1.7), may be calculated through (1.11), (1.13), (2.1), and (2.4).

Satisfying the mixed boundary conditions (1.8), we arrive at

$$\tau_{\gamma\gamma 0}^{(s)} = \tau_{\gamma\gamma}^{+(s)} - \tau_{\gamma\gamma *}^{(s)}(\zeta = 1),$$

$$\tau_{\beta\gamma 0}^{(s)} = \frac{1}{B_{33}} \left(B_{43}(\tau_{\gamma\gamma}^{+(s)} - \tau_{\gamma\gamma *}^{(s)}(\zeta = 1)) + A_{45} V_\alpha^{(s)} - A_{55} V_\beta^{(s)} \right), \tag{2.6}$$

$$\tau_{\alpha\gamma 0}^{(s)} = \frac{1}{B_{33}} \left(B_{53}(\tau_{\gamma\gamma}^{+(s)} - \tau_{\gamma\gamma *}^{(s)}(\zeta = 1)) - A_{44} V_\alpha^{(s)} + A_{54} V_\beta^{(s)} \right),$$

where $\tau_{\gamma\gamma}^{+(s)}$ is defined through (2.3), and $V_\alpha^{(s)}, B_{ik}$ are specified in (2.5).

In case of mixed problem (1.9) imposed at $\gamma = h$ we obtain

$$\tau_{\alpha\gamma 0}^{(s)}(\xi, \eta) = \tau_{\alpha\gamma}^{+(s)} - \tau_{\alpha\gamma *}^{(s)}(\zeta = 1),$$

$$\tau_{\beta\gamma 0}^{(s)}(\xi, \eta) = \tau_{\beta\gamma}^{+(s)} - \tau_{\beta\gamma *}^{(s)}(\zeta = 1), \tag{2.7}$$

$$\tau_{\gamma\gamma 0}^{(s)}(\xi, \eta) = \frac{1}{A_{33}} \left(V_\gamma^{(s)} - A_{34}(\tau_{\beta\gamma}^{+(s)} - \tau_{\beta\gamma *}^{(s)}(\zeta = 1)) \right.$$

$$\left. - A_{35}(\tau_{\alpha\gamma}^{+(s)} - \tau_{\alpha\gamma *}^{(s)}(\zeta = 1)) \right),$$

where

$$V_\gamma^{(s)} = \frac{1}{2} \left(w^{+(s)} - w^{-(s)} + w_*^{(s)}(\zeta = -1) - w_*^{(s)}(\zeta = +1) \right),$$

$$w^{\pm(0)} = w^{\pm}/R, \quad w^{\pm(s)} = 0, \quad s \neq 0, \tag{2.8}$$

$$\tau_{\alpha\gamma}^{+(0)} = \sigma_{\alpha\gamma}^{+}, \quad \tau_{\alpha\gamma}^{+(1)} = r_2 \sigma_{\alpha\gamma}^{+}, \quad \tau_{\alpha\gamma}^{+(s)} = 0, \quad s > 1, \quad (\alpha, \beta; r_2, r_1).$$

The solution of the outer problem associated with (1.5), (1.9), follows from (1.11), (1.13), (2.1), and (2.7).

Let us consider two illustrative examples.

1. Let us study an orthotropic cylindrical shell (Fig. 10.2), having the outer surface rigidly fixed, with constant pressure P acting from the interior. Let α be the angular coordinate and β denoting the axial coordinate. Then

$$A = B = 1, \ r_1 = 0, \ r_2 = 1 \tag{2.9}$$

$$k_\alpha = k_\beta = 0, \ \ 1/R_1 = 0, \ \ R_2 = R$$

where R is the radius of the mid-surface. For the sake of simplicity, the assumptions of constant temperature field and absence of volume forces are adopted here.

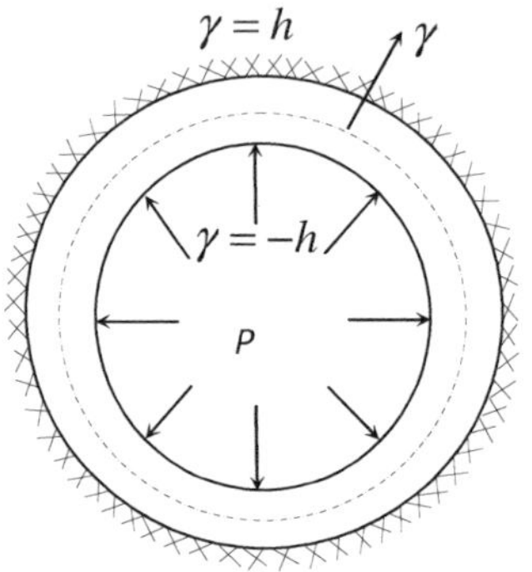

Fig. 10.2

Using the formulae (1.11), (1.13), (2.1), and (2.2), and confining to accuracy $O(\varepsilon^2)$, it is possible to obtain the outer solution in the form

$$\sigma_{\alpha\gamma} = \sigma_{\beta\gamma} = \sigma_{\alpha\beta} = 0, \ \ u_\alpha = u_\beta = 0,$$

$$\sigma_{\gamma\gamma} = -\frac{R-h}{R+h}P - A_{23}\frac{\gamma+h}{R+\gamma}P,$$

$$\sigma_{\alpha\alpha} = -\left(A_{13}(R-h) + A_{13}A_{23}(h+\gamma) - \frac{a_{12}}{\Delta}A_{33}(\gamma-h)\right)\frac{P}{R+\gamma},$$

$$\sigma_{\beta\beta} = -\left(A_{23}(R-h-\gamma) + A_{23}^2(h+\gamma) + \frac{a_{11}}{\Delta}A_{33}(\gamma-h)\right)\frac{P}{R}, \tag{2.10}$$

$$u_\gamma = -A_{33}(\gamma-h)(R-h+hA_{23})\frac{P}{R}$$

$$+\frac{1}{2}A_{33}(\gamma^2-h^2)\frac{P}{R} - A_{23}A_{33}(\gamma-h)\frac{Ph}{R},$$

$$\Delta = a_{11}a_{22} - a_{12}^2.$$

In case of a finite cylindrical shell the solution (2.10) will be valid away from the edges, in the area where the decaying boundary layer solution vanishes. The boundary layer may be constructed in a way analogous to that previously presented in respect of elastic plates.

2. Let the outer surface of a cylindrical shell be rigidly fixed, whereas the internal surface is free of loading, in the absence of the volume forces. Let us consider thermal field variation, which is for simplicity assumed in linear form ($\theta = a\gamma + b$).

Let us restrict ourselves to accuracy of $O(\varepsilon^2)$ which is sufficient for most of the practical applications. Then, according to (1.11), (1.13), (1.14), (2.1), and (2.2), the solution is given by

$$\sigma_{\alpha\gamma} = \sigma_{\beta\gamma} = 0,$$

$$\sigma_{\gamma\gamma} = \frac{\gamma + h}{R + \gamma} C_{22}(a(\gamma - h) + 2b),$$

$$\sigma_{\alpha\alpha} = C_{11}(a\gamma + b) + A_{13}C_{22}\frac{\gamma + h}{2(R + h)}(a(\gamma - h) + 2b)$$

$$-\frac{a_{12}}{2\Delta}C_{33}\frac{\gamma - h}{R + \gamma}(a(\gamma + h) + 2b),$$

$$\sigma_{\beta\beta} = C_{22}(a\gamma + b) + A_{23}C_{22}\frac{\gamma + h}{2R}(a(\gamma - h) + 2b)$$

$$+\frac{a_{12}}{2\Delta}C_{33}\frac{\gamma - h}{R}(a(\gamma + h) + 2b),$$

$$u_\alpha = \frac{1}{2}\alpha_{13}(\gamma - h)(a(\gamma + h) + 2b), \qquad (2.11)$$

$$u_\beta = \frac{1}{2}\alpha_{23}(\gamma - h)(a(\gamma + h) + 2b)$$

$$+\frac{1}{6R}\alpha_{23}\left(a(\gamma^3 - 3h^2\gamma + 2h^3) + 3b(\gamma^2 + 2h\gamma - 3h^2)\right),$$

$$u_\gamma = \frac{1}{2}C_{33}(\gamma - h)(a(\gamma + h) + 2b)$$

$$+\frac{b}{2R}A_{33}C_{22}(\gamma^2 + 2h\gamma - 3h^2) - \frac{b}{2R}A_{23}C_{33}(\gamma - h)^2$$

$$+\frac{a}{6R}(A_{33}C_{22} - A_{23}C_{33})(\gamma^3 - 3h^2\gamma + 2h^3),$$

$$C_{11} = (a_{12}\alpha_{22} - a_{22}\alpha_{11})/\Delta, \qquad (1,2),$$

$$C_{33} = A_{13}\alpha_{11} + A_{23}\alpha_{22} + \alpha_{33}, \quad \Delta = a_{11}a_{22} - a_{12}^2.$$

The solution of more advanced problems may be obtained through the same method.

As follows from the studied examples and more general analysis presented above, the outer solution depends entirely on the face boundary conditions. This is in fact a fundamental difference of the so-called non-classical boundary problems from the classical shell theory. Indeed, within conventional approach of both classical and refined shell theories after fulfilling the face boundary conditions along with other relations, one arrives at differential equations, with the unknowns depending on the tangential surface variables, with solutions involving arbitrary constants to be defined from the edge boundary conditions. This is not the case for the analyzed class of boundary value problems. Another clear distinction from the classical problems is related to the fact that all of the stress components, and all of the displacement components are of the same asymptotic order. Therefore, the main assumption of the classical shell theory becomes meaningless. On the other hand, the assumptions are no longer necessary since the solution process described in Sections 10.1 and 10.2 is relatively straightforward.

The procedures of construction of the boundary layer for both cases are similar, however, the matching procedure is slightly simpler for the non-classical problems, since the constants of the boundary layer solution are determined fully from the edge boundary conditions. Therefore, the derivation of the decay conditions is not necessary.

Thus, by changing the type of face boundary conditions, it has become possible to achieve qualitative change in the type of the stress-strain field of a shell.

10.3 Non-Classical Boundary Value Problems for a Two-Layer Shell

Let us now consider a two-layer anisotropic thermo-elastic shell, with the layers of thickness $h_1 h_2$: $\Omega = \{(\alpha, \beta, \gamma); \alpha, \beta \in \Omega_0, -h_2 \leq \gamma \leq h_1\}$. Let the interfacial surface Ω_0 of the contact between the layers contain the curvature lines α and β, with the γ axis directed perpendicular to the contact surface. The effects of volume forces with components $F_\alpha^{(i)}$, $F_\beta^{(i)}$, $F_\gamma^{(i)}$ and thermal effects are incorporated into the problem statement. According to the Duhamel-Neumann model the temperature effects are described through the function of temperature rate $\theta^{(i)}(\alpha, \beta, \gamma) = T^{(i)}(\alpha, \beta, \gamma) - T_0^{(i)}(\alpha, \beta, \gamma)$. Here and below we denote through $i = 1$ for the upper layer $0 \leq \gamma \leq h_1$, and $i = 2$ for the lower layer $-h_2 \leq \gamma \leq 0$.

Let us study the stress-strain field of such a shell subject to the boundary conditions formulated in terms of displacements at the face $\gamma = -h_2$

$$u_\alpha(-h_2) = u^-(\alpha, \beta), \quad u_\beta(-h_2) = v^-(\alpha, \beta), \quad u_\gamma(-h_2) = w^-(\alpha, \beta) \qquad (3.1)$$

along with one of the following types of boundary conditions at the opposite face $\gamma = h_1$ (written through stresses, displacements or their combination)

$$\sigma_{\alpha\gamma}(h_1) = \varepsilon^{-1}\sigma_{\alpha\gamma}^+, \quad \sigma_{\beta\gamma}(h_1) = \varepsilon^{-1}\sigma_{\beta\gamma}^+, \quad \sigma_{\gamma\gamma}(h_1) = \varepsilon^{-1}\sigma_{\gamma\gamma}^+ \qquad (3.2)$$

or

$$u_\alpha(h_1) = u^+, \quad u_\beta(h_1) = v^+, \quad u_\gamma(h_1) = w^+ \qquad (3.3)$$

or

$$u_\alpha(h_1) = u^+, \quad u_\beta(h_1) = v^+, \quad \sigma_{\gamma\gamma}(h_1) = \varepsilon^{-1}\sigma_{\gamma\gamma}^+ \qquad (3.4)$$

or

$$\sigma_{\alpha\gamma}(h_1) = \varepsilon^{-1}\sigma_{\alpha\gamma}^+, \quad \sigma_{\beta\gamma}(h_1) = \varepsilon^{-1}\sigma_{\beta\gamma}^+, \quad u_\gamma(h_1) = w^+. \qquad (3.5)$$

The edge boundary conditions are adopted in the form of any standard boundary value problem of elasticity. We also require the full contact conditions on the surface $\gamma = 0$, namely

$$u_\alpha^{(1)}(0) = u_\alpha^{(2)}(0), \quad \sigma_{\alpha\gamma}^{(1)}(0) = \sigma_{\alpha\gamma}^{(2)}(0), \quad (\alpha, \beta), \qquad (3.6)$$
$$u_\gamma^{(1)}(0) = u_\gamma^{(2)}(0), \quad \sigma_{\gamma\gamma}^{(1)}(0) = \sigma_{\gamma\gamma}^{(2)}(0).$$

Following a usual procedure, we begin with scaling of variables for each of the layers, transforming to dimensionless variables ξ, η, ζ along with the dimensionless displacements

$$\alpha = R\xi, \quad \beta = R\eta, \quad \gamma = \varepsilon R\zeta, \tag{3.7}$$
$$u_\alpha^{(i)} = Ru^{(i)}, \quad u_\beta^{(i)} = Rv^{(i)}, \quad u_\gamma^{(i)} = Rw^{(i)},$$

where $\varepsilon = h/R$, $h = \max(h_1, h_2)$, where R is a typical length scale of the shell. For the sake of definiteness and without loss of generality, we assume $h_2 > h_1$, hence $\varepsilon = h_2/R$. Clearly, in case of $h_1 > h_2$ the formulae may be easily modified.

As a result of the scaling (3.7), we obtain the following system in terms of $u^{(i)}$, $v^{(i)}$, $w^{(i)}$, $\tau_{\alpha\alpha}^{(i)}$, $\tau_{\alpha\beta}^{(i)}$, (α, β, γ), which is singularly perturbed by a small parameter ε

$$\frac{1}{AB}\frac{\partial}{\partial\xi}\left(B\tau_{\alpha\alpha}^{(i)}\right) - Rk_\beta\tau_{\beta\beta}^{(i)} + \frac{1}{AB}\frac{\partial}{\partial\eta}\left(A\tau_{\beta\alpha}^{(i)}\right)$$

$$+Rk_\alpha\tau_{\alpha\beta}^{(i)} + (1 + \varepsilon r_1\zeta)\varepsilon^{-1}\frac{\partial\tau_{\alpha\gamma}^{(i)}}{\partial\zeta} + 2r_1\tau_{\alpha\gamma}^{(i)}$$

$$+R(1 + \varepsilon\zeta(r_1 + r_2) + \varepsilon^2\zeta r_1 r_2)F_\alpha^{(i)} = 0, \quad (\alpha, \beta; \xi, \eta; 1, 2; A, B),$$

$$\varepsilon^{-1}\frac{\partial\tau_{\gamma\gamma}^{(i)}}{\partial\zeta} - r_1\tau_{\alpha\alpha}^{(i)} - r_2\tau_{\beta\beta}^{(i)} + \frac{1}{A}\frac{\partial\tau_{\alpha\gamma}^{(i)}}{\partial\xi} + \frac{1}{B}\frac{\partial\tau_{\beta\gamma}^{(i)}}{\partial\eta}$$

$$+R(k_\beta\tau_{\alpha\gamma}^{(i)} + k_\alpha\tau_{\beta\gamma}^{(i)}) + R(1 + \varepsilon\zeta(r_1 + r_2) + \varepsilon^2\zeta r_1 r_2)F_\gamma^{(i)} = 0,$$

$$(1 + \varepsilon r_2\zeta)\left(\frac{1}{A}\frac{\partial u^{(i)}}{\partial\xi} + Rk_\alpha v^{(i)} + r_1 w^{(i)}\right) = a_{11}^{(i)}\tau_{\alpha\alpha}^{(i)}$$

$$+a_{12}^{(i)}\tau_{\beta\beta}^{(i)} + a_{13}^{(i)}\tau_{\gamma\gamma}^{(i)} + ... + a_{16}^{(i)}\tau_{\alpha\beta}^{(i)} + \alpha_{11}^{(i)}(1 + \varepsilon\zeta(r_1 + r_2)$$

$$+\varepsilon^2\zeta^2 r_1 r_2)\theta^{(i)} + \varepsilon\zeta(a_{11}^{(i)}r_1\tau_{\alpha\alpha}^{(i)} + a_{12}^{(i)}r_2\tau_{\beta\beta}^{(i)}$$

$$+a_{14}^{(i)}r_2\tau_{\beta\gamma}^{(i)} + a_{15}^{(i)}r_1\tau_{\alpha\gamma}^{(i)} + a_{16}^{(i)}r_1\tau_{\alpha\beta}^{(i)}), \quad (\alpha, \beta; \xi, \eta; 1j, 2j; u, v; A, B),$$

$$(1 + \varepsilon\zeta(r_1 + r_2) + \varepsilon^2\zeta^2 r_1 r_2)\varepsilon^{-1}\frac{\partial w^{(i)}}{\partial\zeta}$$

$$= a_{13}^{(i)}\tau_{\alpha\alpha}^{(i)} + a_{23}^{(i)}\tau_{\beta\beta}^{(i)} + a_{33}^{(i)}\tau_{\gamma\gamma}^{(i)} + a_{34}^{(i)}\tau_{\beta\gamma}^{(i)}$$

$$+a_{35}^{(i)}\tau_{\alpha\gamma}^{(i)} + a_{36}^{(i)}\tau_{\alpha\beta}^{(i)} + \alpha_{33}^{(i)}(1 + \varepsilon\zeta(r_1 + r_2) + \varepsilon^2\zeta^2 r_1 r_2)\theta^{(i)}$$

$$+\varepsilon^2\zeta^2 r_1 r_2)\theta^{(i)} + \varepsilon\zeta(a_{11}^{(i)}r_1\tau_{\alpha\alpha}^{(i)} + a_{12}^{(i)}r_2\tau_{\beta\beta}^{(i)}$$

$$+a_{14}^{(i)}r_2\tau_{\beta\gamma}^{(i)} + a_{15}^{(i)}r_1\tau_{\alpha\gamma}^{(i)} + a_{16}^{(i)}r_1\tau_{\alpha\beta}^{(i)}), \quad (\alpha, \beta; \xi, \eta; 1j, 2j; u, v; A, B),$$

$$(1 + \varepsilon\zeta(r_1 + r_2) + \varepsilon^2\zeta^2 r_1 r_2)\varepsilon^{-1}\frac{\partial w^{(i)}}{\partial\zeta}$$

$$= a_{13}^{(i)}\tau_{\alpha\alpha}^{(i)} + a_{23}^{(i)}\tau_{\beta\beta}^{(i)} + a_{33}^{(i)}\tau_{\gamma\gamma}^{(i)} + a_{34}^{(i)}\tau_{\beta\gamma}^{(i)}$$

$$+a_{35}^{(i)}\tau_{\alpha\gamma}^{(i)} + a_{36}^{(i)}\tau_{\alpha\beta}^{(i)} + \alpha_{33}^{(i)}(1 + \varepsilon\zeta(r_1 + r_2) + \varepsilon^2\zeta^2 r_1 r_2)\theta^{(i)}$$

$$+\varepsilon\zeta(a_{13}^{(i)}r_1\tau_{\alpha\alpha}^{(i)} + a_{23}^{(i)}r_2\tau_{\beta\beta}^{(i)} + a_{34}^{(i)}r_2\tau_{\beta\gamma}^{(i)} + a_{35}^{(i)}r_1\tau_{\alpha\gamma}^{(i)} + a_{36}^{(i)}r_1\tau_{\alpha\beta}^{(i)}),$$

$$(1 + \varepsilon r_1 \zeta) \left(\frac{1}{B} \frac{\partial u^{(i)}}{\partial \eta} - R k_\beta v^{(i)} \right) + (1 + \varepsilon r_2 \zeta) \left(\frac{1}{A} \frac{\partial v^{(i)}}{\partial \xi} - R k_\alpha u^{(i)} \right)$$

$$= a_{16}^{(i)} \tau_{\alpha\alpha}^{(i)} + a_{26}^{(i)} \tau_{\beta\beta}^{(i)} + a_{36}^{(i)} \tau_{\gamma\gamma}^{(i)} + a_{46}^{(i)} \tau_{\beta\gamma}^{(i)} + a_{56}^{(i)} \tau_{\alpha\gamma}^{(i)} + a_{66}^{(i)} \tau_{\alpha\beta}^{(i)}$$

$$+ \alpha_{12}^{(i)} (1 + \varepsilon \zeta (r_1 + r_2) + \varepsilon^2 \zeta^2 r_1 r_2) \theta^{(i)} \qquad (3.8)$$

$$+ \varepsilon \zeta (a_{16}^{(i)} r_1 \tau_{\alpha\alpha}^{(i)} + a_{26}^{(i)} r_2 \tau_{\beta\beta}^{(i)} + a_{46}^{(i)} r_2 \tau_{\beta\gamma}^{(i)} + a_{56}^{(i)} r_1 \tau_{\alpha\gamma}^{(i)} + a_{66}^{(i)} r_1 \tau_{\alpha\beta}^{(i)}),$$

$$(1 + \varepsilon \zeta (r_1 + r_2) + \varepsilon^2 \zeta r_1 r_2) \varepsilon^{-1} \frac{\partial u^{(i)}}{\partial \zeta}$$

$$- (1 + \varepsilon r_2 \zeta) r_1 u^{(i)} + \frac{1}{A} (1 + \varepsilon r_2 \zeta) \frac{\partial w^{(i)}}{\partial \xi}$$

$$= a_{15}^{(i)} \tau_{\alpha\alpha}^{(i)} + a_{25}^{(i)} \tau_{\beta\beta}^{(i)} + a_{35}^{(i)} \tau_{\gamma\gamma}^{(i)} + a_{45}^{(i)} \tau_{\beta\gamma}^{(i)} + a_{55}^{(i)} \tau_{\alpha\gamma}^{(i)} + a_{65}^{(i)} \tau_{\alpha\beta}^{(i)}$$

$$+ \alpha_{13}^{(i)} (1 + \varepsilon \zeta (r_1 + r_2) + \varepsilon^2 \zeta^2 r_1 r_2) \theta^{(i)}$$

$$+ \varepsilon \zeta (a_{15}^{(i)} r_1 \tau_{\alpha\alpha}^{(i)} + a_{25}^{(i)} r_2 \tau_{\beta\beta}^{(i)} + a_{45}^{(i)} r_2 \tau_{\beta\gamma}^{(i)} + a_{55}^{(i)} r_1 \tau_{\alpha\gamma}^{(i)} + a_{56}^{(i)} r_1 \tau_{\alpha\beta}^{(i)}),$$

$$(u, v; \xi, \eta; 13, 23; j5, j4; A, B), \quad i = 1, 2.$$

Here $\tau_{jk}^{(i)}$ are the components of non-symmetric stress tensor. It is also assumed that $\alpha = R\xi$, $\beta = R\eta$ have been substituted into expressions for $A, B, r_1, r_2, k_\alpha, k_\beta$, with the results clearly independent of the small parameter ε.

The solution of the singularly perturbed system (3.8) is composed of the outer solution and the boundary layer solution. Due to important applications associated with the solution of the outer problem, we focus our attention on the latter. We note that the idea of construction of boundary layer solutions is analogous to that presented in details for plates.

The outer solution is sought-for in the form of asymptotic expansion

$$Q^{(i)} = \varepsilon^{\chi_Q + s} Q^{(i,s)}, \qquad s = \overline{0, S}, \ (i = 1, 2) \qquad (3.9)$$

which leads to consistent asymptotic process provided $\chi_u = 0$ for displacements and $\chi_\tau = -1$ for stresses. It is also assumed that the volume forces and function of temperature variation allow the following representation

$$F_\alpha^{(i)} = \varepsilon^{-2+s} F_{\alpha s}^{(i)}, \qquad \theta^{(i)} = \varepsilon^{-1+s} \theta_s^{(i)}, \qquad s = \overline{0, S}, \quad (i = 1, 2) \qquad (3.10)$$

where, as before, summation over repeated index s is performed in the limits of $0, S$. If the intensity of volume forces (which is reasonably large) or thermal effects is not described by (3.10), then the appropriate terms are of next asymptotic order, and will arise at higher order approximations. In this case formal use of (3.10) reveals the appropriate zero coefficients. However, in certain specific problems it is possible to adopt some of the coefficients of (3.10) as vanishing.

Substituting the representations (3.9), (3.10) into the governing system (3.8), one arrives at a system in respect of the coefficients $Q^{(i,s)}$. On satisfying the boundary conditions (3.1) at $\gamma = -h_2$ along with continuity conditions at the contact surface (3.6), we obtain the general solution of the outer problem. It is given below

$$\tau_{\alpha\gamma}^{(i,s)} = \tau_{\alpha\gamma0}^{(s)}(\xi,\eta) + \tau_{\alpha\gamma*}^{(i,s)}(\xi,\eta,\zeta),$$

$$\tau_{\beta\gamma}^{(i,s)} = \tau_{\beta\gamma0}^{(s)}(\xi,\eta) + \tau_{\beta\gamma*}^{(i,s)}(\xi,\eta,\zeta),$$

$$\tau_{\gamma\gamma}^{(i,s)} = \tau_{\gamma\gamma0}^{(s)}(\xi,\eta) + \tau_{\gamma\gamma*}^{(i,s)}(\xi,\eta,\zeta),$$

$$\tau_{\alpha\alpha}^{(i,s)} = A_{13}^{(i)}\tau_{\gamma\gamma0}^{(s)} + A_{14}^{(i)}\tau_{\beta\gamma0}^{(s)} + A_{15}^{(i)}\tau_{\alpha\gamma0}^{(s)} + \tau_{\alpha\alpha*}^{(i,s)}(\xi,\eta,\zeta),$$

$$\tau_{\beta\beta}^{(i,s)} = A_{23}^{(i)}\tau_{\gamma\gamma0}^{(s)} + A_{24}^{(i)}\tau_{\beta\gamma0}^{(s)} + A_{25}^{(i)}\tau_{\alpha\gamma0}^{(s)} + \tau_{\beta\beta*}^{(i,s)}(\xi,\eta,\zeta), \qquad (3.11)$$

$$\tau_{\alpha\beta}^{(i,s)} = A_{63}^{(i)}\tau_{\gamma\gamma0}^{(s)} + A_{64}^{(i)}\tau_{\beta\gamma0}^{(s)} + A_{65}^{(i)}\tau_{\alpha\gamma0}^{(s)} + \tau_{\alpha\beta*}^{(i,s)}(\xi,\eta,\zeta),$$

$$u^{(i,s)} = D_{53}^{(i)}\tau_{\gamma\gamma0}^{(s)} + D_{54}^{(i)}\tau_{\beta\gamma0}^{(s)} + D_{55}^{(i)}\tau_{\alpha\gamma0}^{(s)} + u^{-(s)}$$

$$+u_*^{(i,s)}(\xi,\eta,\zeta) - u_*^{(2,s)}(\zeta = -1),$$

$$v^{(i,s)} = D_{43}^{(i)}\tau_{\gamma\gamma0}^{(s)} + D_{44}^{(i)}\tau_{\beta\gamma0}^{(s)} + D_{45}^{(i)}\tau_{\alpha\gamma0}^{(s)} + v^{-(s)}$$

$$+v_*^{(i,s)}(\xi,\eta,\zeta) - v_*^{(2,s)}(\zeta = -1),$$

$$w^{(i,s)} = D_{33}^{(i)}\tau_{\gamma\gamma0}^{(s)} + D_{34}^{(i)}\tau_{\beta\gamma0}^{(s)} + D_{35}^{(i)}\tau_{\alpha\gamma0}^{(s)} + w^{-(s)}$$

$$+w_*^{(i,s)}(\xi,\eta,\zeta) - w_*^{(2,s)}(\zeta = -1), \quad (i = 1, 2),$$

where

$$D_{ml}^{(i)} = \zeta A_{ml}^{(i)} + A_{ml}^{(2)}, \quad m, l = 3, 4, 5,$$

$$A_{ml}^{(i)} = a_{m1}^{(i)}A_{1l}^{(i)} + a_{m2}^{(i)}A_{2l}^{(i)} + a_{m6}^{(i)}A_{6l}^{(i)} + a_{ml}^{(i)}, \quad A_{ml}^{(i)} \neq A_{lm}^{(i)},$$

$$A_{kl}^{(i)} = -a_{1l}^{(i)}B_{k1}^{(i)} - a_{2l}^{(i)}B_{k2}^{(i)} - a_{6l}^{(i)}B_{k6}^{(i)},$$

$$B_{nj}^{(i)} = \frac{1}{\Delta^{(i)}}(a_{nk}^{(i)}a_{jk}^{(i)} - a_{nj}^{(i)}a_{kk}^{(i)}), \quad B_{nj}^{(i)} = B_{jn}^{(i)},$$

$$B_{kk}^{(i)} = \frac{1}{\Delta^{(i)}}(a_{nn}^{(i)}a_{jj}^{(i)} - (a_{nj}^{(i)})^2), \quad (n \neq j \neq k \neq n), \quad (n, j, k = 1, 2, 6),$$

$$\Delta^{(i)} = a_{11}^{(i)}a_{22}^{(i)}a_{66}^{(i)} + 2a_{12}^{(i)}a_{26}^{(i)}a_{16}^{(i)} - a_{11}^{(i)}(a_{26}^{(i)})^2 - a_{22}^{(i)}(a_{16}^{(i)})^2 - a_{66}^{(i)}(a_{12}^{(i)})^2,$$

$$u^{-(0)} = u^-/R, \quad u^{-(s)} = 0, \quad s > 0, \quad (u, v, w),$$

$$\tau_{\alpha\gamma*}^{(i,s)} = -\int_0^\zeta \left(\frac{1}{AB}\frac{\partial}{\partial\xi}(B\tau_{\alpha\alpha}^{(i,s-1)}) + \frac{1}{AB}\frac{\partial}{\partial\eta}(A\tau_{\beta\alpha}^{(i,s-1)}) \right.$$

$$+R\left(k_\alpha\tau_{\alpha\beta}^{(i,s-1)} - k_\beta\tau_{\beta\beta}^{(i,s-1)}\right) + RF_{\alpha s}^{(i)} + r_1\left(\zeta\frac{\partial\tau_{\alpha\gamma}^{(i,s-1)}}{\partial\zeta} + 2\tau_{\alpha\gamma}^{(i,s-1)}\right)$$

$$\left. +\zeta R(r_1+r_2)F_{\alpha(s-1)}^{(i)} + \zeta^2 Rr_1r_2F_{\alpha(s-2)}^{(i)} \right)d\zeta, \quad (\alpha, \beta; \xi, \eta; r_1, r_2; A, B),$$

$$\sigma_{\gamma\gamma}^{(i,s)} = -\int_0^\zeta \left(\frac{1}{A}\frac{\partial\tau_{\alpha\gamma}^{(i,s-1)}}{\partial\xi} + \frac{1}{B}\frac{\partial\tau_{\beta\gamma}^{(i,s-1)}}{\partial\eta} \right.$$

$$+R\left(k_\alpha\tau_{\beta\gamma}^{(i,s-1)} + k_\beta\tau_{\alpha\gamma}^{(i,s-1)}\right) - r_1\tau_{\alpha\alpha}^{(i,s-1)}$$

$$\left. -r_2\tau_{\beta\beta}^{(i,s-1)} + RF_{\gamma s}^{(i)} + \zeta R(r_1+r_2)F_{\gamma(s-1)}^{(i)} + \zeta^2 Rr_1r_2F_{\gamma(s-2)}^{(i)} \right)d\zeta,$$

$$\tau_{\alpha\alpha*}^{(i,s)} = B_{11}^{(i)} P_1^{(i,s)} + B_{12}^{(i)} P_2^{(i,s)} + B_{16}^{(i)} P_3^{(i,s)}, \quad (\alpha,\beta;1k,2k),$$

$$\tau_{\alpha\beta*}^{(i,s)} = B_{16}^{(i)} P_1^{(i,s)} + B_{26}^{(i)} P_2^{(i,s)} + B_{36}^{(i)} P_3^{(i,s)},$$

$$P_1^{(i,s)} = \frac{1}{A}\frac{\partial u^{(i,s-1)}}{\partial \xi} + Rk_\alpha v^{(i,s-1)} + r_1 w^{(i,s-1)}$$

$$- a_{13}^{(i)}\tau_{\gamma\gamma*}^{(i,s)} - a_{14}^{(i)}\tau_{\beta\gamma*}^{(i,s)} - a_{15}^{(i)}\tau_{\alpha\gamma*}^{(i,s)} - \alpha_{11}^{(i)}\theta_s^{(i)}$$

$$+ \zeta r_2 \left(\frac{1}{A}\frac{\partial u^{(i,s-2)}}{\partial \xi} + Rk_\alpha v^{(i,s-2)} + r_1 w^{(i,s-2)} \right)$$

$$- \zeta(a_{11}^{(i)} r_1 \tau_{\alpha\alpha}^{(i,s-1)} + a_{12}^{(i)} r_2 \tau_{\beta\beta}^{(i,s-1)}$$

$$+ a_{44}^{(i)} r_2 \tau_{\beta\gamma}^{(i,s-1)} + a_{15}^{(i)} r_1 \tau_{\alpha\gamma}^{(i,s-1)} + a_{16}^{(i)} r_1 \tau_{\alpha\beta}^{(i,s-1)})$$

$$- \alpha_{11}^{(i)}\zeta(r_1 + r_2)\theta_{s-1}^{(i)} - \alpha_{11}^{(i)}\zeta^2 r_1 r_2 \theta_{s-2}^{(i)},$$

$$P_2^{(i,s)} = \frac{1}{B}\frac{\partial v^{(i,s-1)}}{\partial \eta} + Rk_\beta u^{(i,s-1)} + r_2 w^{(i,s-1)}$$

$$- a_{23}^{(i)}\tau_{\gamma\gamma*}^{(i,s)} - a_{24}^{(i)}\tau_{\beta\gamma*}^{(i,s)} - a_{25}^{(i)}\tau_{\alpha\gamma*}^{(i,s)} - \alpha_{22}^{(i)}\theta_s^{(i)}$$

$$+ \zeta r_1 \left(\frac{1}{B}\frac{\partial v^{(i,s-2)}}{\partial \eta} + Rk_\beta u^{(i,s-2)} + r_2 w^{(i,s-2)} \right)$$

$$- \zeta(a_{12}^{(i)} r_1 \tau_{\alpha\alpha}^{(i,s-1)} + a_{22}^{(i)} r_2 \tau_{\beta\beta}^{(i,s-1)} \tag{3.12}$$

$$+ a_{24}^{(i)} r_2 \tau_{\beta\gamma}^{(i,s-1)} + a_{25}^{(i)} r_1 \tau_{\alpha\gamma}^{(i,s-1)} + a_{26}^{(i)} r_1 \tau_{\alpha\beta}^{(i,s-1)})$$

$$- \alpha_{22}^{(i)}\zeta(r_1 + r_2)\theta_{s-1}^{(i)} - \alpha_{22}^{(i)}\zeta^2 r_1 r_2 \theta_{s-2}^{(i)},$$

$$P_3^{(i,s)} = \frac{1}{B}\frac{\partial u^{(i,s-1)}}{\partial \eta} - Rk_\beta v^{(i,s-1)} + \frac{1}{A}\frac{\partial v^{(i,s-1)}}{\partial \xi}$$

$$- Rk_\alpha u^{(i,s-1)} - a_{36}^{(i)}\tau_{\gamma\gamma*}^{(s)} - a_{46}^{(i)}\tau_{\beta\gamma*}^{(s)} - a_{56}^{(i)}\tau_{\alpha\gamma*}^{(s)}$$

$$- \alpha_{11}^{(i)}\theta_s^{(i)} + \zeta r_1 \left(\frac{1}{B}\frac{\partial u^{(i,s-2)}}{\partial \eta} - Rk_\beta v^{(i,s-2)} \right)$$

$$+ \zeta r_2 \left(\frac{1}{A}\frac{\partial v^{(i,s-2)}}{\partial \xi} - Rk_\alpha u^{(i,s-2)} \right) - \zeta(a_{16}^{(i)} r_1 \tau_{\alpha\alpha}^{(i,s-1)}$$

$$+ a_{26}^{(i)} r_2 \tau_{\beta\beta}^{(i,s-1)} + a_{46}^{(i)} r_2 \tau_{\beta\gamma}^{(i,s-1)} + a_{56}^{(i)} r_1 \tau_{\alpha\gamma}^{(i,s-1)} + a_{66}^{(i)} r_1 \tau_{\alpha\beta}^{(i,s-1)})$$

$$- \alpha_{12}^{(i)}\zeta(r_1 + r_2)\theta_{s-1}^{(i)} - \alpha_{12}^{(i)}\zeta^2 r_1 r_2 \theta_{s-2}^{(i)},$$

$$u_*^{(i,s)} = \int_0^\zeta \left(a_{15}^{(i)}\tau_{\alpha\alpha*}^{(i,s)} + a_{25}^{(i)}\tau_{\beta\beta*}^{(i,s)} + a_{35}^{(i)}\tau_{\gamma\gamma*}^{(i,s)} \right.$$

$$+ a_{45}^{(i)}\tau_{\beta\gamma*}^{(i,s)} + a_{55}^{(i)}\tau_{\alpha\gamma*}^{(i,s)} + a_{56}^{(i)}\tau_{\alpha\beta*}^{(i,s)} + \alpha_{13}^{(i)}\theta_s^{(i)}$$

$$- \frac{1}{A}\frac{\partial w^{(i,s-1)}}{\partial \xi} + r_1 u^{(i,s-1)} + \zeta \left(a_{15}^{(i)} r_1 \tau_{\alpha\alpha}^{(i,s-1)} + a_{25}^{(i)} r_2 \tau_{\beta\beta}^{(i,s-1)} \right.$$

$$+ a_{45}^{(i)} r_2 \tau_{\beta\gamma}^{(i,s-1)} + a_{55}^{(i)} r_1 \tau_{\alpha\gamma}^{(i,s-1)} + a_{56}^{(i)} r_1 \tau_{\alpha\gamma}^{(i,s-1)}$$

$$+ (r_1 + r_2)\left(\alpha_{13}^{(i)}\theta_{s-1}^{(i)} - \frac{\partial u^{(i,s-1)}}{\partial \zeta} \right) - r_2 \frac{1}{A}\frac{\partial w^{(i,s-2)}}{\partial \xi} + r_1 r_2 u^{(i,s-2)} \right)$$

$$+\zeta^2 r_1 r_2 \left(\alpha_{13}^{(i)}\theta_{s-2}^{(i)} - \frac{\partial u^{(i,s-2)}}{\partial \zeta}\right)\right)d\zeta,$$

$$v_*^{(i,s)} = \int_0^\zeta \left(a_{14}^{(i)}\tau_{\alpha\alpha*}^{(i,s)} + a_{24}^{(i)}\tau_{\beta\beta*}^{(i,s)} + a_{34}^{(i)}\tau_{\gamma\gamma*}^{(i,s)}\right.$$

$$+a_{44}^{(i)}\tau_{\beta\gamma*}^{(i,s)} + a_{45}^{(i)}\tau_{\alpha\gamma*}^{(i,s)} + a_{46}^{(i)}\tau_{\alpha\beta*}^{(i,s)} + \alpha_{23}^{(i)}\theta_s^{(i)}$$

$$-\frac{1}{B}\frac{\partial w^{(i,s-1)}}{\partial \eta} + r_2 v^{(i,s-1)} + \zeta\left(a_{14}^{(i)}r_1\tau_{\alpha\alpha}^{(i,s-1)} + a_{24}^{(i)}r_2\tau_{\beta\beta}^{(i,s-1)}\right.$$

$$+a_{44}^{(i)}r_2\tau_{\beta\gamma}^{(i,s-1)} + a_{45}^{(i)}r_1\tau_{\alpha\gamma}^{(i,s-1)} + a_{46}^{(i)}r_1\tau_{\alpha\beta}^{(i,s-1)}$$

$$+(r_1+r_2)\left(\alpha_{23}^{(i)}\theta_{s-1}^{(i)} - \frac{\partial v^{(i,s-1)}}{\partial \zeta}\right) - r_1\frac{1}{B}\frac{\partial w^{(i,s-2)}}{\partial \eta} + r_1 r_2 v^{(i,s-2)}\right)$$

$$+\zeta^2 r_1 r_2 \left(\alpha_{23}^{(i)}\theta_{s-2}^{(i)} - \frac{\partial v^{(i,s-2)}}{\partial \zeta}\right)\right)d\zeta,$$

$$w_*^{(i,s)} = \int_0^\zeta \left(a_{13}^{(i)}\tau_{\alpha\alpha*}^{(i,s)} + a_{23}^{(i)}\tau_{\beta\beta*}^{(i,s)} + a_{33}^{(i)}\tau_{\gamma\gamma*}^{(i,s)} + a_{34}^{(i)}\tau_{\beta\gamma*}^{(i,s)}\right.$$

$$+a_{35}^{(i)}\tau_{\alpha\gamma*}^{(i,s)} + a_{36}^{(i)}\tau_{\alpha\beta*}^{(i,s)} + \alpha_{33}^{(i)}\theta_s^{(i)} + \zeta(a_{13}^{(i)}r_1\tau_{\alpha\alpha}^{(i,s-1)} + a_{23}^{(i)}r_2\tau_{\beta\beta}^{(i,s-1)}$$

$$+a_{34}^{(i)}r_2\tau_{\beta\gamma}^{(i,s-1)} + a_{35}^{(i)}r_1\tau_{\alpha\gamma}^{(i,s-1)} + a_{36}^{(i)}r_1\tau_{\alpha\beta}^{(i,s-1)})$$

$$+\zeta(r_1+r_2)\left(\alpha_{33}^{(i)}\theta_{s-1}^{(i)} - \frac{\partial w^{(i,s-1)}}{\partial \zeta}\right)$$

$$+\zeta^2 r_1 r_2 \left(\alpha_{33}^{(i)}\theta_{s-2}^{(i)} - \frac{\partial w^{(i,s-2)}}{\partial \zeta}\right)\right)d\zeta,$$

$$r_i = \frac{R}{R_i}, \qquad i = 1, 2.$$

Here the quantities $\tau_{\alpha\gamma0}^{(s)}$, $\tau_{\beta\gamma0}^{(s)}$, $\tau_{\gamma\gamma0}^{(s)}$ are still to be determined from the boundary conditions at $\gamma = h_1$. Some more related details are presented in the following section.

10.4 The Outer Problem for a Two-Layer Shell

Let us now determine the functions $\tau_{\alpha\gamma0}^{(s)}, \tau_{\beta\gamma0}^{(s)}, \tau_{\gamma\gamma0}^{(s)}$ for all of the boundary value problems (3.2)-(3.5) and discuss the procedure for all of the sought-for quantities of the outer solution.

Suppose that the boundary conditions (3.2) formulated in terms of stresses are imposed at the face $\gamma = h_1(\zeta = \zeta_1 = h_1/h_2)$. Then we obtain for the unknown functions above

$$\tau_{\alpha\gamma0}^{(s)}(\xi,\eta) = \tau_{\alpha\gamma}^{+(s)} - \tau_{\alpha\gamma*}^{(1,s)}(\zeta_1),$$

$$\tau_{\beta\gamma0}^{(s)}(\xi,\eta) = \tau_{\beta\gamma}^{+(s)} - \tau_{\beta\gamma*}^{(1,s)}(\zeta_1), \tag{4.1}$$

$$\tau_{\gamma\gamma0}^{(s)}(\xi,\eta) = \tau_{\gamma\gamma}^{+(s)} - \tau_{\gamma\gamma*}^{(1,s)}(\zeta_1),$$

where

$$\tau_{\alpha\gamma}^{+(0)} = \sigma_{\alpha\gamma}^{+}, \quad \tau_{\alpha\gamma}^{+(1)} = \zeta_1 r_2 \sigma_{\alpha\gamma}^{+},$$

$$\tau_{\alpha\gamma}^{+(s)} = 0 \quad s > 1 \quad (\alpha, \beta; r_1, r_2),$$

$$\tau_{\gamma\gamma}^{+(0)} = \sigma_{\gamma\gamma}^{+}, \quad \tau_{\gamma\gamma}^{+(1)} = \zeta_1 (r_1 + r_2)\sigma_{\gamma\gamma}^{+}, \qquad (4.2)$$

$$\tau_{\gamma\gamma}^{+(2)} = \zeta_1^2 r_1 r_2 \sigma_{\gamma\gamma}^{+}, \quad \tau_{\gamma\gamma}^{+(s)} = 0, \quad s > 2.$$

Then the outer stress-strain field of the problem (3.1), (3.2) for a two-layer anisotropic shell is determined by the expressions (3.9), (3.11), (3.12), (4.1), and (4.2).

If the boundary conditions at the surface $\gamma = h_1$ are formulated in terms of displacements (3.3), then

$$\tau_{\alpha\gamma0}^{(s)} = C_{55} V_\alpha^{(s)} + C_{54} V_\beta^{(s)} + C_{53} V_\gamma^{(s)},$$

$$\tau_{\beta\gamma0}^{(s)} = C_{45} V_\alpha^{(s)} + C_{44} V_\beta^{(s)} + C_{43} V_\gamma^{(s)}, \qquad (4.3)$$

$$\tau_{\gamma\gamma0}^{(s)} = C_{35} V_\alpha^{(s)} + C_{34} V_\beta^{(s)} + C_{33} V_\gamma^{(s)},$$

where

$$C_{jk} = (A_{jk} A_{ll} - A_{jl} A_{lk})/\Delta_1,$$

$$C_{ll} = (A_{jk} A_{kj} - A_{jj} A_{kk})/\Delta_1,$$

$$(j \neq k \neq l \neq j, \quad j, k, l = 3, 4, 5),$$

$$A_{mn} = \zeta_1 A_{mn}^{(1)} + A_{mn}^{(2)}, \quad n, m = 3, 4, 5, \quad \zeta_1 = h_1/h_2, \qquad (4.4)$$

$$\Delta_1 = A_{53}(A_{44} A_{35} - A_{34} A_{45}) + A_{54}(A_{45} A_{33} - A_{43} A_{35})$$

$$+ A_{55}(A_{34} A_{43} - A_{33} A_{44}),$$

$$V_\alpha^{(s)} = u^{+(s)} - u^{-(s)} + u_*^{(2,s)}(\zeta = -1) - u_*^{(1,s)}(\zeta_1),$$

$$u^{\pm(0)} = u^{\pm}/R, \quad u^{\pm(s)} = 0, \quad s \neq 0, \quad (\alpha, \beta, \gamma; u, v, w).$$

The resulting outer solution of the problem (3.1), (3.3) is given by the formulae (3.9), (3.11), (3.12), (4.3), and (4.4).

In case of mixed boundary conditions (3.4) it is possible to obtain

$$\tau_{\alpha\gamma0}^{(s)} = \left(C_{53}(\tau_{\gamma\gamma}^{+(s)} - \tau_{\gamma\gamma*}^{(s)}(\zeta_1)) + A_{54} V_\beta^{(s)} - A_{44} V_\alpha^{(s)} \right)/C_{33},$$

$$\tau_{\beta\gamma0}^{(s)} = \left(C_{43}(\tau_{\gamma\gamma}^{+(s)} - \tau_{\gamma\gamma*}^{(s)}(\zeta_1)) - A_{55} V_\beta^{(s)} + A_{45} V_\alpha^{(s)} \right)/C_{33}, \qquad (4.5)$$

$$\tau_{\gamma\gamma0}^{(s)} = \tau_{\gamma\gamma}^{+(s)} - \tau_{\gamma\gamma*}^{(1,s)}(\zeta_1),$$

where

$$C_{33} = A_{45} A_{54} - A_{44} A_{55}, \quad C_{53} = A_{53} A_{44} - A_{54} A_{43},$$

$$C_{43} = A_{43} A_{55} - A_{45} A_{53}, \quad A_{mn} = \zeta_1 A_{mn}^{(1)} + A_{mn}^{(2)},$$

$$\tau_{\gamma\gamma}^{+(0)} = \sigma_{\gamma\gamma}^{+}, \quad \tau_{\gamma\gamma}^{+(1)} = \zeta_1 (r_1 + r_2)\sigma_{\gamma\gamma}^{+}, \qquad (4.6)$$

$$\tau_{\gamma\gamma}^{+(2)} = \zeta_1^2 r_1 r_2 \sigma_{\gamma\gamma}^{+}, \quad \tau_{\gamma\gamma}^{+(s)} = 0, \quad s > 2,$$

$$V_\alpha^{(s)} = u^{+(s)} - u^{-(s)} + u_*^{(2,s)}(\zeta = -1) - u_*^{(1,s)}(\zeta_1),$$

$$u^{\pm(0)} = u^{\pm}/R, \quad u^{\pm(s)} = 0, \quad s \neq 0, \quad (\alpha, \beta; u, v)$$

with the associated outer solution of the boundary value problem (3.1), (3.4) following from (3.9), (3.11), (3.12), (4.5), and (4.6).

Finally, when considering the boundary conditions at the face $\gamma = h_1$ in the mixed form of (3.5), we obtain for the unknown functions $\tau_{\alpha\gamma0}^{(s)}, \tau_{\beta\gamma0}^{(s)}, \tau_{\gamma\gamma0}^{(s)}$

$$\tau_{\alpha\gamma0}^{(s)} = \tau_{\alpha\gamma}^{+(s)} - \tau_{\alpha\gamma*}^{(1,s)}(\zeta_1),$$

$$\tau_{\beta\gamma0}^{(s)} = \tau_{\beta\gamma}^{+(s)} - \tau_{\beta\gamma*}^{(1,s)}(\zeta_1), \quad (4.7)$$

$$\tau_{\gamma\gamma0}^{(s)} = \frac{1}{A_{33}}\left(V_\gamma^{(s)} - A_{34}(\tau_{\beta\gamma}^{+(s)} - \tau_{\beta\gamma*}^{(1,s)}(\zeta_1)) - A_{35}(\tau_{\alpha\gamma}^{+(s)} - \tau_{\alpha\gamma*}^{(1,s)}(\zeta_1))\right),$$

where

$$\tau_{\alpha\gamma}^{+(0)} = \sigma_{\alpha\gamma}^{+}, \quad \tau_{\alpha\gamma}^{+(1)} = \zeta_1 r_2 \sigma_{\alpha\gamma}^{+},$$

$$\tau_{\alpha\gamma}^{+(s)} = 0, \quad s > 1, \quad (\alpha,\beta; r_1, r_2),$$

$$A_{mn} = \zeta_1 A_{mn}^{(1)} + A_{mn}^{(2)}, \quad (4.8)$$

$$V_\gamma^{(s)} = w^{+(s)} - w^{-(s)} + w_*^{(2,s)}(\zeta = -1) - w_*^{(1,s)}(\zeta_1),$$

$$w^{\pm(0)} = w^{\pm}/R, \quad w^{\pm(s)} = 0, \quad s \neq 0.$$

The resulting outer solution of (3.1), (3.5) is given by (3.9), (3.11), (3.12), (4.7), and (4.8).

In order to discuss in more details the proposed approach, let us show several examples of two-layer cylindrical shells.

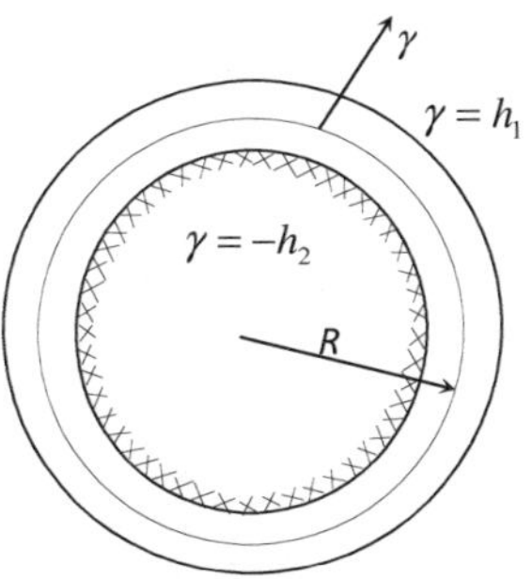

Fig. 10.3

1. Consider a two-layer orthotropic round cylindrical shell with free outer face and fixed interior surface, see Fig. 10.3. The temperature variation is assumed linear throughout the thickness coordinate

$$\theta^{(i)} = a_i \gamma + b, \quad i = 1, 2.$$

After choosing α as length in circular direction, and β as the length along the axial direction, the constants are defined as (2.9). The outer stress-strain field for such a shell may be calculated from the expressions (3.9), (3.11), (3.12), (4.1), and (4.2) up to any desired tolerance. For the sake of brevity let us present the solutions up to $O(\varepsilon^2)$. The result is

$$\sigma_{\alpha\gamma}^{(i)} = \sigma_{\beta\gamma}^{(i)} = 0, \quad \sigma_{\alpha\beta}^{(i)} = -\frac{\alpha_{12}^{(i)}}{a_{66}^{(i)}}(a_i\gamma + b),$$

$$\sigma_{\gamma\gamma}^{(i)} = C_{22}^{(i)}\frac{\gamma(a_i\gamma + 2b)}{2(R+\gamma)} - C_{22}^{(1)}\frac{h_1}{2(R+\gamma)}(a_1h_1 + 2b),$$

$$\sigma_{\alpha\alpha}^{(i)} = C_{11}^{(i)}(a_i\gamma + b) + C_{22}^{(i)}\frac{\gamma}{2(R+\gamma)}(a_i\gamma + 2b)$$

$$-A_{13}^{(i)}C_{22}^{(1)}\frac{h_1}{2(R+\gamma)}(a_1h_1 + 2b)$$

$$-\frac{a_{12}^{(i)}}{\Delta^{(i)}(R+\gamma)}(C_{33}^{(i)}\gamma(a_i\gamma + 2b) - C_{33}^{(2)}h_2(a_2h_2 - 2b)),$$

$$\sigma_{\beta\beta}^{(i)} = C_{22}^{(i)}(a_i\gamma + b) + A_{23}^{(i)}C_{22}^{(i)}\frac{\gamma}{2R}(a_i\gamma + 2b) - A_{23}^{(i)}C_{22}^{(1)}\frac{h_1}{2R}(a_1h_1 + 2b)$$

$$+\frac{a_{11}^{(i)}}{\Delta^{(i)}}(C_{33}^{(i)}\frac{\gamma}{2R}(a_i\gamma + 2b) - C_{33}^{(2)}\frac{h_2}{2R}(a_2h_2 - 2b)), \quad (4.9)$$

$$u_\alpha^{(i)} = \frac{1}{2}\alpha_{13}^{(i)}\gamma(a_i\gamma + 2b) - \frac{1}{2}\alpha_{13}^{(2)}h_2(a_2h_2 - 2b),$$

$$u_\beta^{(i)} = \frac{1}{2}\alpha_{23}^{(i)}\gamma(a_i\gamma + 2b) - \frac{1}{2}\alpha_{23}^{(2)}h_2(a_2h_2 - 2b) + \frac{\gamma^2}{6R}\alpha_{23}^{(i)}(a_i\gamma + 3b)$$

$$-\frac{\gamma h_2}{2R}\alpha_{23}^{(2)}(a_2h_2 - 2b) - \frac{h_2^2}{6R}\alpha_{23}^{(2)}(2a_2h_2 - 3b),$$

$$u_\gamma^{(i)} = \frac{1}{2}C_{33}^{(i)}\gamma(a_i\gamma + 2b) - \frac{1}{2}C_{33}^{(2)}h_2(a_2h_2 - 2b)$$

$$-\frac{h_1}{2R}(A_{33}^{(i)}\gamma + A_{33}^{(2)}h_2)C_{22}^{(1)}(a_1h_1 + 2b)$$

$$+\frac{\gamma h_2}{2R}A_{23}^{(i)}C_{33}^{(2)}(a_2h_2 - 2b) + \frac{\gamma^2}{6R}(A_{33}^{(i)}C_{22}^{(i)} - A_{23}^{(i)}C_{33}^{(i)})(a_i\gamma + 3b)$$

$$+\frac{h_2^2}{6R}A_{33}^{(2)}C_{22}^{(2)}(a_2h_2 - 3b) + \frac{h_2^2}{6R}A_{23}^{(2)}C_{33}^{(2)}(2a_2h_2 - 3b),$$

$$C_{11}^{(i)} = (a_{12}^{(i)}\alpha_{22}^{(i)} - a_{22}^{(i)}\alpha_{11}^{(i)})/(a_{11}^{(i)}a_{22}^{(i)} - (a_{12}^{(i)})^2), \quad (1,2),$$

$$C_{33}^{(i)} = A_{13}^{(i)}\alpha_{11}^{(i)} + A_{23}^{(i)}\alpha_{22}^{(i)} + \alpha_{33}^{(i)}.$$

2. Consider now another example, when the outer surface of a two-layered orthotropic cylindrical shell is rigidly fixed, with constant pressure P acting on the interior of the shell in the absence of volume forces at constant temperature (Fig. 10.4).

Following the same procedure, we obtain

$$\sigma_{\alpha\gamma}^{(i)} = \sigma_{\beta\gamma}^{(i)} = \sigma_{\alpha\beta}^{(i)} = u_\alpha^{(i)} = u_\beta^{(i)} = 0,$$

$$\sigma_{\gamma\gamma}^{(i)} = -\frac{R-h_2}{R+\gamma}P - (A_{23}^{(i)}\gamma + A_{23}^{(2)}h_2)\frac{P}{R+\gamma},$$

$$\sigma_{\alpha\alpha}^{(i)} = -A_{13}^{(i)}(R - h_2 + h_2 A_{23}^{(2)} + \gamma A_{23}^{(i)})\frac{P}{R+\gamma}$$

$$+\frac{a_{12}^{(i)}}{\Delta^{(i)}}(A_{33}^{(i)}\gamma - h_1 A_{33}^{(1)})\frac{P}{R+\gamma},$$

$$\sigma_{\beta\beta}^{(i)} = -A_{23}^{(i)}(R - h_2 - \gamma + h_2 A_{23}^{(2)} + \gamma A_{23}^{(i)})\frac{P}{R}$$

$$-\frac{a_{11}^{(i)}}{\Delta^{(i)}}(A_{33}^{(i)}\gamma - h_1 A_{33}^{(1)})\frac{P}{R}, \qquad (4.10)$$

$$u_\gamma^{(i)} = -(A_{33}^{(i)}\gamma - h_1 A_{33}^{(1)})(R - h_2 + h_2 A_{23}^{(2)})\frac{P}{R}$$

$$-\frac{1}{2}(A_{33}^{(1)}h_1^2 - \gamma^2 A_{33}^{(i)})\frac{P}{R} - A_{33}^{(1)}(A_{23}^{(i)}\gamma - h_1 A_{23}^{(1)})\frac{h_1 P}{R},$$

$$\Delta^{(i)} = a_{11}^{(i)}a_{22}^{(i)} - (a_{12}^{(i)})^2.$$

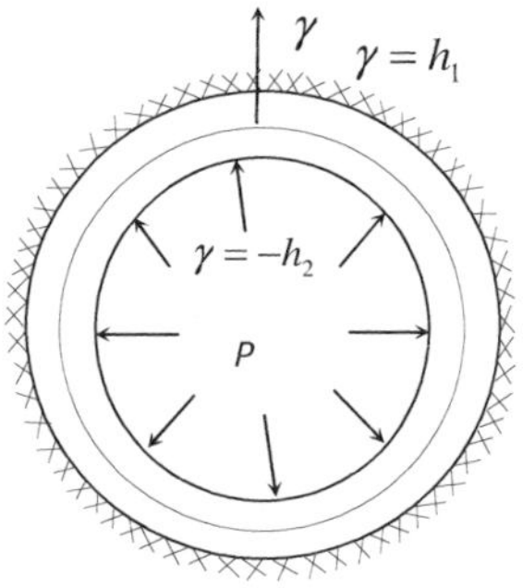

Fig. 10.4

We remark that the solutions of some more advanced problems involving more complicated surface and volume loading, and thermal effects, may be obtained in a similar manner.

The presented asymptotic method could also be applied to new classes of problems for layered beams, plates and shells. As an example, we mention a problem when there is no full contact between the layers, and one has to incorporate dry friction into consideration. In this case the general form of the asymptotic expansion (1.11) will be the same, but the values of the constants χ_u, χ_σ providing a consistent system for the quantities $Q^{(s)}$, will be different. It should be emphasized that the problem of evaluation of such consistent values for χ_u, χ_σ is actually the most challenging problem for each class of problems, though theoretically, the sought-for values always exist. As one of the approaches to rather complicated problems, it is suggested to investigate relatively simple model examples through other analytical methods. Sometimes the sought-for values of χ_u, χ_σ may be guessed from physical insight.

The asymptotic method may also be successfully applied to dynamic problems of elasticity including the problems of interaction of beams, plates and shells with various physical fields.

Chapter 11

Spatial Dynamic Problems for Anisotropic Plates

Introduction

The asymptotic method has also turned out to be rather efficient for tackling novel types of dynamic boundary value problems for thin deformable bodies, including beams, plates and shells. In particular, we mention 3D free and forced vibration problems for single and multi-layered structures, localized and interfacial vibration problems, problems of interaction of thin structures with various physical fields, problems of bio- and geo-physics, seismology, structural engineering etc.

Among the important contributions of the above-mentioned research field we mention Babich V.M., Buldyrev V.S., Bogolubov N.N., Mitropolskii U.A., Goldenweiser A.L., Rogacheva N.N., Kaplunov J.D., Tovstik P.E., Kossovich L.Yu., Nolde E.V., Mikhasev G.I., Prikazchikov D.A., Wilde M.V., Rogerson G.A., and others, for more details see Babich and Buldyrev (1972); Bogolubov and Mitropolskii (1974); Goldenveizer (1980, 1987); Rogacheva (1994); Kaplunov et al. (1998); Kaplunov and Wilde (2000); Tovstik and Smirnov (2001); Mikhasev (1997); Kaplunov et al. (2004, 2005a,b); Prikazchikov et al. (2007); Mikhasev and Tovstik (2009); Wilde et al. (2010); Drozd et al. (2013) and others.

The asymptotic method has also proved to be efficient for analysis of static and dynamic problems of thin walled structures within the framework of non-symmetric elasticity theory, see Sargsyan (2008).

In this chapter we present some of the results for dynamic problems of plates and shells, obtained by the author and his students.

11.1 Formulation of the Forced Vibration Problems for Anisotropic Plates

Motivated by obvious practical applications, let us consider the forced vibration problems for anisotropic plates. The geometry of the plate is given by $D = \{(x,y,z) : 0 \leq x \leq a, 0 \leq y \leq b, -h \leq z \leq h, \min(a,b) = \ell, h \ll \ell\}$, see Fig. 11.1.

309

The following classes of problems may be considered:
a) vibration of plates with fixed rigid foundation:

$$u(x, y, -h) = 0, \quad v(x, y, -h) = 0, \quad w(x, y, -h) = 0 \tag{1.1}$$

with the appropriate stress or displacement components prescribed in the form of time-harmonic load at $z = h$, namely

$$\sigma_{jz}(x, y, h) = \sigma_{jz}^{+}(x, y) \exp(i\Omega t), \quad j = x, y, z \tag{1.2}$$

or

$$u(x, y, h) = u^{+}(x, y) \exp(i\Omega t), \quad (u, v, w); \tag{1.3}$$

b) vibrations induced by the imposed displacements on the face $z = -h$

$$u(x, y, -h) = u^{-}(x, y) \exp(i\Omega t), \quad (u, v, w) \tag{1.4}$$

with the opposite face $z = h$ assumed free of load

$$\sigma_{xz}(x, y, h) = \sigma_{yz}(x, y, h) = \sigma_{zz}(x, y, h) = 0 \tag{1.5}$$

or rigidly fixed

$$u(x, y, h) = v(x, y, h) = w(x, y, h) = 0; \tag{1.6}$$

c) vibrations corresponding to the conditions of the first boundary value problem

$$\sigma_{jz}(x, y, \pm h) = \pm\sigma_{jz}^{\pm}(x, y) \exp(i\Omega t), \quad j = x, y, z \tag{1.7}$$

where $\sigma_{jz}^{\pm}$, $u^{\pm}$, $v^{\pm}$, $w^{\pm}$ are given functions, and Ω is the forcing frequency. The edge boundary conditions are to be specified later, since it may be shown that these correspond to boundary layers in the formulated problems. Let us consider stationary vibrations.

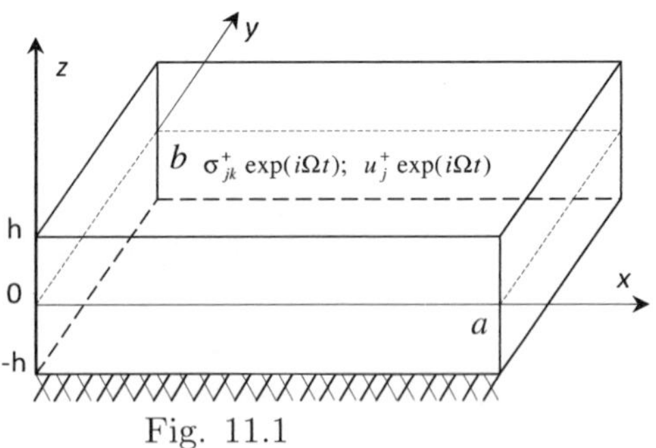

Fig. 11.1

Below we seek for solutions of the 3D elastodynamics, satisfying the appropriate constitutive relations, along with one of the groups of face boundary conditions (1.1)-(1.7). It should be mentioned that the boundary value problem (1.4)-(1.6) models the effect of seismic influence on foundations of buildings and other structures.

11.2 Forced Vibrations of Orthotropic Plates

Since the orthotropic plates are widely used in real-life problems, we begin our consideration with these, see Aghalovyan (2000, 2002, 2004). The equations of motion are taken in conventional form

$$\frac{\partial \sigma_{jx}}{\partial x} + \frac{\partial \sigma_{jy}}{\partial y} + \frac{\partial \sigma_{jz}}{\partial z} = \rho \frac{\partial^2 u}{\partial t^2}, \quad (j = x, y, z; u, v, w). \tag{2.1}$$

The constitutive relations for orthotropic plates are adopted in the form

$$\frac{\partial u}{\partial x} = a_{11}\sigma_{xx} + a_{12}\sigma_{yy} + a_{13}\sigma_{zz},$$

$$\frac{\partial v}{\partial y} = a_{12}\sigma_{xx} + a_{22}\sigma_{yy} + a_{23}\sigma_{zz},$$

$$\frac{\partial w}{\partial z} = a_{13}\sigma_{xx} + a_{23}\sigma_{yy} + a_{33}\sigma_{zz}, \tag{2.2}$$

$$\frac{\partial u}{\partial y} + \frac{\partial v}{\partial x} = a_{66}\sigma_{xy}, \quad \frac{\partial w}{\partial x} + \frac{\partial u}{\partial z} = a_{55}\sigma_{xz}, \quad \frac{\partial w}{\partial y} + \frac{\partial v}{\partial z} = a_{44}\sigma_{yz}.$$

The face boundary conditions are specified in the form (1.1)-(1.7). Since the dynamic process is assumed to be time-harmonic, the solution of (2.1)-(2.2) may be sought in the form

$$\sigma_{\alpha\beta}(x, y, z, t) = \sigma_{jk}(x, y, z)\exp(i\Omega t), \quad \alpha, \beta = x, y, z; j, k = 1, 2, 3 \tag{2.3}$$

$$u(x, y, z, t) = \bar{u}(x, y, z)\exp(i\Omega t), \quad (u, v, w).$$

Substituting (2.3) into the governing equations (2.1)-(2.2) and transforming to dimensionless quantities

$$\xi = x/\ell, \quad \eta = y/\ell, \quad \zeta = z/h, \tag{2.4}$$

$$U = \bar{u}/\ell, \quad V = \bar{v}/\ell, \quad W = \bar{w}/\ell$$

we arrive at a singularly perturbed system

$$\frac{\partial \sigma_{11}}{\partial \xi} + \frac{\partial \sigma_{12}}{\partial \eta} + \varepsilon^{-1}\frac{\partial \sigma_{13}}{\partial \zeta} + \varepsilon^{-2}\Omega_*^2 U = 0,$$

$$\frac{\partial \sigma_{12}}{\partial \xi} + \frac{\partial \sigma_{22}}{\partial \eta} + \varepsilon^{-1}\frac{\partial \sigma_{23}}{\partial \zeta} + \varepsilon^{-2}\Omega_*^2 V = 0,$$

$$\frac{\partial \sigma_{13}}{\partial \xi} + \frac{\partial \sigma_{23}}{\partial \eta} + \varepsilon^{-1}\frac{\partial \sigma_{33}}{\partial \zeta} + \varepsilon^{-2}\Omega_*^2 W = 0,$$

$$\frac{\partial U}{\partial \xi} = a_{11}\sigma_{11} + a_{12}\sigma_{22} + a_{13}\sigma_{33}, \tag{2.5}$$

$$\frac{\partial V}{\partial \eta} = a_{12}\sigma_{11} + a_{22}\sigma_{22} + a_{23}\sigma_{33},$$

$$\varepsilon^{-1}\frac{\partial W}{\partial \zeta} = a_{13}\sigma_{11} + a_{23}\sigma_{22} + a_{33}\sigma_{33}, \quad \frac{\partial V}{\partial \xi} + \frac{\partial U}{\partial \eta} = a_{66}\sigma_{12},$$

$$\frac{\partial W}{\partial \eta} + \varepsilon^{-1}\frac{\partial V}{\partial \zeta} = a_{44}\sigma_{23}, \quad \frac{\partial W}{\partial \xi} + \varepsilon^{-1}\frac{\partial U}{\partial \zeta} = a_{55}\sigma_{13},$$

where $\varepsilon = h/\ell$ is a small parameter, $\Omega_*^2 = \rho h^2 \Omega^2$ and ρ is volume density of mass. The solution of the system (2.5) is composed of the outer solution (I^{out}) and the boundary layer solution (I_b), namely

$$I = I^{out} + I_b. \tag{2.6}$$

Since the analyzed phenomenon is dynamic, that implies the presence of inertial terms in the leading order approximation. Bearing this in mind, we deduce that the consistent asymptotic process may be constructed for the following representation of the sought-for outer solution

$$\sigma_{jk}^{out} = \varepsilon^{-1+s} \sigma_{jk}^{(s)}(\xi, \eta, \zeta), \quad j, k = 1, 2, 3, \quad s = \overline{0, N},$$
$$(U^{out}, V^{out}, W^{out}) = \varepsilon^s (U^{(s)}, V^{(s)}, W^{(s)}). \tag{2.7}$$

It may be inferred from the first three equations of (2.5) that the scaled frequency $\Omega_*^2 = \rho h^2 \Omega^2$ is assumed of order unity. We remark that in case of $\Omega_*^2 = O(\varepsilon)$ the analyzed vibrations are essentially quasi-static, in other words the inertial terms do not appear at leading order approximation of the iterative procedure. For the values $\Omega_*^2 = O(\varepsilon^{-1})$ and higher, the vibrations are of the high-frequency regime. We remark that the case of $\Omega_*^2 = O(\varepsilon^{-1})$ may be described by the system (2.5) through a substitution of $\Omega_*^2 = \rho \ell h^2 \Omega^2$ instead of $\Omega_*^2 = \rho h^2 \Omega^2$ in the governing system (2.5). Similarly, in case of $\Omega_*^2 = O(\varepsilon^{-2})$ the terms $\Omega_*^2 = \rho h^2 \Omega^2$ must be replaced by $\Omega_*^2 = \rho \ell^2 \Omega^2$. However, when $\Omega_*^2 \geq O(\varepsilon^{-3})$, some other novel type of asymptotic process should be applied. Now substituting (2.7) into (2.5) and equating the coefficients at the same orders of ε, we obtain the following consistent system in respect of the coefficients $\sigma_{jk}^{(s)}, U^{(s)}, V^{(s)}, W^{(s)}$

$$\frac{\partial \sigma_{11}^{(s-1)}}{\partial \xi} + \frac{\partial \sigma_{12}^{(s-1)}}{\partial \eta} + \frac{\partial \sigma_{13}^{(s)}}{\partial \zeta} + \Omega_*^2 U^{(s)} = 0,$$

$$\frac{\partial \sigma_{12}^{(s-1)}}{\partial \xi} + \frac{\partial \sigma_{22}^{(s-1)}}{\partial \eta} + \frac{\partial \sigma_{23}^{(s)}}{\partial \zeta} + \Omega_*^2 V^{(s)} = 0,$$

$$\frac{\partial \sigma_{13}^{(s-1)}}{\partial \xi} + \frac{\partial \sigma_{23}^{(s-1)}}{\partial \eta} + \frac{\partial \sigma_{33}^{(s)}}{\partial \zeta} + \Omega_*^2 W^{(s)} = 0,$$

$$\frac{\partial U^{(s-1)}}{\partial \xi} = a_{11} \sigma_{11}^{(s)} + a_{12} \sigma_{22}^{(s)} + a_{13} \sigma_{33}^{(s)}, \tag{2.8}$$

$$\frac{\partial V^{(s-1)}}{\partial \eta} = a_{12} \sigma_{11}^{(s)} + a_{22} \sigma_{22}^{(s)} + a_{23} \sigma_{33}^{(s)},$$

$$\frac{\partial W^{(s)}}{\partial \zeta} = a_{13} \sigma_{11}^{(s)} + a_{23} \sigma_{22}^{(s)} + a_{33} \sigma_{33}^{(s)}, \quad \frac{\partial V^{(s-1)}}{\partial \xi} + \frac{\partial U^{(s-1)}}{\partial \eta} = a_{66} \sigma_{12}^{(s)},$$

$$\frac{\partial W^{(s-1)}}{\partial \eta} + \frac{\partial V^{(s)}}{\partial \zeta} = a_{44} \sigma_{23}^{(s)},$$

$$\frac{\partial W^{(s-1)}}{\partial \xi} + \frac{\partial U^{(s)}}{\partial \zeta} = a_{55} \sigma_{13}^{(s)}, \quad Q^{(m)} \equiv 0 \text{ at } m < 0.$$

Using the system (2.8), the stress components may be expressed as

$$\sigma_{13}^{(s)} = \frac{1}{a_{55}}\left[\frac{\partial U^{(s)}}{\partial \zeta} + \frac{\partial W^{(s-1)}}{\partial \xi}\right], \quad \sigma_{23}^{(s)} = \frac{1}{a_{44}}\left[\frac{\partial V^{(s)}}{\partial \zeta} + \frac{\partial W^{(s-1)}}{\partial \eta}\right],$$

$$\sigma_{12}^{(s)} = \frac{1}{a_{66}}\left[\frac{\partial V^{(s-1)}}{\partial \xi} + \frac{\partial U^{(s-1)}}{\partial \eta}\right],$$

$$\sigma_{11}^{(s)} = -A_{23}\frac{\partial W^{(s)}}{\partial \zeta} + A_{22}\frac{\partial U^{(s-1)}}{\partial \xi} - A_{12}\frac{\partial V^{(s-1)}}{\partial \eta}, \quad (2.9)$$

$$\sigma_{22}^{(s)} = -A_{13}\frac{\partial W^{(s)}}{\partial \zeta} - A_{12}\frac{\partial U^{(s-1)}}{\partial \xi} + A_{33}\frac{\partial V^{(s-1)}}{\partial \eta},$$

$$\sigma_{33}^{(s)} = A_{11}\frac{\partial W^{(s)}}{\partial \zeta} - A_{23}\frac{\partial U^{(s-1)}}{\partial \xi} - A_{13}\frac{\partial V^{(s-1)}}{\partial \eta},$$

where

$$A_{11} = (a_{11}a_{22} - a_{12}^2)/\Delta, \quad A_{22} = (a_{22}a_{33} - a_{23}^2)/\Delta,$$

$$A_{33} = (a_{11}a_{33} - a_{13}^2)/\Delta, \quad A_{13} = (a_{11}a_{23} - a_{12}a_{13})/\Delta, \quad (2.10)$$

$$A_{23} = (a_{22}a_{13} - a_{12}a_{23})/\Delta, \quad A_{12} = (a_{12}a_{33} - a_{13}a_{23})/\Delta,$$

$$\Delta = a_{11}a_{22}a_{33} + 2a_{12}a_{13}a_{23} - a_{11}a_{23}^2 - a_{22}a_{13}^3 - a_{33}a_{12}^2.$$

Then the values of $\sigma_{13}^{(s)}$, $\sigma_{23}^{(s)}$, $\sigma_{33}^{(s)}$ specified in (2.9) may be substituted into the first three equations of (2.8), yielding the three equations for $U^{(s)}, V^{(s)}, W^{(s)}$

$$\frac{\partial^2 U^{(s)}}{\partial \zeta^2} + a_{55}\Omega_*^2 U^{(s)} = R_u^{(s)}, \quad (2.11)$$

$$R_u^{(s)} = -\frac{\partial^2 W^{(s-1)}}{\partial \xi \partial \zeta} - a_{55}\left(\frac{\partial \sigma_{11}^{(s-1)}}{\partial \xi} + \frac{\partial \sigma_{12}^{(s-1)}}{\partial \eta}\right),$$

$$\frac{\partial^2 V^{(s)}}{\partial \zeta^2} + a_{44}\Omega_*^2 V^{(s)} = R_v^{(s)}, \quad (2.12)$$

$$R_v^{(s)} = -\frac{\partial^2 W^{(s-1)}}{\partial \eta \partial \zeta} - a_{44}\left(\frac{\partial \sigma_{12}^{(s-1)}}{\partial \xi} + \frac{\partial \sigma_{22}^{(s-1)}}{\partial \eta}\right),$$

$$A_{11}\frac{\partial^2 W^{(s)}}{\partial \zeta^2} + \Omega_*^2 W^{(s)} = R_w^{(s)}, \quad (2.13)$$

$$R_w^{(s)} = A_{23}\frac{\partial^2 U^{(s-1)}}{\partial \xi \partial \zeta} + A_{13}\frac{\partial^2 V^{(s-1)}}{\partial \eta \partial \zeta} - \left(\frac{\partial \sigma_{13}^{(s-1)}}{\partial \xi} + \frac{\partial \sigma_{23}^{(s-1)}}{\partial \eta}\right).$$

The solutions of (2.11)-(2.13) are written as

$$U^{(s)} = U_0^{(s)}(\xi, \eta, \zeta) + U_\tau^{(s)}(\xi, \eta, \zeta), \quad (U, V, W) \quad (2.14)$$

where the quantities with index "0" and "τ" are solutions of the homogeneous and non-homogeneous equations (2.11)-(2.13), respectively. The homogeneous solutions may be presented in the form

$$U_0^{(s)} = C_1^{(s)}(\xi,\eta)\sin\gamma_1\zeta + C_2^{(s)}(\xi,\eta)\cos\gamma_1\zeta, \quad \gamma_1 = \Omega_*\sqrt{a_{55}},$$

$$V_0^{(s)} = C_3^{(s)}(\xi,\eta)\sin\gamma_2\zeta + C_4^{(s)}(\xi,\eta)\cos\gamma_2\zeta, \quad \gamma_2 = \Omega_*\sqrt{a_{44}}, \qquad (2.15)$$

$$W_0^{(s)} = C_5^{(s)}(\xi,\eta)\sin\gamma_3\zeta + C_6^{(s)}(\xi,\eta)\cos\gamma_3\zeta, \quad \gamma_3 = \frac{\Omega_*}{\sqrt{A_{11}}}.$$

Having determined the displacement components, the stresses may be derived from (2.9). Then the procedure assumes fulfilling the appropriate boundary conditions (1.1)-(1.7), allowing evaluation of the functions $C_i^{(s)}(\xi,\eta)$. The resulting outer solution is therefore given by (2.3), (2.7), (2.9), (2.14), and (2.15). According to (2.9), (2.14), and (2.15) the stresses $\sigma_{13}^{(s)}$, $\sigma_{23}^{(s)}$, $\sigma_{33}^{(s)}$ are

$$\sigma_{13}^{(s)} = \frac{\gamma_1}{a_{55}}(C_1^{(s)}(\xi,\eta)\cos\gamma_1\zeta - C_2^{(s)}(\xi,\eta)\sin\gamma_1\zeta) + \sigma_{13\tau}^{(s)}(\xi,\eta,\zeta),$$

$$\sigma_{23}^{(s)} = \frac{\gamma_2}{a_{44}}(C_3^{(s)}(\xi,\eta)\cos\gamma_2\zeta - C_4^{(s)}(\xi,\eta)\sin\gamma_2\zeta) + \sigma_{23\tau}^{(s)}(\xi,\eta,\zeta),$$

$$\sigma_{33}^{(s)} = \gamma_3 A_{11}(C_5^{(s)}(\xi,\eta)\cos\gamma_3\zeta - C_6^{(s)}(\xi,\eta)\sin\gamma_3\zeta) + \sigma_{33\tau}^{(s)}(\xi,\eta,\zeta), \quad (2.16)$$

$$\sigma_{13\tau}^{(s)} = \frac{1}{a_{55}}\left(\frac{\partial U_\tau^{(s)}}{\partial\zeta} + \frac{\partial W^{(s-1)}}{\partial\xi}\right), \quad \sigma_{23\tau}^{(s)} = \frac{1}{a_{44}}\left(\frac{\partial V_\tau^{(s)}}{\partial\zeta} + \frac{\partial W^{(s-1)}}{\partial\eta}\right),$$

$$\sigma_{33\tau}^{(s)} = A_{11}\frac{\partial W_\tau^{(s)}}{\partial\zeta} - A_{23}\frac{\partial U^{(s-1)}}{\partial\xi} - A_{13}\frac{\partial V^{(s-1)}}{\partial\eta}.$$

Using (2.7), along with (2.14)-(2.16), and satisfying the conditions (1.1)-(1.2) in respect of U and σ_{xz}, we obtain the following system

$$-C_1^{(s)}\sin\gamma_1 + C_2^{(s)}\cos\gamma_1 = -U_\tau^{(s)}(\xi,\eta,-1),$$

$$C_1^{(s)}\cos\gamma_1 - C_2^{(s)}\sin\gamma_1 = \frac{\sqrt{a_{55}}}{\Omega_*}(\sigma_{xz}^{+(s)} - \sigma_{13\tau}^{(s)}(\xi,\eta,1)), \qquad (2.17)$$

where $\sigma_{xz}^{+(0)} = \varepsilon\sigma_{xz}^{+}$, $\sigma_{xz}^{+(s)} = 0$, $s \neq 0$. Therefore,

$$C_1^{(s)} = \frac{1}{\cos 2\gamma_1}\left(\frac{\sqrt{a_{55}}}{\Omega_*}(\sigma_{xz}^{+(s)} - \sigma_{13\tau}^{(s)}(\xi,\eta,1))\cos\gamma_1 - U_\tau^{(s)}(\xi,\eta,-1)\sin\gamma_1\right),$$

$$C_2^{(s)} = \frac{1}{\cos 2\gamma_1}\left(\frac{\sqrt{a_{55}}}{\Omega_*}(\sigma_{xz}^{+(s)} - \sigma_{13\tau}^{(s)}(\xi,\eta,1))\sin\gamma_1 - U_\tau^{(s)}(\xi,\eta,-1)\cos\gamma_1\right). \quad (2.18)$$

Using the obtained values of $C_i^{(s)}$, we get

$$U^{(s)} = \frac{1}{\cos 2\gamma_1}(\frac{\sqrt{a_{55}}}{\Omega_*}(\sigma_{xz}^{+(s)} - \sigma_{13\tau}^{(s)}(\xi,\eta,1))\sin\gamma_1(1+\zeta)$$

$$-U_\tau^{(s)}(\xi,\eta,-1)\cos\gamma_1(1-\zeta)) + U_\tau^{(s)}(\xi,\eta,\zeta),$$

$$\sigma_{13}^{(s)} = \frac{\gamma_1}{a_{55}}(\frac{\sqrt{a_{55}}}{\Omega_*}(\sigma_{xz}^{+(s)} - \sigma_{13\tau}^{(s)}(\xi,\eta,1))\cos\gamma_1(1+\zeta) \qquad (2.19)$$

$$-U_\tau^{(s)}(\xi,\eta,-1)\sin\gamma_1(1-\zeta))\frac{1}{\cos 2\gamma_1} + \sigma_{13\tau}^{(s)}(\xi,\eta,\zeta).$$

Substituting the conditions (1.1)-(1.2) in respect of V and σ_{yz}, it is possible to evaluate the functions $C_3^{(s)} C_4^{(s)}$ and obtain the corresponding solution

$$V^{(s)} = \frac{1}{\cos 2\gamma_2}\left(\frac{\sqrt{a_{44}}}{\Omega_*}(\sigma_{yz}^{+(s)} - \sigma_{23\tau}^{(s)}(\xi,\eta,1))\sin\gamma_2(1+\zeta)\right.$$
$$\left. -V_\tau^{(s)}(\xi,\eta,-1)\cos\gamma_2(1-\zeta)) + V_\tau^{(s)}(\xi,\eta,\zeta),\right. \tag{2.20}$$

$$\sigma_{23}^{(s)} = \frac{\gamma_2}{a_{44}\cos 2\gamma_2}\left(\frac{\sqrt{a_{44}}}{\Omega_*}(\sigma_{yz}^{+(s)} - \sigma_{23\tau}^{(s)}(\xi,\eta,1))\cos\gamma_2(1+\zeta)\right.$$
$$\left. -V_\tau^{(s)}(\xi,\eta,-1)\sin\gamma_2(1-\zeta)) + \sigma_{23\tau}^{(s)}(\xi,\eta,\zeta).\right.$$

Following a similar procedure for W and σ_{zz}, we derive the expressions for $C_5^{(s)}$, $C_6^{(s)}$ and the corresponding solution

$$W^{(s)} = \frac{1}{\cos 2\gamma_3}\left(\frac{1}{\Omega_*\sqrt{A_{11}}}(\sigma_{zz}^{+(s)} - \sigma_{33\tau}^{(s)}(\xi,\eta,1))\sin\gamma_3(1+\zeta)\right.$$
$$\left. -W_\tau^{(s)}(\xi,\eta,-1)\cos\gamma_3(1-\zeta)) + W_\tau^{(s)}(\xi,\eta,\zeta),\right. \tag{2.21}$$

$$\sigma_{33}^{(s)} = \frac{\gamma_3 A_{11}}{\cos 2\gamma_3}\left(\frac{1}{\Omega_*\sqrt{A_{11}}}(\sigma_{zz}^{+(s)} - \sigma_{33\tau}^{(s)}(\xi,\eta,1))\cos\gamma_3(1+\zeta)\right.$$
$$\left. -W_\tau^{(s)}(\xi,\eta,-1)\sin\gamma_3(1-\zeta)) + \sigma_{33\tau}^{(s)}(\xi,\eta,\zeta).\right.$$

The remaining stress components $\sigma_{12}^{(s)}$, $\sigma_{11}^{(s)}$, $\sigma_{22}^{(s)}$ are then determined from (2.9). The derived solution (2.19)-(2.21) is finite provided

$$\cos 2\gamma_1 \neq 0, \quad \cos 2\gamma_2 \neq 0, \quad \cos 2\gamma_3 \neq 0. \tag{2.22}$$

According to the expressions (2.15) for γ_i the conditions (2.22) are violated for the following values of the forcing frequency

$$\Omega_n^{xz} = (2n+1)\frac{\pi}{4h}\frac{1}{\sqrt{\rho a_{55}}} = (2n+1)\frac{\pi}{4h}\sqrt{\frac{G_{13}}{\rho}}, \quad n \in N,$$

$$\Omega_n^{yz} = (2n+1)\frac{\pi}{4h}\frac{1}{\sqrt{\rho a_{44}}} = (2n+1)\frac{\pi}{4h}\sqrt{\frac{G_{23}}{\rho}}, \tag{2.23}$$

$$\Omega_n^{zz} = (2n+1)\frac{\pi}{4h}\sqrt{\frac{A_{11}}{\rho}},$$

where G_{ik} is the shear modulus, with $V_s^{xz} = \sqrt{\frac{G_{13}}{\rho}}$, $V_s^{yz} = \sqrt{\frac{G_{23}}{\rho}}$ denoting the shear wave speeds, and $V_p^{zz} = \sqrt{\frac{A_{11}}{\rho}}$ represents the longitudinal wave speed in the direction of the OZ axis, given explicitly as

$$V_p^{zz} = \sqrt{\frac{A_{11}}{\rho}} = \sqrt{\frac{E_3}{\rho}\frac{1 - \nu_{12}\nu_{21}}{1 - \nu_{12}\nu_{21} - \nu_{31}(\nu_{12}\nu_{23} + \nu_{13}) - \nu_{32}(\nu_{21}\nu_{13} + \nu_{23})}}. \tag{2.24}$$

It is worth noting that a well-known result for longitudinal speed in isotropic media

$$V_p = \sqrt{\frac{A_{11}}{\rho}} = \sqrt{\frac{E}{\rho}\frac{1-\nu}{(1+\nu)(1-2\nu)}} \tag{2.25}$$

may be deduced from (2.24).

The values (2.23) may be rewritten as

$$\Omega_n^{xz} = (2n+1)\frac{\pi}{4h}V_s^{xz}, \quad \Omega_n^{yz} = (2n+1)\frac{\pi}{4h}V_s^{yz},$$

$$\Omega_n^{zz} = (2n+1)\frac{\pi}{4h}V_p^{zz}, \quad n \in N. \tag{2.26}$$

We remark that the obtained frequencies (2.26) coincide with the leading order of the eigenfrequencies of an orthotropic plate subject to the following boundary conditions

$$\sigma_{xz} = \sigma_{yz} = \sigma_{xz} = 0 \quad \text{at} \quad z = h, \quad u = v = w = 0 \quad \text{at} \quad z = -h, \tag{2.27}$$

see Aghalovyan (2002, 2004, 2008a); Aghalovyan and Aghalovyan (2005). It should also be noted that the eigenfrequency of the plate depends on the small parameter $\varepsilon = h/\ell$, with the leading order value following from the associated leading order approximation $s = 0$ of the problem. The next correction is typically of order $O(\varepsilon^2)$, and may therefore be neglected for a number of practical applications, for more details see Aghalovyan (2002); Aghalovyan and Aghalovyan (2005); Aghalovyan and Gulgazaryan (2006).

The obtained results allow some qualitative analysis. For example, we may conclude that the outer solution is fully determined from the face boundary conditions at $z = \pm h$. It is also clear that the resonant effects arise provided that the forcing frequency coincides with the eigenfrequency of the plate. In addition, it is shown below that in case of loading functions appearing in the boundary conditions (1.2)-(1.7) specified as algebraic polynomials in tangential coordinates ξ, η, the iterative process involves finite number of steps, leading to a mathematically exact outer solution (of a spatial problem for a layer).

It follows from (2.19)-(2.21), that in case of orthotropic plate there arise two types of shear and longitudinal vibrations. In view of $R_u^{(0)} = 0$, $R_v^{(0)} = 0$, $R_w^{(0)} = 0$, at leading order these vibrations are independent, however, at higher orders $s \geq 1$ the influence between these types of vibrations emerges, therefore, one type of vibrations may cause the opposite type of vibrations.

Let us now consider the conditions (1.1), (1.3). Using (2.14) and (2.15) and satisfying the face boundary conditions, we obtain the following solution

$$U^{(s)} = \frac{1}{\sin 2\gamma_1}((u^{+(s)} - U_\tau^{(s)}(\xi,\eta,1))\sin\gamma_1(1+\zeta)$$

$$-U_\tau^{(s)}(\xi,\eta,-1)\sin\gamma_1(1-\zeta)) + U_\tau^{(s)}(\xi,\eta,\zeta), \tag{2.28}$$

$$u^{+(0)} = u^+/\ell, \quad u^{+(s)} = 0, \quad s \neq 0, \quad (u,v,w).$$

The results for $V^{(s)}$, $W^{(s)}$ may be obtained by cyclic permutation of $(U, V, W; \gamma_1, \gamma_2, \gamma_3; u^+, v^+, w^+)$. The solution is finite provided

$$\sin 2\gamma_1 \neq 0, \ \sin 2\gamma_2 \neq 0, \ \sin 2\gamma_3 \neq 0, \tag{2.29}$$

hence

$$\Omega \neq \frac{\pi n}{2h}V_s^{xz}, \ \frac{\pi n}{2h}V_s^{yz}, \ \frac{\pi n}{2h}V_p^{zz}, \quad n \in N, \tag{2.30}$$

otherwise resonant effects arise.

Satisfying the conditions (1.4), (1.5) leads to the following solution

$$U^{(s)} = \frac{1}{\cos 2\gamma_1}((u^{-(s)} - U_\tau^{(s)}(\xi, \eta, -1))\cos\gamma_1(1 - \zeta)$$

$$- \frac{a_{55}}{\gamma_1}\sigma_{13\tau}^{(s)}(\xi, \eta, 1)\sin\gamma_1(1 + \zeta)) + U_\tau^{(s)}(\xi, \eta, \zeta),$$

$$(u, v; a_{55}, a_{44}; 1, 2), \qquad (2.31)$$

$$W^{(s)} = \frac{1}{\cos 2\gamma_3}((w^{-(s)} - W_\tau^{(s)}(\xi, \eta, -1))\cos\gamma_3(1 - \zeta)$$

$$- \frac{1}{\gamma_3 A_{11}}\sigma_{33\tau}^{(s)}(\xi, \eta, 1)\sin\gamma_3(1 + \zeta)) + W_\tau^{(s)}(\xi, \eta, \zeta),$$

$$u^{-(0)} = u^-/\ell, \quad u^{-(s)} = 0, \quad s \neq 0, \quad (u, v, w).$$

After the displacement components have been determined, the stresses may be evaluated from (2.9).

The solution associated with conditions (1.4), (1.6) takes the form

$$U^{(s)} = \frac{1}{\sin 2\gamma_1}((u^{-(s)} - U_\tau^{(s)}(\xi, \eta, -1))\sin\gamma_1(1 - \zeta)$$

$$- U_\tau^{(s)}(\xi, \eta, 1)\sin\gamma_1(1 + \zeta)) + U_\tau^{(s)}(\xi, \eta, \zeta),$$

$$(u, v; a_{55}, a_{44}; 1, 2), \qquad (2.32)$$

$$W^{(s)} = \frac{1}{\sin 2\gamma_3}((w^{-(s)} - W_\tau^{(s)}(\xi, \eta, -1))\sin\gamma_3(1 - \zeta)$$

$$- W_\tau^{(s)}(\xi, \eta, 1)\sin\gamma_3(1 + \zeta)) + W_\tau^{(s)}(\xi, \eta, \zeta),$$

$$u^{-(0)} = u^-/\ell, \quad u^{-(s)} = 0, \quad s \neq 0, \quad (u, v, w).$$

The resonant phenomena may be anticipated provided that any of the following conditions occur

$$\sin 2\gamma_1 = 0, \quad \sin 2\gamma_2 = 0, \quad \sin 2\gamma_3 = 0. \qquad (2.33)$$

The resulting solution for each of the cases is delivered through (2.3), (2.7), (2.9), (2.14), and (2.15).

The asymptotic approximation (2.7) provides results of reasonable accuracy for elastodynamic problem, with the boundary conditions of the first type specified on the faces of the plate $z = \pm h$. There is a principal difference between this dynamic asymptotic approximation and the one previously obtained for static first boundary value problem (3.2.5). Therefore, the dynamic effects could lead to rapid variation of the stress-strain field. Let us delve into this subject.

Using (2.16) for $\sigma_{13}^{(s)}$, we may satisfy the conditions (1.7) for σ_{xz}, resulting in the following system in respect of the unknown functions $C_1^{(s)}(\xi, \eta)$ and $C_2^{(s)}(\xi, \eta)$

$$C_1^{(s)}\cos\gamma_1 - C_2^{(s)}\sin\gamma_1 = \frac{a_{55}}{\gamma_1}(\sigma_{xz}^{+(s)} - \sigma_{13\tau}^{(s)}(\xi, \eta, 1)), \qquad (2.34)$$

$$C_1^{(s)}\cos\gamma_1 + C_2^{(s)}\sin\gamma_1 = -\frac{a_{55}}{\gamma_1}(\sigma_{xz}^{-(s)} + \sigma_{13\tau}^{(s)}(\xi, \eta, -1)),$$

from which

$$C_1^{(s)}(\xi,\eta) = \frac{a_{55}}{2\gamma_1 \cos\gamma_1}\left(\sigma_{xz}^{+(s)} - \sigma_{13\tau}^{(s)}(\xi,\eta,1) - (\sigma_{xz}^{-(s)} + \sigma_{13\tau}^{(s)}(\xi,\eta,-1))\right), \qquad (2.35)$$

$$C_2^{(s)}(\xi,\eta) = -\frac{a_{55}}{2\gamma_1 \sin\gamma_1}\left(\sigma_{xz}^{+(s)} - \sigma_{13\tau}^{(s)}(\xi,\eta,1) + \sigma_{xz}^{-(s)} + \sigma_{13}^{-(s)} + \sigma_{13\tau}^{(s)}(\xi,\eta,-1)\right).$$

Substituting these expressions into (2.14)-(2.16), one obtains

$$U^{(s)} = -\frac{a_{55}}{\gamma_1 \sin 2\gamma_1}((\sigma_{xz}^{+(s)} - \sigma_{13\tau}^{(s)}(\xi,\eta,1))\cos\gamma_1(1+\zeta)$$

$$+(\sigma_{xz}^{-(s)} + \sigma_{13\tau}^{(s)}(\xi,\eta,-1))\cos\gamma_1(1-\zeta)) + U_\tau^{(s)}(\xi,\eta,\zeta),$$

$$\sigma_{13}^{(s)} = \frac{1}{\sin 2\gamma_1}((\sigma_{xz}^{+(s)} - \sigma_{13\tau}^{(s)}(\xi,\eta,1))\sin\gamma_1(1+\zeta) \qquad (2.36)$$

$$-(\sigma_{xz}^{-(s)} + \sigma_{13\tau}^{(s)}(\xi,\eta,-1))\sin\gamma_1(1-\zeta)) + \sigma_{13\tau}^{(s)}(\xi,\eta,\zeta),$$

$$\sigma_{xz}^{\pm(0)} = \varepsilon\sigma_{xz}^{\pm}, \qquad \sigma_{xz}^{\pm(s)} = 0, \quad s \neq 0, \quad (x,y,z).$$

Then, satisfying the conditions (1.7) in respect of σ_{yz}, it is possible to determine $C_3^{(s)}$, $C_4^{(s)}$ in a similar manner. The solution may be obtained from (2.36) by changing from $(U, a_{55}, 13, xz, \gamma_1)$ to $(V, a_{44}, 23, yz, \gamma_2)$. Finally, satisfying conditions (1.7) in respect of σ_{zz}, the functions $C_5^{(s)}(\xi,\eta)$, $C_6^{(s)}(\xi,\eta)$ may be determined along with the corresponding solution

$$W^{(s)} = -\frac{1}{A_{11}\gamma_3 \sin 2\gamma_3}((\sigma_{zz}^{+(s)} - \sigma_{33\tau}^{(s)}(\xi,\eta,1))\cos\gamma_3(1+\zeta)$$

$$+(\sigma_{zz}^{-(s)} + \sigma_{33\tau}^{(s)}(\xi,\eta,-1))\cos\gamma_3(1-\zeta)) + W_\tau^{(s)}(\xi,\eta,\zeta), \qquad (2.37)$$

$$\sigma_{33}^{(s)} = \frac{1}{\sin 2\gamma_3}((\sigma_{zz}^{+(s)} - \sigma_{33\tau}^{(s)}(\xi,\eta,1))\sin\gamma_3(1+\zeta)$$

$$-(\sigma_{zz}^{-(s)} + \sigma_{33\tau}^{(s)}(\xi,\eta,-1))\sin\gamma_3(1-\zeta)) + \sigma_{33\tau}^{(s)}(\xi,\eta,\zeta).$$

We note that any of the relations $\sin 2\gamma_1 = 0$, $\sin 2\gamma_2 = 0$, $\sin 2\gamma_3 = 0$ could lead to resonant effects. The associated values of Ω are principal values of the eigenfrequencies corresponding to the free face boundary conditions $\sigma_{jz}(\xi,\eta,\pm h) = 0$, $j = x,y,z$.

11.3   Mathematically Exact Solutions

As before, it should be emphasized that in case of the loading functions $\sigma_{jz}^{\pm}$, $u^{\pm}$, $v^{\pm}$, $w^{\pm}$ appearing in the boundary conditions (1.2)-(1.7) specified as polynomials in tangential coordinates ξ,η, the proposed asymptotic procedure involves finite number of iterations depending on the highest degree of the appropriate polynomial, leading to a mathematically exact outer solution (solution for a spatial layer). Let us present some results for certain types of boundary conditions.

Let us first consider the case of $\sigma_{jz}^{\pm} = $ const for problem (1.1)-(1.2). The asymptotic process then reduces to leading order approximation only. In view of

(2.3)-(2.4), denoting any of the components of displacement vector or stress tensor through Q leads to a general representation

$$Q(x, y, z, t) = \tilde{Q}(\xi, \eta, \zeta) \exp(i\Omega t). \tag{3.1}$$

For the problem under consideration the solution follows from (2.7), (2.19)-(2.21). The expressions of (3.1) may be written explicitly as

$$\tilde{u} = \frac{h a_{55} \sigma_{xz}^+}{\gamma_1 \cos 2\gamma_1} \sin \gamma_1 (1 + \zeta), \quad \gamma_1 = \Omega_* \sqrt{a_{55}},$$

$$\tilde{\sigma}_{xz} = \frac{\sigma_{xz}^+}{\cos 2\gamma_1} \cos \gamma_1 (1 + \zeta), \quad \gamma_2 = \Omega_* \sqrt{a_{44}}, \tag{3.2}$$

$$(u, v; a_{55} a_{44}; \gamma_1, \gamma_2; x, y),$$

$$\tilde{w} = \frac{h \sigma_{zz}^+}{A_{11} \gamma_3 \cos 2\gamma_3} \sin \gamma_3 (1 + \zeta), \quad \gamma_3 = \frac{\Omega_*}{\sqrt{A_{11}}},$$

$$\tilde{\sigma}_{zz} = \frac{\sigma_{zz}^+}{\cos 2\gamma_3} \cos \gamma_3 (1 + \zeta), \quad \Omega_*^2 = \rho h^2 \Omega^2.$$

From the analysis of the outer solution (3.1)-(3.2) it may be deduced that in case of $\sigma_{jz}^+ = \text{const}$ there arise two types of shear and longitudinal motions in the plate, which are independent. At the same time the displacement w depends significantly on the transverse coordinate ζ, which means that the Kirchhoff-Love hypotheses are non-applicable here.

Let us now illustrate the above-stated comment provided the external forcing depends on the tangential coordinates ξ, η, one type of forced vibrations could excite the opposite type of vibrations. Consider the problem (1.1)-(1.2) with

$$\sigma_{zz}^+ = -(a_1 + a_2 \xi + a_3 \eta), \quad \sigma_{xz}^+ = \sigma_{yz}^+ = 0, \quad a_i = \text{const.} \tag{3.3}$$

Then, according to formulae (2.9), (2.19)-(2.21), at leading order $s = 0$ we obtain

$$U^{(0)} = 0, \quad \sigma_{13}^{(0)} = 0, \quad V^{(0)} = 0, \quad \sigma_{23}^{(0)} = 0, \quad \sigma_{12}^{(0)} = 0,$$

$$W^{(0)} = \frac{1}{A_{11} \gamma_3 \cos 2\gamma_3} \sigma_{zz}^{+(0)} \sin \gamma_3 (1 + \zeta), \quad \sigma_{zz}^{+(0)} = \varepsilon \sigma_{zz}^+ = -\varepsilon (a_1 + a_2 \xi + a_3 \eta),$$

$$\sigma_{33}^{(0)} = \frac{1}{\cos 2\gamma_3} \sigma_{zz}^{+(0)} \cos \gamma_3 (1 + \zeta), \tag{3.4}$$

$$\sigma_{11}^{(0)} = -A_{23} \frac{\partial W^{(0)}}{\partial \zeta}, \quad \sigma_{22}^{(0)} = -A_{13} \frac{\partial W^{(0)}}{\partial \zeta}.$$

The iterative procedure stops at next order approximations ($s = 1$), resulting in

$$U^{(1)} = -\frac{1}{\cos 2\gamma_1}\left(\frac{a_{55}}{\gamma_1}\sigma_{13\tau}^{(1)}(\xi,\eta,1)\sin\gamma_1(1+\zeta) + b_1\cos\gamma_1(1-\zeta)\right)$$

$$+ b_1\cos\gamma_3(1+\zeta),$$

$$\sigma_{13}^{(1)} = -\frac{1}{\cos 2\gamma_1}\left(\sigma_{13\tau}^{(1)}(\xi,\eta,1)\cos\gamma_1(1+\zeta) + \frac{b_1\gamma_1}{a_{55}}\sin\gamma_1(1-\zeta)\right)$$

$$+ \sigma_{13\tau}^{(1)}(\xi,\eta,\zeta),$$

$$V^{(1)} = -\frac{1}{\cos 2\gamma_2}\left(\frac{a_{44}}{\gamma_2}\sigma_{23\tau}^{(1)}(\xi,\eta,1)\sin\gamma_2(1+\zeta) + b_3\cos\gamma_2(1-\zeta)\right)$$

$$+ b_3\cos\gamma_3(1+\zeta),$$

$$\sigma_{23}^{(1)} = -\frac{1}{\cos 2\gamma_2}\left(\sigma_{23\tau}^{(1)}(\xi,\eta,1)\cos\gamma_2(1+\zeta) + \frac{b_3\gamma_2}{a_{44}}\sin\gamma_2(1-\zeta)\right)$$

$$+ \sigma_{23\tau}^{(1)}(\xi,\eta,\zeta),$$

$$W^{(1)} = 0, \quad \sigma_{33}^{(1)} = 0, \quad \sigma_{11}^{(1)} = 0, \quad \sigma_{22}^{(1)} = 0, \quad \sigma_{12}^{(1)} = 0, \qquad (3.5)$$

$$b_1 = \frac{(1 - a_{55}A_{23})(\varepsilon a_2)}{\Omega_*^2(a_{55}A_{11} - 1)\cos 2\gamma_3}, \quad b_3 = \frac{(1 - a_{44}A_{13})(\varepsilon a_3)}{\Omega_*^2(a_{44}A_{11} - 1)\cos 2\gamma_3},$$

$$b_2 = \frac{1}{A_{11}\gamma_3} + \frac{\gamma_3(1 - a_{55}A_{23})}{\Omega_*^2(a_{55}A_{11} - 1)}, \quad b_4 = \frac{1}{A_{11}\gamma_3} + \frac{\gamma_3(1 - a_{44}A_{13})}{\Omega_*^2(a_{44}A_{11} - 1)},$$

$$\sigma_{13\tau}^{(1)} = -\frac{\varepsilon a_2 b_2 \sin\gamma_3(1+\zeta)}{a_{55}\cos 2\gamma_3}, \quad \sigma_{23\tau}^{(1)} = -\frac{\varepsilon a_3 b_4 \sin\gamma_3(1+\zeta)}{a_{44}\cos 2\gamma_3}.$$

Then, the formulae (2.7), (3.1), (3.4), and (3.5) imply the following exact solution of the outer problem for a layer

$$\tilde{u} = hU^{(1)}, \quad \tilde{\sigma}_{xz} = \sigma_{13}^{(1)}, \quad \tilde{v} = hV^{(1)}, \quad \tilde{\sigma}_{yz} = \sigma_{23}^{(1)},$$

$$\tilde{w} = \ell W^{(0)} = -\frac{h(a_1 + a_2\xi + a_3\eta)}{A_{11}\gamma_3 \cos 2\gamma_3}\sin\gamma_3(1+\zeta), \qquad (3.6)$$

$$\tilde{\sigma}_{zz} = \varepsilon^{-1}\sigma_{33}^{(0)} = -\frac{(a_1 + a_2\xi + a_3\eta)}{\cos 2\gamma_3}\cos\gamma_3(1+\zeta),$$

$$\tilde{\sigma}_{xx} = \varepsilon^{-1}\sigma_{11}^{(0)}, \quad \tilde{\sigma}_{yy} = \varepsilon^{-1}\sigma_{22}^{(0)}, \quad \tilde{\sigma}_{12} = 0.$$

Let us now present the results of the boundary value problem (1.1), (1.3) for $u^+, v^+, w^+ = \text{const}$. The iterative procedure contains only leading order approximation $s = 0$, with the expressions (2.11)-(2.15), (2.28), (3.1) giving the following exact outer solution

$$\tilde{u} = \frac{u^+}{\sin 2\gamma_1}\sin(1+\zeta)\gamma_1, \quad \tilde{\sigma}_{xz} = \frac{1}{a_{55}}\frac{u^+}{h}\frac{\gamma_1}{\sin 2\gamma_1}\cos(1+\zeta)\gamma_1,$$

$$(u, v; x, y; a_{55}, a_{44}; \gamma_1, \gamma_2),$$

$$\tilde{w} = \frac{w^+}{\sin 2\gamma_3}\sin(1+\zeta)\gamma_3, \quad \tilde{\sigma}_{zz} = A_{11}\frac{w^+}{h}\frac{\gamma_3}{\sin 2\gamma_3}\cos(1+\zeta)\gamma_3, \qquad (3.7)$$

$$\tilde{\sigma}_{xx} = -A_{23}\frac{w^+}{h}\frac{\gamma_3}{\sin 2\gamma_3}\cos(1+\zeta)\gamma_3,$$

$$\tilde{\sigma}_{yy} = -A_{13}\frac{w^+}{h}\frac{\gamma_3}{\sin 2\gamma_3}\cos(1+\zeta)\gamma_3, \quad \tilde{\sigma}_{xy} = 0.$$

In case of the boundary value problem (1.4), (1.5) for $u^-, v^-, w^- = $ const the asymptotic process also involves leading order only. Using (2.9), (2.11)-(2.13), and (2.31), the exact solution (3.1) takes the form

$$\tilde{u} = \frac{u^-}{\cos 2\gamma_1}\cos(1-\zeta)\gamma_1, \quad \tilde{\sigma}_{xz} = \frac{1}{a_{55}}\frac{u^-}{h}\frac{\gamma_1}{\cos 2\gamma_1}\sin(1-\zeta)\gamma_1,$$

$$(u, v; x, y; a_{55}, a_{44}; \gamma_1, \gamma_2),$$

$$\tilde{w} = \frac{w^-}{\cos 2\gamma_3}\cos(1-\zeta)\gamma_3, \quad \tilde{\sigma}_{zz} = A_{11}\frac{w^-}{h}\frac{\gamma_3}{\cos 2\gamma_3}\sin(1-\zeta)\gamma_3, \qquad (3.8)$$

$$\tilde{\sigma}_{xx} = -A_{23}\frac{w^-}{h}\frac{\gamma_3}{\cos 2\gamma_3}\sin(1-\zeta)\gamma_3,$$

$$\tilde{\sigma}_{yy} = -A_{13}\frac{w^-}{h}\frac{\gamma_3}{\cos 2\gamma_3}\sin(1-\zeta)\gamma_3, \quad \tilde{\sigma}_{xy} = 0.$$

Consider now the conditions (1.4)-(1.6) for $u^-, v^-, w^- = $ const. The associated exact solution is given by

$$\tilde{u} = \frac{u^-}{\sin 2\gamma_1}\sin(1-\zeta)\gamma_1, \quad \tilde{\sigma}_{xz} = \frac{1}{a_{55}}\frac{u^-}{h}\frac{\gamma_1}{\sin 2\gamma_1}\cos(1-\zeta)\gamma_1,$$

$$(u, v; x, y; a_{55}, a_{44}; \gamma_1, \gamma_2),$$

$$\tilde{w} = \frac{w^-}{\sin 2\gamma_3}\sin(1-\zeta)\gamma_3, \quad \tilde{\sigma}_{zz} = -A_{11}\frac{w^-}{h}\frac{\gamma_3}{\sin 2\gamma_3}\cos(1-\zeta)\gamma_3, \quad (3.9)$$

$$\tilde{\sigma}_{xx} = A_{23}\frac{w^-}{h}\frac{\gamma_3}{\sin 2\gamma_3}\cos(1-\zeta)\gamma_3, \quad \tilde{\sigma}_{yy} = A_{13}\frac{w^-}{h}\frac{\gamma_3}{\sin 2\gamma_3}\cos(1-\zeta)\gamma_3,$$

$$\tilde{\sigma}_{xy} = 0.$$

One more case, which is of particular interest, is related to conditions (1.7) for the first boundary value problem of elasticity, since this case was problematic in static formulation. As previously, in case of polynomial loading functions $\sigma_{xz}^\pm$, $\sigma_{yz}^\pm$, $\sigma_{zz}^\pm$ in ξ, η the exact solution may be obtained after finite number of iterations. In case of $\sigma_{xz}^\pm$, $\sigma_{yz}^\pm$, $\sigma_{zz}^\pm = $ const according to (2.36), (2.37), and (3.1), the solution may be written as

$$\tilde{u} = \ell U^{(0)} = -\frac{ha_{55}}{\gamma_1 \sin 2\gamma_1}(\sigma_{xz}^+ \cos(1+\zeta)\gamma_1 + \sigma_{xz}^- \cos(1-\zeta)\gamma_1),$$

$$\tilde{\sigma}_{xz} = \varepsilon^{-1}\sigma_{13}^{(0)} = \frac{1}{\sin 2\gamma_1}(\sigma_{xz}^+ \sin(1+\zeta)\gamma_1 - \sigma_{xz}^- \sin(1-\zeta)\gamma_1),$$

$$(u, v; x, y; a_{55}a_{44}; \gamma_1, \gamma_2), \quad (3.10)$$

$$\tilde{w} = \ell W^{(0)} = -\frac{1}{A_{11}}\frac{h}{\gamma_3 \sin 2\gamma_3}(\sigma_{zz}^+ \cos(1+\zeta)\gamma_3 + \sigma_{zz}^- \cos(1-\zeta)\gamma_3),$$

$$\tilde{\sigma}_{zz} = \varepsilon^{-1}\sigma_{33}^{(0)} = \frac{1}{\sin 2\gamma_3}(\sigma_{zz}^+ \sin(1+\zeta)\gamma_3 - \sigma_{zz}^- \sin(1-\zeta)\gamma_3).$$

We note that the cases of higher order polynomials may be treated similar using the formulae (2.19)-(2.21), (2.28), (2.31), (2.32), (2.36), and (2.37).

11.4 Boundary Layer of an Orthotropic Plate

As a rule, the outer solution found in Sections 11.2 and 11.3 does not satisfy the edge boundary conditions, therefore, this discrepancy may be resolved through boundary layer solution. Let us construct the boundary layer near the edge $x = 0$. We begin with introducing the scaling $\gamma = \xi/\varepsilon$ and assigning the sought-for quantities with an index "b" (corresponding to "boundary"). Then the governing equations are rewritten as

$$\varepsilon^{-1}\frac{\partial \sigma_{11b}}{\partial \gamma} + \frac{\partial \sigma_{12b}}{\partial \eta} + \varepsilon^{-1}\frac{\partial \sigma_{13b}}{\partial \zeta} + \varepsilon^{-2}\Omega_*^2 U_b = 0,$$

$$\varepsilon^{-1}\frac{\partial \sigma_{12b}}{\partial \gamma} + \frac{\partial \sigma_{22b}}{\partial \eta} + \varepsilon^{-1}\frac{\partial \sigma_{23b}}{\partial \zeta} + \varepsilon^{-2}\Omega_*^2 V_b = 0,$$

$$\varepsilon^{-1}\frac{\partial \sigma_{13b}}{\partial \gamma} + \frac{\partial \sigma_{23b}}{\partial \eta} + \varepsilon^{-1}\frac{\partial \sigma_{33b}}{\partial \zeta} + \varepsilon^{-2}\Omega_*^2 W_b = 0,$$

$$\varepsilon^{-1}\frac{\partial U_b}{\partial \gamma} = a_{11}\sigma_{11b} + a_{12}\sigma_{22b} + a_{13}\sigma_{33b}, \quad (4.1)$$

$$\frac{\partial V_b}{\partial \eta} = a_{12}\sigma_{11b} + a_{22}\sigma_{22b} + a_{23}\sigma_{33b},$$

$$\varepsilon^{-1}\frac{\partial W_b}{\partial \zeta} = a_{13}\sigma_{11b} + a_{23}\sigma_{22b} + a_{33}\sigma_{33b}, \quad \varepsilon^{-1}\frac{\partial V_b}{\partial \gamma} + \frac{\partial U_b}{\partial \eta} = a_{66}\sigma_{12b},$$

$$\varepsilon^{-1}\frac{\partial W_b}{\partial \gamma} + \varepsilon^{-1}\frac{\partial U_b}{\partial \zeta} = a_{55}\sigma_{13b}, \quad \frac{\partial W_b}{\partial \eta} + \varepsilon^{-1}\frac{\partial V_b}{\partial \zeta} = a_{44}\sigma_{23b}.$$

Since the outer solution satisfies the non-homogeneous boundary conditions (1.2)-(1.7), the conditions for the boundary layer solution are homogeneous

$$U_b(\gamma, \eta, -1) = 0, \quad V_b(\gamma, \eta, -1) = 0, \quad W_b(\gamma, \eta, -1) = 0, \quad (4.2)$$

$$\sigma_{j3b}(\gamma, \eta, 1) = 0, \quad (j = 1, 2, 3)$$

$$U_b(\gamma, \eta, \pm 1) = 0, \quad V_b(\gamma, \eta, \pm 1) = 0, \quad W_b(\gamma, \eta, \pm 1) = 0, \quad (4.3)$$

$$\sigma_{j3b}(\gamma, \eta, \pm 1) = 0, \quad j = 1, 2, 3. \tag{4.4}$$

The boundary layer solution should obviously satisfy the decay conditions away from the edge $x = 0$ in the interior of the plate. The solution is sought in the form

$$\sigma_{jkb} = \varepsilon^{-1+s} \sigma_{jkb}^{(s)}(\eta, \zeta) \exp(-\lambda\gamma), \quad j, k = 1, 2, 3, \quad s = \overline{0, N},$$

$$(U_b, V_b, W_b) = \varepsilon^s (U_b^{(s)}(\eta, \zeta), V_b^{(s)}(\eta, \zeta), W_b^{(s)}(\eta, \zeta)) \exp(-\lambda\gamma), \tag{4.5}$$

where λ is a constant to be determined indicating the velocity of attenuation of the boundary layer solution. Substitution (4.5) into (4.1) leads to

$$-\lambda\sigma_{11b}^{(s)} + \frac{\partial\sigma_{12b}^{(s-1)}}{\partial\eta} + \frac{\partial\sigma_{13b}^{(s)}}{\partial\zeta} + \Omega_*^2 U_b^{(s)} = 0,$$

$$-\lambda\sigma_{12b}^{(s)} + \frac{\partial\sigma_{22b}^{(s-1)}}{\partial\eta} + \frac{\partial\sigma_{23b}^{(s)}}{\partial\zeta} + \Omega_*^2 V_b^{(s)} = 0,$$

$$-\lambda\sigma_{13b}^{(s)} + \frac{\partial\sigma_{23b}^{(s-1)}}{\partial\eta} + \frac{\partial\sigma_{33b}^{(s)}}{\partial\zeta} + \Omega_*^2 W_b^{(s)} = 0,$$

$$-\lambda U_b^{(s)} = a_{11}\sigma_{11b}^{(s)} + a_{12}\sigma_{22b}^{(s)} + a_{13}\sigma_{33b}^{(s)}, \tag{4.6}$$

$$\frac{\partial V_b^{(s-1)}}{\partial\eta} = a_{12}\sigma_{11b}^{(s)} + a_{22}\sigma_{22b}^{(s)} + a_{23}\sigma_{33b}^{(s)},$$

$$\frac{\partial W_b^{(s)}}{\partial\zeta} = a_{13}\sigma_{11b}^{(s)} + a_{23}\sigma_{22b}^{(s)} + a_{33}\sigma_{33b}^{(s)}, \quad -\lambda V_b^{(s)} + \frac{\partial U_b^{(s-1)}}{\partial\eta} = a_{66}\sigma_{12b}^{(s)},$$

$$-\lambda W_b^{(s)} + \frac{\partial U_b^{(s)}}{\partial\zeta} = a_{55}\sigma_{13b}^{(s)}, \quad \frac{\partial W_b^{(s-1)}}{\partial\eta} + \frac{\partial V_b^{(s)}}{\partial\zeta} = a_{44}\sigma_{23b}^{(s)}.$$

The stresses may be expressed through displacements from (4.6) as

$$\sigma_{12b}^{(s)} = \frac{1}{a_{66}}\left(-\lambda V_b^{(s)} + \frac{\partial U_b^{(s-1)}}{\partial\eta}\right), \quad \sigma_{13b}^{(s)} = \frac{1}{a_{55}}\left(-\lambda W_b^{(s)} + \frac{\partial U_b^{(s)}}{\partial\zeta}\right),$$

$$\sigma_{23b}^{(s)} = \frac{1}{a_{44}}\left(\frac{\partial V_b^{(s)}}{\partial\zeta} + \frac{\partial W_b^{(s-1)}}{\partial\eta}\right),$$

$$\sigma_{11b}^{(s)} = -\lambda A_{22} U_b^{(s)} - A_{23}\frac{\partial W_b^{(s)}}{\partial\zeta} - A_{12}\frac{\partial V_b^{(s-1)}}{\partial\eta}, \tag{4.7}$$

$$\sigma_{22b}^{(s)} = \lambda A_{12} U_b^{(s)} - A_{13}\frac{\partial W_b^{(s)}}{\partial\zeta} + A_{33}\frac{\partial V_b^{(s-1)}}{\partial\eta},$$

$$\sigma_{33b}^{(s)} = \lambda A_{23} U_b^{(s)} + A_{11}\frac{\partial W_b^{(s)}}{\partial\zeta} - A_{13}\frac{\partial V_b^{(s-1)}}{\partial\eta},$$

where the quantities A_{ik} are defined through (2.10). Substituting the expressions for $\sigma_{12b}^{(s)}$, $\sigma_{23b}^{(s)}$ into the second equation of (4.6), we arrive at the following equation for $V_b^{(s)}$

$$\frac{\partial^2 V_b^{(s)}}{\partial \zeta^2} + \left(\frac{a_{44}}{a_{66}} \lambda^2 + \Omega_*^2 a_{44} \right) V_b^{(s)} = f_{vb}^{(s)},$$

$$f_{vb}^{(s)} = \frac{a_{44}}{a_{66}} \lambda \frac{\partial U_b^{(s-1)}}{\partial \eta} - a_{44} \frac{\partial \sigma_{22b}^{(s-1)}}{\partial \eta} - \frac{\partial^2 W_b^{(s-1)}}{\partial \zeta \partial \eta}. \tag{4.8}$$

Then, inserting the values of $\sigma_{11b}^{(s)}$, $\sigma_{13b}^{(s)}$, $\sigma_{33b}^{(s)}$ into the first and third equations of (4.6), we obtain a system in respect of $U_b^{(s)}$, $W_b^{(s)}$, namely

$$\frac{\partial^2 U_b^{(s)}}{\partial \zeta^2} + a_{55}(\lambda^2 A_{22} + \Omega_*^2) U_b^{(s)} + \lambda(A_{23} a_{55} - 1) \frac{\partial W_b^{(s)}}{\partial \zeta} = f_{ub}^{(s)},$$

$$A_{11} a_{55} \frac{\partial^2 W_b^{(s)}}{\partial \zeta^2} + (\lambda^2 + a_{55}\Omega_*^2) W_b^{(s)} + \lambda(A_{23} a_{55} - 1) \frac{\partial U_b^{(s)}}{\partial \zeta} = f_{wb}^{(s)}, \tag{4.9}$$

where

$$f_{ub}^{(s)} = -a_{55} \left((\lambda A_{12} \frac{\partial V_b^{(s-1)}}{\partial \eta} + \frac{\partial \sigma_{12b}^{(s-1)}}{\partial \eta} \right),$$

$$f_{wb}^{(s)} = a_{55} \left(A_{13} \frac{\partial^2 V_b^{(s-1)}}{\partial \eta \partial \zeta} - \frac{\partial \sigma_{23b}^{(s-1)}}{\partial \eta} \right). \tag{4.10}$$

Using (4.9), it is possible to express $W_b^{(s)}$ through $U_b^{(s)}$ as

$$W_b^{(s)} = C_1 \left(A_{11} a_{55} \frac{\partial^3 U_b^{(s)}}{\partial \zeta^3} + C_2 \frac{\partial U_b^{(s)}}{\partial \zeta} + f_{uwb}^{(s)} \right), \tag{4.11}$$

where

$$C_1 = \frac{1}{\lambda(\lambda^2 + a_{55}\Omega_*^2)(A_{23} a_{55} - 1)},$$

$$C_2 = A_{11} a_{55}^2 (\lambda^2 A_{22} + \Omega_*^2) - \lambda^2 (A_{23} a_{55} - 1)^2, \tag{4.12}$$

$$f_{uwb}^{(s)} = -A_{11} a_{55} \frac{\partial f_{ub}^{(s)}}{\partial \zeta} + \lambda(A_{23} a_{55} - 1) f_{wb}^{(s)}.$$

The quantity $U_b^{(s)}$ is determined from the equation

$$\frac{\partial^4 U_b^{(s)}}{\partial \zeta^4} + B_{11} \frac{\partial^2 U_b^{(s)}}{\partial \zeta^2} + B_{22} U_b^{(s)} = \Psi_U^{(s)} \tag{4.13}$$

with

$$B_{11} = \frac{\lambda^2}{A_{11}} (a_{55}(A_{11} A_{22} - A_{23}^2) + 2A_{23}) + \frac{\Omega_*^2}{A_{11}} (1 + A_{11} a_{55}),$$

$$B_{22} = \frac{1}{A_{11}} (\lambda^2 A_{22} + \Omega_*^2)(\lambda^2 + \Omega_*^2 a_{55}), \tag{4.14}$$

$$\Psi_U^{(s)} = \frac{1}{A_{11} a_{55}} (\lambda^2 + \Omega_*^2 a_{55}) f_{ub}^{(s)} - \frac{1}{A_{11} a_{55}} \frac{\partial f_{uwb}^{(s)}}{\partial \zeta}.$$

The solution of (4.13) may be presented in the form

$$U_b^{(s)} = A_1^{(s)}\Psi_1 + A_2^{(s)}\Psi_2 + A_3^{(s)}\Psi_3 + A_4^{(s)}\Psi_4 + U_{\tau b}^{(s)} \tag{4.15}$$

where $U_{\tau b}^{(s)}$ is a particular solution of equation (4.13), and

$$\Psi_1 = \mathrm{ch}k_1\zeta, \quad \Psi_2 = \mathrm{sh}k_1\zeta, \quad \Psi_3 = \mathrm{ch}k_3\zeta, \quad \Psi_4 = \mathrm{sh}k_3\zeta,$$

$$k_{1,3} = \sqrt{\frac{-B_{11} \pm \sqrt{B_{11}^2 - 4B_{22}}}{2}}. \tag{4.16}$$

Using (4.7), (4.11), (4.15), and (4.16), it is possible to determine $\sigma_{13b}^{(s)}$, $\sigma_{33b}^{(s)}$, $W_b^{(s)}$ and satisfy the conditions (4.2)-(4.4) in respect of U_b, W_b, σ_{13b}, σ_{33b}. At $s = 0$ the conditions (4.2)-(4.4) imply an algebraic system of homogeneous linear equations, yielding non-trivial solution provided that the associated determinant is zero. This transcendental equation gives the roots $Re\lambda_n > 0$, which are typically complex conjugate pairs. We note that $Re\lambda_n > 0$ is characterizing the velocity of attenuation of the boundary layer solution. For conditions (4.4) in respect of σ_{13b}, σ_{33b} the derivation of the associated transcendental equation leads to two cases:

a)

$$(C_3 - C_4)\mathrm{sh}(k_1 + k_3) + (C_3 + C_4)\mathrm{sh}(k_1 - k_3) = 0 \tag{4.17}$$

b)

$$(C_3 - C_4)\mathrm{sh}(k_1 + k_3) - (C_3 + C_4)\mathrm{sh}(k_1 - k_3) = 0 \tag{4.18}$$

where

$$C_3 = (C_1\lambda A_{11}a_{55}k_1^2 + C_1C_2 - 1)(\lambda A_{23} + A_{11}^2 C_1 a_{55}k_3^4 + C_1C_2 A_{11}k_3^2)k_1, \tag{4.19}$$
$$C_4 = (C_1\lambda A_{11}a_{55}k_3^2 + C_1C_2 - 1)(\lambda A_{23} + A_{11}^2 C_1 a_{55}k_1^4 + C_1C_2 A_{11}k_1^2)k_3.$$

If λ is the root of equation (4.17), then $A_2^{(0)} \equiv 0$, $A_4^{(0)} \equiv 0$ in (4.15), hence $A_3^{(0)}$ may be expressed through $A_1^{(0)}$, with the solution written as

$$U_b^{(0)} = A_1^{(0)}\left(\Psi_1 + b_1\frac{\mathrm{sh}k_1}{\mathrm{sh}k_3}\Psi_3\right),$$

$$b_1 = -\frac{(C_1\lambda A_{11}a_{55}k_1^2 + C_1C_2 - 1)k_1}{(C_1\lambda A_{11}a_{55}k_3^2 + C_1C_2 - 1)k_3}. \tag{4.20}$$

On the other hand, if λ is the root of equation (4.18), then $A_1^{(0)} \equiv 0$, $A_3^{(0)} \equiv 0$, hence $A_4^{(0)}$ is expressed through $A_2^{(0)}$, with the solution taking the form

$$U_b^{(0)} = A_2^{(0)}\left(\Psi_2 + b_1\frac{\mathrm{ch}k_1}{\mathrm{ch}k_3}\Psi_4\right). \tag{4.21}$$

The solutions (4.20), (4.21) correspond to the plane symmetric (extension/compression), and anti-symmetric (bending) dynamic boundary layers, respectively. As before, only the roots λ_n satisfying the decay condition $Re\lambda_n > 0$ are of interest for construction of the boundary layer solutions. This procedure may be used to formulate the transcendental equations and the corresponding solutions satisfying the conditions (4.2), (4.3).

The solution of (4.8) may be expressed as

$$V_b^{(s)} = B_1^{(s)} \sin m\zeta + B_2^{(s)} \cos m\zeta + V_{\tau b}^{(s)}(\eta, \zeta),$$

$$m = \sqrt{\frac{a_{44}}{a_{66}}\lambda^2 + \Omega_*^2 a_{44}}. \tag{4.22}$$

Using relations (4.7), (4.22), the quantity $\sigma_{23b}^{(s)}$ is evaluated, and conditions (4.2)-(4.4) are satisfied in respect of V_b and σ_{23b}. At $s = 0$ these result in

$$\sigma_{23b}^{(0)} = \frac{1}{a_{44}}\frac{\partial V_b^{(0)}}{\partial \zeta} = \frac{m}{a_{44}}(B_1^{(0)} \cos m\zeta - B_2^{(0)} \sin m\zeta). \tag{4.23}$$

The conditions (4.2) in respect of $V_b^{(0)}$, $\sigma_{23b}^{(0)}$ imply

$$\cos 2m = 0 \quad \Rightarrow \quad m = (2n+1)\frac{\pi}{4}, \quad n \in N \tag{4.24}$$

which leads to the following roots (in view of (4.22))

$$\lambda_{an} = -\sqrt{\frac{a_{66}}{a_{44}}\left((2n+1)^2\frac{\pi^2}{16} - \Omega_*^2 a_{44}\right)}, \quad n > \frac{2}{\pi}\Omega_*\sqrt{a_{44}} - \frac{1}{2}, \quad n \in N \tag{4.25}$$

along with the solution given by

$$V_b^{(0)} = B_2^{(0)}(\tan m \sin m\zeta + \cos m\zeta). \tag{4.26}$$

Consider now conditions (4.3). The two cases arise

$$a) \qquad \cos m = 0 \quad \Rightarrow \quad m = (2n+1)\frac{\pi}{2}, \ V^{(0)} = B_2^{(0)} \cos m\zeta \tag{4.27}$$

$$b) \qquad \sin m = 0 \quad \Rightarrow \quad m = \pi n, \ V^{(0)} = B_1^{(0)} \sin m\zeta \tag{4.28}$$

$$\lambda_{an} = -\sqrt{\frac{a_{66}}{a_{44}}(m^2 - \Omega_*^2 a_{44})}, \quad n > \frac{\Omega_*}{\pi}\sqrt{a_{44}}. \tag{4.29}$$

On satisfying the conditions (4.4) for $\sigma_{23b}^{(0)}$ we deduce

$$a) \qquad \cos m = 0 \quad \Rightarrow \quad m = (2n+1)\frac{\pi}{2}, \ V^{(0)} = B_1^{(0)} \sin m\zeta \tag{4.30}$$

$$b) \qquad \sin m = 0 \quad \Rightarrow \quad m = \pi n, \ V^{(0)} = B_2^{(0)} \cos m\zeta. \tag{4.31}$$

It may be inferred from the solutions presented above that the roots λ_n corresponding to conditions for $V_b^{(0)}$, $\sigma_{23b}^{(0)}$, are real, which correspond to anti-plane vibrations, with the remaining conditions of (4.2)-(4.4) associated with plane vibrations localized near the edge $x = 0$. At leading order ($s = 0$) these plane and anti-plane vibrations are independent, however, it may be shown from the next order analysis ($s \geq 1$) that they become linked, and that one type of vibrations may excite the opposite type as well. For example, it may be observed from relations (4.8), (4.10) that $f_{vb}^{(1)}$ depends on $U_b^{(0)}$, $\sigma_{22b}^{(0)}$, $W_b^{(0)}$, i.e. on the quantities of the plane boundary layer, whereas the quantities $f_{ub}^{(1)}$, $f_{wb}^{(1)}$ depend on the values of $V_b^{(0)}$, $\sigma_{23b}^{(0)}$ associated with the anti-plane boundary layer solution. However, the amplitudes of the next order terms are clearly small compared to those of the leading order quantities, therefore in a variety of practical problems it is sufficient to reduce the analysis to leading order only. In general, two types of solutions may be pointed out, $\overset{p}{Q}{}^{(s)}$ and $\overset{a}{Q}{}^{(s)}$, corresponding to the values of λ_n for plane (λ_{pn}) and anti-plane (λ_{an}) boundary layers. It is clear that

$$
\overset{a}{\sigma}{}_{11b}^{(0)} = \overset{a}{\sigma}{}_{22b}^{(0)} = \overset{a}{\sigma}{}_{13b}^{(0)} = \overset{a}{\sigma}{}_{33b}^{(0)} = 0, \quad \overset{a}{U}{}_b^{(0)} = \overset{a}{W}{}_b^{(0)} = 0, \tag{4.32}
$$

$$
\overset{p}{\sigma}{}_{12b}^{(0)} = \overset{p}{\sigma}{}_{23b}^{(0)} = 0, \quad \overset{p}{V}{}_b^{(0)} = 0.
$$

Taking into account that the roots λ_{pn} of equations (4.17), (4.18) are complex conjugate pairs (similarly to the conditions (4.2), (4.3) for the plane boundary layer), the resulting solution is real. Indeed, in view of (4.20)

$$
A_1^{(0)} = \frac{1}{2}\left(A_{1n}^{(0)} - iA_{2n}^{(0)} \right), \tag{4.33}
$$

$$
\tilde{U}_{bn} = u_{bn}\exp(-\lambda_{pn}\gamma),
$$

where u_{bn} is a coefficient of $A_1^{(0)}$ in (4.20). Hence, the plane boundary layer solution may be expressed as

$$
\begin{aligned}
U_{bn}^{(0)} &= A_{1n}^{(0)}\mathrm{Re}\tilde{U}_{bn} + A_{2n}^{(0)}\mathrm{Im}\tilde{U}_{bn}, \\
W_{bn}^{(0)} &= A_{1n}^{(0)}\mathrm{Re}\tilde{W}_{bn} + A_{2n}^{(0)}\mathrm{Im}\tilde{W}_{bn}, \\
\sigma_{jkbn}^{(0)} &= A_{1n}^{(0)}\mathrm{Re}\tilde{\sigma}_{jkbn}^{(0)} + A_{2n}^{(0)}\mathrm{Im}\tilde{\sigma}_{jkbn}^{(0)},
\end{aligned} \tag{4.34}
$$

$$
\tilde{\sigma}_{jkbn}^{(0)} = \sigma_{jkbn}\exp(-\lambda_{pn}\gamma), \quad j,k = 1,3.
$$

We remark that the solution for dynamic boundary layer for equations (4.1) subject to homogeneous boundary conditions (4.2)-(4.4) is defined up to a constant factor. Therefore, if $\overset{p}{Q}$ is a solution associated with the values $\lambda_n = \lambda_{pn}$ in (4.6) along with the corresponding anti-plane boundary layer, the quantity $\varepsilon^\mu \overset{p}{Q}$ is also a solution. Similarly, $\varepsilon^\chi \overset{a}{Q}$ is also a solution provided that $\overset{a}{Q}$ is the appropriate solution corresponding to the roots of (4.6) as $\lambda_n = \lambda_{an}$. The values of whole numbers μ, χ are to be determined from the procedure of matching the boundary layer with the outer solution taking place during analysis of the edge boundary conditions. Thus, the general representation of the solution is given by

$$
I = I^{out} + \varepsilon^\mu \overset{p}{Q} + \varepsilon^\chi \overset{a}{Q}. \tag{4.35}
$$

11.4.1 *Matching of the Outer Solution and the Boundary Layer*

Let us present in more details the matching procedure for the outer solution and the boundary layer for several types of typical boundary conditions imposed on the edge $x = 0$

$$\sigma_{xx} = \sigma_{xy} = \sigma_{xz} = 0, \tag{4.36}$$

$$u = v = w = 0, \tag{4.37}$$

$$\sigma_{xx} = \sigma_{xy} = 0, \quad w = 0. \tag{4.38}$$

Other types of boundary conditions along with other edges $x = a$; $y = 0, b$ may be considered in an analogous manner. In view of (2.3), (2.4), (2.7), (4.5), and (4.35) the conditions (4.36) take the form

$$\varepsilon^{-1+s}\sigma_{11}^{(s)}(\xi = 0, \eta, \zeta) + \varepsilon^{-1+\mu+s}\overset{p}{\sigma}\,_{11b}^{(s)}(\gamma = 0, \eta, \zeta),$$
$$+\varepsilon^{-1+\chi+s}\overset{a}{\sigma}\,_{11b}^{(s)}(\gamma = 0, \eta, \zeta) = 0, \tag{4.39}$$
$$\varepsilon^{-1+s}\sigma_{13}^{(s)}(\xi = 0, \eta, \zeta) + \varepsilon^{-1+\mu+s}\overset{p}{\sigma}\,_{13b}^{(s)}(\gamma = 0, \eta, \zeta)$$
$$+\varepsilon^{-1+\chi+s}\overset{a}{\sigma}\,_{13b}^{(s)}(\gamma = 0, \eta, \zeta) = 0,$$

$$\varepsilon^{-1+s}\sigma_{12}^{(s)}(\xi = 0, \eta, \zeta) + \varepsilon^{-1+\mu+s}\overset{p}{\sigma}\,_{12b}^{(s)}(\gamma = 0, \eta, \zeta)$$
$$+\varepsilon^{-1+\chi+s}\overset{a}{\sigma}\,_{12b}^{(s)}(\gamma = 0, \eta, \zeta) = 0. \tag{4.40}$$

Since the outer solution is known, from the compatibility conditions we deduce that $\mu = \chi = 0$. On the other hand, taking into account that $\overset{a}{\sigma}\,_{11b}^{(0)} = \overset{a}{\sigma}\,_{13b}^{(0)} = 0$, along with $\overset{p}{\sigma}\,_{12b}^{(0)} = 0$, it follows from (4.39) for the plane boundary layer that

$$\overset{p}{\sigma}\,_{11b}^{(s)}(\gamma = 0, \eta, \zeta) = -\sigma_{11}^{(s)}(\xi = 0, \eta, \zeta) - \overset{a}{\sigma}\,_{11b}^{(s)}(\gamma = 0, \eta, \zeta), \tag{4.41}$$
$$\overset{p}{\sigma}\,_{13b}^{(s)}(0, \eta, \zeta) = -\sigma_{13}^{(s)}(0, \eta, \zeta) - \overset{a}{\sigma}\,_{13b}^{(s)}(0, \eta, \zeta)$$

whereas for the anti-plane boundary layer a similar condition takes the form

$$\overset{a}{\sigma}\,_{12b}^{(s)}(\gamma = 0, \eta, \zeta) = -\sigma_{12}^{(s)}(\xi = 0, \eta, \zeta) - \overset{p}{\sigma}\,_{12b}^{(s)}(\gamma = 0, \eta, \zeta). \tag{4.42}$$

According to (4.34), the conditions (4.41) at leading order ($s = 0$) may be rewritten as

$$A_{1n}^{(0)}Re\tilde{\sigma}_{11bn}^{(0)}(0, \eta, \zeta) + A_{2n}^{(0)}Im\tilde{\sigma}_{11bn}^{(0)}(0, \eta, \zeta) = -\sigma_{11}^{(0)}(0, \eta, \zeta), \quad n = \overline{1, N}, \tag{4.43}$$
$$A_{1n}^{(0)}Re\tilde{\sigma}_{13bn}^{(0)}(0, \eta, \zeta) + A_{2n}^{(0)}Im\tilde{\sigma}_{13bn}^{(0)}(0, \eta, \zeta) = -\sigma_{13}^{(0)}(0, \eta, \zeta)$$

with summation assumed over the repeated index n in the limits from 1 to N, where N is a number of roots λ_n chosen for computations, sorted in increasing order of their real parts. The constants $A_{1n}^{(0)}, A_{2n}^{(0)}$ are evaluated from (4.43) through one of the following methods: collocation, Fourier or least squares. We note that higher order approximations $s \geq 1$ affect only the right-hand sides of (4.43).

Depending on the specified boundary value problem (1.1), (1.2)-(1.7) and the corresponding anti-plane boundary layer solution (4.26)-(4.31), the condition (4.40) is used for the Fourier method to determine the unique value of the real constant $B_i^{(s)}$, and, hence, the resulting anti-plane boundary layer solution.

In case of relations (4.37) from the consistency of the procedure we deduce $\mu = 0$, $\chi = 0$. The associated conditions for the plane boundary layer solution are written as

$$\overset{p}{U}{}^{(s)}_{b}(\gamma = 0, \eta, \zeta) = -U^{(s)}(\xi = 0, \eta, \zeta) - \overset{a}{U}{}^{(s)}_{b}(\gamma = 0, \eta, \zeta), \qquad (4.44)$$

$$\overset{p}{W}{}^{(s)}_{b}(\gamma = 0, \eta, \zeta) = -W^{(s)}(\xi = 0, \eta, \zeta) - \overset{a}{W}{}^{(s)}_{b}(\gamma = 0, \eta, \zeta).$$

Bearing in mind (4.34), we obtain at leading order $s = 0$

$$A_{1n}^{(0)}\mathrm{Re}\tilde{U}_{bn}(\gamma = 0, \eta, \zeta) + A_{2n}^{(0)}\mathrm{Im}\tilde{U}_{bn}(\gamma = 0, \eta, \zeta)$$
$$= -U^{(0)}(\xi = 0, \eta, \zeta), \quad n = \overline{0, N}, \quad (4.45)$$
$$A_{1n}^{(0)}\mathrm{Re}\tilde{W}_{bn}(\gamma = 0, \eta, \zeta) + A_{2n}^{(0)}\mathrm{Im}\tilde{W}_{bn}(\gamma = 0, \eta, \zeta) = -W^{(0)}(\xi = 0, \eta, \zeta)$$

from which the constants $A_{1n}^{(0)}, A_{2n}^{(0)}$ may be computed through one of the above-mentioned methods, leading to the resulting approximation of $s = 0$. Once again, at higher orders $s \geq 1$ the only changes happen at the right-hand sides of (4.45). The condition corresponding to $V = 0$ is given by

$$\overset{a}{V}{}^{(s)}_{b}(\gamma = 0, \eta, \zeta) = -V^{(s)}(\xi = 0, \eta, \zeta) - \overset{p}{V}{}^{(s)}_{b}(\gamma = 0, \eta, \zeta). \qquad (4.46)$$

Then, using the Fourier method in application to (4.26)-(4.31) the constant $B_i^{(s)}$ is determined, thus leading to the sought-for anti-plane boundary layer solution.

In case of conditions (4.38) the consistent values are $\mu = 0$, $\chi = 0$. The boundary conditions in respect of σ_{xx}, W leads to

$$\overset{p}{\sigma}{}^{(s)}_{11b}(\gamma = 0, \eta, \zeta) = -\sigma_{11}^{(s)}(\xi = 0, \eta, \zeta) - \overset{a}{\sigma}{}^{(s)}_{11b}(\gamma = 0, \eta, \zeta), \qquad (4.47)$$

$$\overset{p}{W}{}^{(s)}_{b}(\gamma = 0, \eta, \zeta) = -W^{(s)}(\xi = 0, \eta, \zeta) - \overset{a}{W}{}^{(s)}_{b}(\gamma = 0, \eta, \zeta).$$

At $s = 0$ we obtain

$$A_{1n}^{(0)}\mathrm{Re}\tilde{\sigma}^{(0)}_{11bn}(0, \eta, \zeta) + A_{2n}^{(0)}\mathrm{Im}\tilde{\sigma}^{(0)}_{11bn}(0, \eta, \zeta) = -\sigma_{11}^{(0)}(0, \eta, \zeta), \quad n = \overline{1, N}, \quad (4.48)$$
$$A_{1n}^{(0)}\mathrm{Re}\tilde{W}_{bn}(0, \eta, \zeta) + A_{2n}^{(0)}\mathrm{Im}\tilde{W}_{bn}(0, \eta, \zeta) = -W^{(0)}(0, \eta, \zeta)$$

from which the constants $A_{1n}^{(0)}, A_{2n}^{(0)}$ are evaluated, providing the resulting leading order plane boundary layer solution ($s = 0$). As before, the analysis of higher orders is very similar, with only changes occurring on the right-hand sides of (4.48). We remark that the second condition of (4.38) implies fulfilling of (4.42).

11.5　Forced Vibrations of Layered Orthotropic Plates

The asymptotic approach may be extended to dynamic boundary value problems for layered media. Let us illustrate it by considering a two-layered plate $D = \{(x, y, z) : 0 \leq x \leq a, 0 \leq y \leq b, -h_2 \leq z \leq h_1, h = h_1 + h_2, \ell = \min(a, b), h \ll \ell\}$ for several types of boundary conditions (Aghalovyan, 2002, 2008a; Aghalovyan and Zakaryan, 2011). Consider a plate composed of two orthotropic layers of thicknesses h_1, h_2. The quantities associated with the first and second layers are denoted by indices I and II, respectively, with k denoting arbitrary layer. Let us investigate the cases of the following boundary conditions imposed on the face surfaces $z = h_1, -h_2$
a)

$$u^{II}(-h_2) = u^-(\xi, \eta)\exp(i\Omega t), \quad (u, v, w) \tag{5.1}$$

and

$$u^I(h_1) = v^I(h_1) = w^I(h_1) = 0 \tag{5.2}$$

or

$$\sigma_{xz}^I(h_1) = \sigma_{yz}^I(h_1) = \sigma_{zz}^I(h_1) = 0 \tag{5.3}$$

b)

$$\sigma_{jz}^I(h_1) = \sigma_{jz}^+(\xi, \eta)\exp(i\Omega t), \quad j = x, y, z \tag{5.4}$$

$$\sigma_{jz}^{II}(-h_2) = -\sigma_{jz}^-(\xi, \eta)\exp(i\Omega t), \quad \xi = x/\ell, \ \zeta = z/h. \tag{5.5}$$

The boundary value problem (5.1), (5.2) models seismic action over the basement-foundation structures. The continuity conditions are assumed over the contact area

$$u^I(x, y, 0, t) = u^{II}(x, y, 0, t), \quad (u, v, w), \tag{5.6}$$
$$\sigma_{jz}^I(x, y, 0, t) = \sigma_{jz}^{II}(x, y, 0, t), \quad j = x, y, z.$$

We begin the consideration from assigning an index "k" to all of the quantities of the equations of motion (2.1) and the constitutive relations (2.2). The solution is then sought in the form

$$\sigma_{\alpha\beta}^k(x, y, z, t) = \sigma_{jm}^k(x, y, z)\exp(i\Omega t), \quad \alpha, \beta = x, y, z; \ j, m = 1, 2, 3,$$
$$u^k(x, y, z, t) = \bar{u}^k(x, y, z)\exp(i\Omega t), \quad (u, v, w), \quad k = I, II. \tag{5.7}$$

Transforming to dimensionless variables and displacements (2.4), equations (2.1) are rewritten as

$$\frac{\partial \sigma_{11}^k}{\partial \xi} + \frac{\partial \sigma_{12}^k}{\partial \eta} + \varepsilon^{-1}\frac{\partial \sigma_{13}^k}{\partial \zeta} + \varepsilon^{-2}\rho_k\Omega_*^2 U^k = 0, \quad \Omega_* = h\Omega,$$

$$\frac{\partial \sigma_{12}^k}{\partial \xi} + \frac{\partial \sigma_{22}^k}{\partial \eta} + \varepsilon^{-1}\frac{\partial \sigma_{23}^k}{\partial \zeta} + \varepsilon^{-2}\rho_k\Omega_*^2 V^k = 0, \tag{5.8}$$

$$\frac{\partial \sigma_{13}^k}{\partial \xi} + \frac{\partial \sigma_{23}^k}{\partial \eta} + \varepsilon^{-1}\frac{\partial \sigma_{33}^k}{\partial \zeta} + \varepsilon^{-2}\rho_k\Omega_*^2 W^k = 0 \quad k = I, II.$$

Clearly, the form of the constitutive relations is very similar to (2.5). Once again, the presence of a singularly perturbed system allows decomposition of the solution into the outer solution and the boundary layer $I = I^{out} + I_b$. The outer solution I^{out} is sought in the form

$$\sigma_{jm}^{k,out} = \varepsilon^{-1+s}\sigma_{jm}^{(k,s)}(\xi,\eta,\zeta), \quad j,m = 1,2,3, \quad s = \overline{0,N},$$

$$(U^{k,out}, V^{k,out}, W^{k,out}) = \varepsilon^s(U^{(k,s)}, V^{(k,s)}, W^{(k,s)}), \quad k = I, II. \tag{5.9}$$

Substituting (5.9) into (5.8) and the constitutive relations, and using the relations (2.9), (2.10) it is possible to express the stresses through displacement components. The governing system for $U^{(k,s)}, V^{(k,s)}, W^{(k,s)}$ is then given by

$$\frac{\partial^2 U^{(k,s)}}{\partial\zeta^2} + a_{55}^k \rho_k \Omega_*^2 U^{(k,s)} = R_u^{(k,s)} \quad \Omega_*^2 = h^2\Omega^2,$$

$$\frac{\partial^2 V^{(k,s)}}{\partial\zeta^2} + a_{44}^k \rho_k \Omega_*^2 V^{(k,s)} = R_v^{(k,s)}, \tag{5.10}$$

$$A_{11}^k \frac{\partial^2 W^{(k,s)}}{\partial\zeta^2} + \rho_k \Omega_*^2 W^{(k,s)} = R_w^{(k,s)}, \quad k = I, II,$$

where $R_u^{(k,s)}$, $R_v^{(k,s)}$, $R_w^{(k,s)}$ are defined after (2.11)-(2.13) with the index "k" added to all of the appropriate quantities. The solution of (5.10) may be presented as

$$U^{(k,s)} = U_0^{(k,s)}(\xi,\eta,\zeta) + U_\tau^{(k,s)}(\xi,\eta,\zeta), \quad (U,V,W) \tag{5.11}$$

where

$$U_0^{(k,s)} = C_1^{(k,s)} \sin a_1^k \Omega_* \zeta + C_2^{(k,s)} \cos a_1^k \Omega_* \zeta,$$

$$V_0^{(k,s)} = C_3^{(k,s)} \sin a_2^k \Omega_* \zeta + C_4^{(k,s)} \cos a_2^k \Omega_* \zeta, \tag{5.12}$$

$$W_0^{(k,s)} = C_5^{(k,s)} \sin a_3^k \Omega_* \zeta + C_6^{(k,s)} \cos a_3^k \Omega_* \zeta,$$

$$a_1^k = \sqrt{a_{55}^k \rho_k}, \ a_2^k = \sqrt{a_{44}^k \rho_k}, \ a_3^k = \sqrt{\frac{\rho_k}{A_{11}^k}}, \quad k = I, II,$$

and $U_\tau^{(k,s)}$, $V_\tau^{(k,s)}$, $W_\tau^{(k,s)}$ are particular solutions of (5.10). Then, using the solutions (5.11)-(5.12) along with relations (2.9) (for both layers), it is possible to get the stress components satisfying the conditions (5.1)-(5.5). Thus, the resulting solutions are obtained for each type of the boundary conditions. Satisfying the conditions (5.1), (5.2), and (5.6), we have

$$U^{(k,s)} = C_1^{(k,s)}(\xi,\eta)\sin a_1^k \Omega_* \zeta + C_2^{(k,s)}(\xi,\eta)\cos a_1^k \Omega_* \zeta$$
$$+ U_\tau^{(k,s)}(\xi,\eta,\zeta),$$

$$C_1^{(I,s)} = \sqrt{\frac{a_{55}^I \rho_{II}}{a_{55}^{II} \rho_I}}\, \frac{f_1^{(s)}}{\Delta_1} + \sqrt{\frac{a_{55}^I}{\rho_I}}\, \frac{1}{\Omega_*}(b_{13}^{(II,s)}(\zeta=0) - b_{13}^{(I,s)}(\zeta=0)),$$

$$C_2^{(I,s)} = \frac{f_2^{(s)}}{\Delta_1} + U_\tau^{(II,s)}(\zeta=0) - U_\tau^{(I,s)}(\zeta=0),$$

$$C_1^{(II,s)} = \frac{f_1^{(s)}}{\Delta_1}, \quad C_2^{(II,s)} = \frac{f_2^{(s)}}{\Delta_1},$$

$$f_1^{(s)} = -(u^{-(s)} - U_\tau^{(II,s)}(\zeta=-\zeta_2))\sqrt{\frac{\rho_I}{a_{55}^I}}\,\Omega_* \cos a_1^I \Omega_* \zeta_1$$

$$+ (b_{13}^{(I,s)}(\zeta=0) - b_{13}^{(II,s)}(\zeta=0))\sin a_1^I \Omega_* \zeta_1 \cos a_1^{II}\Omega_* \zeta_2$$

$$- \sqrt{\frac{\rho_I}{a_{55}^I}}\,\Omega_*((U_\tau^{(II,s)}(\zeta=0) - U_\tau^{(I,s)}(\zeta=0))\cos a_1^I \Omega_* \zeta_1$$

$$+ U_\tau^{(I,s)}(\zeta=\zeta_1))\cos a_1^{II}\Omega_* \zeta_2, \quad \zeta_1 = \frac{h_1}{h}, \quad \zeta_2 = \frac{h_2}{h}, \qquad (5.13)$$

$$f_2^{(s)} = (u^{-(s)} - U_\tau^{(II,s)}(\zeta=-\zeta_2))\sqrt{\frac{\rho_{II}}{a_{55}^{II}}}\,\Omega_* \sin a_1^I \Omega_* \zeta_1$$

$$+ (b_{13}^{(I,s)}(\zeta=0) - b_{13}^{(II,s)}(\zeta=0))\sin a_1^I \Omega_* \zeta_1 \sin a_1^{II}\Omega_* \zeta_2$$

$$- \sqrt{\frac{\rho_I}{a_{55}^I}}\,\Omega_*((U_\tau^{(II,s)}(\zeta=0) - U_\tau^{(I,s)}(\zeta=0))\cos a_1^I \Omega_* \zeta_1$$

$$+ U_\tau^{(I,s)}(\zeta=\zeta_1))\sin a_1^{II}\Omega_* \zeta_2,$$

$$b_{13}^{(k,s)} = \frac{1}{a_{55}^k}\left(\frac{\partial U_\tau^{(k,s)}}{\partial \zeta} + \frac{\partial W^{(k,s-1)}}{\partial \xi}\right),$$

$$u^{-(0)} = \frac{u^-}{\ell}, \quad u^{-(s)} = 0, \quad s \neq 0, \quad (u,v,w), \quad \Omega_* = h\Omega,$$

$$\Delta_1 = \left(\sqrt{\frac{\rho_I}{a_{55}^I}}\cos a_1^I \Omega_* \zeta_1 \sin a_1^{II}\Omega_* \zeta_2\right.$$

$$\left. + \sqrt{\frac{\rho_{II}}{a_{55}^{II}}}\sin a_1^I \Omega_* \zeta_1 \cos a_1^{II}\Omega_* \zeta_2\right)\Omega_*,$$

$$V^{(k,s)} = C_3^{(k,s)}(\xi,\eta)\sin a_2^k \Omega_* \zeta + C_4^{(k,s)}(\xi,\eta)\cos a_2^k \Omega_* \zeta$$
$$+ V_\tau^{(k,s)}(\xi,\eta,\zeta),$$

$$C_3^{(I,s)} = \sqrt{\frac{a_{44}^I \rho_{II}}{a_{44}^{II}\rho_I}}\,\frac{f_3^{(s)}}{\Delta_2} + \sqrt{\frac{a_{44}^I}{\rho_I}}\,\frac{1}{\Omega_*}(b_{23}^{(II,s)}(\zeta=0) - b_{23}^{(I,s)}(\zeta=0)),$$

$$C_4^{(I,s)} = \frac{f_4^{(s)}}{\Delta_2} + V_\tau^{(II,s)}(\zeta=0) - V_\tau^{(I,s)}(\zeta=0),$$

$$C_3^{(II,s)} = \frac{f_3^{(s)}}{\Delta_2}, \quad C_4^{(II,s)} = \frac{f_4^{(s)}}{\Delta_2},$$

$$f_3^{(s)} = -(v^{-(s)} - V_\tau^{(II,s)}(\zeta = -\zeta_2))\sqrt{\frac{\rho_I}{a_{44}^I}}\Omega_* \cos a_2^I \Omega_* \zeta_1$$

$$+(b_{23}^{(I,s)}(\zeta = 0) - b_{23}^{(II,s)}(\zeta = 0))\sin a_2^I \Omega_* \zeta_1 \cos a_2^{II} \Omega_* \zeta_2$$

$$-\sqrt{\frac{\rho_I}{a_{44}^I}}\Omega_*((V_\tau^{(II,s)}(\zeta = 0) - V_\tau^{(I,s)}(\zeta = 0))\cos a_2^I \Omega_* \zeta_1$$

$$+V_\tau^{(I,s)}(\zeta = \zeta_1))\cos a_2^{II} \Omega_* \zeta_2,$$

$$f_4^{(s)} = (v^{-(s)} - V_\tau^{(II,s)}(\zeta = -\zeta_2))\sqrt{\frac{\rho_{II}}{a_{44}^{II}}}\Omega_* \sin a_2^I \Omega_* \zeta_2$$

$$+(b_{23}^{(I,s)}(\zeta = 0) - b_{23}^{(II,s)}(\zeta = 0))\sin a_2^I \Omega_* \zeta_1 \sin a_2^{II} \Omega_* \zeta_2$$

$$-\sqrt{\frac{\rho_I}{a_{44}^I}}\Omega_*((V_\tau^{(II,s)}(\zeta = 0) - V_\tau^{(I,s)}(\zeta = 0))\cos a_2^I \Omega_* \zeta_1$$

$$+V_\tau^{(I,s)}(\zeta = \zeta_1))\sin a_2^{II} \Omega_* \zeta_2, \quad (5.14)$$

$$b_{23}^{(k,s)} = \frac{1}{a_{44}^k}\left(\frac{\partial V_\tau^{(k,s)}}{\partial \zeta} + \frac{\partial W^{(k,s-1)}}{\partial \eta}\right),$$

$$\Delta_2 = \left(\sqrt{\frac{\rho_I}{a_{44}^I}}\cos a_2^I \Omega_* \zeta_1 \sin a_2^{II} \Omega_* \zeta_2\right.$$

$$\left.+\sqrt{\frac{\rho_{II}}{a_{44}^{II}}}\sin a_2^I \Omega_* \zeta_1 \cos a_2^{II} \Omega_* \zeta_2\right)\Omega_*,$$

$$W^{(k,s)} = C_5^{(k,s)}(\xi, \eta)\sin a_3^k \Omega_* \zeta$$

$$+C_6^{(k,s)}(\xi, \eta)\cos a_3^k \Omega_* \zeta + W_\tau^{(k,s)}(\xi, \eta, \zeta),$$

$$C_5^{(I,s)} = \sqrt{\frac{A_{11}^{II}\rho_{II}}{A_{11}^I \rho_I}}\frac{f_5^{(s)}}{\Delta_3} + \frac{1}{\sqrt{A_{11}^I \rho_I}}\frac{1}{\Omega_*}(b_{33}^{(II,s)}(\zeta = 0) - b_{33}^{(I,s)}(\zeta = 0)),$$

$$C_6^{(I,s)} = \frac{f_6^{(s)}}{\Delta_3} + W_\tau^{(II,s)}(\zeta = 0) - W_\tau^{(I,s)}(\zeta = 0),$$

$$C_5^{(II,s)} = \frac{f_5^{(s)}}{\Delta_3}, \quad C_6^{(II,s)} = \frac{f_6^{(s)}}{\Delta_3}, \quad (5.15)$$

$$f_5^{(s)} = -(w^{-(s)} - W_\tau^{(II,s)}(\zeta = -\zeta_2))\Omega_*\sqrt{A_{11}^I \rho_I}\cos a_3^I \Omega_* \zeta_1$$

$$+(b_{33}^{(I,s)}(\zeta = 0) - b_{33}^{(II,s)}(\zeta = 0))\sin a_3^I \Omega_* \zeta_1 \sin a_3^{II} \Omega_* \zeta_2$$

$$+\sqrt{A_{11}^I \rho_I}\Omega_*\left((W_\tau^{(I,s)}(\zeta = 0)\right.$$

$$\left.-W_\tau^{(II,s)}(\zeta = 0)\right)\cos a_3^I \Omega_* \zeta_1 \cos a_3^{II} \Omega_* \zeta_2,$$

$$f_6^{(s)} = (w^{-(s)} - W_\tau^{(II,s)}(\zeta = -\zeta_2))\sqrt{A_{11}^{II}\rho_{II}}\Omega_* \sin a_3^I \Omega_* \zeta_1$$

$$+(b_{33}^{(I,s)}(\zeta = 0) - b_{33}^{(II,s)}(\zeta = 0))\sin a_3^I \Omega_* \zeta_1 \cos a_3^{II} \Omega_* \zeta_2$$

$$+\sqrt{A_{11}^I \rho_I}\Omega_*((W_\tau^{(I,s)}(\zeta = 0) - W_\tau^{(II,s)}(\zeta = 0))\cos a_3^I \Omega_* \zeta_1 \sin a_3^{II} \Omega_* \zeta_2,$$

$$b_{33}^{(k,s)} = A_{11}^k \frac{\partial W_\tau^{(k,s)}}{\partial \zeta} - A_{23}^k \frac{\partial U_\tau^{(k,s-1)}}{\partial \xi} - A_{13}^k \frac{\partial V_\tau^{(k,s-1)}}{\partial \eta}$$

and

$$\Delta_3 = \left(\sqrt{A^I_{11}\rho_I} \cos a^I_3 \Omega_* \zeta_1 \sin a^{II}_3 \Omega_* \zeta_2 \sqrt{A^{II}_{11}\rho_{II}} \sin a^I_3 \Omega_* \zeta_1 \cos a^{II}_3 \Omega_* \zeta_2 \right) \Omega_*.$$

The stress tensor components may be calculated through expressions (2.9) with the assigned index "k". The solutions (5.13)-(5.16) are finite provided that $\Delta_1 \neq 0$, $\Delta_2 \neq 0$, $\Delta_3 \neq 0$, otherwise the resonant phenomena occur. It may be shown that $\Delta_i = 0$ corresponds to the case of Ω being a natural frequency.

We remark once again that if the functions $u^-(\xi,\eta)$, $v^-(\xi,\eta)$, $w^-(\xi,\eta)$ are polynomials in ξ,η, then the asymptotic process gives mathematically exact solution for a layer in a finite number of iterations. In particular, if $u^- = v^- = w^- = \text{const}$, the solution is given by:

for the first layer $(0 \leq \zeta \leq \zeta_1)$

$$U^I = \sqrt{\frac{\rho_{II}}{a^{II}_{55}}} \frac{u^-}{\ell} \frac{\Omega_*}{\Delta_1} \sin a^I_1 \Omega_*(\zeta_1 - \zeta),$$

$$V^I = \sqrt{\frac{\rho_{II}}{a^{II}_{44}}} \frac{v^-}{\ell} \frac{\Omega_*}{\Delta_2} \sin a^I_2 \Omega_*(\zeta_1 - \zeta),$$

$$W^I = \sqrt{A^{II}_{11}\rho_{II}} \frac{w^-}{\ell} \frac{\Omega_*}{\Delta_3} \sin a^I_3 \Omega_*(\zeta_1 - \zeta),$$

$$\sigma^I_{13} = \frac{1}{a^I_{55}} \frac{\partial U^I}{\partial \zeta}, \quad \sigma^I_{23} = \frac{1}{a^I_{44}} \frac{\partial V^I}{\partial \zeta}, \quad \sigma^I_{33} = A^I_{11} \frac{\partial W^I}{\partial \zeta}, \qquad (5.16)$$

$$\sigma^I_{12} = 0, \quad \sigma^I_{11} = -A^I_{23} \frac{\partial W^I}{\partial \zeta}, \quad \sigma^I_{22} = -A^I_{13} \frac{\partial W^I}{\partial \zeta},$$

for the second layer $(-\zeta_2 \leq \zeta \leq 0)$

$$U^{II} = \left(\sqrt{\frac{\rho_{II}}{a^{II}_{55}}} \sin a^I_1 \Omega_* \zeta_1 \cos a^{II}_1 \Omega_* \zeta \right.$$

$$\left. - \sqrt{\frac{\rho_I}{a^I_{55}}} \cos a^I_1 \Omega_* \zeta_1 \sin a^{II}_1 \Omega_* \zeta \right) \frac{u^-}{\ell} \frac{\Omega_*}{\Delta_1},$$

$$V^{II} = \left(\sqrt{\frac{\rho_{II}}{a^{II}_{44}}} \sin a^I_2 \Omega_* \zeta_1 \cos a^{II}_2 \Omega_* \zeta \right.$$

$$\left. - \sqrt{\frac{\rho_I}{a^I_{44}}} \cos a^I_2 \Omega_* \zeta_1 \sin a^{II}_2 \Omega_* \zeta \right) \frac{v^-}{\ell} \frac{\Omega_*}{\Delta_2},$$

$$W^{II} = \left(\sqrt{A^{II}_{11}\rho_{II}} \sin a^I_3 \Omega_* \zeta_1 \cos a^{II}_3 \Omega_* \zeta \right.$$

$$\left. - \sqrt{A^I_{11}\rho_I} \cos a^I_3 \Omega_* \zeta_1 \sin a^{II}_3 \Omega_* \zeta \right) \frac{w^-}{\ell} \frac{\Omega_*}{\Delta_3}, \qquad (5.17)$$

$$\sigma^{II}_{13} = \frac{1}{a^{II}_{55}} \frac{\partial U^{II}}{\partial \zeta}, \quad \sigma^{II}_{23} = \frac{1}{a^{II}_{44}} \frac{\partial V^{II}}{\partial \zeta}, \quad \sigma^{II}_{33} = A^{II}_{11} \frac{\partial W^{II}}{\partial \zeta},$$

$$\sigma^{II}_{12} = 0, \quad \sigma^{II}_{11} = -A^{II}_{23} \frac{\partial W^{II}}{\partial \zeta}, \quad \sigma^{II}_{22} = -A^{II}_{13} \frac{\partial W^{II}}{\partial \zeta}.$$

In view of (2.4), (5.7), and (5.9) the resulting solution is written as

$$\sigma_{\alpha\beta}^{k}(x,y,z,t) = \varepsilon^{-1}\sigma_{jm}^{k}\exp(i\Omega t), \quad \alpha,\beta = x,y,z, \quad j,m = 1,2,3,$$

$$u^{k}(x,y,z,t) = \ell U^{k}\exp(i\Omega t), \quad (u,v,w), \quad k = I, II. \qquad (5.18)$$

The solution of the problem (5.1), (5.3), and (5.6) may be obtained in a similar manner.

For the sake of brevity we present below the solution for $u^{-} = v^{-} = w^{-} = \text{const}$ (with the asymptotic process containing leading order approximation only).

For the first layer ($0 \le \zeta \le \zeta_{1}$)

$$U^{I} = \frac{b_{1}}{\delta_{1}}\frac{u^{-}}{\ell}\cos a_{1}^{I}\Omega_{*}(\zeta_{1} - \zeta),$$

$$V^{I} = \frac{b_{2}}{\delta_{2}}\frac{v^{-}}{\ell}\cos a_{2}^{I}\Omega_{*}(\zeta_{1} - \zeta),$$

$$W^{I} = \frac{b_{3}}{\delta_{3}}\frac{w^{-}}{\ell}\cos a_{3}^{I}\Omega_{*}(\zeta_{1} - \zeta),$$

$$\sigma_{13}^{I} = \frac{a_{1}^{II}}{a_{55}^{II}}\frac{\Omega_{*}}{\delta_{1}}\frac{u^{-}}{\ell}\sin a_{1}^{I}\Omega_{*}(\zeta_{1} - \zeta), \qquad (5.19)$$

$$\sigma_{23}^{I} = \frac{a_{2}^{II}}{a_{44}^{II}}\frac{\Omega_{*}}{\delta_{2}}\frac{v^{-}}{\ell}\sin a_{2}^{I}\Omega_{*}(\zeta_{1} - \zeta),$$

$$\sigma_{33}^{I} = \frac{A_{1}^{II}a_{3}^{II}}{\delta_{3}}\Omega_{*}\frac{w^{-}}{\ell}\sin a_{3}^{I}\Omega_{*}(\zeta_{1} - \zeta),$$

$$\sigma_{12}^{I} = 0, \quad \sigma_{11}^{I} = -A_{23}^{I}\frac{\partial W^{I}}{\partial\zeta}, \quad \sigma_{22}^{I} = -A_{13}^{I}\frac{\partial W^{I}}{\partial\zeta},$$

$$b_{1} = \frac{a_{55}^{I}a_{1}^{II}}{a_{55}^{II}a_{1}^{I}}, \quad b_{2} = \frac{a_{44}^{I}a_{2}^{II}}{a_{44}^{II}a_{2}^{I}}, \quad b_{3} = \frac{A_{11}^{II}a_{3}^{II}}{A_{11}^{I}a_{2}^{I}},$$

where

$$\delta_{1} = b_{1}\cos a_{1}^{I}\Omega_{*}\zeta_{1}\cos a_{1}^{II}\Omega_{*}\zeta_{2} - \sin a_{1}^{I}\Omega_{*}\zeta_{1}\sin a_{1}^{II}\Omega_{*}\zeta_{2},$$

$$\delta_{2} = b_{2}\cos a_{2}^{I}\Omega_{*}\zeta_{1}\cos a_{2}^{II}\Omega_{*}\zeta_{2} - \sin a_{2}^{I}\Omega_{*}\zeta_{1}\sin a_{2}^{II}\Omega_{*}\zeta_{2}, \qquad (5.20)$$

$$\delta_{3} = b_{3}\cos a_{3}^{I}\Omega_{*}\zeta_{1}\cos a_{3}^{II}\Omega_{*}\zeta_{2} - \sin a_{3}^{I}\Omega_{*}\zeta_{1}\sin a_{3}^{II}\Omega_{*}\zeta_{2}.$$

For the second layer ($-\zeta_{2} \le \zeta \le 0$)

$$U^{II} = \frac{u^{-}}{\delta_{1}\ell}\left(b_{1}\cos a_{1}^{I}\Omega_{*}\zeta_{1}\cos a_{1}^{II}\Omega_{*}\zeta + \sin a_{1}^{I}\Omega_{*}\zeta_{1}\sin a_{1}^{II}\Omega_{*}\zeta\right),$$

$$V^{II} = \frac{v^{-}}{\delta_{2}\ell}\left(b_{2}\cos a_{2}^{I}\Omega_{*}\zeta_{1}\cos a_{2}^{II}\Omega_{*}\zeta + \sin a_{2}^{I}\Omega_{*}\zeta_{1}\sin a_{2}^{II}\Omega_{*}\zeta\right),$$

$$W^{II} = \frac{w^{-}}{\delta_{3}\ell}\left(b_{3}\cos a_{3}^{I}\Omega_{*}\zeta_{1}\cos a_{3}^{II}\Omega_{*}\zeta + \sin a_{3}^{I}\Omega_{*}\zeta_{1}\sin a_{3}^{II}\Omega_{*}\zeta\right), \qquad (5.21)$$

$$\sigma_{13}^{II} = \frac{1}{a_{55}^{II}}\frac{\partial U^{II}}{\partial\zeta}, \quad \sigma_{23}^{II} = \frac{1}{a_{44}^{II}}\frac{\partial V^{II}}{\partial\zeta}, \quad \sigma_{33}^{II} = A_{11}^{II}\frac{\partial W^{II}}{\partial\zeta},$$

$$\sigma_{12}^{II} = 0, \quad \sigma_{11}^{II} = -A_{23}^{II}\frac{\partial W^{II}}{\partial\zeta}, \quad \sigma_{22}^{II} = -A_{13}^{II}\frac{\partial W^{II}}{\partial\zeta}.$$

The final solution is given by (5.18). Clearly, the solution is finite provided that $\delta_1 \neq 0$, $\delta_2 \neq 0$, $\delta_3 \neq 0$, otherwise resonant effects may be expected. It may be shown that the values of the frequency of external forcing Ω for which $\delta_i = 0$, coincide with the natural frequencies of the two-layered plate, for more details see Aghalovyan (2008a).

Consider now the conditions (5.4)-(5.6). From the analysis of the latter the constants $C_j^{(k,s)}(\xi, \eta)$ are evaluated, giving

$$C_1^{(I,s)} = b_1^{(s)} + \sqrt{\frac{\rho_{II} a_{55}^I}{\rho_I a_{55}^{II}}} \frac{d_1^{(s)}}{\delta_1}, \quad C_1^{(II,s)} = \frac{d_1^{(s)}}{\delta_1},$$

$$C_2^{(I,s)} = b_2^{(s)} + \frac{d_2^{(s)}}{\delta_1}, \quad C_2^{(II,s)} = \frac{d_2^{(s)}}{\delta_1},$$

$$b_1^{(s)} = \frac{1}{\Omega_*} \sqrt{\frac{a_{55}^I}{\rho_I}} (\sigma_{13\tau}^{(II,s)}(\xi, \eta, 0) - \sigma_{13\tau}^{(I,s)}(\xi, \eta, 0)),$$

$$b_2^{(s)} = U_\tau^{(II,s)}(\xi, \eta, 0) - U_\tau^{(I,s)}(\xi, \eta, 0),$$

$$b_3^{(s)} = \frac{1}{\Omega_*} \sqrt{\frac{a_{55}^{II}}{\rho_{II}}} (-\sigma_{xz}^{-(s)} - \sigma_{13\tau}^{(II,s)}(\xi, \eta, -\zeta_2)), \qquad (5.22)$$

$$b_4^{(s)} = \frac{1}{\Omega_*} \sqrt{\frac{a_{55}^I}{\rho_I}} (\sigma_{xz}^{+(s)} - \sigma_{13\tau}^{(I,s)}(\xi, \eta, \zeta_1)),$$

$$d_1^{(s)} = b_3^{(s)} \sin a_1^I \zeta_1 + (b_4^{(s)} - b_1^{(s)} \cos a_1^I \zeta_1 + b_2^{(s)} \sin a_1^I \zeta_1) \sin a_1^{II} \zeta_2,$$

$$d_2^{(s)} = b_3^{(s)} \sqrt{\frac{\rho_{II} a_{55}^I}{\rho_I a_{55}^{II}}} \cos a_1^I \zeta_1$$

$$- (b_4^{(s)} - b_1^{(s)} \cos a_1^I \zeta_1 + b_2^{(s)} \sin a_1^I \zeta_1) \cos a_1^{II} \zeta_2,$$

$$\delta_1 = \sin a_1^I \zeta_1 \cos a_1^{II} \zeta_2 + \sqrt{\frac{\rho_{II} a_{55}^I}{\rho_I a_{55}^{II}}} \cos a_1^I \zeta_1 \sin a_1^{II} \zeta_2,$$

$$C_3^{(I,s)} = b_5^{(s)} + \sqrt{\frac{\rho_{II} a_{44}^I}{\rho_I a_{44}^{II}}} \frac{d_3^{(s)}}{\delta_2}, \quad C_3^{(II,s)} = \frac{d_3^{(s)}}{\delta_2},$$

$$C_4^{(I,s)} = b_6^{(s)} + \frac{d_4^{(s)}}{\delta_2}, \quad C_4^{(II,s)} = \frac{d_4^{(s)}}{\delta_2},$$

$$b_5^{(s)} = \frac{1}{\Omega_*} \sqrt{\frac{a_{44}^I}{\rho_I}} (\sigma_{23\tau}^{(II,s)}(\xi, \eta, 0) - \sigma_{23\tau}^{(I,s)}(\xi, \eta, 0)),$$

$$b_6^{(s)} = V_\tau^{(II,s)}(\xi, \eta, 0) - V_\tau^{(I,s)}(\xi, \eta, 0), \qquad (5.23)$$

$$b_7^{(s)} = \frac{1}{\Omega_*} \sqrt{\frac{a_{44}^{II}}{\rho_{II}}} (-\sigma_{yz}^{-(s)} - \sigma_{23\tau}^{(II,s)}(\xi, \eta, -\zeta_2)),$$

$$b_8^{(s)} = \frac{1}{\Omega_*}\sqrt{\frac{a_{44}^I}{\rho_I}}(\sigma_{yz}^{+(s)} - \sigma_{23\tau}^{(I,s)}(\xi,\eta,\zeta_1)),$$

$$d_3^{(s)} = b_7^{(s)}\sin a_2^I\zeta_1 + (b_8^{(s)} - b_5^{(s)}\cos a_2^I\zeta_1 + b_6^{(s)}\sin a_2^I\zeta_1)\sin a_2^{II}\zeta_2,$$

$$d_4^{(s)} = b_7^{(s)}\sqrt{\frac{\rho_{II}a_{44}^I}{\rho_I a_{44}^{II}}}\cos a_2^I\zeta_1$$

$$-(b_8^{(s)} - b_5^{(s)}\cos a_2^I\zeta_1 + b_6^{(s)}\sin a_2^I\zeta_1)\cos a_2^{II}\zeta_2,$$

$$\delta_2 = \cos a_2^{II}\zeta_2\sin a_2^I\zeta_1 + \sqrt{\frac{\rho_{II}a_{44}^I}{\rho_I a_{44}^{II}}}\cos a_2^I\zeta_1\sin a_2^{II}\zeta_2,$$

$$C_5^{(I,s)} = b_9^{(s)} + \sqrt{\frac{\rho_{II}A_{11}^{II}}{\rho_I A_{11}^I}}\frac{d_5^{(s)}}{\delta_3}, \quad C_5^{(II,s)} = \frac{d_5^{(s)}}{\delta_3},$$

$$C_6^{(I,s)} = b_{10}^{(s)} + \frac{d_6^{(s)}}{\delta_3}, \quad C_6^{(II,s)} = \frac{d_6^{(s)}}{\delta_3},$$

$$b_9^{(s)} = \frac{1}{\Omega_*}\sqrt{\frac{1}{\rho_I A_{11}^I}}(\sigma_{33\tau}^{(II,s)}(\xi,\eta,0) - \sigma_{33\tau}^{(I,s)}(\xi,\eta,0)),$$

$$b_{10}^{(s)} = W_\tau^{(II,s)}(\xi,\eta,0) - W_\tau^{(I,s)}(\xi,\eta,0),$$

$$b_{11}^{(s)} = \frac{1}{\Omega_*}\sqrt{\frac{1}{\rho_{II}A_{11}^{II}}}(-\sigma_{zz}^{-(s)} - \sigma_{33\tau}^{(II,s)}(\xi,\eta,-\zeta_2)), \qquad (5.24)$$

$$b_{12}^{(s)} = \frac{1}{\Omega_*}\sqrt{\frac{1}{\rho_I A_{11}^I}}(\sigma_{zz}^{+(s)} - \sigma_{33\tau}^{(I,s)}(\xi,\eta,\zeta_1)),$$

$$d_5^{(s)} = b_{11}^{(s)}\sin a_3^I\zeta_1 + (b_{12}^{(s)} - b_9^{(s)}\cos a_3^I\zeta_1 + b_{10}^{(s)}\sin a_3^I\zeta_1)\sin a_3^{II}\zeta_2,$$

$$d_6^{(s)} = b_{11}^{(s)}\sqrt{\frac{\rho_{II}A_{11}^{II}}{\rho_I A_{11}^I}}\cos a_3^I\zeta_1$$

$$-(b_{12}^{(s)} - b_9^{(s)}\cos a_3^I\zeta_1 + b_{10}^{(s)}\sin a_3^I\zeta_1)\cos a_3^{II}\zeta_2,$$

$$\delta_3 = \cos a_3^{II}\zeta_2\sin a_3^I\zeta_1 + \sqrt{\frac{\rho_{II}A_{11}^{II}}{\rho_I A_{11}^I}}\cos a_3^I\zeta_1\sin a_3^{II}\zeta_2,$$

where $\sigma_{jz}^{\pm(0)} = \varepsilon\sigma_{jz}^\pm$, $\sigma_{jz}^{\pm(s)} = 0$, $s \neq 0$, $j = x,y,z$; $\sigma_{m3\tau}^{(k,s)}$, $m = 1,2,3$ are defined after (2.16) with index "k", $k = I, II$ added to all the appropriate quantities. Similarly to the previous case, the solution (5.22)-(5.24) is finite provided that $\delta_1 \neq 0$, $\delta_2 \neq 0$, $\delta_3 \neq 0$, with resonant effects occurring at $\delta_i = 0$, with the corresponding frequencies Ω coinciding with the natural frequencies of the plate with free faces.

Once again, we remark that in case of polynomial loading functions $\sigma_{jz}^\pm$ (in ξ,η) the iterative process involves finite number of steps providing a mathematically exact outer solution, for more details the reader is referred to a recent contribution of Aghalovyan and Zakaryan (2011). As an example, let us consider

$$\sigma_{zz}^{+} = -P_1 = \text{const}, \quad \sigma_{zz}^{-} = P_2 = \text{const}, \tag{5.25}$$

$$\sigma_{xz}^{\pm} = \sigma_{yz}^{\pm} = 0.$$

It follows from (5.22)-(5.24) as

$$C_1^{(k,0)} = C_2^{(k,0)} = C_3^{(k,0)} = C_4^{(k,0)} = 0, \quad k = I, II,$$

$$C_5^{(I,0)} = -\frac{\varepsilon}{\delta_3 \Omega_*} \left(P_1 \frac{1}{\rho_I A_{11}^I} \sqrt{\rho_{II} A_{11}^{II}} \sin a_3^{II} \zeta_2 + P_2 \sqrt{\frac{1}{\rho_I A_{11}^I}} \sin a_3^I \zeta_1 \right), \tag{5.26}$$

$$C_5^{(II,0)} = -\frac{\varepsilon}{\delta_3 \Omega_*} \left(P_1 \sqrt{\frac{1}{\rho_I A_{11}^I}} \sin a_3^{II} \zeta_2 + P_2 \sqrt{\frac{1}{\rho_{II} A_{11}^{II}}} \sin a_3^I \zeta_1 \right),$$

$$C_6^{(I,0)} = C_6^{(II,0)} = -\frac{\varepsilon}{\delta_3 \Omega_*} \sqrt{\frac{1}{\rho_I A_{11}^I}} (P_2 \cos a_3^I \zeta_1 - P_1 \cos a_3^{II} \zeta_2),$$

therefore

$$U^{(k,0)} = 0, \quad V^{(k,0)} = 0, \quad \sigma_{13}^{(k,0)} = \sigma_{23}^{(k,0)} = \sigma_{12}^{(k,0)} = 0,$$

$$W^{(k,0)} = C_5^{(k,0)} \sin a_3^k \zeta + C_6^{(k,0)} \cos a_3^k \zeta,$$

$$\sigma_{33}^{(I,0)} = -\frac{\varepsilon}{\delta_3} \left((P_2 \sin a_3^I \zeta_1 + P_1 \sqrt{\frac{\rho_{II} A_{11}^{II}}{\rho_I A_{11}^I}} \sin a_3^{II} \zeta_2) \cos a_3^I \zeta \right.$$

$$\left. -(P_2 \cos a_3^I \zeta_1 - P_1 \cos a_3^{II} \zeta_2) \sin a_3^I \zeta \right), \tag{5.27}$$

$$\sigma_{33}^{(II,0)} = -\frac{\varepsilon}{\delta_3} \left((P_2 \sin a_3^I \zeta_1 + P_1 \sqrt{\frac{\rho_{II} A_{11}^{II}}{\rho_I A_{11}^I}} \sin a_3^{II} \zeta_2) \cos a_3^{II} \zeta \right.$$

$$\left. -\sqrt{\frac{\rho_{II} A_{11}^{II}}{\rho_I A_{11}^I}} (P_2 \cos a_3^I \zeta_1 - P_1 \cos a_3^{II} \zeta_2) \sin a_3^{II} \zeta \right),$$

$$\sigma_{11}^{(k,0)} = -A_{23}^k \frac{\partial W^{(k,0)}}{\partial \zeta}, \quad \sigma_{22}^{(k,0)} = -A_{13}^k \frac{\partial W^{(k,0)}}{\partial \zeta},$$

$$\delta_3 = \cos a_3^{II} \zeta_2 \sin a_3^I \zeta_1 + \sqrt{\frac{\rho_{II} A_{11}^{II}}{\rho_I A_{11}^I}} \cos a_3^I \zeta_1 \sin a_3^{II} \zeta_2.$$

The higher order approximations ($s \geq 1$) reveal zero components of the stress tensor and displacement vector, thus leading to the following resulting outer solution (a mathematically exact solution for two layers)

$$u^k = 0, \quad v^k = 0, \quad \sigma_{xy}^k = \sigma_{xz}^k = \sigma_{yz}^k = 0, \quad k = I, II,$$

$$w^k = \ell w^{(k,0)} \exp(i\Omega t), \quad \sigma_{xx}^k = \varepsilon^{-1} \sigma_{11}^{(k,0)} \exp(i\Omega t), \tag{5.28}$$

$$\sigma_{yy}^k = \varepsilon^{-1} \sigma_{22}^{(k,0)} \exp(i\Omega t), \quad \sigma_{zz}^k = \varepsilon^{-1} \sigma_{33}^{(k,0)} \exp(i\Omega t).$$

In the boundary value problems (5.1)-(5.3), (5.6), and (5.4)-(5.6) considered above the outer solution is fully determined by the face boundary conditions. As a rule, the outer solution does not satisfy the edge boundary conditions, with the discrepancy treated through the boundary layer solution. The asymptotic approach may be extended to similar problems for three-layer and multi-layer plates and shells, see Aghalovyan (2008b); Aghalovyan and Hovhannisyan (2008). It is worth noting that in the above cited contributions it has been shown that in case of a boundary value problem similar to (5.1)-(5.3) in extension to a three-layer plate, when the tangential force is acting on the lower layer (analogous to (5.1)), which is modeling the real seismic action, and the middle layer is relatively weak (composed of a rubber-like material), the effect of the load on the upper layer is reduced. Thus, the necessity of seismic isolators use in structures has been mathematically justified.

11.6 Forced Vibrations of Plates of General Anisotropy

Below we generalize the results of forced vibration problems for plates of general anisotropy (involving 21 material constants). The formulated boundary value problems (1.1)-(1.7) for plates of general anisotropy may be analyzed through asymptotic method, see Aghalovyan and Aghalovyan (2005); Aghahovyan M. (2007). The governing system of equations involves (2.1) along with the following constitutive relations

$$
\begin{aligned}
\frac{\partial u}{\partial x} &= a_{11}\sigma_{xx} + a_{12}\sigma_{yy} + a_{13}\sigma_{zz} + a_{14}\sigma_{yz} + a_{15}\sigma_{xz} + a_{16}\sigma_{xy}, \\
\frac{\partial v}{\partial y} &= a_{12}\sigma_{xx} + a_{22}\sigma_{yy} + a_{23}\sigma_{zz} + a_{24}\sigma_{yz} + a_{25}\sigma_{xz} + a_{26}\sigma_{xy}, \\
\frac{\partial w}{\partial z} &= a_{13}\sigma_{xx} + a_{23}\sigma_{yy} + a_{33}\sigma_{zz} + a_{34}\sigma_{yz} + a_{35}\sigma_{xz} + a_{36}\sigma_{xy}, \\
\frac{\partial v}{\partial z} + \frac{\partial w}{\partial y} &= a_{14}\sigma_{xx} + a_{24}\sigma_{yy} + a_{34}\sigma_{zz} + a_{44}\sigma_{yz} + a_{45}\sigma_{xz} + a_{46}\sigma_{xy}, \quad (6.1) \\
\frac{\partial u}{\partial z} + \frac{\partial w}{\partial x} &= a_{15}\sigma_{xx} + a_{25}\sigma_{yy} + a_{35}\sigma_{zz} + a_{45}\sigma_{yz} + a_{55}\sigma_{xz} + a_{56}\sigma_{xy}, \\
\frac{\partial u}{\partial y} + \frac{\partial v}{\partial x} &= a_{16}\sigma_{xx} + a_{26}\sigma_{yy} + a_{36}\sigma_{zz} + a_{46}\sigma_{yz} + a_{56}\sigma_{xz} + a_{66}\sigma_{xy}.
\end{aligned}
$$

Each type of the boundary conditions (1.1)-(1.7) may be considered. The solution is then sought in the form (2.3). After transformation to dimensionless quantities (2.4), one arrives at a singularly perturbed system of the form (2.6). The outer solution may be found as (2.7), leading to the following equations in respect of $\sigma_{jk}^{(s)}$, $u^{(s)}, v^{(s)}, \mathrm{w}^{(s)}$

$$\frac{\partial \sigma_{11}^{(s-1)}}{\partial \xi} + \frac{\partial \sigma_{12}^{(s-1)}}{\partial \eta} + \frac{\partial \sigma_{13}^{(s)}}{\partial \zeta} + \Omega_*^2 U^{(s)} = 0,$$

$$\frac{\partial \sigma_{12}^{(s-1)}}{\partial \xi} + \frac{\partial \sigma_{22}^{(s-1)}}{\partial \eta} + \frac{\partial \sigma_{23}^{(s)}}{\partial \zeta} + \Omega_*^2 V^{(s)} = 0, \tag{6.2}$$

$$\frac{\partial \sigma_{13}^{(s-1)}}{\partial \xi} + \frac{\partial \sigma_{23}^{(s-1)}}{\partial \eta} + \frac{\partial \sigma_{33}^{(s)}}{\partial \zeta} + \Omega_*^2 W^{(s)} = 0$$

and

$$\frac{\partial U^{(s-1)}}{\partial \xi} = a_{11}\sigma_{11}^{(s)} + a_{12}\sigma_{22}^{(s)} + a_{13}\sigma_{33}^{(s)} + a_{14}\sigma_{23}^{(s)} + a_{15}\sigma_{13}^{(s)} + a_{16}\sigma_{12}^{(s)},$$

$$\frac{\partial V^{(s-1)}}{\partial \eta} = a_{12}\sigma_{11}^{(s)} + a_{22}\sigma_{22}^{(s)} + a_{23}\sigma_{33}^{(s)} + a_{24}\sigma_{23}^{(s)} + a_{25}\sigma_{13}^{(s)} + a_{26}\sigma_{12}^{(s)},$$

$$\frac{\partial W^{(s)}}{\partial \zeta} = a_{13}\sigma_{11}^{(s)} + a_{23}\sigma_{22}^{(s)} + a_{33}\sigma_{33}^{(s)} + a_{34}\sigma_{23}^{(s)} + a_{35}\sigma_{13}^{(s)} + a_{36}\sigma_{12}^{(s)}, \tag{6.3}$$

$$\frac{\partial V^{(s)}}{\partial \zeta} + \frac{\partial W^{(s-1)}}{\partial \eta} = a_{14}\sigma_{11}^{(s)} + a_{24}\sigma_{22}^{(s)} + a_{34}\sigma_{33}^{(s)}$$
$$+ a_{44}\sigma_{23}^{(s)} + a_{45}\sigma_{13}^{(s)} + a_{46}\sigma_{12}^{(s)},$$

$$\frac{\partial W^{(s-1)}}{\partial \xi} + \frac{\partial U^{(s)}}{\partial \zeta} = a_{15}\sigma_{11}^{(s)} + a_{25}\sigma_{22}^{(s)} + a_{35}\sigma_{33}^{(s)}$$
$$+ a_{45}\sigma_{23}^{(s)} + a_{55}\sigma_{13}^{(s)} + a_{56}\sigma_{12}^{(s)},$$

$$\frac{\partial V^{(s-1)}}{\partial \xi} + \frac{\partial U^{(s-1)}}{\partial \eta} = a_{16}\sigma_{11}^{(s)} + a_{26}\sigma_{22}^{(s)} + a_{36}\sigma_{33}^{(s)}$$
$$+ a_{46}\sigma_{23}^{(s)} + a_{56}\sigma_{13}^{(s)} + a_{66}\sigma_{12}^{(s)}.$$

The system (6.3) may be treated as algebraic in terms of the six components of the stress tensor $\sigma_{jk}^{(s)}$, which are then expressed through $\dfrac{\partial U^{(s)}}{\partial \zeta}, \dfrac{\partial V^{(s)}}{\partial \zeta}, \dfrac{\partial W^{(s)}}{\partial \zeta}$.

As follows from the generalized Hooke's law, the determinant of the system Δ is non-zero, hence, the stresses are uniquely expressed through strains and vice versa (Lekhnitskii, 1977). Using the Cramer's rule, we obtain

$$\sigma_{13}^{(s)} = \frac{1}{\Delta}\left(A_{55}\frac{\partial U^{(s)}}{\partial \zeta} + A_{45}\frac{\partial V^{(s)}}{\partial \zeta} + A_{35}\frac{\partial W^{(s)}}{\partial \zeta} + \sigma_{13*}^{(s)} \right),$$

$$\sigma_{23}^{(s)} = \frac{1}{\Delta}\left(A_{54}\frac{\partial U^{(s)}}{\partial \zeta} + A_{44}\frac{\partial V^{(s)}}{\partial \zeta} + A_{34}\frac{\partial W^{(s)}}{\partial \zeta} + \sigma_{23*}^{(s)} \right),$$

$$\sigma_{33}^{(s)} = \frac{1}{\Delta}\left(A_{53}\frac{\partial U^{(s)}}{\partial \zeta} + A_{43}\frac{\partial V^{(s)}}{\partial \zeta} + A_{33}\frac{\partial W^{(s)}}{\partial \zeta} + \sigma_{33*}^{(s)} \right),$$

$$\sigma_{11}^{(s)} = \frac{1}{\Delta}\left(A_{15}\frac{\partial U^{(s)}}{\partial \zeta} + A_{14}\frac{\partial V^{(s)}}{\partial \zeta} + A_{13}\frac{\partial W^{(s)}}{\partial \zeta} + \sigma_{11*}^{(s)}\right),$$

$$\sigma_{22}^{(s)} = \frac{1}{\Delta}\left(A_{25}\frac{\partial U^{(s)}}{\partial \zeta} + A_{24}\frac{\partial V^{(s)}}{\partial \zeta} + A_{23}\frac{\partial W^{(s)}}{\partial \zeta} + \sigma_{22*}^{(s)}\right), \qquad (6.4)$$

$$\sigma_{12}^{(s)} = \frac{1}{\Delta}\left(A_{56}\frac{\partial U^{(s)}}{\partial \zeta} + A_{46}\frac{\partial V^{(s)}}{\partial \zeta} + A_{36}\frac{\partial W^{(s)}}{\partial \zeta} + \sigma_{12*}^{(s)}\right),$$

where A_{ij} are cofactors of the elements a_{ij} of the matrix of the system (6.3), with

$$\sigma_{13*}^{(s)} = \frac{\partial}{\partial \xi}\left(A_{15}U^{(s-1)} + A_{65}V^{(s-1)} + A_{55}W^{(s-1)}\right)$$
$$+ \frac{\partial}{\partial \eta}\left(A_{65}U^{(s-1)} + A_{25}V^{(s-1)} + A_{45}W^{(s-1)}\right),$$

$$\sigma_{23*}^{(s)} = \frac{\partial}{\partial \xi}\left(A_{14}U^{(s-1)} + A_{64}V^{(s-1)} + A_{54}W^{(s-1)}\right)$$
$$+ \frac{\partial}{\partial \eta}\left(A_{64}U^{(s-1)} + A_{24}V^{(s-1)} + A_{44}W^{(s-1)}\right),$$

$$\sigma_{33*}^{(s)} = \frac{\partial}{\partial \xi}\left(A_{13}U^{(s-1)} + A_{63}V^{(s-1)} + A_{53}W^{(s-1)}\right)$$
$$+ \frac{\partial}{\partial \eta}\left(A_{63}U^{(s-1)} + A_{23}V^{(s-1)} + A_{43}W^{(s-1)}\right), \qquad (6.5)$$

$$\sigma_{jj*}^{(s)} = \frac{\partial}{\partial \xi}\left(A_{j1}U^{(s-1)} + A_{j6}V^{(s-1)} + A_{j5}W^{(s-1)}\right)$$
$$+ \frac{\partial}{\partial \eta}\left(A_{j6}U^{(s-1)} + A_{j2}V^{(s-1)} + A_{j4}W^{(s-1)}\right),$$
$$(j = 1, 2),$$

$$\sigma_{12*}^{(s)} = \frac{\partial}{\partial \xi}\left(A_{16}U^{(s-1)} + A_{66}V^{(s-1)} + A_{56}W^{(s-1)}\right)$$
$$+ \frac{\partial}{\partial \eta}\left(A_{66}U^{(s-1)} + A_{26}V^{(s-1)} + A_{46}W^{(s-1)}\right).$$

Substituting the values of $\sigma_{13}^{(s)}, \sigma_{23}^{(s)}, \sigma_{33}^{(s)}$ from (6.4) into (6.2) we result in the following system for $U^{(s)}, V^{(s)}, W^{(s)}$

$$A_{55}\frac{\partial^2 U^{(s)}}{\partial \zeta^2} + A_{45}\frac{\partial^2 V^{(s)}}{\partial \zeta^2} + A_{35}\frac{\partial^2 W^{(s)}}{\partial \zeta^2} + \Delta\Omega_*^2 U^{(s)} = R_u^{(s)}(\xi, \eta, \zeta),$$

$$A_{54}\frac{\partial^2 U^{(s)}}{\partial \zeta^2} + A_{44}\frac{\partial^2 V^{(s)}}{\partial \zeta^2} + A_{34}\frac{\partial^2 W^{(s)}}{\partial \zeta^2} + \Delta\Omega_*^2 V^{(s)} = R_v^{(s)}(\xi, \eta, \zeta), \qquad (6.6)$$

$$A_{53}\frac{\partial^2 U^{(s)}}{\partial \zeta^2} + A_{43}\frac{\partial^2 V^{(s)}}{\partial \zeta^2} + A_{33}\frac{\partial^2 W^{(s)}}{\partial \zeta^2} + \Delta\Omega_*^2 W^{(s)} = R_w^{(s)}(\xi, \eta, \zeta),$$

where

$$R_u^{(s)} = -\Delta \left(\frac{\partial \sigma_{11}^{(s-1)}}{\partial \xi} + \frac{\partial \sigma_{12}^{(s-1)}}{\partial \eta} \right) - \frac{\partial \sigma_{13*}^{(s)}}{\partial \zeta},$$

$$R_v^{(s)} = -\Delta \left(\frac{\partial \sigma_{12}^{(s-1)}}{\partial \xi} + \frac{\partial \sigma_{22}^{(s-1)}}{\partial \eta} \right) - \frac{\partial \sigma_{23*}^{(s)}}{\partial \zeta}, \tag{6.7}$$

$$R_w^{(s)} = -\Delta \left(\frac{\partial \sigma_{13}^{(s-1)}}{\partial \xi} + \frac{\partial \sigma_{23}^{(s-1)}}{\partial \eta} \right) - \frac{\partial \sigma_{33*}^{(s)}}{\partial \zeta}.$$

It is clear that $R_u^{(0)} = R_v^{(0)} = R_w^{(0)} \equiv 0$.

Depending on the type of anisotropy the system (6.6) could be simplified. For example, in case of orthotropic plates

$$A_{35} = A_{45} = A_{34} = A_{54} = A_{43} = A_{53} = 0 \tag{6.8}$$

hence, the system (6.6) is split into three, which are independent at leading order ($s = 0$), coinciding with (2.11)-(2.13). The presence of plane of symmetry (involving 13 material constants) implies $a_{14} = a_{24} = a_{34} = a_{46} = a_{15} = a_{25} = a_{35} = a_{56} = 0$, therefore the matrix of (6.1) becomes

$$\begin{pmatrix} a_{11} & a_{12} & a_{13} & 0 & 0 & a_{16} \\ a_{12} & a_{22} & a_{23} & 0 & 0 & a_{26} \\ a_{13} & a_{23} & a_{33} & 0 & 0 & a_{36} \\ 0 & 0 & 0 & a_{44} & a_{45} & 0 \\ 0 & 0 & 0 & a_{45} & a_{55} & 0 \\ a_{16} & a_{26} & a_{36} & 0 & 0 & a_{66} \end{pmatrix}. \tag{6.9}$$

Therefore,

$$A_{35} = A_{34} = A_{53} = A_{43} = 0,$$

$$A_{55} = a_{44}\Delta_1, \quad A_{45} = -a_{45}\Delta_1, \quad A_{33} = \Delta_2 \Delta_3,$$

$$\Delta_2 = a_{11}a_{22}a_{66} + 2a_{12}a_{16}a_{26} - a_{11}a_{26}^2 - a_{22}a_{16}^2 - a_{66}a_{12}^2, \tag{6.10}$$

$$\Delta_1 = \begin{vmatrix} a_{11} & a_{12} & a_{13} & a_{16} \\ a_{12} & a_{22} & a_{23} & a_{26} \\ a_{13} & a_{23} & a_{33} & a_{36} \\ a_{16} & a_{26} & a_{36} & a_{66} \end{vmatrix}, \quad \Delta_3 = a_{44}a_{55} - a_{45}^2$$

and the system (6.6) is rewritten as

$$a_{44}\frac{\partial^2 U^{(s)}}{\partial \zeta^2} - a_{45}\frac{\partial^2 V^{(s)}}{\partial \zeta^2} + \Delta_3 \Omega_*^2 U^{(s)} = \frac{1}{\Delta_1} R_u^{(s)},$$

$$-a_{45}\frac{\partial^2 U^{(s)}}{\partial \zeta^2} + a_{55}\frac{\partial^2 V^{(s)}}{\partial \zeta^2} + \Delta_3 \Omega_*^2 V^{(s)} = \frac{1}{\Delta_1} R_v^{(s)}, \tag{6.11}$$

$$\frac{\partial^2 W^{(s)}}{\partial \zeta^2} + \frac{\Delta_1}{\Delta_2} \Omega_*^2 W^{(s)} = \frac{1}{\Delta_2 \Delta_3} R_w^{(s)}.$$

It is worth noting that at $s = 0$ the right-hand sides of (6.11) are equal to zero, with the first two equations describing tangential vibrations, whereas the third equation corresponding to transverse vibrations. In other words, the presence of the plane of symmetry leads to separation of plane and anti-plane vibrations at leading order. We remark however, that at higher orders the interaction between those occur.

The first two equations of (6.11) may be represented as

$$\ell_{11}U^{(s)} + \ell_{12}V^{(s)} = \frac{1}{\Delta_1}R_u^{(s)},$$

$$\ell_{12}U^{(s)} + \ell_{22}V^{(s)} = \frac{1}{\Delta_1}R_v^{(s)}, \tag{6.12}$$

where the operators

$$\ell_{11} = a_{44}\frac{\partial^2}{\partial\zeta^2} + \Delta_3\Omega_*^2, \quad \ell_{12} = -a_{45}\frac{\partial^2}{\partial\zeta^2}, \tag{6.13}$$

$$\ell_{22} = a_{55}\frac{\partial^2}{\partial\zeta^2} + \Delta_3\Omega_*^2.$$

The last system (6.12) is transformed to an equation for $U^{(s)}$, namely

$$(\ell_{11}\ell_{22} - \ell_{12}^2)U^{(s)} = \frac{1}{\Delta_1}(\ell_{22}R_u^{(s)} - \ell_{12}R_v^{(s)}). \tag{6.14}$$

When (6.14) is solved, the function $V^{(s)}$ is determined by

$$V^{(s)} = -\frac{a_{44}a_{55} - a_{45}^2}{a_{45}\Delta_3\Omega_*^2}\frac{\partial^2 U^{(s)}}{\partial\zeta^2} - \frac{a_{55}}{a_{45}}U^{(s)} + \frac{1}{\Delta_1\Delta_3\Omega_*^2}(\frac{a_{55}}{a_{45}}R_u^{(s)} + R_v^{(s)}). \tag{6.15}$$

The solution of the third equation of (6.11) is given by

$$W^{(s)} = C_1^{(s)}(\xi,\eta)\sin\sqrt{\frac{\Delta_1}{\Delta_3}}\Omega_*\zeta + C_2^{(s)}(\xi,\eta)\cos\sqrt{\frac{\Delta_1}{\Delta_3}}\Omega_*\zeta + W_\tau^{(s)}(\xi,\eta,\zeta). \tag{6.16}$$

Equation (6.14) is an ordinary differential equation of the fourth order, which could be solved. Then, using the formulae (6.4), (6.15), and (6.16), the stress components are determined, and the boundary conditions (1.1)-(1.7) are satisfied.

Consider now the general case assuming that all of the coefficients A_{ij} of the system (6.6) are non-zero. Excluding the second derivative $\dfrac{\partial^2 W^{(s)}}{\partial\zeta^2}$ from the first two equations, one arrives at the following equation in respect of $U^{(s)}$, $V^{(s)}$

$$\ell_1 U^{(s)} + \ell_2 V^{(s)} = R_{uv}^{(s)}. \tag{6.17}$$

Now if one excludes $\dfrac{\partial^2 W^{(s)}}{\partial\zeta^2}$ from the first and third equations of (6.6), then $W^{(s)}$ is expressed through $U^{(s)}$ and $V^{(s)}$ as

$$W^{(s)} = \frac{b_1}{\Delta\Omega_*^2}\frac{\partial^2 U^{(s)}}{\partial\zeta^2} + \frac{b_2}{\Delta\Omega_*^2}\frac{\partial^2 V^{(s)}}{\partial\zeta^2} + b_3 U^{(s)} - C_w^{(s)}, \tag{6.18}$$

where

$$\ell_1 = b_4\frac{\partial^2}{\partial\zeta^2} + A_{34}\Delta\Omega_*^2, \quad \ell_2 = b_5\frac{\partial^2}{\partial\zeta^2} - A_{35}\Delta\Omega_*^2,$$

$$C_w^{(s)} = \frac{1}{A_{35}\Delta\Omega_*^2}(A_{33}R_u^{(s)} - A_{35}R_w^{(s)}),$$

$$b_1 = \frac{1}{A_{35}}(A_{55}A_{33} - A_{53}A_{35}), \quad b_2 = \frac{1}{A_{35}}(A_{45}A_{33} - A_{43}A_{35}), \qquad (6.19)$$

$$b_3 = \frac{A_{33}}{A_{35}}, \quad b_4 = A_{55}A_{34} - A_{54}A_{35}, \quad b_5 = A_{45}A_{34} - A_{44}A_{35},$$

$$R_{uv}^{(s)} = A_{34}R_u^{(s)} - A_{35}R_v^{(s)}.$$

Substituting the value of $W^{(s)}$ into the second equation of (6.6), we obtain one more equation for $U^{(s)}$ and $V^{(s)}$, namely

$$\ell_3 U^{(s)} + \ell_4 V^{(s)} = R_{vw}^{(s)}, \qquad (6.20)$$

where

$$\ell_3 = \frac{A_{34}b_1}{\Delta\Omega_*^2}\frac{\partial^4}{\partial\zeta^4} + (A_{54} + A_{34})\frac{\partial^2}{\partial\zeta^2}, \quad \ell_4 = \frac{A_{34}b_2}{\Delta\Omega_*^2}\frac{\partial^4}{\partial\zeta^4} + A_{44}\frac{\partial^2}{\partial\zeta^2} + \Delta\Omega_*^2,$$

$$R_{vw}^{(s)} = R_v^{(s)} + A_{34}\frac{\partial^2 C_w^{(s)}}{\partial\zeta^2}. \; (6.21)$$

Equations (6.17) and (6.20) may now be used to exclude $V^{(s)}$ and arrive at

$$(\ell_3\ell_2 - \ell_1\ell_4)U^{(s)} = \ell_2 R_{vw}^{(s)} - \ell_4 R_{uv}^{(s)}), \qquad (6.22)$$

which may be rewritten in more explicit form as

$$B_1\frac{\partial^6 U^{(s)}}{\partial\zeta^6} + B_2\Delta\Omega_*^2\frac{\partial^4 U^{(s)}}{\partial\zeta^4} + B_3\Delta^2\Omega_*^4\frac{\partial^2 U^{(s)}}{\partial\zeta^2} + B_4\Delta^3\Omega_*^6 = \psi_u^{(s)} \qquad (6.23)$$

with

$$B_1 = A_{34}(b_1 b_5 - b_2 b_4),$$

$$B_2 = b_5(A_{54} + b_3 A_{34}) - A_{34}A_{35}b_1 - A_{44}b_4 - A_{34}^2 b_2,$$

$$B_3 = -(A_{35}(A_{54} + b_3 A_{34}) + b_4 + A_{44}A_{34}), \quad B_4 = A_{34}, \qquad (6.24)$$

$$\psi_u^{(s)} = (\ell_2 R_{vw}^{(s)} - \ell_4 R_{uv}^{(s)})\Delta\Omega_*^2.$$

The quantity $U^{(s)}$ is determined from equation (6.23), then $V^{(s)}$ may be obtained from (6.17) and (6.20) as

$$V^{(s)} = \frac{1}{A_{34}(A_{35}^2 b_2 - A_{45}b_5)\Delta\Omega_*^2}\left(\frac{A_{34}b_5}{\Delta\Omega_*^2}(b_1 b_5 - b_2 b_4)\frac{\partial^4 U^{(s)}}{\partial\zeta^4}\right.$$

$$+((A_{54} + b_3 A_{34})b_5^2 - A_{34}b_2(A_{34}b_5 + A_{35}b_4) - A_{44}b_4 b_5)\frac{\partial^2 U^{(s)}}{\partial\zeta^2}$$

$$\left. -A_{34}\Delta\Omega_*^2(A_{34}A_{35}b_2 + A_{44}b_5)U^{(s)} - b_5^2\bar{R}_{vw}^{(s)}\right), \qquad (6.25)$$

$$\bar{R}_{vw}^{(s)} = R_{vw}^{(s)} - \frac{A_{44}}{b_5}R_{uv}^{(s)} - \frac{A_{34}b_2 A_{35}}{b_5^2}R_{uv}^{(s)} - \frac{A_{34}b_2}{b_5\Delta\Omega_*^2}\frac{\partial^2 R_{uv}^{(s)}}{\partial\zeta^2}$$

with $W^{(s)}$ found from (6.18). The stress components are then calculated through (6.4)-(6.5).

The solution of equation (6.23) may be presented in the form

$$U^{(s)} = U_0^{(s)} + U_\tau^{(s)}(\xi, \eta, \zeta), \tag{6.26}$$

where $U_0^{(s)}$ is a solution of homogeneous, and $U_\tau^{(s)}$ is a particular solution of a non-homogeneous equation (6.23). The solution of a homogeneous equation may be sought as

$$U_0^{(s)} = \exp \lambda\zeta, \tag{6.27}$$

leading to the following auxiliary equation for λ after substitution of (6.27) into (6.23)

$$B_1\lambda^6 + B_2\Delta\Omega_*^2\lambda^4 + B_3\Delta^2\Omega_*^4\lambda^2 + B_4\Delta^3\Omega_*^6 = 0 \tag{6.28}$$

which may be rewritten as

$$B_1k^3 + B_2k^2 + B_3k + B_4k = 0, \quad k = \left(\frac{\lambda}{\Omega_*\sqrt{\Delta}}\right)^2. \tag{6.29}$$

It follows from above that if λ is a root of (6.28), its opposite value $-\lambda$ will also be a solution. It may be deduced from a straightforward analysis of the cubic equation (6.29) that it possesses at least one real root, corresponding to a pair of real or purely imaginary λ, and it could possess a pair of complex conjugate roots k and $\bar{k}$. Hence, the solution $U_0^{(s)}$ may be written as

$$U_0^{(s)} = C_1^{(s)}(\xi, \eta) \exp \lambda_1\zeta + C_2^{(s)}(\xi, \eta) \exp(-\lambda_1\zeta) + C_3^{(s)}(\xi, \eta) \exp \lambda_2\zeta \tag{6.30}$$
$$+ C_4^{(s)}(\xi, \eta) \exp(-\lambda_2\zeta) + C_5^{(s)}(\xi, \eta) \exp \lambda_3\zeta + C_6^{(s)}(\xi, \eta) \exp(-\lambda_3\zeta).$$

Denoting

$$C_1^{(s)} = \frac{D_1^{(s)} + D_2^{(s)}}{2}, \quad C_2^{(s)} = \frac{D_1^{(s)} - D_2^{(s)}}{2}, \quad \dots , \tag{6.31}$$
$$C_5^{(s)} = \frac{D_5^{(s)} + D_6^{(s)}}{2}, \quad C_6^{(s)} = \frac{D_5^{(s)} - D_6^{(s)}}{2}$$

enables the following representation of (6.30)

$$U_0^{(s)} = D_1^{(s)}(\xi, \eta)\cosh\lambda_1\zeta + D_2^{(s)}(\xi, \eta)\sinh\lambda_1\zeta + D_3^{(s)}(\xi, \eta)\cosh\lambda_2\zeta$$
$$+ D_4^{(s)}(\xi, \eta)\sinh\lambda_2\zeta + D_5^{(s)}(\xi, \eta)\cosh\lambda_3\zeta + D_6^{(s)}(\xi, \eta)\sinh\lambda_3\zeta. \tag{6.32}$$

Depending on the values of λ_j the solution (6.32) may always be expressed through real functions. In general, it is given by

$$U_0^{(s)} = D_1^{(s)}(\xi, \eta)\varphi_1(\zeta) + D_2^{(s)}(\xi, \eta)\varphi_2(\zeta) + \cdots + D_6^{(s)}(\xi, \eta)\varphi_6(\zeta). \tag{6.33}$$

In case of a positive real root $k_1 > 0$ of equation (6.29) it follows that $\lambda_1 = \Omega_*\sqrt{k_1\Delta}$, hence

$$\varphi_1 = \cosh\lambda_1\zeta, \quad \varphi_2 = \sinh\lambda_1\zeta \tag{6.34}$$

whereas if $k_1 < 0$, then

$$\varphi_1 = \cos\Omega_* \sqrt{-k_1\Delta}\zeta, \qquad \varphi_2 = \sin\Omega_* \sqrt{-k_1\Delta}\zeta. \tag{6.35}$$

In case of a complex root of (6.29) $k = k_2$, the corresponding four roots for λ may be presented as

$$\lambda_{3,4} = \alpha_1 \pm \beta_1 i, \quad \lambda_{5,6} = -\alpha_1 \pm \beta_1 i, \tag{6.36}$$

therefore the corresponding functions are given by

$$\varphi_3 = \cosh\lambda_1\zeta \sin\beta_1\zeta, \qquad \varphi_4 = \cosh\lambda_1\zeta \cos\beta_1\zeta,$$
$$\varphi_5 = \sinh\lambda_1\zeta \sin\beta_1\zeta, \qquad \varphi_6 = \sinh\lambda_1\zeta \cos\beta_1\zeta, \tag{6.37}$$

with α_1, β_1 being expressed through the material constants. The functions φ_i involved in (6.33), will be of the form (6.34), (6.35), and (6.37). The solution (6.26) is then written as

$$U^{(s)} = D_1^{(s)}(\xi,\eta)\varphi_1(\zeta) + \cdots + D_6^{(s)}(\xi,\eta)\varphi_6(\zeta) + U_\tau^{(s)}(\xi,\eta,\zeta). \tag{6.38}$$

Now once the function $U^{(s)}$ is determined, the values of $V^{(s)}$ and $W^{(s)}$ may be obtained through (6.18) and (6.25), and therefore, the stress components may be calculated using the formulae (6.4)-(6.5), which enables consideration of the boundary conditions (1.1)-(1.7) for the plate of general anisotropy.

The presented asymptotic method may also be used for analysis of forced vibrations of anisotropic shells, see Aghalovyan and Gulgazaryan (2009); Aghalovyan (2011a).

The method has also allowed successful analysis of free vibrations of plates and shells, in particular, evaluation of natural frequencies, see Aghalovyan and Aghalovyan (2005); Aghalovyan and Gulgazaryan (2006); Aghalovyan (2008a, 2010).

The same method has been applied to investigation of interaction between the plates and shells and various physical fields, including the thermal field (Aghalovyan and Gevorkyan, 2011) and electro-elastic field (Aghalovyan et al., 2011).

We also mention a review of investigations performed by the asymptotic methods (Aghalovyan, 2011b). The prospectives of using asymptotic analysis for certain classes of static and dynamic boundary value problems for thin bodies including beams, plates, and shells have been discussed in Aghalovyan (2012a).

The proposed method has also enabled analysis of a novel class of problems of seismology and earthquake resistant structures, for more details see Aghalovyan and Aghalovyan (2012).

It has been established by now that 95% of earthquakes occur due to tectonic motion of the lithospheric plates. Using the data of the seismic stations network along with that of the GPS, it is possible to apply the proposed asymptotic method to analyze the stress-strain field arising in Earth lithospheric plates, to investigate the variation of the appropriate quantities in time and therefore reveal the areas which are currently of increased seismic risks, which could be used to predict strong earthquakes (Aghalovyan, 2012b, 2013).

Bibliography

Abovsky, N. P., Andreev, N. P. and Deruga, A. P. (1978). *Variational Principles of the Elasticity Theory and the Shell Theory*, (Nayka, Moscow), in Russian.

Abramyan, B. L. (1957). On the plane problem of elasticity for a rectangle. *J. Appl. Math. Mech. (PMM)* **21**, 1, pp. 89–100, in Russian.

Aghalovyan, L. A. (1965). On the refinement of the classical theory of bending of anisotropic plates. *Izv. AN Arm. SSR. Ser. of Phys. and Math. Scien. (Ser. F.M.N.)* **15**, 5, pp. 16–30, in Russian.

Aghalovyan, L. A. (1966a). Application of the method of asymptotic integration to the construction of an approximate theory of anisotropic shells. *J. Appl. Math. Mech. (PMM)* **30**, 2, pp. 388–398.

Aghalovyan, L. A. (1966b). On the theory of bending of orthotropic plates. *Izv. AN SSSR. Rigid Body Mechanics (MTT)* **15**, 6, pp. 114–121, in Russian.

Aghalovyan, L. A. (1966c). On the boundary conditions for bending of anisotropic plates. *Izv. AN Arm. SSR. Mechanics* **19**, 4, pp. 13–27, in Russian.

Aghalovyan, L. A. (1970). On the bending equations of anisotropic plates. *Proc. of 7th All-Union Conf. on the Theory of Shells and Plates* pp. 17–21, (Nayka, Moscow), in Russian.

Aghalovyan, L. A. (1972a). On the boundary layer of plates. *Dokl. AN Arm. SSR* **45**, 3, pp. 149–155, in Russian.

Aghalovyan, L. A. (1972b). Some relations of the classical linear theory of anisotropic shells and possibilities for their refinement. *Izv. AN SSSR. Rigid Body Mechanics (MTT)* **1**, pp. 109–120, in Russian.

Aghalovyan, L. A. (1973a). On the boundary layer of orthotropic plates. *Izv. AN Arm. SSR. Mechanics* **26**, 2, pp. 27–43, in Russian.

Aghalovyan, L. A. (1973b). On the accounting of transverse shear in calculation of orthotropic shells. *Proc. of 8th All-Union Conf. on the Theory of Shells and Plates*, pp. 7–13, (Nayka, Moscow), in Russian.

Aghalovyan, L. A. (1977). On the interaction between the boundary layer with the outer stress-strain state of a strip. *Izv. AN Arm. SSR. Mechanics* **30**, 5, pp. 48–62, in Russian.

Aghalovyan, L. A. (1978a). On the question of bringing the boundary conditions for three-dimensional problem to two-dimensional in the theory of anisotropic plates. *Ychenie zapiski, Erevan Univ. of Nat. Scien.* **138**, 2, pp. 20–27, in Russian.

Aghalovyan, L. A. (1978b). On the boundary conditions in the theory of anisotropic plates. *Ychenie zapiski, Erevan Univ. of Nat. Scien.* **139**, 3, pp. 21–30, in Russian.

Aghalovyan, L. A. (1982). On the structure of solution of a class of plane problems if anistropic elastic solids. *Mezhvuz. Sb., EGU, Mechanics* **2**, pp. 7–12, in Russian.

Aghalovyan, L. A. (1983). Determining the stress-strain state of a two-layer strip and validity of the Winkler hypothesis. *Proc. of 13th All-Union Conf. on the Theory of Plates and Shells* **1**, pp. 13–18, (Tallin), in Russian.

Aghalovyan, L. A. (1984). The elastic boundary layer for a class of plane problems. *Mechanics, Erevan State Univ.* **3**, pp. 51–58, in Russian.

Aghalovyan, L. A. (2000). On one approach of studying the free and forced vibrations of bases and fundaments of structures. In *Earthquake Hazard and Seismic Risk Reduction*, (Springer, Dordrecht), pp. 395–402.

Aghalovyan, L. A. (2002). On one class of problems on forced vibrations of anisotropic plates. In book: *Problems of Mechanics of Thin Deformable Bodies*, (Yerevan), pp. 9–19.

Aghalovyan, L. A. (2004). On asymptotic method in the solution of static and dynamic boundary value problems. *Proc. of NAS of Armenia, Mechanics* **57**, 4, pp. 3–14.

Aghalovyan, L. A. (2008a). An asymptotic method for solving three-dimensional boundary value problems of statics and dynamics of thin bodies. *Proc. of the IUTAM Symposium on the Relations of Shell, Plate, Beam, and 3D Models.* Springer, pp. 1–20.

Aghalovyan, L. A. (2008b). On one method of solution of three-dimensional dynamic problems for layered elastic plates and applications in seismology and seismosteady construction. *Proc. of Fourth European Conf. of Structures Control*, St. Petersburgh, pp. 25–33.

Aghalovyan, L. A. (2010). An asymptotic method of boundary-value problems solution of elasticity theory for thin bodies. *Proc. of the Symposium: Recent Advances in Mechanics.* Athens, pp. 9–26.

Aghalovyan, L. A. (2011a). Non-classical spatial boundary value problems of statics and dynamics of shells and the asymptotic method of their solution. In *Shell-like Structures, Non-classical Theories and Applications. Advanced Structured Materials, Vol 15*, (Springer), pp. 3–14.

Aghalovyan, L. A. (2011b). On the classes of problems for deformable one-layer and multilayer thin bodies, solvable by the asymptotic method. *Mechanics of Composite Materials* **47**, 1, pp. 59–72.

Aghalovyan, L. A. (2012a). On possibilities of the asymptotic method in the questions of theory of plates and shells. In *Problems of Mechanics of Deformable Solid Body* (Yerevan), pp. 33–41, in Russian.

Aghalovyan, L. A. (2012b). The research of lithospheric plates tectonics of the earth on the base of data seismostations, GPS systems, the solutions of problems of elasticity theory and the earthquakes prediction. *Proc. of the 15th World Conference on Earthquake Engineering*, Portugal, pp. 2012–3008.

Aghalovyan, L. A. (2013). On some classes of space boundary-value problems of static and dynamics of plates and shells. *Proc. Intern. Sci. Conf.: Shell and Membrane Theories in Mechanics and Biology*, Minsk, pp. 5–7.

Aghalovyan, L. A. and Adamyan, S. H. (1986). On the stress strain state of a two-layer strip - rectangle with variable elastic characteristics. *Izv. AN Arm. SSR. Mechanics.* **39**, 5, pp. 3–15, in Russian.

Aghalovyan, L. A. and Adamyan, S. H. (1987). On the coefficient of foundation for bases with variable elastic characteristics. *Dokl. AN Arm. SSR.* **84**, 3, pp. 115–118, in Russian.

Aghalovyan, L. A. and Aghalovyan, M. L. (2005). Asymptotics of free vibrations of anisotropic elastic plates fastened with an absolutely rigid base. In *Modern Problems of Deformable Bodies Mechanics. Collection of Papers Dedicated to the Memory of Prof. Pericles S. Theocaris, Vol I*, Yerevan, pp. 8–19.

Aghalovyan, L. A. and Aghalovyan, M. L. (2012). On dynamic conduct of lithospheric plates of the earth on the base of the data of seismic stations and GPS systems. In *Topical Problems of Continuum Mechanics, Vol.1*, Yerevan, pp. 42–46, In Russian.

Aghalovyan, L. A., Azatyan, G. Z., Gevorkyan, R. S. and Poghosyan, H. M. (2011). On asymptotic solution of the dynamic 3D problem for rectangular piezoceramical plate. *Reports NAS RA* **111**, 2, pp. 129–137, in Russian.

Aghalovyan, L. A. and Gevorkyan, R. S. (1984). Non-classical boundary-value problems for plates with general anisotropy. *Proc. of the 4th All-Union Symposium on the Mechanics of Composite Material Structures*, pp. 105–110, (Novosibirsk, Nauka), in Russian.

Aghalovyan, L. A. and Gevorkyan, R. S. (1986). On the asymptotic solution of mixed three-dimensional problems for double-layer anisotropic plates. *J. Appl. Math. Mech. (PMM)* **50**, 2, pp. 202–208.

Aghalovyan, L. A. and Gevorkyan, R. S. (1989). On the asymptotic solution of nonclassical boundary-value problems for two-layer anisotropic thermoelastic shells. *Izv. AN Arm. SSR. Mechanics* **42**, 3, pp. 28–36, in Russian.

Aghalovyan, L. A. and Gevorkyan R.S. (2011). Asymptotic solution of coupled dynamic problems of thermoelasticity for thin bodies of anisotropic inhomogeneous in-plane materials. *J. Appl. Math. Mech. (PMM)* **75**, 5, pp. 601–611.

Aghalovyan, L. A. and Gulgazaryan L. G. (2006). Asymptotic solutions of non-classical boundary value problems of the natural vibrations of orthotropic shells. *J. Appl. Math. Mech. (PMM)* **70**, 1, pp. 102–115.

Aghalovyan, L. A. and Gulgazaryan L.G. (2009). Non-classical boundary-value problems of the forced vibrations of orthotropic shells *International Appl. Mechanics* **45**, 8, pp. 105–122.

Aghalovyan, L. A. and Hovhannisyan, R. Zh. (2008). Theoretical proof of seismoisolator application necessity. *Proc. Int. Workshop Base Isolat. High-Rise Build.* Yerevan, pp. 185–199.

Aghalovyan, L. A. and Khachatryan, S. M. (1975). On general Papkovich orthogonality and conditions of existence of decaying solutions in a plain-strain problem of elasticity for an orthtropic semi-strip. *Dokl. AN Arm. SSR. Mechanics* **50**, 3, pp. 157–163, in Russian.

Aghalovyan, L. A. and Khachatryan, S. M. (1977a). Asymptotic analysis of the stress-strain state of an orthotropic strip. *Uch. Zap. Erevan Univ. of Nat. Sci.* **134**, 1, pp. 22–30, in Russian.

Aghalovyan, L. A. and Khachatryan, S. M. (1977b). Plain strain problems for an orthotropic strip. *Uch. Zap. Erevan Univ. of Nat. Sci.* **135**, 1, pp. 20–26, in Russian.

Aghalovyan, L. A. and Zakaryan T. V. (2011). The asymptotic solution of the first dynamic boundary value problem of the theory of elasticity for two-layered orthotropic plate. *Proc. of NAS of Armenia, Mechanics* **64**, 2, pp. 15–25.

Aghalovyan, M. L. (2007). On the character of forced vibrations of plates for general anisotropy. *Proceedings of the International Conference: On Models of Continuum Mechanics, Saratov University Press.* Saratov, pp. 14–18, in Russian.

Ainola, L. Ya. (1965). Nonlinear Timoshenko type theory of elastic shells. *Izv. AN Estonskoi SSR. Ser. Fiz.-Mat. Tekhn. Nauk.* **14**, 3, pp. 337–344, in Russian.

Ainola, L. Ya. (1967). Stress and moment equations in a Timoshenko-type elastic-shell theory. *Izv. AN Estonskoi SSR. Ser. Fiz.-Mat. Tekhn. Nauk.* **16**, 4, pp. 463–465, in Russian.

Aksentyan, O. K. and Vorovich I. I. (1963). Stress state of a thin plate. *J. Appl. Math. Mech. (PMM)* **27**, 6, pp. 1057–1074.

Aksentyan, O. K. (1967). Singularities of stress-strain state of plate near edge. *J. Appl. Math. Mech. (PMM)* **31**, 1, pp. 178–186.

Aksentyan, O. K. and Luschik, O. N. (1978). Conditions for bounded stress at the edge of a compound wedge. *Izv. AN SSSR Mech. Solids (MTT)* **5**, pp. 102–108.

Alfutov, N. A. (1992). On some paradoxes of the theory of thin elastic plates, *Izv. RAN (MTT)* **3**, pp. 65–72, in Russian.

Altenbach, H. and Meenen, J. (2008). On the different possibilities to derive plate and shell theories. *Proc. of hte IUTAM Symposium on the Relations of Shell, Plate, Beam and 3D Models.* Springer, pp. 37–47.

Alumyae, N. A. (1972). Theory of elastic shells and plates. *Fifty Years of Mechanics in the USSR* pp. 227–266, (Nayka, T.3.M.), in Russian.

Ambartsumyan, S. A. (1957). On the calculation of two-layer orthotropic shells. *Izv. AN SSSR. OTN* **7**, pp. 57–64, in Russian.

Ambartsumyan, S. A. (1958). On the theory of bending of anisotropic plates. *Izv. AN SSSR. OTN* **5**, pp. 67–77, in Russian.

Ambartsumyan, S. A. (1961). *Theory of Anisotropic Shells*, (Fizmatgiz, Moscow), in Russian.

Ambartsumyan, S. A. (1964). Certain questions of the theory of anisotropic layered shells. *Izv. AN Arm. SSR. Ser. of Phys. and Math. Scien. (Ser. F.M.N.)* **17**, 3, pp. 55–77, in Russian.

Ambartsumyan, S. A. (1967). *Theory of Anisotropic Plates*, (Nayka, Moscow), in Russian.

Ambartsumyan, S. A. (1968). Specific features of the shell theory for modern materials. *Izv. AN Arm. SSR. Mechanics* **21**, 4, pp. 3–19, in Russian.

Ambartsumyan, S. A. (1974). *General Theory of Anisotropic Shells*, (Nayka, Moscow), in Russian.

Andreev, A. N. and Nemirovskii, U. V. (1977). Theory of elastic multilayer anisotropic shells. *Izv. AN SSSR. Rigid Body Mechanics (MTT)* **5**, pp. 77–96, in Russian.

Arutyunyan, N. Kh. and Abramyan, B. L. (1968). Torsion of the rods. *Durability, Stability and Vibrations* **1**, pp. 239–286, (Mashinostroenie, Moscow) in Russian.

Avetisyan, A. G. and Chobanyan, K. S. (1972). On character of stresses in the vicinity of rigidly fixed bound of juction surface for compound body, loaded under plan elasticity problem conditions. *Izv. AN Arm. SSR. Mechanics* **25**, 6, pp. 13–25, in Russian.

Babenkova, E. and Kaplunov, J. (2005). Radiation conditions for a semi-infinite elastic strip. *Proc. Roy. Soc. Lond. A* **461**, pp. 1163–1179.

Babich, V. M. and Buldyrev, V. S. (1972). *Asymptotic Methods in Problems of Diffraction of Short Waves*, (Nauka, Moscow), in Russian.

Babich, V. M. and Buldyrev, V. S. (1977). The art of asymptotics. *Vestn. Leningr. Univer.* **13**, 3, pp. 5–12, in Russian.

Babloyan, A. A. and Gulkanyan, N. O. (1969). On the mixed problem for a rectangle. *Izv. AN Arm. SSR. Mechanics* **22**, 1, pp. 3–16, in Russian.

Babloyan, A. A. and Mkrtchyan, A. M. (1972). Solution of the mixed problem for a rectangle. *Izv. AN Arm. SSR. Mechanics* **25**, 2, pp. 3–14, in Russian.

Babuska, J. and Prager, M. (1960). Reissnerian Algorithmus in the Theory of Elasticity. *Bulletin de l'Academie Polonaise des Sciences, Srie des sciences techniques* **8**, 8, pp. 411–417.

Baikov, V. N. and Strongin, S. G. (1980). *Building Constructions*, (Stroiizdat, Moscow), in Russian.

Barancev, R. G. (1976). Asimptology. *Vestn. Leningr. Univer.* **1**, 1, pp. 69–71, in Russian.

Berdichevsky, V. L. (1972). Variational methods of constructing models of shells. *J. Appl. Math. Mech. (PMM)* **36**, 5, pp. 788–804.

Berdichevsky, V. L. (1974). On the proof of the Saint-Venant principle for bodies of arbitrary shape. *J. Appl. Math. Mech. (PMM)* **38**, 5, pp. 851–864.

Berdichevsky, V. L. (1975). Dynamic equations of the theory of anisotropic plates. *DAN SSSR* **224**, 1, pp. 54–57.

Berdichevsky, V. L. (1978). Energy methods in some problems of decay of solutions. *J. Appl. Math. Mech. (PMM)* **42**, 1, pp. 140–156.

Berdichevsky, V. L. (1983). *Variational Principles of Continuum Mechanics*, (Nauka, Moscow), in Russian.

Bogolubov, N. N. and Mitropolskii, U. A. (1974). *Asymptotic Methods in the Theory of Nonlinear Oscillations*, (Nauka, Moscow), in Russian.

Bogy, D. B. (1968). Edge bonded dissimilar ortbogonal elastic wedges under normal and shear loading. *Trans ASME J. Appl. Mech.* **35**, pp. 460–466.

Bogy, D. B. (1970). On the problem of edge-bonded elastic quarter-planes loaded at the boundary. *Int. J. Solids Structures* **6**, pp. 1287–1313.

Bogy, D. B. (1971). Two edge-bonded elastic wedges of different materials and wedge angles under surface tractions. *Trans ASME J. Appl. Mech.* **38**, pp. 377–386.

Bolotin, V. V. (1963). On the theory of layered slabs. *Izv. AN SSSR, Mechanics and Mechanical Engineering* **3**, pp. 65–72.

Bolotin, V. V. and Novichkov, U. N. (1980). *Mechanics of Multilayer Structures*, (Mashinostroenie, Moscow), in Russian.

Cherepanov, G. P. (1970). On singular solutions in elasticity theory. *Problems of Solid Mechanics*, (Sudostroenie, Leningrad), in Russian.

Chernishev, G. N. (1966). Asymptotic method in the theory of shells (concentrated loads). *Proc. of the 6th All-Union Conf. on Theory of Shells and Plates* pp. 799–810, (Nauka, Moscow), in Russian.

Chernishev, G. N. (1970). Character of the solutions of equations of zero curvature shells with concentrated exposures. *Proc. of the 7th All-Union Conf. on Theory of Shells and Plates* pp. 597–600, (Nauka, Moscow), in Russian.

Chernykh, K. F. (1962). *The Linear Theory of Shells*, (LSU), in Russian.

Chjen, C. (1970). Mechanical properties of anisotropic fiber composite material. *J. Appl. Mech. Trans. ASME*, **37**, 1, pp. 197–199.

Chobanyan, K. S. (1987). *Stresses in Compound Elastic Solids*, (Arm. Ac. Sci. Erevan), in Russian.

Chobanyan, K. S. and Aleksanyan, R. K. (1971). Thermoelastic stresses in the vicinity of the edge of the joining surface in compound solids. *Izv. AN Arm. SSR. Mechanics* **24**, 3, pp. 22–32, in Russian.

Choi, I. and Horgan, G. O. (1977). Saint-Venant's principle and end effects in anisotropic elasticity. *J. Appl. Mech. Trans. ASME* **44**, pp. 424–430.

Chulkov, P. P. and Ivanov, A. V. (1969). Calculation of transverse deformation of filler in stability problems of three-layer panels with different supporting layers. *AN SSSR, Izvestiya, Mehanika Tverdogo Tela* **6**, pp. 101–107.

Ciarlet, P. G. (1997). Mathematical elasticity. *Theory of Plates.* North-Holland, **2**.

Darevskii, V. M. (1961). On basic relations in the theory of thin shells. *J. Appl. Math. Mech. (PMM)* **25**, 3, pp. 768–790.

Darevskii, V. M. (1995). On static boundary conditions in classical plates and shells theory, *Izv. RAN (MTT)* **4**, pp. 129–132.

Drozd, E. S., Mikhasev, G. I., Botogova, M. G. and Chizhik, S. A. (2013). Nano-scale shell theory-based estimation of the elastic characteristic of biological cells. *Proc. of Int. Sci. Conf. Shells and Membrane Theories* pp. 137–138, (Munsk, Belarus).

Feodos'ev, V. I. (1970). *Strength of Materials*, (Moscow, Nauka), in Russian.

Filin, A. P. (1975). *Applied Mechanics of Deformable Bodies*, (Nauka, Moscow), in Russian.

Filin, A. P. (1987). *Elements of the Theory of Shells*, (Stroiizdat, Leningrad), in Russian.

Finikova, V. O and Stolyar, A. M. (2011). Asymptotic integration of one narrow plate problem. *Advanced Structural Materials*, Springer, *15*, pp. 53–62.

Friedrichs, K. O. (1950). *Kirchhoff's Boundary Conditions and the Edge Effect for Elastic Plates*, (Proc. Symp., Appl. Math. 3 Amer. Math. Soc., N.Y.).

Friedrichs, K. O. (1955). Asymptotic Phenomena in Mathematical Physics. *Bull. Amer. Math. Soc.* **61**, p. 485.

Friedrichs, K. O. and Dressler R. F. (1961). Boundary-Layer Theory for Elastic Plates. *Comm. Pure and Appl. Math.* **14**, 1.

Galimov, K. Z. (1976). To non-linear theory of thin shells of Timoshenko type. *Izv. AN SSSR. Rigid Body Mechanics (MTT)*, **4**, pp. 155–166, in Russian.

Galimov, K. Z. and Surkin, R. G. (1967). On the works of Kazan Scientists on the Theory of Plates and Shells, *Sb. Research on the Theory of Plates and Shells*, **5**, (Izd. Kazan. Univ., Kazan), in Russian.

Galin'sh, A. K. (1967). Calculation of plates and shells on the specified theories. *Sb. Research on the Theory of Plates and Shells*, **5, 6**, (Izd. Kazan. Univ., Kazan), in Russian.

Gersevanov, N. M. and Polshin D. E. (1948). *Theoretical Foundations of Soil Mechanics*, (Pergamon Press), (Gosstroiizdat, Moscow), in Russian.

Gevorkyan, S. Kh. (1968). On the specific features of solutions of some problems of anisotropic elasticity. *Izv. AN Arm. SSR. Mechanics* **21**, 4, pp. 30–39, in Russian.

Gevorkyan, R. S. (1984). Asymptotics of the boundary layer for a class of problems for anisotropic plates. *Izv. AN Arm. SSR. Mechanics* **37**, 6, pp. 3–15, in Russian.

Goldenveizer, A. L. (1958). Theory of bending of Reissner plates. *Izv. AN SSSR (OTN)* **4**, 1, pp. 102–109.

Goldenveizer, A. L. (1959). Asymptotic integration of linear partial differential equations with a small principal part. *J. Appl. Math. Mech.(PMM)* **23**, 1, pp. 35–57.

Goldenveizer, A. L. (1960). Some mathematical problems in the linear theory of thin elastic shells. *Russian Mathematical Surveys (YMN)* **15**, 5, pp. 1–73, in Russian.

Goldenveizer, A. L. (1961). *Theory of Elastic Thin Shells*, Nauka, Moscow, in Russian.

Goldenveizer, A. L. (1962). An application of asymptotic integration of the equations of elasticity to derive an approximate theory for plate bending. *J. Appl. Math. Mech.(PMM)* **26**, pp. 668–686.

Goldenveizer, A. L. (1963). Derivation an approximate theory of shells by means of asymptotic integration of the equations of the elasticity theory. *J. Appl. Math. Mech.(PMM)* **27**, 4, pp. 593–608.

Goldenveizer, A. L. (1968a). *On the Two-dimensional Equations of the General Linear Theory of Elastic Shells*, (Nauka, Moscow), in Russian.

Goldenveizer, A. L. (1968b). Methods for justifying and refining the theory of shells. *J. Appl. Math. Mech.(PMM)* **32**, 4, pp. 684–695.

Goldenveizer, A. L. (1969). Boundary layer and its interaction with the interior state of stress of an elastic thin shell. *J. Appl. Math. Mech.(PMM)* **33**, 6, pp. 996–1028.

Goldenveizer, A. L. (1976). *Theory of Elastic Thin Shells*,(second edition), Nauka, Moscow, in Russian.

Goldenveizer, A. L. (1980). Asymptotic methods in the theory of shells. *Proc. 15th Intern. Cong. Theory Appl. Mech.* Toronto, pp. 91–104, (North-Holland, Amsterdam).

Goldenveizer, A. L. (1987). On the forced harmonic vibrations of shells. *Izv. AN SSSR Mech. Solids (MTT)* **5**, pp. 168–177, in Russian.

Goldenveizer, A. L. (1994). On algorithms of asymptotic derivation of two-dimensional shell theory and the Saint-Venant principle. *J. Appl. Math. Mech.(PMM)* **58**, 6, pp. 96–108.

Goldenveizer, A. L., Kaplunov, J. D. and Nolde, E. V. (1990). Asymptotic analysis and refinement of Timoshenko-Reissner type shell and plate theories. *Izv AN SSSR, Rigid Body Mechanics (MTT)* **6**, pp. 124–138, in Russian.

Goldenveizer, A. L., Kaplunov, J. D. and Nolde, E. V. (1993). On Timoshenko-Reissner type theories of plates and shells. *Intern. J. Solids and Struct.* **30(5)**, pp. 675–694.

Goldenveizer, A. L. and Kolos, A. V. (1965). On the derivation of two-dimensional equations in the theory of thin elastic plates. *J. Appl. Math. Mech.(PMM)* **29**, 1, pp. 151–166.

Goldenveizer, A. L., Lidskii, V. B. and Tovstik, P. E. (1979). *Free Oscillations of Thin Elastic Shells*, (Nauka, Moscow), in Russian.

Gorbunov-Posadov, M. I., Malikova T.A. and Solomin V.I. (1984). *Calculations of Structures on Elastic Foundation*, (Stroyizdat).

Green, A. E. (1962a). On the linear theory of thin elastic shells. *Proc. Roy. Soc. Ser. A* **266**, 1325.

Green, A. E. (1962b). Boundary layer equations in the linear theory of thin elastic shells. *Proc. Roy. Soc. Ser. A* **269**, 1339.

Green, A. E. and Zerna, W. (1954). *Theoretical Elasticity*, (Oxford: Clarendons Press).

Grigolyuk, E. I. (1957). Equation for three-layer shells with lightweight aggregate. *Izv. AN SSSR. OTN* **1**, pp. 77–84, in Russian.

Grigolyuk, E. I. (1958). Finite deflections of three-layer shells with rigid filler. *Izv. AN SSSR. OTN* **1**, pp. 77–84, in Russian.

Grigolyuk, E. I. and Chulkov, P. P. (1965). Nonlinear equations of thin elastic laminated anisotropic shallow shells with rigid filler. *Izv. AN SSSR. Mechanics* **5**, pp. 68–80, in Russian.

Grigolyuk, E. I. and Kogan, F. A. (1972). Modern theory of multilayer shells. *Appl. Mech.* **8**, 6, pp. 3–17, in Russian.

Grigolyuk, E. I. and Selezov, I. T. (1973). *Nonclassical Theory of Vibrations of Rods, Plates and Shells*, (Results of science. Solid Mechanics. VINITI.), **5**, in Russian.

Grigorenko, Y. H. (1973). *Isotropic and Anisotropic Layered Shells of Revolution of Variable Stiffness*, (Naukova Dumka, Kiev), in Russian.

Grigorenko, Y. H. and Vasilenko, A. T. (1981). *Theory of Shells of Variable Stiffness*, (Naukova Dumka, Kiev), in Russian.

Grinberg, G. A. (1953). On the approach of P.F. Papkovich for plane problem of elasticity for a rectangle and for bending of a rectangular plate with two fixed edges. *J. Appl. Math. Mech. (PMM)* **17**, 2, pp. 211–228.

Gusein-Zade, M. I. (1965). On the necessary and sufficient conditions for existence of decaying solutions of plain-strain problem of elasticity for a semi-strip. *J. Appl. Math. Mech. (PMM)* **29**, 4, pp. 752–759.

Gusein-Zade, M. I. (1966). Development of a bending theory of sandwich plates. *Proc. of the 5th All-Union Conf. on Theory of Shells and Plates* pp. 333–343, (Moscow, Nauka), in Russian.

Gusein-Zade, M. I. (1970). Stressed State of a Boundary Layer for Layered Plates. *Proc. of the 7th All-Union Conf. on Plates and Shells Theory* pp. 638–643, in Russian.

Gusein-Zade, M. I. (1974). Asymptotic analysis of three-dimensional dynamic equations for a thin plate. *J. Appl. Math. Mech. (PMM)* **38**, 6, pp. 1072–1078.

Gusein-Zade, M. I. (1978). Asymptotic analysis of bouhdart and initial conditions in the dynamics of thin plates. *J. Appl. Math. Mech. (PMM)* **42**, 5, pp. 899–907.

Guz', A. N. (1971). *Stability of Three-Dimensional Deformable Bodies*, (Naukova Dumka, Kiev), in Russian.

Guz', A. N. and Nemish, Y. N. (1982). *Perturbation Methods in Three-Dimensional Problems of the Theory of Elasticity*, (Vishcha Shkola, Kiev), in Russian.

Hillman, A. P. and Salzer, H. E. (1943). Roots of $\mathrm{Sin}z = z$. *Phil. Mag.* **34**, pp. 575–576.

Kalandiya, I. A. (1969). Notes on the singularity of elastic solutions near corners. *J. Appl. Math. Mech. (PMM)* **27**, 1, pp. 132–135.

Kaplunov, J. D., Kossovich L.Yu. and Nolde E.V. (1998). *Dynamics of Thinwalled Elastic Bodies*, (Academic Press).

Kaplunov, J. D. and Wilde, M. V. (2000). Edge and interfacial vibrations in elastic shells of evolution. *J. Appl. Math. Phys. (ZAMP)* **51**, pp. 29–48.

Kaplunov, J. D., Nolde E. V. and Rogerson G. A. (2002). An asymptotically consistent model for long-wave high-frequency motion in a pre-stressed elastic plate. *Mathematics and Mechanics of Solids* **7(6)**, pp. 581–606.

Kaplunov, J. D., Nolde E. V. and Shorr B. F. (2005). A perturbation approach for evaluating natural frequencies of moderately thick elliptic plates. *J. Sound Vib.* **281**, pp. 905–919.

Kaplunov J. D., Prikazchikov, D. A. and Rogerson, G. A. (2004). Edge vibration of a prestressed semi-infinite strip with traction free edge and mixed face boundary conditions. *ZAMP* **55**, 4, pp. 701–719.

Kaplunov J. D., Prikazchikov, D. A. and Rogerson, G. A. (2005). On three dimensional edge waves in semi-infinite isotropic plates subject to mixed face boundary conditions. *J. Acoust. Soc. Am.* **118**, 5, pp. 2975–2983.

Kaplunov J. D., Rogerson, G. A. and Tovstik, P. E. (2005). Localized vibration in elastic structures with slowly varying thickness. *Q. J. Mech. Appl. Math.* **58**, 4, pp. 645–664.

Keldysh, M. V. (1951). On eigenvalues and eigenfunctions of some classes of non-self-adjoint equations. *Doklady AN SSSR* **77**, 1, pp. 11–14, in Russian.

Khachatryan, S. M. (1976). On determination of stress-strain state for anisotropic strip. *Izv. AN Arm. SSR. Mech.* **29**, 6, pp. 19–32, in Russian.

Khachatryan, T. T. (1963). On the theory of bending and compression of thick plates. *Izv. AN Arm. SSR. Ser. FMN* **16**, 6, in Russian.

Khoma, I. Y. (1975). Certain aspects of the theory of anisotropic plates and shells. *Materials of the 1st All-Union Conf. on Theory and Numerical Methods in the Design of Plates and Shells* (Izd. Tbilis. Univ., Tbilisi), pp. 409–420, in Russian.

Kienzler, R. (2002). On consistent shell theories. *Arch. Appl. Mech.* **72**, pp. 229–247.

Kilchevskii, N. A. (1939). Generalization of the modern theory of shells. *J. Appl. Math. Mech. (PMM)* **2**, 4, in Russian.

Kilchevskii, N. A. (1962). Analysis of various methods of bringing three-dimensional problems of elasticity theory for two-dimensional and research formulation of boundary problems in the theory of shells. *Proc. of the 2nd All-Union Conf. on Theory of Shells and Plates* pp. 58–69, in Russian.

Kilchevskii, N. A. (1963). *Fundamentals of Analytical Mechanics of Shells*, (AN USSR, Kiev), in Russian.

Koiter, W. (1966). On the nonlinear theory of thin elastic shells. *Proc. Koninklijke Nederlandse Akademie van Wetenschappen, Series B* **69**, 1, pp. 1–54.

Koiter, W. (1970). On the foundation of linear theory of thin elastic shells. *Proc. Koninklijke Nederlandse Akademie van Wetenschappen, Series B* **73**, 3, p. 169.

Koiter, W. (1971). On the mathematical foundation of shell theory. *Proc. Int. Congr. of*

Mathematics, Nice. **3**, pp. 123–130.

Koiter, W. and Simmonds, J. G. (1973). Foundations of shell theory. Springer Berlin Heidelberg, pp. 150–176.

Kolos, A. V. (1964). On a refinement of the classical theory of bending of circular plates. *J. Appl. Math. Mech. (PMM)* **28**, 3, pp. 718–726.

Kolos, A. V. (1965). Methods of refining the classical theory of bending and extension of plates. *J. Appl. Math. Mech. (PMM)* **29**, 4, pp. 914–925.

Kolos, A. V. (1966). On the application of approximate theories of plate bending type of Reissner theory. *Proc. of the 6th All-Union Conf. on Theory of Shells and Plates,* (Nauka, Moscow), in Russian.

Kornishin, M. S. (1964). *Nonlinear Problems of Plate and Flat Shell Theories and Methods of Solutions,* (Nauka, Moscow), in Russian.

Korolev, V. I. (1965). *Laminated Anisotropic Plates and Shells Made of Reinforced Plastics,* (Mashinostroenie, Moscow), in Russian.

Kostyuchenko, A. G. and Orazov, M. B. (1975). Certain properties of the roots of a self-adjoint quadratic pencil. *Functional Analysis and Its Applications* **9**, 4, pp. 295–305.

Kostyuchenko, A. G. and Orazov, M. B. (1977). On the completeness of the root vectors of certain self-adjoint quadratic pencils. *Functional Analysis and Its Applications* **11**, 4, p. 85–87.

Krasnoselsky, M. I., Vainikko G. M., Zabreiko P. P., Rutitsky Y. B. and Stecenko V. Y. (1969). *Approximate Methods of Solution for Operator Equations,* (Nauka, Moscow), in Russian.

Krein, M. G. and Langer, H. (1965) Certain mathematical principles of the linear theory of damped vibrations of continua. *Appl. Theory of Functions in Continuum Mechanics. Proc. Internat. Sympos.* (Tbilisi), pp. 283–322, in Russian.

Kupradze, V. D., Gegelia, T. G., Basheleishvili, M. O. and Burchuladze, T. V. (1976) *Three-dimensional Problems of the Mathematical Theory of Elasticity and Thermoelasticity,* (Nauka, Moscow), in Russian.

Lantsosh, K. (1961). *Practical Methods of Applied Analysis,* (Fizmatgiz, Moscow), in Russian.

Leibenzon, L. S. (1947). *Course in the Theory of Elasticity,* (Gostekhizdat, Moscow-Leningrad), in Russian.

Lekhnitskii, S. G. (1962). The elastic equilibrium of a transversely isotropic layer and a thick plate, *J. Appl. Math. Mech. (PMM)* **26**, 4, pp. 687–696.

Lekhnitskii, S. G. (1968). *Anisotropic Plates,* (Gordon and Breach Science Publishers, New York).

Lekhnitskii, S. G. (1977). *Elasticity Theory of Anisotropic Bodies,* (Nauka, Moscow), in Russian.

Lomov, S. A. (1981). *Introduction to General Theory of Singular Perturbations,* (Nauka, Moscow), in Russian.

Lourier, A. I. (1940). The general theory of thin elastic shells. *J. Appl. Math. Mech. (PMM)* **4**, 2, pp. 7–34, in Russian.

Lourier, A. I. (1942). Thick plate theory revisited. *J. Appl. Math. Mech. (PMM)* **6**, pp. 151–169, in Russian.

Lourier, A. I. (1947). *Statics of Thin Elastic Shells,* (Moscow-Leningrad: Gostekhizdat), in Russian.

Lourier, A. I. (1955). *Spatial Problems of Elastic Theory,* (Moscow-Leningrad: Gostekhizdat), in Russian.

Lourier, A. I. (1970). *Theory of Elasticity,* (Nauka, Moscow).

Markus, A. S. (1962) On the expansion of the root vectors of weakly perturbed self-adjoint operator. *DAN SSSR* **142**, 3.

Mikhasev, G. I. (1997). Free and parametric vibrations of cylindrical shells under static and periodic axial loads. *Technische Mechanik* **17**, 3, pp. 209–216.

Mikhasev, G. I. and Tovstik, P. E. (2009). *Localized Vibrations and Waves in Thin Shells*, (FizMatLit, Moscow), in Russian.

Mikhlin, S. G. (1964). *Variational Methods in Mathemetical Physics*, (Pergamon Press).

Mittelman, B. S. and Hillman, A. P. (1946). Zeros of sinz+z. *Math. Tables and other Aids to Comp.* **2**, 11, p. 60.

Morgenstern, D. (1959a). *Herleitung der Plattentheorie aus der Dreidimensionalen Elastizittstheorie*, (Arch. Rational Mech. Analysis), **4**.

Morgenstern, D. (1959b). Bernoullische Hypothesen bei Balken und Plattentheorie. *J. of Appl. Math. and Mech. (ZAMM)*, **39**, 9-11, pp. 420–422.

Mushtari, K. M. (1947). Domain of applicability of approximate theory of Kirchhoff-Love shells. *J. Appl. Math. Mech. (PMM)* **11**, 5, pp. 517–520.

Mushtari, K. M. and Galimov, K. Z. (1957). *Nonlinear Elastic Shell Theory*, (Tatknigoizdat, Kazan), in Russian.

Mushtari, K. M. and Teregulov, I. G. (1959). The theory of hollow orthotropic shells of average thickness. *Izv. AN SSSR, Mekhanika i Mashinostroenie* **128**, 6, pp. 60–67.

Naghdi, P. M. (1956). On the theory of thin elastic plates. *Quart. Appl. Math.* **14**, 4.

Nayfeh, A. H. (1973). *Perturbation Methods*, (John Wiley and Sons).

Nemirovskii, Y. V. (1970). Stability of reinforced shells and plates beyond the elastic limit. *Izv. AN SSSR, Mekh. Tverd. Tela.* **2**, 2, pp. 67–74, in Russian.

Nigul, U. K. (1963). Application of the Lur'e symbolic method to stress analysis and to the two-dimensional theory of elastic plates. *J. Appl. Math. Mech. (PMM)* **27**, 3, pp. 583–588.

Nolde E. V., Prikazchikova L. A. and Rogerson G. A. (2004). Dispersion of small amplitude waves in a pre-stressed, compressible elastic plate *J. Elast.* **75**, pp. 1–29.

Novatsky, V. (1970). Dynamic problems in thermoelasticity, (Mir, Moscow).

Novotn B. (1970). On the asymptotic integration of the three-dimensional non-linear equations of thin elastic shells and plates. *International Journal of Solids and Structures* **6**, 4, pp. 433–451.

Novotn B. (1972). *Methods of Asymptotic Integration in the Theory of Thin Shells of Variable Thickness*, (Stavebnicky casopis, Rocnik 10. Ciclo 3. Bratislawa).

Novozhilov V. V. (1962). *The Theory of Thin Shells*, (Sudpromgiz, Leningrad), in Russian.

Novozhilov, V. V. and Finkelstein, R. M. (1943). On the error of Kirchhoff assumptions in shell theory. *J. Appl. Math. Mech. (PMM)* **7**, 2, pp. 331–340.

Nuller, B. M. (1969). On the generalized orthogonality conditions of P.A. Schiff. *J. Appl. Math. Mech. (PMM)* **33**, 2, pp. 376–387.

Obraztsov, I. F. and Onanov, G. G. (1973). *Structural Mechanics of Tapered Thin-Walled Systems*, (Mashinostroenie, Moscow), in Russian.

Ogibalov P. M. (1958). *Bending, Stability and Vibrations of Plates*, (Izd. MGU), in Russian.

Oleynik, O. A. and Iosif'yan G. A. (1976). An Analogue of Saint-Venant's principle and the uniqueness of solutions of boundary value problems for parabolic equations in unbounded domains. *Russian Mathematical Surveys* **31**, 6.

Oleynik, O. A. and Iosif'yan G. A. (1977a). On energy estimates of generalized solutions of second order boundary value problems for elliptic equations and their applications. *Doklady AN SSSR* **232**, 6, pp. 1257–1260, in Russian.

Oleynik, O. A. and Iosif'yan G. A. (1977b). On the Saint-Venant principle for a mixed problem of elasticity and its applications. *Doklady AN SSSR* **233**, 5, pp. 824–827, in Russian.

Orazov M. B. (1976). The completeness of the eigenvectors and associated vectors of a self-adjoint quadratic bundle. *Funct. Anal. Appl.* **10**, 2, pp. 153–155, in Russian.

Paimushin V. N. (1978). Relations of the theory of thin shells of Timoshenko type in curvilinear coordinates of the reference surface. *J. Appl. Math. Mech. (PMM)* **42**, 4, pp. 753–758, in Russian.

Paimushin V. N. (1980). On the problem of analyzing plates and shells with a complex contour. *Prikl. Mekh.* **16**, pp. 63–70, in Russian.

Palant Y. A. (1961). On test for completeness of a system of eigenvectors and adjoint vectors of a polynomial bundle of operators. *Dokl. Akad. Nauk SSSR*, **141**, 3, pp. 558–562, in Russian.

Papkovich, P. F. (1940). On a form of solution of plane problem of elasticity for a rectangular strip. *Dokl. AN SSSR* **27**, 4, pp. 335–339, in Russian.

Papkovich, P. F. (1941). On the two issues of thin plate bending. *J. Appl. Math. Mech. (PMM)* **5**, 3, pp. 359–374, in Russian.

Pelekh, B. L. (1973). *Theory of Shells with Finite Shear Stiffness*, (Naukova Dumka, Kiev), in Russian.

Pichugin A. V. and Rogerson G. A. (2002). An asymptotic membrane-like theory for long-wave motion in a pre-stressed elastic plate *Proc. R. Soc. A* **458**, pp. 1447–1468

Pietraszkiewicz, W., Chroscielewski, J. and Makowski J. (2005). On dynamically and kinematically exact theory of shells. In *Shell Structures: Theory and Applications*, (Taylor and Francis. London), pp. 163–167.

Podio-Guidugli, P. (2008). Validation of classical beam and plate models by variational convergence. *Proc. of the IUTAM Symp. on the Relations of Shell, Plate, Beam and 3D Models*, Springer, pp. 177–188.

Ponyatovsky, V. V. (1964). Toward a theory of the bending of anisotropic plates. *J. Appl. Math. Mech.(PMM)* **28**, 6, pp. 1033–1039.

Ponyatovsky, V. V. (1965). *Equations of the Theory of Anisotropic Plates*, (The collection: Studies on the elasticity and plasticity. 4. LSU), in Russian.

Ponyatovsky, V. V. (1968). On asymptotic integration of the problem of equilibrium for a thin rod with arbitrary surface loading. *Izv. AN SSR. Mechanics (MTT)* **5**, pp. 139–143, in Russian.

Prikazchikov, D. A., Rogerson, G. A., and Sandiford, K. J. (2007). On localized vibrations in pre-stressed incompressible transversely isotropic elastic solids. *J. Sound. Vib.* **301**, pp. 701–717.

Prokopov, V. K. (1965). Application of the symbolic method to the derivation of the equations of the theory of plates. *J. Appl. Math. Mech.(PMM)* **29**, 5, pp. 1064–1083.

Prokopov, V. K. (1966). Homoheneous solution of elasticity and applications to thin plate theories. *Proc. 2nd USSR Congr. Theor. and Appl. Mech.* **3**, pp. 253–259, (Nauka, Moscow), in Russian.

Prokopov, V. K. (1967). Review of studies on homogeneous solutions in the theory of elasticity and their applications. *Tr. Leningr. Politekh. Inst.* **279**, pp. 31–46.

Prokopov, V. K. and Gruzdev, I. A. (1968). Multi-moment theory of equilibrium of thick plates. *J. Appl. Math. Mech.(PMM)* **32**, 2, pp. 342–351.

Rabotnov, Y. N. (1962). *Strength of Materials*, (Fizmatgiz, Moscow), in Russian.

Reissner, E. (1944). On the theory of bending of elastic plates. *J. Math. and Phys.* **23**, pp. 184–191.

Reissner, E. (1945). The effect of transverse shear deformation on the bending of elastic plates. *J. of Appl. Mech.* **12**, pp. 69–77.

Rogacheva, N. N. (1974). Reissner-Naghdi elasticity relations (shell theory equations. *Prik-*

ladnaia Matematika i Mekhanika **38**, pp. 1063–1071.

Rogacheva, N. N. (1975). Refined theory of thermoelastic shells. *Proc. 10th All-Union Conf. on Shell and Plate Theory* (Tbilisi), **1**, pp. 251–259.

Rogacheva, N. N. (1994). *The Theory of Piezoelectric Plates and Shells*, (CRC Press, London).

Romenskaya, G. I. and Shlenev, M. A. (1973). Asymptotic method for solving three-dimensional problem of a transversely isotropic plate. *Proc. 11th All-Union Conf. on Shell and Plate Theory* (Nauka, Moscow), in Russian.

Romenskaya, G. I. and Shlenev, M. A. (1976). *On the roots of the equation* $\sin \omega z = -\omega \sin z$, (In the collection: calculation of shells and plates, Rostov-na-Dony), in Russian.

Rutten H. S. (1971). *Asymptotic approximation in the three-dimensional theory of thin and thick elastic shells*, (Nederlandse Boekdruk Industrie NV).

Sapondjyan O. M. (1975). *Bending of Thin Elastic Plates*, (Erevan, Aiastan), in Russian.

Sargsyan, S. H. (2008). Boundary value problems of asymmetric theory of elasticity for thin plates. *J. Appl. Math. Mech. (PMM)* **72**, 1, pp. 77–86.

Sarkisyan V. S. (1976). *Some Problems of the Mathematical Theory of Elasticity of an Anisotropic Body*, (Erevan State University, Erevan), in Russian.

Schiff, P. A. (1883). Sur L'equilibre d'un cilindre elastique. *Journal Mathematiques pures et Appliquees*, **3**, III.

Sedov, L. I. (1965). Mathematical methods for constructing new models of continuous media. *Russ. Math. Surv.* **20**, 5, pp. 123–182.

Shariyat, M. (2010). A generalized high-order global-local plate theory for nonlinear bending and buckling analyses of imperfect sandwich plates subjected to thermo-mechanical loads. *Composite Struct.* **92**, 1, pp. 130–143.

Shoykhet, B. A., (1973). On asymptotically precise equations of thin plates of complicated structure. *J. Appl. Math. Mech.(PMM)* **38**, 5, pp. 914–924.

Sternberg, E. (1954). On Saint-Venants principle. *Quart. Appl. Math.* **11**, pp. 393–402.

Tamarkin, J. D. (1917). *About Certain General Problems of Theory of Ordinary Linear Differential Equations and about Expansions of Derivative Functions into Series*, (Petrograd), in Russian.

Teregulov I. G. (1962). On the formulation of refined theories of plates and shells. *J. Appl. Math. Mech.(PMM)* **26**, 2, pp. 495–502.

Thai, H.-T. and Choi, D.-H. (2013). Size-dependent functionally graded Kirchhoff and Mindlin plate models based on a modified couple stress theory. *Composite Struct.* **95**, 1, pp. 142–153.

Timoshenko, S. P. (1957). *History of Strength of Materials*, (IL, Moscow), in Russian.

Timoshenko, S. P. and Goodier J. N. (1970). *Elasticity Theory*, (3rd Edn, McGraw-Hill, New York).

Toupin, R. A. (1965). Saint-Venants principle. *Arch. Rat. Mech. Anal.* **18**, pp. 83–96.

Tovstik, P. E. (1975). Low-frequency oscillations of convex shell of revolution. *Mechanics of Solids* **10**, 6, pp. 95–100.

Tovstik, P. E. and Smirnov, A. L. (2001). *Asymptotic Methods in the Buckling Thenry of Elastic Shells*, (World Scientific, Singapore).

Trenogin, V. A. (1970). Development and applications of Vishik-Lyusternik's asymptotic method. *Usp. Mat. Nauk.* **25**, 4, pp. 123–156.

Uflyand, Ya. S. (1967). *Integral Transforms in Problems of Elasticity Theory*, (Nauka, Leningrad), in Russian.

Ustinov, Y. A. and Iudovich, V. I. (1973). The completeness of a system of elementary solutions to a biharmonic equation in a half-strip. *J. Appl. Math. Mech.(PMM)* **37**,

4, pp. 706–714.

Ustinov, Y. A. (1974). Some properties of homogeneous solutions of inhomogeneous plates. *Sov. Phys. Dokl.* **216**, 4, pp. 123–156.

Ustinov, Y. A. (1976). On the structure of the boundary layer in laminar slabs. *Sov. Phys. Dokl.* **229**, 2, pp. 325–328.

Vasiliev, V. V. (1992). On the theory of thin plates, *Izv. RAN (MTT)* **3**, pp. 26–47.

Vasiliev, V. V. (1995). To discussion on the theory of thin plates, *Izv. RAN (MTT)* **4**, pp. 140–150,.

Vasilieva, A. B. and Butuzov V. F. (1973). *Asymptotic Expansions of Solutions of Singularly Perturbed Equations*, (Nauka, Moscow), in Russian.

Vazov, V. (1968). *Asymptotic Expansions of Solutions of Ordinary Differential Equations*, (Mir, Moscow), in Russian.

Vekua, I. N. (1955). On a method of computing prismatic shells. *Trudy Tbiliss. Mat. Inst. Razmadze* **21**, pp. 191–259, in Russian.

Vekua, I. N. (1975). On two methods of constructing a consistent theory of shells. *1st All-Union School on the Theory and Numerical Methods of Calculation of Shells and and Plates* pp. 5–50, (Izd. Tbiliss. Univ., Tbilissi), in Russian.

Vekua, I. N. (1982). *Some General Methods of Constructing Different Variants of Shell Theories*, (Nayka, Moscow), in Russian.

Vishik, M. I. and Lusternik L. A. (1957). Regular degeneration and boundary layer for linear differential equations with small parameter. *Usp. Math. Nauk* **12**, 5, pp. 3–122, in Russian.

Vishik, M. I. and Lyusternik, L. A. (1960). Solvability of some problems concerning perturbations in case of matrices and selfadjoint and non-selfadjoint differential equations. *Usp. Math. Nauk* **15**, 3, pp. 3–80, in Russian.

Vizitei, V. N. and Markus, A. S. (1965). On convergence of multiple expansions in the eigenvectors and associated vectors of an operator bundle. *Sb. Mathematics* **66.108**, 2, pp. 287–320, in Russian.

Vlasov, V. Z. (1949). *General Shell Theory and Its Engineering Applications*, (Gostekhizdat, Moscow-Leningrad), in Russian.

Vlasov, V. V. (1975). *The Method of Initial Functions in Problems of the Theory of Elasticity and Structural Mechanics*, (Stroiizdat, Moscow), in Russian.

Vlasov, V. Z. and Leont'ev N. N. (1960). *Plates and Shells on an Elastic Foundation*, (Fizmatgiz, Moscow), in Russian.

Volkov, A. N. (1971). Theory of thick shells on the basis of the method of initial functions. *International Applied Mechanics*, **7**, 10, pp. 1093–1097.

Volokh, K. Yu. (1994). On classical plate theory. *J. Appl. Math. Mech.(PMM)* **58**, 6, pp. 156–165.

Von Mises, R. (1945). On Saint-Venant's principle. *Bull. Amer. Math. Soc.* **51**, 8, pp. 555–562.

Vorovich, I. I. (1966a). Some mathematical aspects of plates and shells theory. *Proc. 2nd All-Union Conf. on Theor. and Appl. Mech.*, pp. 116–136, (Moscow, Nauka), in Russian.

Vorovich, I. I. (1966b). General problems in the theory of plates and shells. *Proc. of the 6th All-Union Conf. on Theory of Shells and Plates*, pp. 896–903, (Moscow, Nauka), in Russian.

Vorovich, I. I. (1975). Some results and problems of the asymptotic theory of plates and shells. *Proc. of the 1st All-Union School on Theory and Numerical Methods in Shell and Plate Theory*, pp. 51–149, (Izd. Tbiliss. Univ., Tbilissi), in Russian.

Vorovich, I. I. Aleksandrov, V. M. and Babeshko, V. A. (1974). *Nonclassical Mixed Prob-*

lems in the Theory of Elasticity, (Moscow, Nauka), in Russian.

Vorovich, I. I. and Bazarenko, N. A. (1965). Asymptotic solution of the elasticity problem for a hollow, finite length, thin cylinder. *J. Appl. Math. Mech.(PMM)*, **29**, 6, pp. 1219–1238.

Vorovich, I. I. and Kadomtsev, I. G. (1970). Qualitative investigation of the stress-strain state of a sandwich plate. *J. Appl. Math. Mech.(PMM)*, **34**, 5, pp. 830–836.

Vorovich, I. I., Kadomtsev, I. G. and Ustinov, Yu. A. (1975). To the theory of plates inhomogeneous in thickness. *Izv. AN SSSR. Rigid Body Mechanics (MTT)*, **3**, pp. 119–129.

Vorovich, I. I. and Kopasenko, V. V. (1966). On problems of elasticity theory for a semi-strip. *J. Appl. Math. Mech.(PMM)* **30**, 1, pp. 109–115.

Vorovich, I. I. and Malkina, O. S. (1966). Asymptotic method for solving elasticy problems in case of a thick plate. *Proc. VI All-Union Conf. on PLates and Shells*, pp. 251–254 (Moscow, Nauka), in Russian.

Vorovich, I. I. and Shlenev, M. A. (1963). *Plates and Shells. Results of Science. Mechanics*, (Izd. AN SSSR), in Russian.

Vorovich, I. I. and Vilenskaia, T. V. (1966). Asymptotic behavior of the solution of the problem of elasticity for a thin spherical shell. *J. Appl. Math. Mech.(PMM)*, **30**, 2, pp. 342–361.

Wang, Y. H., Tham L. G., and Cheung, Y. K. (2005). Beams and plates on elastic foundations: a review. *Progr. Srruct. Engng. Mater.*, **7**, pp. 174–182.

Widera, O. E. (1969). An asymptotic theory for the moderately large deflections of anisotropic plates. *Journal of Engineering Mathematics* **3**, 3, pp. 239–244.

Wilde, M. V., Kaplunov, J. D. and Kossovich, L. Yu. (2010). *Boundary and Interfacial Rassonance Phenomena in Elastic Solids*, (Fizmatlit, Moscow), in Russian.

Williams, M. L. (1952). Stress Singularities Resulting from Various Boundary Conditions in Angular Corners of Plates in Extension. *J. Appl. Mech.* **19**, pp. 526–528.

Williams, M. L. (1959). The Stresses Around a Fault or a Crack in Dissimilar Media. *Bull. Seismol. Soc. Am.* **49**, pp. 199–204.

Zhilin, P. A. (1992). On the plates theories of Poisson and Kirchhoff from the prospective of modern plate theories, *Izv. RAN (MTT)* **3**, pp. 48–64.

Zhilin, P. A. (1995). On the classical plate theory and Kelvin-Tate transformation, *Izv. RAN (MTT)* **4**, pp. 133–139.